Edited by Kathy Lüdge

Nonlinear Laser Dynamics

Related Titles

Okhotnikov, O. G. (ed.)

Semiconductor Disk Lasers
Physics and Technology

2010
ISBN: 978-3-527-40933-4

Harrison, P.

Quantum Wells, Wires and Dots
Theoretical and Computational Physics of Semiconductor Nanostructures

2009
ISBN: 978-0-470-77097-9

Kane, D., Shore, A. (eds.)

Unlocking Dynamical Diversity
Optical Feedback Effects on Semiconductor Lasers

2005
ISBN: 978-0-470-85619-2

Edited by Kathy Lüdge

Nonlinear Laser Dynamics

From Quantum Dots to Cryptography

WILEY-VCH

WILEY-VCH Verlag GmbH & Co. KGaA

The Editor

Dr. Kathy Lüdge
Technische Universität Berlin
Inst. für Theoretische Physik
Hardenbergstr. 36
10623 Berlin
Germany

Cover
The picture on the cover page was designed by Benjamin Lingnau.

All books published by **Wiley-VCH** are carefully produced. Nevertheless, authors, editors, and publisher do not warrant the information contained in these books, including this book, to be free of errors. Readers are advised to keep in mind that statements, data, illustrations, procedural details or other items may inadvertently be inaccurate.

Library of Congress Card No.: applied for

British Library Cataloguing-in-Publication Data
A catalogue record for this book is available from the British Library.

Bibliographic information published by the Deutsche Nationalbibliothek
The Deutsche Nationalbibliothek lists this publication in the Deutsche Nationalbibliografie; detailed bibliographic data are available on the Internet at <http://dnb.d-nb.de>.

© 2012 Wiley-VCH Verlag & Co. KGaA, Boschstr. 12, 69469 Weinheim, Germany

All rights reserved (including those of translation into other languages). No part of this book may be reproduced in any form – by photoprinting, microfilm, or any other means – nor transmitted or translated into a machine language without written permission from the publishers. Registered names, trademarks, etc. used in this book, even when not specifically marked as such, are not to be considered unprotected by law.

Typesetting Laserwords Private Limited, Chennai, India
Printing and Binding Fabulous Printers Pte Ltd, Singapore
Cover Design Adam-Design, Weinheim

Printed in Singapore
Printed on acid-free paper

Print ISBN: 978-3-527-41100-9
ePDF ISBN: 978-3-527-63984-7
oBook ISBN: 978-3-527-63982-3
ePub ISBN: 978-3-527-63983-0
Mobi ISBN: 978-3-527-63985-4

In honor of Prof. Dr. Eckehard Schöll's 60th birthday.

Contents

Preface *XV*
List of Contributors *XVII*

Part I Nanostructured Devices *1*

1 **Modeling Quantum-Dot-Based Devices** *3*
  Kathy Lüdge
1.1 Introduction *3*
1.2 Microscopic Coulomb Scattering Rates *4*
1.2.1 Carrier–Carrier Scattering *5*
1.2.2 Detailed Balance *8*
1.3 Laser Model with Ground and Excited States in the QDs *9*
1.3.1 Temperature Effects *14*
1.3.2 Impact of Energy Confinement *15*
1.3.3 Eliminating the Excited State Population Dynamics *17*
1.4 Quantum Dot Switching Dynamics and Modulation Response *18*
1.4.1 Inhomogeneous Broadening *19*
1.4.2 Temperature-Dependent Losses in the Reservoir *20*
1.4.3 Comparison to Experimental Results *20*
1.5 Asymptotic Analysis *21*
1.5.1 Consequences of Optimizing Device Performance *25*
1.6 QD Laser with Doped Carrier Reservoir *26*
1.7 Model Reduction *28*
1.8 Comparison to Quantum Well Lasers *29*
1.9 Summary *30*
  Acknowledgment *30*
  References *30*

2 **Exploiting Noise and Polarization Bistability in Vertical-Cavity Surface-Emitting Lasers for Fast Pulse Generation and Logic Operations** *35*
  Jordi Zamora-Munt and Cristina Masoller
2.1 Introduction *35*

2.2	Spin-Flip Model 39
2.3	Polarization Switching 40
2.4	Pulse Generation Via Asymmetric Triangular Current Modulation 44
2.5	Influence of the Noise Strength 48
2.6	Logic Stochastic Resonance in Polarization-Bistable VCSELs 49
2.7	Reliability of the VCSEL-Based Stochastic Logic Gate 52
2.8	Conclusions 53
	Acknowledgment 54
	References 54

3	**Mode Competition Driving Laser Nonlinear Dynamics** 57
	Marc Sciamanna
3.1	Introduction 57
3.2	Mode Competition in Semiconductor Lasers 58
3.3	Low-Frequency Fluctuations in Multimode Lasers 61
3.4	External-Cavity Mode Beating and Bifurcation Bridges 64
3.5	Multimode Dynamics in Lasers with Short External Cavity 65
3.6	Polarization Mode Hopping in VCSEL with Time Delay 67
3.6.1	Polarization Switching Induced by Optical Feedback 67
3.6.2	Polarization Mode Hopping with Time-Delay Dynamics 69
3.6.3	Coherence Resonance in a Bistable System with Time Delay 71
3.7	Polarization Injection Locking Properties of VCSELs 73
3.7.1	Optical Injection Dynamics 74
3.7.2	Polarization and Transverse Mode Switching and Locking: Experiment 76
3.7.3	Bifurcation Picture of a Two-Mode Laser 81
3.8	Dynamics of a Two-Mode Quantum Dot Laser with Optical Injection 83
3.9	Conclusions 85
	Acknowledgments 86
	References 86

4	**Quantum Cascade Laser: An Emerging Technology** 91
	Andreas Wacker
4.1	The Essence of QCLs 92
4.1.1	Semiconductor Heterostructures 92
4.1.2	Electric Pumping 94
4.1.3	Cascading 94
4.2	Different Designs 96
4.2.1	Optical Transition and Lifetime of the Upper State 96
4.2.2	Effective Extraction from the Lower Laser Level 96
4.2.3	Injection 97
4.3	Reducing the Number of Levels Involved 98
4.4	Modeling 100

4.5	Outlook *103*	
	Acknowledgments *104*	
4.6	Appendix: Derivation of Eq. (4.1) *104*	
	References *105*	
5	**Controlling Charge Domain Dynamics in Superlattices** *111*	
	Mark T. Greenaway, Alexander G. Balanov, and T. Mark Fromhold	
5.1	Model of Charge Domain Dynamics *112*	
5.2	Results *117*	
5.2.1	Drift Velocity Characteristics for $\theta = 0°, 25°$, and $40°$ *118*	
5.2.2	Current–Voltage Characteristics for $\theta = 0°, 25°$, and $40°$ *119*	
5.2.3	$I(t)$ Curves for $\theta = 0°, 25°$, and $40°$ *120*	
5.2.4	Charge Dynamics for $\theta = 0°, 25°$, and $40°$ *122*	
5.2.5	Stability and Power of $I(t)$ Oscillations for $0° < \theta < 90°$ *128*	
5.2.6	Frequency of $I(t)$ for $0° < \theta < 90°$ *130*	
5.3	Conclusion *132*	
	Acknowledgment *132*	
	References *132*	
	Part II **Coupled Laser Device** *137*	
6	**Quantum Dot Laser Tolerance to Optical Feedback** *139*	
	Christian Otto, Kathy Lüdge, Evgeniy Viktorov, and Thomas Erneux	
6.1	Introduction *139*	
6.2	QD Laser Model with One Carrier Type *141*	
6.3	Electron-Hole Model for QD Laser *142*	
6.3.1	Similar Scattering Times τ_e and τ_h *143*	
6.3.2	Different Scattering Times τ_e and τ_h *144*	
6.3.3	Small Scattering Lifetime of the Holes $a = O(1)$ *144*	
6.3.4	Very Small Scattering Lifetime of the Holes $a = O(\gamma^{-1/2})$ *144*	
6.4	Summary *145*	
	Acknowledgment *146*	
6.5	Appendix A: Rate Equations for Quantum Well Lasers *146*	
6.6	Appendix B: Asymptotic Analysis for a QD Laser Model with One Carrier Type *148*	
6.7	Appendix C: Asymptotic Analysis for a QD Laser Model with Two Carrier Types *153*	
	References *158*	
7	**Bifurcation Study of a Semiconductor Laser with Saturable Absorber and Delayed Optical Feedback** *161*	
	Bernd Krauskopf and Jamie J. Walker	
7.1	Introduction *161*	
7.2	Bifurcation Analysis of the SLSA *164*	

7.3	Equilibria of the DDE and Their Stability 168
7.4	Bifurcation Study for Excitable SLSA 171
7.5	Bifurcation Study for Nonexcitable SLSA 173
7.6	Dependence of the Bifurcation Diagram on the Gain Pump Parameter 176
7.7	Conclusions 178
	References 179
8	**Modeling of Passively Mode-Locked Semiconductor Lasers** 183
	Andrei G. Vladimirov, Dmitrii Rachinskii, and Matthias Wolfrum
8.1	Introduction 183
8.2	Derivation of the Model Equations 184
8.3	Numerical Results 189
8.4	Stability Analysis for the ML Regime in the Limit of Infinite Bandwidth 197
8.4.1	New's Stability Criterion 197
8.4.2	Slow Stage 199
8.4.3	Fast Stage 199
8.4.4	Laser Without Spectral Filtering 200
8.5	The Q-Switching Instability of the ML Regime 203
8.5.1	Laser Without Spectral Filtering 204
8.5.2	Weak Saturation Limit 207
8.5.3	Variational Approach 209
8.6	Conclusion 212
	Acknowledgments 213
	References 213
9	**Dynamical and Synchronization Properties of Delay-Coupled Lasers** 217
	Cristina M. Gonzalez, Miguel C. Soriano, M. Carme Torrent, Jordi Garcia-Ojalvo, and Ingo Fischer
9.1	Motivation: Why Coupling Lasers? 217
9.2	Dynamics of Two Mutually Delay-Coupled Lasers 218
9.2.1	Dynamical Instability 218
9.2.2	Instability of Isochronous Solution 220
9.3	Properties of Leader–Laggard Synchronization 224
9.3.1	Emergence of Leader–Laggard Synchronization 224
9.3.2	Control of Lag Synchronization 226
9.4	Dynamical Relaying as Stabilization Mechanism for Zero-Lag Synchronization 228
9.4.1	Laser Relay 228
9.4.2	Mirror Relay 230
9.5	Modulation Characteristics of Delay-Coupled Lasers 231
9.5.1	Periodic Modulation 231
9.5.2	Noise Modulation 235

9.5.3	Application: Key Exchange Protocol *238*
9.6	Conclusion *240*
	Acknowledgments *240*
	References *241*

10	**Complex Networks Based on Coupled Two-Mode Lasers** *245*
	Andreas Amann
10.1	Introduction *245*
10.2	Complex Networks on the Basis of Two-Mode Lasers *246*
10.3	The Design Principles of Two-Mode Lasers *248*
10.4	The Dynamics of Two-Mode Lasers Under Optical Injection *253*
10.4.1	The Model Equations *253*
10.4.2	The $\epsilon = 0$ Case *254*
10.4.3	The Finite ϵ Case *257*
10.5	Conclusions *264*
	Acknowledgments *265*
	References *265*

Part III   Synchronization and Cryptography *269*

11	**Noise Synchronization and Stochastic Bifurcations in Lasers** *271*
	Sebastian M. Wieczorek
11.1	Introduction *271*
11.2	Class-B Laser Model and Landau–Stuart Model *272*
11.3	The Linewidth Enhancement Factor and Shear *274*
11.4	Detection of Noise Synchronization *275*
11.5	Definition of Noise Synchronization *278*
11.6	Synchronization Transitions via Stochastic d-Bifurcation *280*
11.6.1	Class-B Laser Model Versus Landau–Stuart Equations *282*
11.7	Noise-Induced Strange Attractors *285*
11.8	Conclusions *289*
	References *290*

12	**Emergence of One- and Two-Cluster States in Populations of Globally Pulse-Coupled Oscillators** *293*
	Leonhard Lücken and Serhiy Yanchuk
12.1	Introduction *293*
12.1.1	Pulse-Coupled Oscillators *294*
12.1.2	Phase-Response Curve as a Parameter *295*
12.1.3	System Description *298*
12.2	Numerical Results *300*
12.3	Appearance and Stability Properties of One-Cluster State *302*
12.3.1	Inadequacy of the Linear Stability Analysis *302*
12.3.2	One-Cluster State is a Saddle Point *302*
12.3.2.1	Existence of a Local Unstable Direction *302*

12.3.2.2	Existence of a Local Stable Direction 303
12.3.2.3	Other Stable and Unstable Local Directions 304
12.3.3	Stable Homoclinic Orbit to One-Cluster State 305
12.4	Two-Cluster States 306
12.4.1	Stability of Two-Cluster States 308
12.5	Intermediate State for Symmetric PRC with $\beta = 0.5$ 309
12.6	Conclusions 310
12.7	Appendix: Existence of a Homoclinic Orbit 310
	References 315

13 Broadband Chaos 317
Kristine E. Callan, Lucas Illing, and Daniel J. Gauthier

13.1	Introduction 317
13.2	Optoelectronic Oscillators 318
13.3	Instability Threshold 323
13.4	Transition to Broadband Chaos 325
13.5	Asymptotic Analysis 327
13.6	Summary and Outlook 330
	Acknowledgments 331
	References 331

14 Synchronization of Chaotic Networks and Secure Communication 333
Ido Kanter and Wolfgang Kinzel

14.1	Introduction 333
14.2	Unidirectional Coupling 334
14.3	Transmission of Information 335
14.4	Bidirectional Coupling 336
14.5	Mutual Chaos Pass Filter 339
14.5.1	Protocol 342
14.6	Private Filters 345
14.7	Networks 346
14.8	Outlook 350
	References 350

15 Desultory Dynamics in Diode-Lasers: Drift, Diffusion, and Delay 355
K. Alan Shore

15.1	Introduction 355
15.2	Carrier Diffusion in Diode Lasers 357
15.3	Intersubband Laser Dynamics 359
15.4	Carrier Diffusion Effects in VCSELs 362
15.4.1	Transverse Mode Competition and Secondary Pulsations 362
15.4.2	VCSEL Polarization Selection 363
15.4.3	Nanospin VCSELs 363
15.5	Delayed Feedback and Control of VCSEL Polarization 364
15.6	VCSEL Chaos and Synchronization and Message Transmission 365

15.7	Delay Deletion: Nullified Time of Flight	*369*
15.8	Chaos Communications: Optimization and Robustness	*371*
15.9	Conclusion	*372*
	Acknowledgments	*373*
	References	*373*
	Further Reading	*380*

Index *381*

Preface

Lasers are paradigmatic examples of nonlinear systems and have played a decisive role in the development of nonlinear dynamics into a cross disciplinary subject over the past 40 years. Already a free running laser represents a nontrivial nonlinear system, but even more interesting phenomena arise when lasers are subjected to feedback or coupled to build large networks. Some of these phenomena already found their way to industrial applications, for example, the creation of ultrashort pulses with the mode locking technique by using integrated multisection devices or the stabilization of laser outputs with optical injection. The technological advances in semiconductor processing technologies also allow to produce a variety of lasers with nanostructured active regions that give rise to interesting physics and allow designing new innovative devices.

Nowadays, nonlinear laser dynamics is a still growing field of active research, and this book focuses and reviews recent advances in this area. In an interdisciplinary approach, it will concentrate on mathematical, physical, as well as experimental aspects. By discussing problems such as the modeling of integrated devices, the creation of networks, exploitation of chaotic lasers for secure communication, and the use of nanostructured lasers for logic gates and memory elements, it will enter innovative grounds and hopefully inspire future research on that topic.

On the occasion of the sixtieth birthday of Prof. Eckehard Schöll, this book is also intended to recognize the work during his scientific career, as he always enforced the connection between rigorous mathematical analysis and physical modeling. For this reason, the contributors are former and future collaborators of Prof. Eckehard Schöll.

The book is separated into three parts. Within the first part, "Nanostructured devices", the dynamic properties and modeling aspects of Quantum Dot Lasers, Vertical Cavity Surface Emitting Lasers, and Quantum Cascade Lasers are reviewed, while the second part "Coupled Laser Devices" focusses on the complex dynamics and bifurcations induced by self coupling, delay coupling, or mode coupling of lasers. The third part, "Synchronization and Cryptography", discusses the chaotic dynamics of excitable systems and their application for secure communication or for the generation of synchronized cluster states in networks.

I am grateful to the group of Prof. Schöll for their enduring support during the compilation of this volume and to the staff from Wiley VCH for their excellent help.

Berlin, February 2011 *Kathy Lüdge*

List of Contributors

Andreas Amann
Tyndall National Institute
University College Cork
Lee Maltings
Cork
Ireland

Alexander G. Balanov
Department of Physics
Loughborough University
Loughborough
LE11 3TU
UK

Kristine E. Callan
Department of Physics
Duke University
Durham
North Carolina 27708
USA

Thomas Erneux
Université Libre de Bruxelles
Optique Nonlinéaire Théorique
Campus Plaine C.P. 231
1050 Bruxelles
Belgium

Ingo Fischer
Instituto de Física Interdisciplinar
y Sistemas Complejos, IFISC
(UIB-CSIC)
Campus Universitat de les Illes
Balears
07122 Palma de Mallorca
Spain

T. Mark Fromhold
School of Physics and Astronomy
University of Nottingham
University Park
Nottingham
NG7 2RD
UK

Jordi Garcia-Ojalvo
Departament de Fisica i
Enginyeria Nuclear
Universitat Politecnica de
Catalunya
08222 Terrassa
Barcelona
Spain

Daniel J. Gauthier
Department of Physics
Duke University
Durham
North Carolina 27708
USA

Cristina M. Gonzalez
Centre for Sensors, Instruments
and Systems Development
Universitat Politecnica de
Catalunya
08222 Terrassa
Barcelona
Spain

Mark T. Greenaway
School of Physics and Astronomy
University of Nottingham
University Park
Nottingham
NG7 2RD
UK

Lucas Illing
Department of Physics
Reed College
Portland
Oregon 97202
USA

Ido Kanter
Department of physics
Bar-Ilan University
Ramat-Gan 52900
Israel

Wolfgang Kinzel
Universität Würzburg
Theoretische Physik
Am Hubland
97074 Würzburg
Germany

Bernd Krauskopf
University of Bristol
Department of Engineering
Mathematics
Queen's Building
University Walk
Bristol BS8 1TR
UK

Leonhard Lücken
Institute of Mathematics
Humboldt University of Berlin
Unter den Linden 6
10099 Berlin
Germany

Kathy Lüdge
TU Berlin
Institut für Theoretische Physik
Fakultät Mathematik und
Naturwissenschaften
Hardenbergerstr. 36
10632 Berlin
Germany

Cristina Masoller
Departament de Física i
Enginyeria Nuclear
Universitat Politécnica de
Catalunya
Edifici Gaia, Rambla de Sant
Nebridi s/n
08222 Terrassa
Barcelona
Spain

Christian Otto
TU Berlin
Institut für Theoretische Physik
Fakultät Mathematik und
Naturwissenschaften
Hardenbergerstr. 36
10632 Berlin
Germany

Dmitrii Rachinskii
University College Cork
Department of Applied
Mathematics
Cork
Ireland

Marc Sciamanna
Supélec Campus de Metz
2 Rue Edouard Belin
57070 Metz
France

K. Alan Shore
Bangor University
School of Electronic Engineering
Dean Street
Bangor Gwynedd
LL57 1UT
Wales
UK

Miguel C. Soriano
Instituto de Física Interdisciplinar
y Sistemas Complejos, IFISC
(UIB-CSIC)
Campus Universitat de les Illes
Balears
07122 Palma de Mallorca
Spain

M. Carme Torrent
Departament de Fisica i
Enginyeria Nuclear
Universitat Politecnica de
Catalunya
08222 Terrassa
Barcelona
Spain

Evgeniy Viktorov
Université Libre de Bruxelles
Optique Nonlinéaire Théorique
Campus Plaine C.P. 231
1050 Bruxelles
Belgium

Andrei G. Vladimirov
Weierstrass Institute
Mohrenstr. 39
10117 Berlin
Germany

Andreas Wacker
Lund University
Division of Mathematical Physics,
Department of Physics
Box 118
22100 Lund
Sweden

Jamie J. Walker
University of Bristol
Department of Engineering
Mathematics, Queen's Building
University Walk, Bristol BS8 1TR
UK

Sebastian M. Wieczorek
Mathematics Research Institute
University of Exeter
EX4 4QF Exeter
UK

Matthias Wolfrum
Weierstrass Institute
Mohrenstr. 39
10117 Berlin
Germany

Serhiy Yanchuk
Institute of Mathematics
Humboldt University of Berlin
Unter den Linden 6
10099 Berlin
Germany

Jordi Zamora-Munt
Departament de Física i
Enginyeria Nuclear
Universitat Politècnica de
Catalunya
Edifici Gaia, Rambla de Sant
Nebridi s/n
08222 Terrassa
Barcelona
Spain

Part I
Nanostructured Devices

1
Modeling Quantum-Dot-Based Devices
Kathy Lüdge

1.1
Introduction

During the past decades, the performance of semiconductor lasers has been dramatically improved from a laboratory curiosity to a broadly used light source. Owing to their small size and low costs, they can be found in many commercial applications ranging from their use in DVD players to optical communication networks. The rapid progress in epitaxial growth techniques allows to design complex semiconductor laser devices with nanostructured active regions and, therefore, interesting dynamical properties. Future high-speed data communication applications demand devices that are insensitive to temperature variations and optical feedback effects, and provide features such as high modulation bandwidth and low chirp, as well as error-free operation. Currently, self-organized semiconductor quantum dot (QD) lasers are promising candidates for telecommunication applications [1]. For an introduction to QD-based devices, their growth process, and their optical properties, see, for example, [2, 3].

This review focuses on the modeling of these QD laser devices and on the discussion of their dynamic properties. It uses a microscopically based rate equation model that assumes a classical light field but includes microscopically calculated scattering rates for the collision terms in the carrier rate equations, as introduced in [4–8]. Following the hierarchy of different semiconductor modeling approaches (for an overview, see [9]), this model aims to be sophisticated enough to permit a quantitative modeling of the QD laser dynamics but still allows an analytic treatment of the dynamics. Different levels of complexity will be explored to enable comprehensive insights into the underlying processes.

In order to reduce the numeric effort and still allow for analytic insights, a variety of effects have been neglected. This way, a different approach has to be chosen if, for example, the photon statistics of the emitted light [10] or changes in the emission wavelength due to Coulomb enhancement effects [11, 12] are to be of interest. For the analysis of ultrafast phenomena, as, for example, the gain recovery in QD-based optical amplifiers [13], coherent effects resulting from the dynamics of the microscopic polarization become important, and the model has to be extended

Nonlinear Laser Dynamics: From Quantum Dots to Cryptography, First Edition. Edited by Kathy Lüdge.
© 2012 Wiley-VCH Verlag GmbH & Co. KGaA. Published 2012 by Wiley-VCH Verlag GmbH & Co. KGaA.

to semiconductor Bloch equations. This has been intensively studied in [14, 15] in good agreement with experimental results [16], but it will not be discussed in this review. Note that later on in this book, the experimental results obtained with QD lasers under optical injection are presented in Chapter 3 (by Sciamanna [17]), and the results regarding the sensitivity of QD lasers to optical feedback [18] are discussed in Chapter 6 by Erneux et al. [19].

After a detailed introduction to the microscopical modeling aspects in Section 1.2, the turn-on and switching dynamics of a QD laser with two confined levels is discussed in Sections 1.3 and 1.4, and temperature effects are analyzed in Section 3.1. In Section 1.5, the results of an asymptotic analysis of the rate equation systems are presented, which allows to give analytic expression to relaxation oscillation (RO) frequency and damping of the turn-on dynamics, and thus allows to predict the modulation properties of the laser. Resulting from the analytic predictions, the effect of using a doped carrier reservoir on the laser dynamics is investigated in Section 1.6. At the end, in Section 1.7, the results are discussed and compared to quantum well (QW) laser devices.

1.2
Microscopic Coulomb Scattering Rates

A schematic view of the QD laser structure is shown in Figure 1.1a. The active area of the p–n heterojunction is a dot-in-a-well (DWELL) structure that consists of several InGaAs QW layers that have a height of about 4 nm, and contain embedded QDs that are confined in all three dimensions having a size of approximately 4 nm × 18 nm × 18 nm. During laser operation, an electric current is injected into

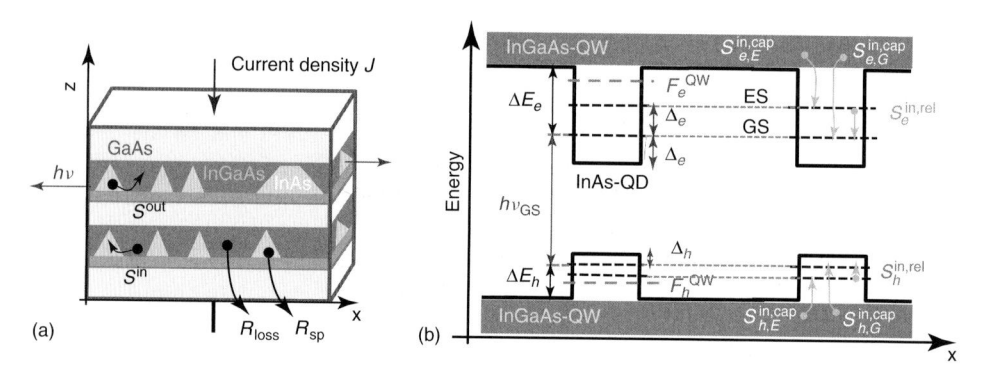

Figure 1.1 (a) Schematic illustration of the QD laser. (b) Energy diagram of the band structure across two QDs in the electron–hole picture. $h\nu_{GS}$ labels the ground state (GS) lasing energy. ΔE_e and ΔE_h mark the distance of the GS from the band edge of the 2D carrier reservoir (QW) for electrons and holes, respectively. Δ_e and Δ_h denote the distance to the bottom of the QD, which is equal to the energetic distances between the GS and the excited state (ES) in the QD. F_e^{QW} and F_h^{QW} are the quasi-Fermi levels for electrons and holes in the QW, respectively. The different processes of direct electron and hole capture ($S^{in,cap}$), as well as relaxation ($S^{in,rel}$) into the QD states, are indicated with gray arrows.

the QW layers. They form the carrier reservoir where carrier–carrier scattering events take place because of Coulomb interaction and lead to a filling (or depletion) of the confined QD levels. As a result, carrier inversion is reached first between the lowest confined QD levels in the conduction band and its counterpart in the valence band. Since the size and the composition of the zero-dimensional QD structures determine the energetic position of the QD levels, it is possible to design lasers with different emission wavelengths. The lasers discussed here have a ground state (GS) emission wavelength of 1.3 μm, as needed for optical data communication.

For high carrier densities in the reservoir, that is, during electrical pumping, the Coulomb interaction (carrier–carrier Auger scattering) will dominate the scattering rate into (and out of) the QDs, whereas the scattering events resulting from carrier–phonon interaction are negligible [20]. Inside the QD, two confined energy levels are modeled. Thus, direct capture processes for electrons ($b = e$) and holes ($b = h$) into or out of the GS labeled as $S_{b,G}^{cap}$, into or out of the excited state (ES) labeled as $S_{b,E}^{cap}$, and relaxation processes between GSs and ESs named S_b^{rel} are considered as depicted in Figure 1.1b, where gray arrows indicate the in-scattering events.

Section 1.2.1 systematically describes and quantifies the different Auger processes before they are incorporated into the dynamic rate equation model in Section 1.3. Note that although phonon scattering between the carrier reservoir (QW) and the QDs is neglected, the fast phonon- assisted carrier relaxation processes within the QW states are taken into account by assuming a quasi-Fermi distribution with quasi-Fermi levels F_e^{QW} and F_h^{QW} for electrons in the conduction band and holes in the valence band of the QW, respectively.

1.2.1
Carrier–Carrier Scattering

If the Coulomb interaction is treated in the second-order Born approximation in the Markov limit up to second order in the screened Coulomb potential [21, 22], a Boltzmann equation for the collision terms, which describe the change in the occupation probability in the QD states, can be derived, and subsequently easily incorporated into laser rate equation models (for details, see also [15]). The striking difference from the standard rate equation models is that there are no constant relaxation times. Instead, the detailed modeling of the scattering events inside the reservoir leads to scattering times that are nonlinearly dependent on the carrier densities in the reservoir.

Figure 1.2 gives a systematic overview of all processes leading to in-scattering into the QD electron levels. The gray arrows denote electron transitions of the scattering partners. Panels I and III show pure $e-e$ processes, while panels II and IV display mixed $e-h$ processes. The corresponding processes for in-scattering into the QD hole levels are obtained by exchanging all electron and hole states. The out-scattering processes are obtained by inverting all arrows of the electron transitions. The exchange processes of pure $e-e$ capture processes contributing to the scattering rates are not shown, since there is no qualitative difference from that of the direct processes. In case of mixed $e-h$ processes (II, IV), the exchange processes lead to

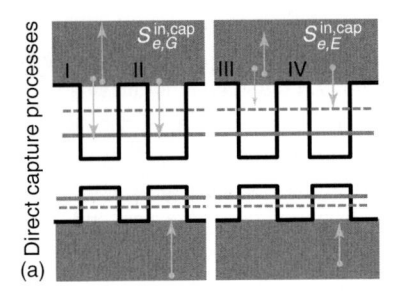

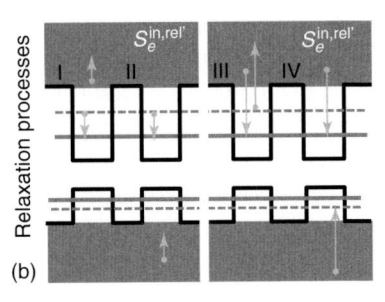

Figure 1.2 Electron transitions during Auger scattering processes (gray arrows denote electron transitions): (a) direct electron capture from the 2D carrier reservoir to the QD ground state (I, II) and first excited state (III, IV). Panels I and III and panels II and IV show pure e–e and mixed e–h scattering processes, respectively. (b) QW-assisted intradot electron relaxation to the QD ground state.

transitions across the band gap, which are neglected since they are unlikely to occur. Note that the process shown in panel III of Figure 1.2b is the exchange process of the one in panel I. In the following, the scattering events shown in Figure 1.2 are decomposed into contributions originating from direct carrier capture from the QW into the QD levels R_m^{cap} (Figure 1.2a) and relaxation processes between the QD states with one and two intra-QD transitions $R^{rel'}$ and $R^{rel''}$, respectively (Figure 1.2b). Processes involving three QD states are neglected. Thus, the collision term in the Boltzmann equation for the carrier occupation probability in the QD states ρ_b^m, where m labels the quantum number of the 2D angular momentum of the confined QD states ($m = E$ for the first ES; $m = G$ for the GS) reads:

$$\frac{\partial \rho_b^m}{\partial t}\bigg|_{col} = R_{b,m}^{cap} + R_b^{rel'} + R_b^{rel''} \tag{1.1}$$

The contribution to Eq. (1.1) from direct capture processes (Figure 1.2a) can be expressed as

$$R_{b,m}^{cap} = S_{b,m}^{in,cap}(1 - \rho_b^m) - S_{b,m}^{out,cap}\rho_b^m \tag{1.2}$$

where the direct capture Coulomb scattering rates for in- $\left(S_{b,m}^{in,cap}\right)$ and out-scattering $\left(S_{b,m}^{out,cap}\right)$ are defined as

$$S_{b,m}^{in,cap} = \sum_{k_1 k_2 k_3, b'} W_{k_1^b k_3^{b'} k_2^{b'} m}^b f_{k_1^b} f_{k_3^{b'}} \left(1 - f_{k_2^{b'}}\right) \quad (k_1 \to m, k_3 \to k_2), \tag{1.3}$$

$$S_{b,m}^{out,cap} = \sum_{k_1 k_2 k_3, b'} W_{m k_2^{b'} k_3^{b'} k_1^b}^b \left(1 - f_{k_1^b}\right)\left(1 - f_{k_3^{b'}}\right) f_{k_2^{b'}} \quad (m \to k_1, k_2 \to k_3). \tag{1.4}$$

States in the QW are labeled by the in-plane carrier momentum k_i^b ($b = e$ and $b = h$ indicate conduction and valence band states, respectively). For both bands in the QW, $f_{k_i^b}$ indicates the electron occupation probability. The transition probability $W_{k_1 k_2 k_3 m'}^b$ for a process where two carriers scatter from initial states k_1 and k_3 to the final states m and k_2, respectively, ($k_1 \to m$, $k_3 \to k_2$) contains the screened Coulomb matrix elements for direct and exchange interactions, and the energy-conserving δ-function [6, 15]. Owing to the microscopic reversibility of the Coulomb matrix elements, the transition probability is equal for reversed direction $W^b_{k^b_1 k^{b'}_3 k^{b'}_2 m} = W^b_{mk^{b'}_2 k^{b'}_3 k^b_1}$.

The relaxation processes shown in Figure 1.2b describe a redistribution of carriers within the intra-QD levels. The contribution from processes I and II to Eq. (1.1) is given by

$$R^{\text{rel}'}_b = S^{\text{in,rel}'}_b \rho^E_b (1 - \rho^G_b) - S^{\text{out,rel}'}_b (1 - \rho^E_b) \rho^G_b. \tag{1.5}$$

The relaxation in-scattering rate is given by

$$S^{\text{in,rel}'}_b = \sum_{k_2 k_3, b'} W^b_{E k^{b'}_3 k^{b'}_2 G} \left(1 - f_{k^{b'}_2}\right) f_{k^{b'}_3} \qquad (E \to G, k_3 \to k_2). \tag{1.6}$$

The dynamical equations for the processes III and IV ($R^{\text{rel}''}$) in Figure 1.2b can be obtained in a similar manner as in Eq. (1.5).

For the calculation of the Coulomb scattering rates, a quasiequilibrium *within* the QW states (fast phonon scattering inside one band) but nonequilibrium *between* the QW electrons and the QD electrons, the QW holes, and the QD holes is assumed. As a result, the electron occupation probability f_{kb} in the conduction ($b = e$) and valence band ($b = h$) of the QW can be expressed by a quasi-Fermi distribution given by

$$f_{kb} = \left[\exp\left(\frac{E_k - F^{\text{QW}}_b}{kT}\right) + 1\right]^{-1} \qquad (b = e, h). \tag{1.7}$$

The quasi-Fermi levels F^{QW}_b are determined by the total carrier density in the respective band via the relation given in Eq. (1.8), as shown in [7, 23],

$$F^{\text{QW}}_b(w_b) = E^{\text{QW}}_b \pm kT \ln\left[\exp\left(\frac{w_b}{D_b kT}\right) - 1\right] \tag{1.8}$$

where the + and − signs correspond to electrons and holes, respectively. Furthermore, $D_b = m_b/(\pi \hbar^2)$ is the 2D density of states, with the effective masses m_b of electrons ($b = e$) and holes ($b = h$), respectively. E^{QW}_b are the QW band edges of conduction and the valence band, respectively. Note that the analytic expression Eq. (1.8) is only valid for a 2D electron gas, where the integrals

$$w_e = \int_{E^{\text{QW}}_e}^{\infty} dE_k D_e f_{ke} \quad \text{and} \quad w_h = \int_{-\infty}^{E^{\text{QW}}_h} dE_k D_h (1 - f_{kh}) \tag{1.9}$$

can be solved. As a result, the quasi-Fermi distributions f_{ke} and f_{kh} are determined by the QW carrier densities w_e and w_h, and thus, the scattering rates given in Eqs. (1.3) and (1.6) are calculated as functions of w_e and w_h. Besides that, the scattering rates parametrically depend on the effective masses of the carriers in the QW bands and on the band structure given by the energetic distances ΔE_b and Δ_b, as indicated in Figure 1.1b. The resulting rates are shown in Figure 1.3 as a function of w_e along the line $w_h/w_e = 1.5$. For the relaxation rates, the sum of all relaxation processes

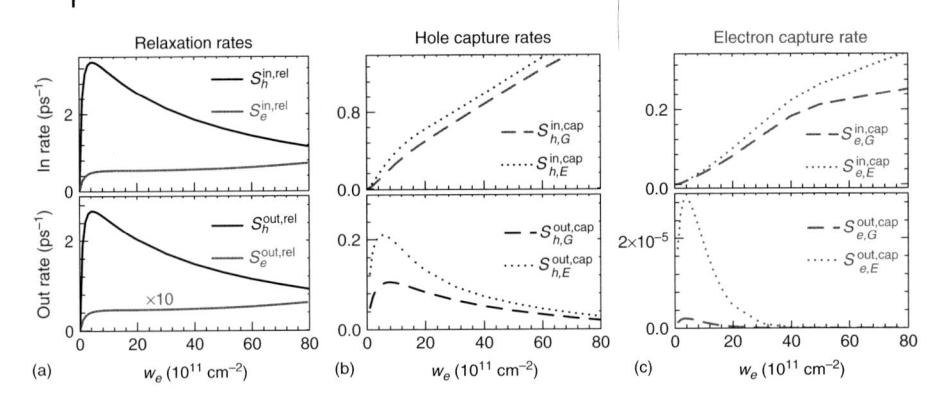

Figure 1.3 Coulomb scattering rates of the QDs-in-a-well system versus QW electron density w_e ($w_h/w_e = 1.5$). (a) Intra-QD relaxation rates for electrons (gray) and holes (black); (b) and (c) direct capture rates into the GS (dashed line) and ES (dotted line) for holes and electrons, respectively. Top and bottom panels show in- and out-scattering rates, respectively. Parameters as in Table 1.1.

is plotted but note that the rates involving a transition within the QD accompanied by a QW transition (rel′) are much larger than the rates involving two QW–QD transitions (rel″). The relaxation rates are characterized first by a sharp increase and later by a decrease in higher carrier densities because of the effect of Pauli blocking. These relaxation scattering events are on a ps time scale, whereas the direct capture rates plotted in Figure 1.3b,c for holes and electrons are an order of magnitude smaller for small carrier densities. Owing to their small effective mass, the rate for electron capture is much smaller, although the dependence on w_e is similar to that of the hole rate. For small electron densities inside the QW, the capture rates increase quadratically with w_e, which is expected from mass action kinetics.

1.2.2 Detailed Balance

In thermodynamic equilibrium, there is a detailed balance between the in- and out-scattering rates of the QD level. This allows one to relate the rate coefficients of in- and out-scattering even for nonequilibrium carrier densities [24].

For a single scattering process between two carriers of type b and b', the in-scattering rate for capture into the GS ($m = \text{G}$) or ES ($m = \text{E}$) is defined in Eq. (1.3), and can be rewritten as

$$W^b_{k_1^b k_3^{b'} k_2^{b'} m} f_{k_1^b} f_{k_3^{b'}} (1 - f_{k_2^{b'}}) \tag{1.10}$$

$$= W^b_{k_1^b k_3^{b'} k_2^{b'} m} (1 - f_{k_1^b})(1 - f_{k_3^{b'}}) f_{k_2^{b'}} \frac{f_{k_1^b}}{1 - f_{k_1^b}} \frac{f_{k_3^{b'}}}{1 - f_{k_3^{b'}}} \frac{1 - f_{k_2^{b'}}}{f_{k_2^{b'}}} \tag{1.11}$$

$$= W^b_{k_1^b k_3^{b'} k_2^{b'} m} (1 - f_{k_1^b})(1 - f_{k_3^{b'}}) f_{k_2^{b'}} \exp\left[\frac{F_e^{QW} - E_{k_1^b} - E_{k_3^{b'}} + E_{k_2^{b'}}}{kT}\right] \tag{1.12}$$

if the quasi-Fermi distribution given in Eq. (1.7) is used, which leads to $f_k/(1-f_k) = \exp\left[(F_e^{QW} - E_k)/(kT)\right]$. Inserting the energy conservation of final and initial states, $E_{k_2'} - E_{k_3'} - E_{k_1} + E_{b,m}^{QD} = 0$, where $E_{b,m}^{QD}$ is the confined QD energy ($m = G, E$) for electrons ($b = e$) or holes ($b = h$), and summing overall initial and final states in k-space gives

$$S_{b,m}^{\text{in,cap}} = S_{b,m}^{\text{out,cap}} e^{\pm\left(F_b^{QW} - E_{b,m}^{QD}\right)/kT} = S_{b,m}^{\text{out,cap}} e^{\frac{\Delta E_{b,m}}{kT}}\left[e^{\frac{w_b}{D_b kT}} - 1\right] \quad (1.13)$$

where $\Delta E_{b,G} = \pm(E_b^{QW} - E_{b,G}^{QD})$ and $\Delta E_{b,E} = \pm(E_b^{QW} - E_{b,E}^{QD})$ are the energetic distances from the QW band edge to the GS and the ES of the QD, respectively, and the $+$ and $-$ signs correspond to electrons and holes, respectively.

Note that Eq. (1.13) holds for the mixed e–h Auger capture process ($b \neq b'$) as well as for the e–e and h–h processes. Thus, besides the Boltzmann factor $e^{\frac{\Delta E}{kT}}$ that is valid for a discrete two-level system with energy difference ΔE, the ratio ($S_b^{\text{in}}/S_b^{\text{out}}$) for Auger scattering between the 2D electron gas of the QW and the discrete QD level also depends on the quasi-Fermi levels F_b^{QW}, and thereby on the carrier density in the QW. As a result of this carrier-density-dependent factor in the detailed balance relation, the out-scattering rates show a pronounced maximum around the degeneracy concentration $D_b kT$, as can be seen in the bottom panel of Figure 1.3b,c.

In contrast to that, the ratio between the in- and out-scattering relaxation rates is a constant factor since both involved levels are indeed localized. For the positively defined energy difference Δ_b ($b = e, h$) between ES and GS (Figure 1.1b), the relation reads:

$$S_b^{\text{in,rel}} = S_b^{\text{out,rel}} e^{\frac{\Delta_b}{kT}} \quad (1.14)$$

1.3
Laser Model with Ground and Excited States in the QDs

Using the microscopic scattering rates defined in the last section, an eight-variable rate equation system can be formulated, which contains the Boltzmann collision terms for the direct capture processes, $R_{b,m}^{\text{cap}}$, defined in Eq. (1.2), and those for relaxation into the GS, R_b^{rel}, defined in Eq. (1.5). As used earlier for the scattering contributions in Section 1.2, carrier densities in the GS and ES have the index G and E, respectively, and the index b labels the carrier type. Further, the photon densities n_{ph}^G and n_{ph}^E are introduced, which result from the GS and ES transition in the QD, respectively. Starting from the occupation probability of the confined QD levels $\rho_b^{G,E}$, the carrier densities in the QD are defined by $n_b^{G,E} = N^{QD} \nu_{G,E} \rho_b^{G,E}$. N^{QD} denotes twice the QD density of the lasing subgroup (the factor of 2 accounts for spin degeneracy), and $\nu_{G,E}$ is the degeneracy of the states ($\nu_G = 1, \nu_E = 2$).

The induced processes of absorption and emission at the GS wavelength are modeled by a linear gain $R_{\text{ind}}^{G} = WA(n_{e}^{G} + n_{h}^{G} - N^{\text{QD}}) n_{\text{ph}}^{G}$, where W is the Einstein coefficient that can be determined from a full quantum mechanical approach of the light matter interaction [9], and A is the in-plane area of the QW. Analogous to the simple two-level system, the model introduced above yields positive gains if the occupation probability of electrons in the localized conduction band level $f_{e}^{C} = \rho_{e}^{G} = n_{e}^{G}/N^{\text{QD}}$ of the QDs is higher than the occupation probability of electrons in their localized valence band level $f_{e}^{V} = 1 - \rho_{h}^{G}$. Thus, the linear gain term $R_{\text{ind}} = WAN^{\text{QD}}(f_{e}^{C} - f_{e}^{V}) n_{\text{ph}} = WAN^{\text{QD}}(f_{e}^{C}(1 - f_{e}^{V}) - f_{e}^{V}(1 - f_{e}^{C})) n_{\text{ph}}$ corresponds to the standard net rate of stimulated emission minus absorption [25]. The rate of induced emission at the ES wavelength is obtained analogously, but by assuming a different Einstein coefficient $\overline{W}$; thus, $R_{\text{ind}}^{E} = \overline{W}A(n_{e}^{E} + n_{h}^{E} - 2N^{\text{QD}}) n_{\text{ph}}^{E}$. As a result of the size distribution and material composition fluctuations of the QDs, only a subgroup (QD density N^{QD}) of all QDs (N^{sum}) matches the mode energies for lasing. The QD density N^{sum} is twice the total QD density as given by experimental surface imaging (again, the factor of 2 accounts for spin degeneracy). As discussed below, N^{QD} is not a constant but can increase with increasing pump current if the number of longitudinal modes in the laser output is increased (see Figure 1.9a for experimental lasing spectra).

The nonlinear rate equations (Eqs. (1.15)–(1.19)) describe the dynamics of the charge carrier densities in the GS and ES of the QDs, n_{b}^{G} and n_{b}^{E}, respectively, the carrier densities in the QW, w_{b}, and the photon density emitted from the GS and the ES, n_{ph}^{G} and n_{ph}^{E}, respectively.

$$\dot{n}_{b}^{E} = N^{\text{QD}}(2R_{b,E}^{\text{cap}} - R_{b}^{\text{rel}}) - \overline{W}A(n_{e}^{E} + n_{h}^{E} - 2N^{\text{QD}})n_{\text{ph}}^{E} - \overline{W}n_{e}^{E}\rho_{h}^{E}, \quad (1.15)$$

$$\dot{n}_{b}^{G} = N^{\text{QD}}(R_{b,G}^{\text{cap}} + R_{b}^{\text{rel}}) - WA(n_{e}^{G} + n_{h}^{G} - N^{\text{QD}})n_{\text{ph}}^{G} - Wn_{e}^{G}\rho_{h}^{G} \quad (1.16)$$

$$\dot{w}_{b} = \eta\frac{J(t)}{e_{0}} - N^{\text{sum}}\left[R_{b,G}^{\text{cap}} + 2R_{b,E}^{\text{cap}}\right] - Bw_{e}w_{h}, \quad (1.17)$$

$$\dot{n}_{\text{ph}}^{G} = -2\kappa n_{\text{ph}}^{G} + \Gamma WA(n_{e}^{G} + n_{h}^{G} - N^{\text{QD}})n_{\text{ph}}^{G} + \beta Wn_{e}^{G}\rho_{h}^{G} \quad (1.18)$$

$$\dot{n}_{\text{ph}}^{E} = -2\kappa n_{\text{ph}}^{E} + \Gamma\overline{W}A(n_{e}^{E} + n_{h}^{E} - 2N^{\text{QD}})n_{\text{ph}}^{E} + \beta\overline{W}n_{e}^{E}\rho_{h}^{E} \quad (1.19)$$

The spontaneous emission in each level of the QDs is approximated by bimolecular recombination using $R_{\text{sp}}^{G}(n_{e}^{G}, n_{h}^{G}) = Wn_{e}^{G}n_{h}^{G}/N^{\text{QD}}$ and $R_{\text{sp}}^{E}(n_{e}^{E}, n_{h}^{E}) = \overline{W}n_{e}^{E}n_{h}^{E}/(2N^{\text{QD}})$. The loss rate $R_{\text{loss}}^{b} = Bw_{e}w_{h}$, accounting for carrier losses in the QW, is a sum of the spontaneous band–band recombination and Auger-related losses inside the QW [26]. This loss rate determines the lifetime τ_{w}^{b} of carriers in the QW ($R_{\text{loss}}^{b} \equiv w_{b}/\tau_{w}^{b}$), which is on the order of several nanoseconds and decreases with the carrier densities w_{b}. β is the spontaneous emission coefficient, and $\Gamma = \Gamma_{g}N^{\text{QD}}/N^{\text{sum}}$ is the optical confinement factor. Γ is the product of the geometric confinement factor Γ_{g} (i.e., the ratio of the volume of all QDs and the mode volume), and the ratio $N^{\text{QD}}/N^{\text{sum}}$ (accounting for reduced gain since only a subgroup of all QDs matches the mode energy for lasing because of the size distribution and material composition fluctuations of the QDs). The coefficient $2\kappa = (c/\sqrt{\varepsilon_{\text{bg}}})[\kappa_{\text{int}} - \ln(R_{1}R_{2})/2L]$ expresses the total cavity loss [2], where L is the cavity length, and R_{1} and R_{2} are the facet reflectivities, and κ_{int} are the internal losses [6]. J is the injection current density, e_o is the elementary charge, and $\eta = 1 - w_e/N^{QW}$ is the current injection efficiency that accounts for the fact that the injection into the QW is blocked if the QW is already filled (maximum density inside the QW: $w_e = N^{QW}$). Note that within the model the carriers are directly injected into the QW, leading, of course, to an underestimation of the experimentally realized current densities. Therefore, only current densities relative to the threshold value J_{th} are considered for comparisons between theory and experiment. The values of parameters used for the simulations are listed in Table 1.1, if not stated otherwise.

The steady-state characteristics and the turn-on dynamics for the QD laser as resulting from the nonlinear rate equations (Eqs. (1.15)–(1.19)) are depicted in Figure 1.4. The input–output curves in Figure 1.4b show that with increasing pump current, the GS first reaches inversion and starts lasing at the GS threshold current J_{th}^G. By further increasing the pump current, the ES reaches its lasing threshold J_{th}^E and the laser emits light at both wavelengths. As can be seen in Figure 1.4b, the GS efficiency is reduced as soon as the ES lasing sets in. The turn-on dynamics observed before reaching the steady states is shown in Figure 1.4a,c. For currents above J_{th}^G but far below J_{th}^E, highly damped ROs are found for the GS turn-on trajectories (Figure 1.4a) in accordance with experimental results [26]. Above J_{th}^E, the ES turns on with very short turn-on delay times and damped ROs (gray line in Figure 1.4c), while the GS shows overdamped turn-on behavior (black line in Figure 1.4c). The overdamped behavior is due to the high current needed to invert the ES levels, which is accompanied by high carrier densities in the reservoir and thus by high scattering rates into the GS (see Section 1.5 for analytic discussions of the damping rate, which depends on the carrier lifetimes).

The ratio of the threshold currents of the two modes, J_{th}^E/J_{th}^G, depends on the values of the carrier capture and relaxation rates and can be changed by varying the band structure of the QD–QW system. A system where the QW band edge is very close to the ES leads to a faster filling of the ES and thus to a smaller J_{th}^E (compare Figure 1.4b and Figure 1.6c that show the input–output curves for different confinement energies). Besides this microscopic effects, the ratio J_{th}^E/J_{th}^G also depends on the device length. Length-dependent measurements of this ratio

Table 1.1 Numerical parameters used in the simulation, unless stated otherwise.

Symbol	Value	Symbol	Value	Symbol	Value
W	0.7 ns^{-1}	A	4×10^{-5} cm^2	ΔE_e	210 meV
$\overline{W}$	0.88 ns^{-1}	N^{QD}	0.6×10^{10} cm^{-2}	ΔE_h	50 meV
B	0.5 nm^2 ps^{-1}	N^{sum}	6×10^{10} cm^{-2}	Δ_e	64 meV
Γ_g	0.06	N^{QW}	1×10^{12} cm^{-2}	Δ_h	6 meV
2κ	0.16 ps^{-1}	β	5×10^{-6}	$D_{e/h}$	$m_{e/h}/(\pi \hbar^2)$
R_1, R_2	0.32	L	1 mm	m_e	$0.043\, m_0$
ε_{bg}	14.2	κ_{int}	650 m^{-1}	m_h	$0.45\, m_0$

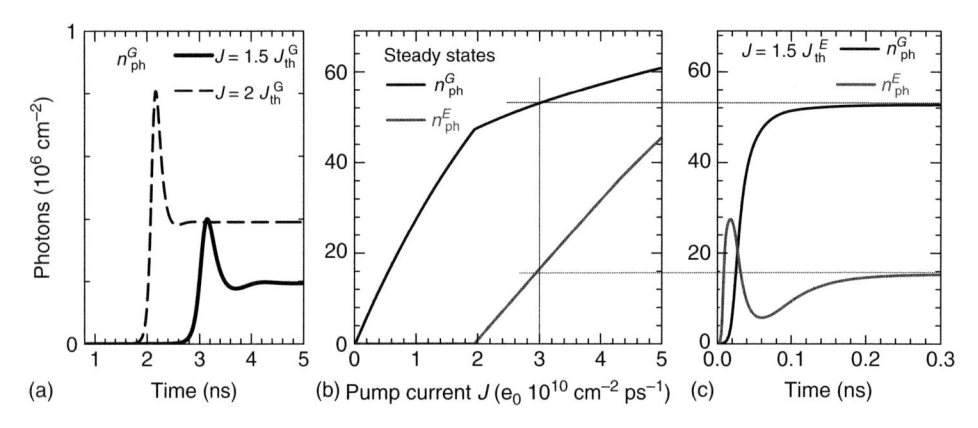

Figure 1.4 Simulated laser turn-on dynamics at the (a) GS lasing wavelength for two different small currents below the ES threshold J_{th}^E and (c) ES and GS lasing above J_{th}^E at $J = 1.5 J_{th}^E$; (b) steady-state photon output from the ES (gray) and GS (black) transition. $J_{th}^E = 1.95 \times 10^8 e_0$ cm^{-2} ps^{-1}; $J_{th}^G = 1.5 \times 10^{10} e_0$ cm^{-2} ps^{-1}. Parameters as in Table 1.1 but $\Delta E_e = 134$ meV, $\Delta E_h = 30$ meV and $B = 0.2$ nm^2 ps^{-1}.

can be found in [27], showing that the shorter the device the smaller is J_{th}^E, while J_{th}^G increases. This is in good agreement with our simulations.

The threshold current J_{th}^G can be obtained from Eqs. (1.15)–(1.19) by deriving the steady-state characteristics of the laser. By neglecting spontaneous emission and photons from the ES transition ($\beta = 0$; $n_{ph}^E = 0$), this leads to

$$n_{ph}^{G\,*}(J) = J_{th}^G \frac{\Gamma N^{QD}}{2\kappa N^{sum}} \left(\frac{J}{J_{th}^G} - 1 \right) \tag{1.20}$$

$$J_{th}^G = B w_{e|th} w_{h|th} + \frac{N^{sum}}{(N^{QD})^2} \left(W n_{e|th}^G n_{h|th}^G + \overline{W} n_{e|th}^E n_{h|th}^E \right). \tag{1.21}$$

Eq. (1.21) shows that J_{th}^G depends upon the loss terms in the rate equations. Thus, the parameters B, W, and N^{sum}, as well as the carrier densities at threshold, labeled with the subscript $|_{th}$ in Eq. (1.21), determine J_{th}^G. The threshold carrier densities are determined by the different scattering contributions (they do not depend on B and N^{sum}). However, owing to the nonlinear dependence of the Auger scattering rates upon the QW carrier densities, it is not possible to give closed analytic expression (see [7] for approximations). The ES threshold current J_{th}^E also depends on the photon density in the GS, which depends on the pump current and the differential gain. The analytic expression reads:

$$J_{th}^E = J_{th}^G + \frac{2\kappa N^{sum}}{\Gamma N^{QD}} n_{ph}^G. \tag{1.22}$$

Since the microscopic model allows for a separate treatment of electron and hole dynamics, the transient behavior of both species will be investigated. Figure 1.5a,b shows the trajectories of the turn-on process projected onto the (n_b^G, n_{ph}^G)-planes. The familiar anticlockwise rotation can be seen for the electron as well as for the

1.3 Laser Model with Ground and Excited States in the QDs

hole density. Nonetheless, their shape is different. The black stars in Figure 1.5a,b denote the steady state values of the electron and hole concentration in the GS levels, n_e^{G*} and n_h^{G*}, respectively, for increasing pump currents. It is interesting to note that the value of n_e^{G*} decreases with J. This is anomalous because the carrier concentration for conventional lasers is clamped at the threshold value (saturation of inversion). Nevertheless, the inversion of the QDs is saturated as the total number of carriers, namely, the threshold density $n_t^G = n_e^{G*} + n_h^{G*}$, is a constant that depends only on the material parameters and not on the pump current. n_t^G can be obtained by neglecting spontaneous emission in Eq. (1.18) and setting $\dot{n}_{ph}^G = 0$:

$$n_t^G = n_e^{G*} + n_h^{G*} = \frac{2\kappa}{\Gamma W A} + N^{QD}. \tag{1.23}$$

Figure 1.5 also reveals that the steady-state values of n_e^{G*} and n_h^{G*} differ a lot. While in the steady state most of the QDs are occupied by an electron, only every fifth hole state is filled. This effect is due to the high out-scattering rates for the holes, which inhibits effective filling of the states. As known from the microscopic scattering rates plotted in Figure 1.3, the hole out-scattering rate decreases with the carrier density in the reservoir and thus with the pump current. This leads to higher n_h^{G*} for higher currents (see [7] for detailed steady-state analysis of a QD laser with one confined level).

As can be seen in the phase portrait of Figure 1.5c, the turn-on process projected onto the (n_e, n_h)-plane deviates from a straight line (which corresponds to the synchronized behavior $n_e \sim n_h$) and instead performs a spiral ending in the fixed point (steady state). This desynchronization between electron and hole dynamics is due to the different carrier lifetimes that stem from the different effective masses and the resulting different energy separation between QW band edge and confined QD level (ΔE_e and ΔE_h in Figure 1.1b).

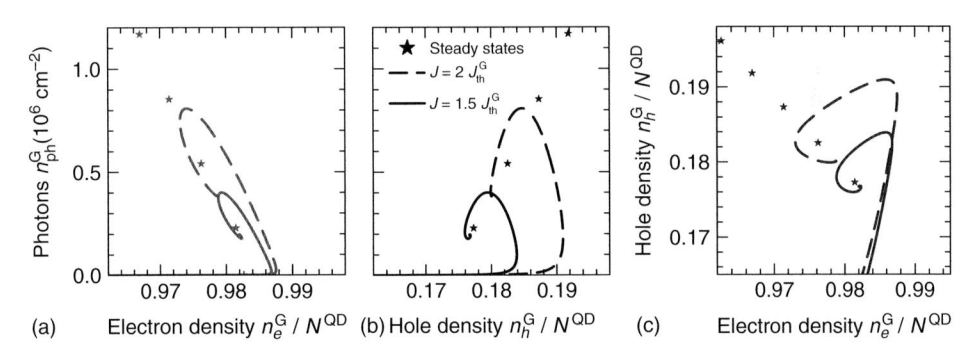

Figure 1.5 Turn-on trajectories for two different pump currents projected onto the (a) (n_e^G, n_{ph}^G)-plane, (b) (n_h^G, n_{ph}^G)-plane, and (c) (n_e^G, n_h^G)-plane; solid and dashed lines correspond to pump currents $J = 1.5 J_{th}^G$ and $J = 2 J_{th}^G$; stars represent the steady-state values for increasing pump currents. Parameters as in Table 1.1 but $B = 0.2$ nm^2 ps^{-1}, $\Delta E_e = 134$ meV and $\Delta E_h = 30$ meV.

1.3.1
Temperature Effects

So far, all simulations have been performed at a constant temperature of 300 K. In this section, the model is refined to account for carrier heating during laser operation. For the shift in the device temperature inside an electrically pumped optical amplifier (with identical active region as the laser diode considered here) Gomis-Bresco et al. [13] found values of $\Delta T = 60$ K at a pump current of $I = 150$ mA, which is about 10 times J_{th}^G. Their measurement suggests a functional relationship of $\Delta T(J) \sim J^2$, which is adapted by implementing $\Delta T(w_e) \sim (w_e)^2$ (see Eq. (1.24)). It is noted that the steady-state relation $w_e^*(J)$ plotted for the discussed model in Figure 1.6f depends on the microscopic details of the scattering processes and is thus different for a QD laser with only one GS as in [26]. The physical reason for the carrier heating lies first in the facts that the carriers are injected into higher k-states during the electrical pumping. Second, the Auger scattering processes between QD and QW lead to scattering into high energy states inside the QW. Both effects change the carrier distribution and if the carriers in the reservoir do not have time to cool down to the lattice temperature, their temperature stays increased. (See [28] for a detailed kinetic modeling of the relaxation processes that allow to determine the carrier temperature from their distribution in k-space and [29] for microscopic calculations of the carrier heating in the low-density limit.) Consequently, the temperature entering the scattering rates is actually not the lattice temperature but the temperature of the carriers inside the QW that surrounds the QDs.

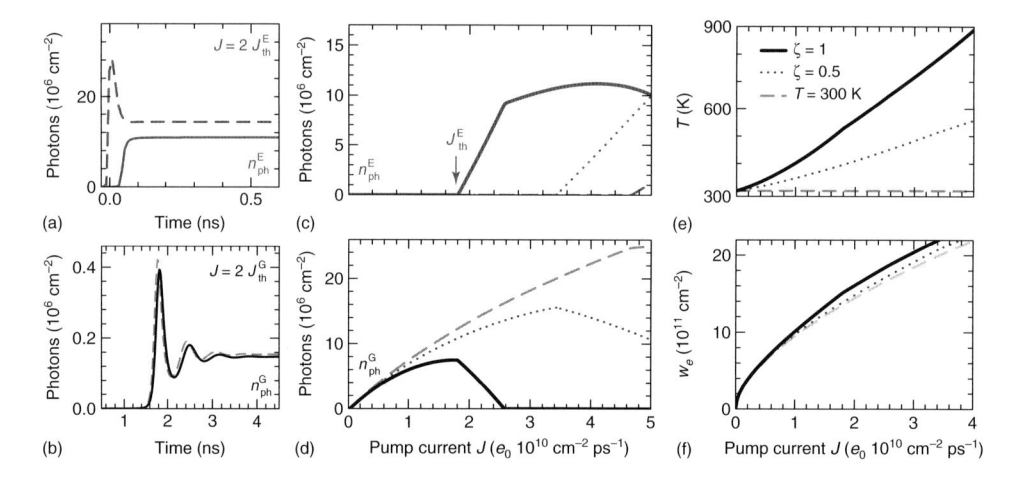

Figure 1.6 Simulated turn-on dynamics for (a) GS lasing ($J = 2J_{th}^G$) and (b) ES lasing ($J = 2J_{th}^E$). (c) and (d) show steady-state characteristics for ES and GS lasing, respectively. (e) and (f) show T and w_e as a function of the pump current J, respectively. In all plots, solid, dotted, and dashed lines represent simulations with increasing T according to Eq. (1.24) with $\zeta = 1$; $\zeta = 0.5$; and $\zeta = 0$. Parameters as in Table 1.1.

$$T = 300 \text{ K} + \zeta \cdot \left(w_e/(w_{e|\text{th}})\right)^2 \tag{1.24}$$

Note that an approach to directly implement $T(J)$, as in [30], suffers from the problem that a large signal modulation of the current leads to unphysical instantaneous switching in the temperature. An alternative approach is to determine the carrier heating from an additional rate equation for the energy density of the carriers in the reservoir [31].

Since the temperature enters the quasi-Fermi distribution that is assumed inside the QW, the microscopic scattering rates were calculated for several temperatures and implemented into the numeric simulation by approximated analytic expressions:

$$S_e^{\text{in}}(T, w_e, w_h) = (1 + 0.22(T - 300 \text{ K})/100 \text{ K}) \cdot S_e^{\text{in}}(300 \text{ K}, w_e, w_h)$$
$$S_h^{\text{in}}(T, w_e, w_h) = (1 + 0.26(T - 300 \text{ K})/100 \text{ K}) \cdot S_h^{\text{in}}(300 \text{ K}, w_e, w_h)$$

The out-scattering rates are related to the in-scattering rates by the detailed balance relations derived in Eqs. (1.13) and (1.14). Figure 1.6c,d shows the changes in the laser turn-on and steady-state dynamics if a dynamic temperature given by Eq. (1.24) is implemented for different constants ζ. The carrier temperature T and the carrier density in the reservoir w_e for the three cases are plotted in Figure 1.6e,f, respectively, as a function of J. Below the ES threshold current, the increasing temperature leads to a reduction in the differential efficiency of the GS steady-state characteristics as can be seen in Figure 1.6d. Furthermore, it reduces the ES threshold J_{th}^E, which results in two-state lasing at smaller pump currents (Figure 1.6c). In contrast to the case with constant T, the GS lasing is reduced as soon as the light is emitted from the ES. For high values of ζ, the GS lasing is completely suppressed. The turn-on dynamics of the ES is also affected by the high temperature (Figure 1.6a). Mainly due to the increased scattering rates at high T, the turn-on process becomes overdamped without a pronounced relaxation peak. At low currents close to the GS threshold J_{th}^G, the temperature does not change much and, thus, the turn-on process is also nearly unchanged (Figure 1.6b).

If these results are compared to two-state lasing experiments, a good agreement can be found. The suppression of the GS emission is indeed observed in experiments done by Wu *et al.* [32] on InP devices or by Ji *et al.* [30] on GaAs QD devices.

1.3.2
Impact of Energy Confinement

The energy diagram of the QD laser structure along the in-plane direction is shown in Figure 1.1. In principle, the sum $\Delta E_h + \Delta E_e$ can be determined from photoluminescence experiments that measure the energy of the GS emission of the QD ($h\nu_{\text{GS}}$) and the wavelength of the QW emission. However, for the devices used here, there is a large uncertainty for the position of the QW band edge. Increasing the distance to the QD confined levels reduces the capture rates but does not have a large effect on the relaxation rates between the QD states. Simulations of the QD laser with larger confinement energies ΔE_e and ΔE_h show two major changes in the dynamics. At first, the ratio between the threshold currents J_{th}^E/J_{th}^G increases, as can be seen by comparing the steady-state characteristics shown in Figure 1.4b with the simulations for larger confinement energies in Figure 1.6c. The effect can be explained with the reduced capture rates into the ES, which inhibit an effective filling of the ES.

Another change in the dynamics that results from changes in the confinement energies is the reduced damping of the ROs. This can be seen if the GS turn-on in Figure 1.6b for higher $\Delta E_e(\Delta E_h) = 210(50)$ meV is compared to the turn-on with smaller $\Delta E_e(\Delta E_h) = 134(30)$ meV in Figure 1.4a. Similar to the case of a damped harmonic oscillator, the damping of the turn-on dynamics determines the response of the laser to a pump current that is modulated with a certain frequency and a small modulation amplitude. Modulation response curves obtained for different pump currents (close to the GS threshold) as a function of the modulating frequency are plotted in Figure 1.7a,c for the two different confinement energies discussed so far. Note that the parameter for the losses in the reservoir, B, is different in both cases to yield equal threshold currents of $J_{th} = 3.4 \times 10^8 e_0$ cm^{-2} ps^{-1} and thus, according to Eq. (1.39), an RO frequency that is also observed in experiments (Figure 1.7b). The modulation response for the less damped case shown in Figure 1.7a shows a pronounced maximum at the frequency of the ROs, whereas it disappears for the strongly damped case in Figure 1.7c. The explanation for the impact of the confinement energies on the damping rate is given later on by using asymptotic methods in Section 1.5. There it is shown that the scattering rates, that is, the carrier lifetimes, determine the damping of the turn-on process, and increasing the lifetimes (decreasing the rates) of the smaller species (electrons) reduces the damping. The total lifetimes (including all capture and relaxation processes) of the GS levels are plotted in Figure 1.8a as a function of J for the two different cases discussed above. Obviously the lifetimes are decreased by decreasing ΔE_h but in

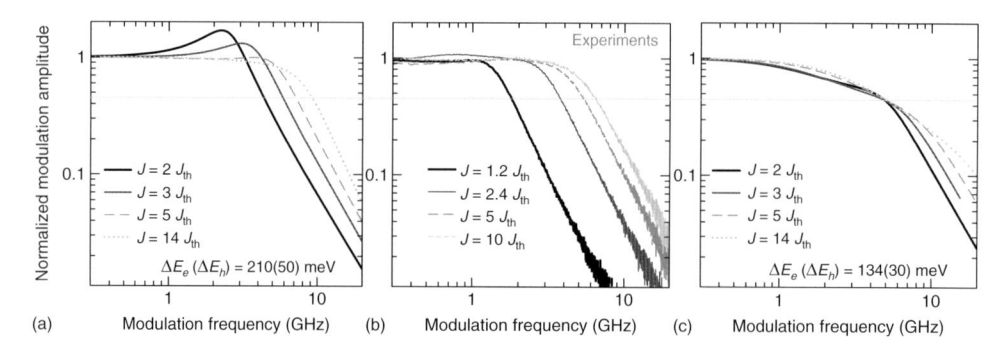

Figure 1.7 Small signal modulation response curves for different pump currents J simulated for confinement energies of (a) $\Delta E_e(\Delta E_h) = 210(50)$ meV ($B = 1.5$ nm^2 ps^{-1}) and (c) $\Delta E_e(\Delta E_h) = 134(30)$ meV ($B = 0.5$ nm^2 ps^{-1}). Other parameters as in Table 1.1. Measured curves taken from [6] are plotted in (b).

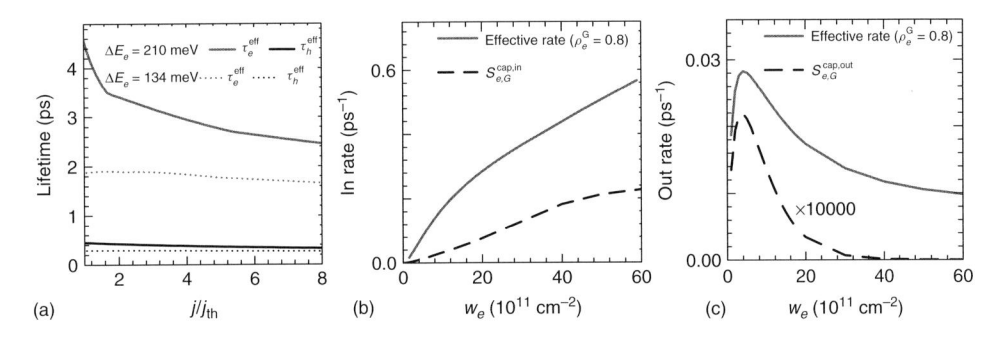

Figure 1.8 (a) Effective GS lifetimes τ_b^{eff} of electrons (gray) and holes (black) as a function of J for $\Delta E_e(\Delta E_h) = 210(50)$ meV (solid) and $\Delta E_e(\Delta E_h) = 134(30)$ meV (dotted). (b) and (c) show effective in- and out-scattering capture rates, respectively, resulting from adiabatic elimination of the ES variables according to Eq. (1.26).

both cases, the hole lifetimes are an order of magnitude smaller than those of the electrons.

1.3.3
Eliminating the Excited State Population Dynamics

One way to simplify the eight-variable rate equation system for the case where light is only emitted from the GS is to eliminate the ES carrier populations. This can be done by adiabatic elimination, which assumes a fast relaxation of the ES variables to their steady-state values and thus assumes $\dot{n}_b^E = 0$. Using Eq. (1.15) gives

$$n_b^E(w_e, w_h, n_b^G) = \frac{2N^{QD} S_{b,E}^{\text{cap,in}} + S_b^{\text{rel,out}} n_b^G}{S_{b,E}^{\text{cap,in}} + S_{b,E}^{\text{cap,out}} + (2N^{QD})^{-1}\left[S_b^{\text{rel,in}}(N^{QD} - n_b^G) + S_b^{\text{rel,out}} n_b^G\right]}. \quad (1.25)$$

Rewriting the remaining equations for w_b and n_b^G leads to

$$\dot{n}_b^G = (N^{QD} - n_b^G)\left[n_b^E \frac{S_b^{\text{rel,in}}}{2N^{QD}} + S_{b,G}^{\text{cap,in}}\right] - R_{\text{ind}}^G - R_{\text{sp}}^G$$

$$- n_b^G \left[(2N^{QD} - n_b^E)\frac{S_b^{\text{rel,out}}}{2N^{QD}} + S_{b,G}^{\text{cap,out}}\right], \quad (1.26)$$

$$\dot{w}_b = \eta \frac{J(t)}{e_0} - N^{\text{sum}} R_{b,G}^{\text{cap}} - \left[Bw_e w_h + 2N^{\text{sum}} R_{b,E}^{\text{cap}}\right], \quad (1.27)$$

Together with the unchanged equation for the photon density, Eq. (1.18), these equations resemble the five-variable case of a QD laser with one confined level if the terms in square brackets in Eq. (1.26) are interpreted as effective in- and out-scattering rates. They are, of course, different if compared to the values resulting from the pure GS scattering rates. The presence of the ES increases the scattering rates because of the possibility of in- or out-scattering via the relaxation cascade.

A comparison of the effective GS capture rates with the pure GS capture rates is shown in Figure 1.8b,c.

As already expected from the analytic expression in Eq. (1.26), the rates increase because of the relaxation rates that have to be added, but the overall nonlinear dependence on the QW carrier densities stays unchanged. The electron in-scattering rates increase by a factor of about 5, while a dramatic increase of 5×10^4 is observed for the electron out-scattering. Nevertheless the out-scattering rates stay an order of magnitude smaller than the in-scattering rates. For the holes the situation is different. The much higher GS out-scattering rate is comparable to the out-relaxation rate and for a large range of operation conditions the effective out-scattering rate is higher than the effective in-scattering rate. The resulting effective lifetimes of the GS levels for both carrier types, $\tau_b^{\text{eff}} = (S_b^{\text{in,eff}} + S_b^{\text{out,eff}})^{-1}$, are plotted in Figure 1.8a.

Another effect that results from the presence of an ES in the QD laser system is an increased loss rate in the equation for the QW carrier density (term in square brackets in Eq. (1.27)). This leads to higher threshold currents and to a speedup of the device.

To get a further insight into the correlations between the scattering rates and the turn-on dynamics, Section 1.5 discusses the analytic approximation for frequency and damping of the ROs of the QD laser. Before doing that, experimental results of QD lasers will be compared to numeric results obtained with the reduced five-variable rate-equation system.

1.4
Quantum Dot Switching Dynamics and Modulation Response

This section aims to discuss the modulation response and switching dynamics in comparison with experimental results [26]. Because the experimental results were obtained on a laser that showed only GS lasing, the reduced five-variable rate equation system of Section 1.3.3 is used for the simulations. As all quantities now refer to the GS, the superscript G is omitted in the following. The nonlinear rate equations (Eqs. (1.28)–(1.32)) describe the dynamics of the charge carrier densities in the QD GS, n_e and n_h, the carrier densities in the QW, w_e and w_h (e and h stand for electrons and holes, respectively), and the photon density n_{ph} of the GS transition.

$$\dot{n}_e = S_e^{\text{in}}(N^{\text{QD}} - n_e) - S_e^{\text{out}} n_e - WA(n_e + n_h - N^{\text{QD}})n_{\text{ph}} - R_{\text{sp}}, \quad (1.28)$$

$$\dot{n}_h = S_h^{\text{in}}(N^{\text{QD}} - n_h) - S_h^{\text{out}} n_h - WA(n_e + n_h - N^{\text{QD}})n_{\text{ph}} - R_{\text{sp}}, \quad (1.29)$$

$$\dot{w}_e = \eta \frac{J(t)}{e_0} - \frac{N^{\text{sum}}}{N^{\text{QD}}} \left[S_e^{\text{in}}(N^{\text{QD}} - n_e) - S_e^{\text{out}} n_e \right] - B(w_e) w_e w_h, \quad (1.30)$$

$$\dot{w}_h = \eta \frac{J(t)}{e_0} - \frac{N^{\text{sum}}}{N^{\text{QD}}} \left[S_h^{\text{in}}(N^{\text{QD}} - n_h) - S_h^{\text{out}} n_h \right] - B(w_e) w_e w_h, \quad (1.31)$$

$$\dot{n}_{\text{ph}} = -2\kappa n_{\text{ph}} + \Gamma WA(n_e + n_h - N^{\text{QD}})n_{\text{ph}} + \beta R_{\text{sp}}. \quad (1.32)$$

1.4 Quantum Dot Switching Dynamics and Modulation Response

Here, the scattering rates S_b^{in} and S_b^{out} ($b = e, h$) used for the following simulations result from microscopic calculations that do not consider a second ES in the QDs. Their values as a function of the carrier densities w_b can be found in [8, 26]. Nevertheless, it is noted that similar results can been obtained with the full system discussed in Section 1.3.

1.4.1
Inhomogeneous Broadening

The spectral properties of the laser output are not addressed in the model, as the photon density is an average of all longitudinal modes inside the cavity. However, changes in the number of longitudinal modes are taken into account by changes in the active QD density N^{QD}, which basically changes the gain of the active medium. With a given QD size distribution p_i (where i is the index for a certain longitudinal mode frequency v_i), the QD density participating in the emission at a given frequency v_i is $N_i^{QD} = p_i N^{sum}$. Thus, the density of all active QDs is given by $N^{QD} = \sum_k p_k N^{sum}$ (the index k denotes the lasing longitudinal modes). The mode spacing inside the cavity ($L = 1$ mm) is $\Delta h\nu = 0.17$ meV($\Delta\lambda = 0.22$ nm), while the standard deviation of the QD size distribution [2] is about $\sigma_{inh} = 65$ meV $= 380\ \Delta h\nu$. Thus, 70% of all QDs are active ($N^{QD} = 0.7\ N^{sum}$) if the laser emits light at 380 longitudinal modes and only 3% ($N^{QD} = 0.03\ N^{sum}$) for a laser linewidth of 3.5 nm. On the basis of the experimental lasing spectra that show an increase in the lasing linewidth with increasing pump current (Figure 1.9), the pump-current-dependent spectral properties of the active QDs are

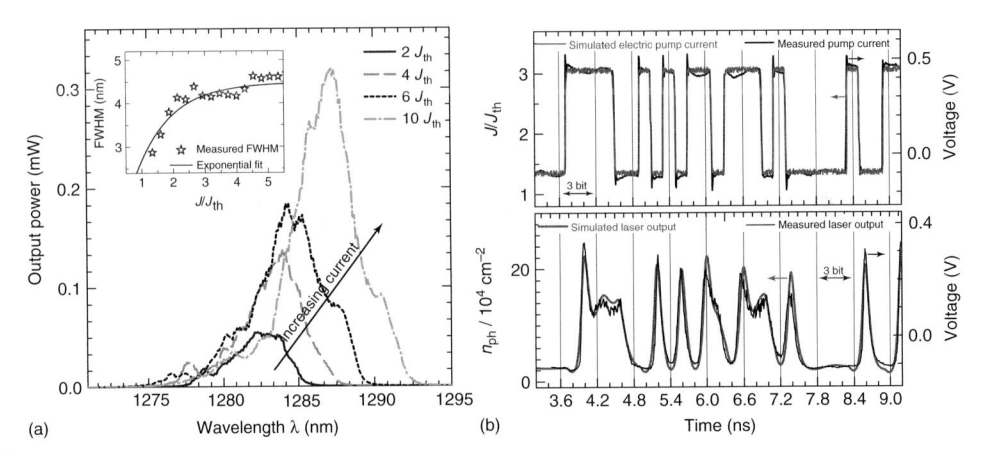

Figure 1.9 (a) Experimental lasing spectra for increasing pump currents illustrating the increasing number of longitudinal modes. Inset: Exponential fit (solid line) and FWHM of the measured spectra (stars), (b) Comparison between simulated (gray line) and measured data (black line) of electrical input signal (upper panel) and optical output (lower panel) versus time. Vertical lines show the separation into 3-bit sequences. (Reprinted from [26].)

taken into account and N^{QD} is implemented as a function of the QW carrier density [26] (Eq. (1.33)).

$$\frac{N^{QD}}{10^{-4}\,\text{nm}^{-2}} = 0.75 - 0.74\exp\left(-\frac{10^6}{1.75}w_e^2\right), \quad (1.33)$$

A more rigorous way to implement inhomogeneous broadening, which occurs in real devices because of fluctuations in QD size and material composition, and directly affects the energy levels, is accounted for by assuming a Gaussian size distribution around a central GS transition frequency ω_0 with standard deviation δ_ω. The spectral QD density is then given by $N(\omega) = \frac{N^{QD}}{\sqrt{2\pi}\delta_\omega}\exp\left[-\frac{(\omega-\omega_0)^2}{2\delta_\omega^2}\right]$ and the total QD density N^{QD} is approximated by a sum over a finite number of subensembles $N^{QD} = \sum_j N^j = \sum_j N(\omega_j)\Delta\omega$, where $\Delta\omega$ denotes the spectral width of the QD subgroups. Subsequently, a separate rate equation is used for each subensemble. For details, see, for example, [15, 33].

1.4.2
Temperature-Dependent Losses in the Reservoir

In addition to the temperature dependence of the in- and out-scattering rates discussed in Section 1.3.1, the carrier losses inside the reservoir will also be modeled as a function of T. The effect of these T-dependent losses will be most prominent for the large signal response of the laser while its effect on the turn-on dynamics and modulation response is small. The rate $R_\text{loss} = Bw_e w_h$ that accounts for these losses is a sum of the spontaneous bimolecular band–band recombination and Auger-related losses inside the QW given by $B_A w_e w_e w_h$. The Auger coefficient B_A has been shown [34] to depend significantly on the temperature T, and is therefore implemented such that it leads to a doubling of the rate for a temperature change of 60 K (Eq. (1.24)) as found in [34]. Thus, $B_A = 305\,\text{nm}^4\,\text{ps}^{-1}\left(\frac{T}{300\,\text{K}}\right)^4$ is used as given in [26]. Keep in mind that in this section a laser with only GS levels in the QDs is modeled. Within the extended model described in Section 1.3, the Auger scattering processes into the ES are already taken into account microscopically, which results in a different B_A for the remaining Auger processes within the QW. An alternative approach to model temperature characteristics is described in [35] by assuming nonradiative losses in the reservoir, which are modeled by capture processes from the reservoir to a midgap defect level.

1.4.3
Comparison to Experimental Results

The laser diode used for the experiments was a ridge waveguide InAs/InGaAs QD laser diode. The diode incorporates 15 stacks of QD layers having a DWELL structure [36]. The ridge is etched through the active layer to reduce current spreading [37] and to enhance wave guiding. The width of the ridge is 4 μm, while the length is 1 mm. To use the diode in high-frequency modulation schemes, top p- and n-contacts in a ground-signal-ground (GSG) configuration, allowing the use of high-speed, low-loss probe heads, have been processed. The threshold current density J_{th}^{exp} at room temperature is 380 A cm^{-2} with an emission wavelength close to 1.3 µm (Figure 1.9a). For pump currents, no ES lasing is found. Both facets of the laser are as cleaved. The diode is mounted on a copper heat sink and the light output is coupled to a standard single-mode fiber. A fiber-based isolator is used to prevent any feedback from influencing the laser diode. Eye diagrams have been obtained with an Agilent ParBert System, which creates an electrical pseudorandom binary sequence (PRBS) in a nonreturn to zero configuration. Here, a PRBS 5 (length: $2^5 - 1$ bit) is used to make the results comparable to theoretical calculations.

Figure 1.9b shows the optical response of the laser to an electrical PRBS 5 signal switching between two levels (1.5 J_{th} and 3 J_{th}) of continuous wave (cw) operation. Simulated and experimentally determined input signals (electrical words) are shown in the upper panel of Figure 1.9b. Owing to the experimental setup (e.g., influence of cables and divider, oscilloscope noise), the measured pump-current signal (black line) is not as flat as the simulated time trace (gray line). Despite this small deviation, the measured optical response (black line, lower panel) matches the simulated laser output (gray line, lower panel) very well. Note that this agreement could only be achieved by including the dynamic parameters discussed in Section 1.4.1. For constant B, the relaxation peak that appears in the photon output after switching to higher currents (Figure 1.9b lower panel at $t = 4$ ns) could not be modeled because the long lifetime of the carriers in the QW inhibits fast changes in the QW carrier densities.

By superposing every 3-bit sequences of the laser output shown in Figure 1.9b, an eye diagram [38] is generated. These eye diagrams can be seen in Figure 1.10a,b, which shows measured and simulated eye patterns, respectively, for switching between two different current levels (left column: $J_{th} \mapsto 3 J_{th}$ and right column: $4 J_{th} \mapsto 6 J_{th}$) and for three different pulse repetition frequencies (2.5, 5 and 10 GHz). Exact agreement in the shape (overshoots, trace, and extinction ratio) of the calculated and measured diagrams is found. Comparing the laser response for the different current levels it can be concluded that in order to improve the eye pattern diagrams, it is better to use higher current levels, as the relaxation peaks are thereby suppressed. The cutoff frequency of this QD laser – which is related to its RO frequency of 7 GHz – leads to a closing of the eyes already at 10 GHz. This can be improved by using higher pump currents; however, the modeling predicts that there is a trade-off since at the same time device heating results in further reduction of the RO frequency.

1.5
Asymptotic Analysis

As discussed in the previous sections, the solution to the QD laser equations exhibits different time scales that require accurate simulations. This section discusses an alternative to computationally expensive studies by using asymptotic methods. They

22 | *1 Modeling Quantum-Dot-Based Devices*

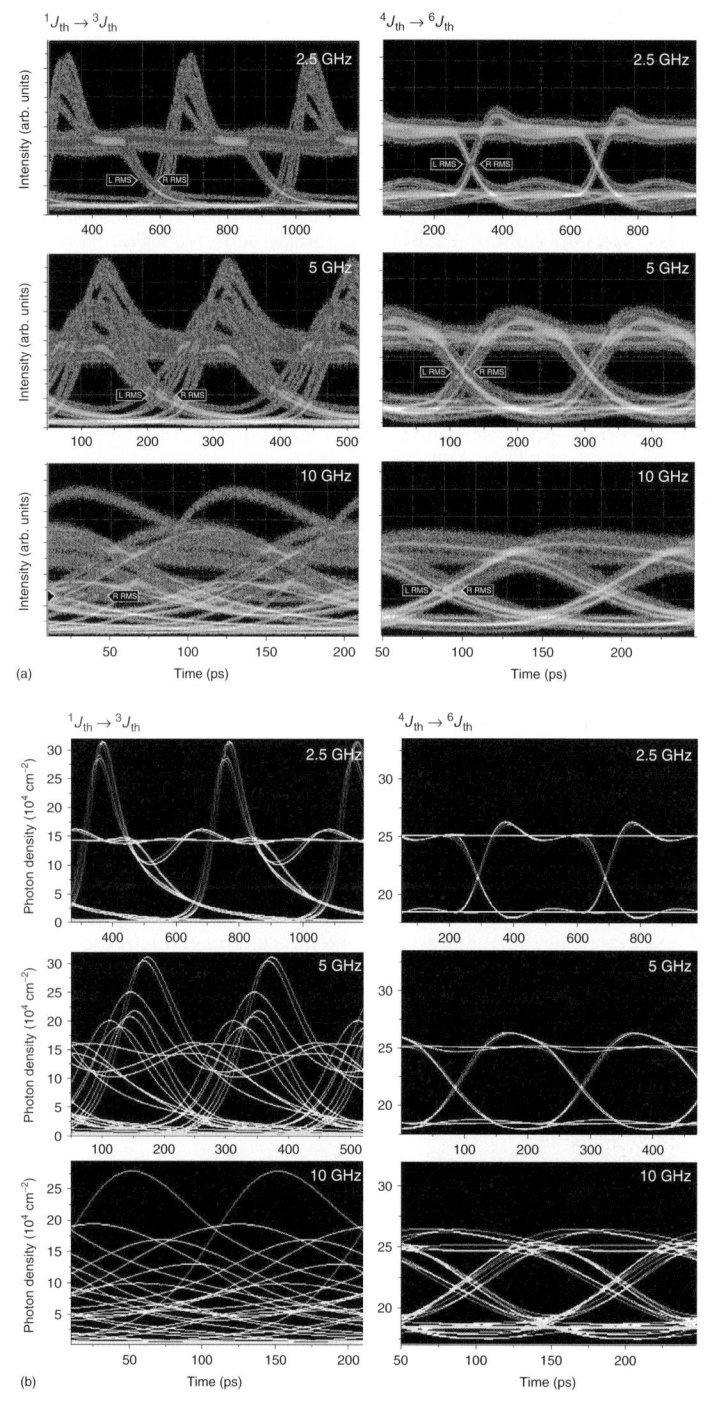

Figure 1.10 (a) Measured and (b) simulated eye diagrams for pump-current switching between $1J_{th}$ and $3J_{th}$ (left column) and between $4J_{th}$ and $6J_{th}$ (right column). Bit repetition frequency varies between 2.5 GHz (first line) and 10 GHz (third line). (Reprinted from [26].)

are motivated by the observation of quite different lifetimes between the carriers and the photons in the cavity ($\gamma = W/(2\kappa) \approx 7 \times 10^{-3}$). In order to simplify the rate equations, it is recalled that semiconductor lasers admit the properties of class B lasers. By applying approximation techniques appropriate for this class of lasers [39], it is possible to expand the full rate equation system in orders of $\sqrt{\gamma}$. The asymptotic techniques used for the analysis are described in detail in Chapter 6 [19], while the main results, that is, the analytic expressions for damping and frequency of the relaxation oscillations (ROs) of the laser, are discussed in the following. Note that the asymptotic results are valid only for scattering lifetimes that are on the order of several picoseconds or larger. For faster carriers the dynamics approaches the one of QW lasers and different scalings have to be used for the asymptotic analysis (see the limit large B in [46]).

The frequency ω_{RO} of the ROs can be obtained from the leading order problem of the expansion in powers of $\sqrt{\gamma}$ (see [40] for details) and is also valid far away from the fixed point. However, it is mathematically more convenient to determine the damping of the RO from the linearized problem including both $O(1)$ and $O(\sqrt{\gamma})$ terms. Thus, the damping rate of the ROs equals the real part of the eigenvalue λ of the characteristic polynomial of the linearized problem. To point out the effect of the scattering lifetimes τ_e and τ_h on the damping rate, the eigenvalues λ are computed as a function of the dimensionless parameter a_e and a_h defined as

$$a_e^{-1} = (S_e^{in} + S_e^{out})^{-1} \cdot \omega_{RO} = \tau_e \cdot \omega_{RO} \quad \text{and} \quad a_h^{-1} = \tau_h \cdot \omega_{RO}. \quad (1.34)$$

The values of Re(λ) and Im(λ) are plotted in Figure 1.11a,b. It is striking that for constant and small a_e, the real part of λ first increases with a_h before it starts to decrease again. Thus, there is an optimal value for the carrier lifetimes if large damping is required. The parameter space for the carrier lifetimes explored

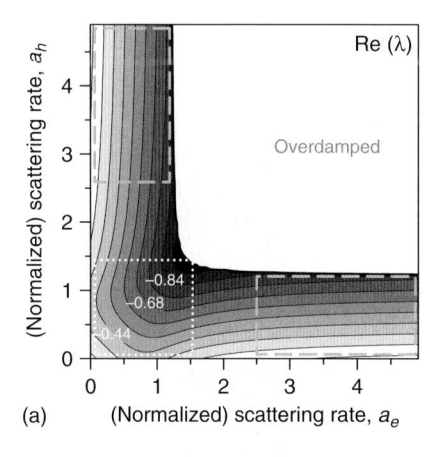

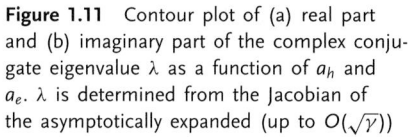

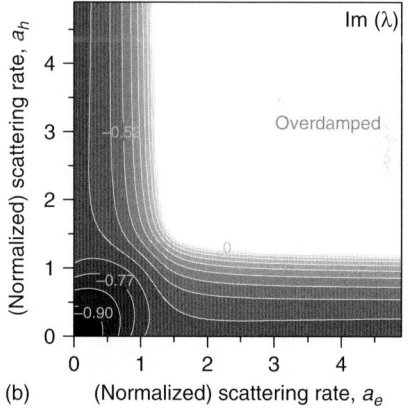

Figure 1.11 Contour plot of (a) real part and (b) imaginary part of the complex conjugate eigenvalue λ as a function of a_h and a_e. λ is determined from the Jacobian of the asymptotically expanded (up to $O(\sqrt{\gamma})$) and linearized laser problem. The white area marks the overdamped case with Im(λ) = 0, while the squared, dotted, and rectangular dashed area mark "case S" and "case D", respectively.

in Figure 1.11 can be separated into three different areas. The "overdamped case" without ROs in the turn-on dynamics if both a parameters are large (small lifetimes); "Case S" having equal lifetimes for both species and "Case D" showing large timescale separation and, therefore, one large and one small parameter of a_e and a_h.

For the QD laser modeled in Section 1.4, the microscopic calculations yield fast hole scattering rates with lifetimes that are in the range of picoseconds. This leads to high values of a_h ($a_h > 5$), while the slower electrons with their small scattering rates are characterized by $a_e \approx 0$. Consequently, following the analytic approach for "case D," the RO frequency and damping rate could be obtained, which are given by Lüdge et al. [40]

$$\omega_{RO}^{Da} = 2\pi f_{RO}^{Da} = \sqrt{An_{ph}^* W 2\kappa}, \tag{1.35}$$

$$= \sqrt{\frac{\Gamma W A N^{QD}}{N^{sum}} J_{th}^G \left(\frac{J}{J_{th}^G} - 1\right)} \tag{1.36}$$

$$\Gamma_{RO}^{Da} = \kappa An_{ph}^* W \tau_h + \frac{1}{2\tau_e} + \frac{W}{2}(An_{ph}^* + \frac{n_h^*}{N^{QD}}) \tag{1.37}$$

$$\approx \frac{1}{2}(\omega_{RO}^{Da})^2 \left((2\kappa)^{-1} + \tau_h\right) + \frac{1}{2\tau_e} \tag{1.38}$$

where the superscript Da means case D with a_h large. The analytic solutions shown in Eqs. (1.35) and (1.37) for frequency and damping of the ROs have been compared to numerically obtained data in Figure 1.12b,c. Note the good agreement between the numeric values (symbols) and the analytic expressions (lines). The numeric values for ω_{RO} and Γ_{RO} have been obtained by fitting the function $n_{ph}(t) \simeq C \sin(\omega_{RO} t + \phi) \exp(-\Gamma_{RO} t)$ to the turn-on transients. Equation (1.37) for the damping rate can be further simplified by omitting the smallest term (the one containing n_h^*), leading to Eq. (1.38). This reveals that the K-factor (ratio between damping rate and frequency squared [41, 42]) depends on three contributions. The smallest results from the scattering processes of the fast species (τ_h), the intermediate contribution is proportional to the cavity lifetime ($(2\kappa)^{-1}$), while the dominating effect scales with the scattering rates of the slow species. The RO frequency instead does not explicitly depend on the scattering rates, but is determined by J_{th}^G and the differential gain $\Gamma W A$ (Eq. (1.36) obtained by inserting the steady-state relation n_{ph}^* from Eq. (1.20)).

A different scaling is found for "case S" (see [40]) with small scattering rates for both carrier types (this can be achieved by changing the band structure and increasing the hole confinement energy).

$$\omega_{RO}^S = \sqrt{2An_{ph}^* W 2\kappa} \tag{1.39}$$

$$\Gamma_{RO}^S = \frac{W}{2}[2An_{ph}^* + 1] + \frac{\kappa}{\Gamma A N^{QD}} + \frac{1}{4\tau_e} + \frac{1}{4\tau_h} \tag{1.40}$$

$$\approx \frac{1}{2}(\omega_{RO}^S)^2 (2\kappa)^{-1} + \frac{1}{4\tau_e} + \frac{1}{4\tau_h}. \tag{1.41}$$

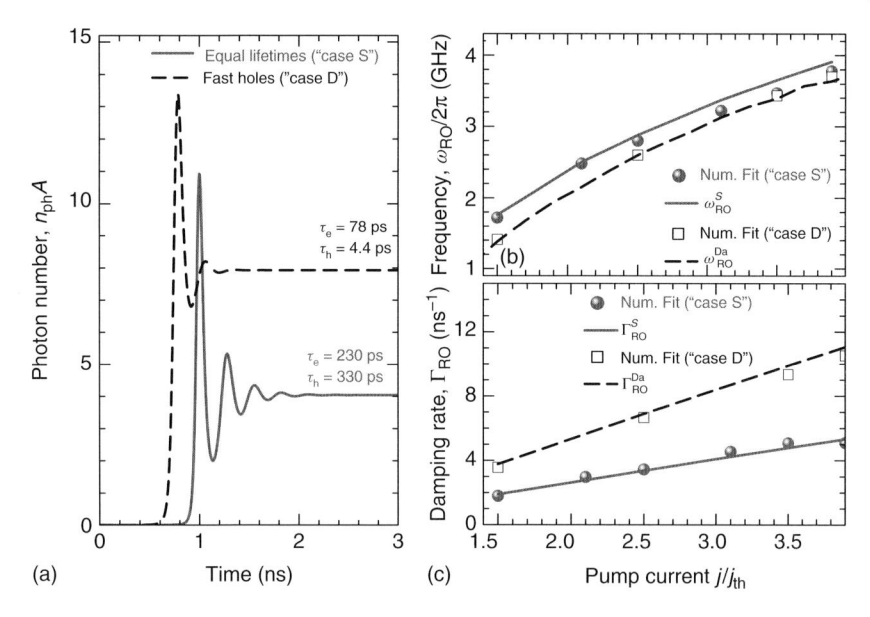

Figure 1.12 (a) Turn-on dynamics of the QD laser for "case S" (similar lifetimes) (gray solid) and "case D" (different lifetimes) (black dashed) at $J = 3.7 J_{th}$. (b) and (c) show comparisons between numerically fitted data (symbols) and analytical data obtained from Eqs. (1.35)–(1.40) (lines) for the RO frequency and RO damping rate, respectively. $(n_{ph}(t) \simeq C \sin(\omega_{RO} t + \phi) \exp(-\Gamma_{RO} t)$ is used to extract ω_{RO} and Γ_{RO} from the numerical simulation of $n_{ph}(t)$).

For this "case S," the expression of the RO frequency in Eq. (1.39) is the same as the one for the conventional semiconductor laser [39]. However, the expression of the damping rate is different. It contains the familiar term $\frac{W}{2}[2An^*_{ph}+1]$ that is found for the damping rate of QW lasers [19], but as already known from "case D" the dominating contribution stems from the scattering rates between QD and the reservoir.

By changing the confinement energy in the numeric calculations of the scattering rates it is possible to obtain small scattering rates and thus a "case S" like behaviour of the QD laser. The results of this simulation are plotted in Figure 1.12a showing weakly damped ROs. The analytic expressions accurately predict this behavior (see lines in Figure 1.12b,c), which makes them a powerful analytical tool for designing QD laser devices with optimal operation conditions.

1.5.1
Consequences of Optimizing Device Performance

The analytical expressions for the RO frequency and RO damping rate for the different parameter ranges show that the RO frequency ω_{RO} does not explicitly depend on the carrier–carrier scattering between QW and QD. It strongly depends on the cavity lifetime $(2\kappa)^{-1}$ and radiative recombination lifetime W^{-1} and on the threshold current of the QD laser, which are determined by the gain, the ratio between in- and out-scattering rates, and the losses in the QW.

The damping rate Γ_{RO}, instead, is crucially affected by the carrier–carrier scattering rates. For equal lifetimes of electrons and holes, the damping increases with decreasing lifetimes τ_e and τ_h. If both carrier types have different lifetimes, only the slowest species (for the chosen QD–QW system, these are the electrons) determines the damping rate, whereas the effect of the fast species is negligible.

In the next section, numeric simulations of QD lasers with doped carrier reservoir are discussed. Owing to the density-dependent Coulomb scattering rates, the doping modifies the carrier lifetimes in a controlled way and thus it is a good tool on the one hand for testing the analytics and on the other hand to optimize the device performance.

1.6
QD Laser with Doped Carrier Reservoir

A doped QW can be implemented by choosing different initial conditions for electron and hole densities in the QW. Without doping, the following initial conditions have been used, that is, $n_e^0 = 0$, $n_h^0 = 0$, $w_e^0 = 10^{-2} D_e kT$, and $w_h^0 = 10^{-2} D_e kT$. Note that charge conservation is contained in the five-variable rate equation system Eqs. (1.28)–(1.32), thus leading to only four independent dynamic variables that are related by

$$N^{\mathrm{sum}}(\dot{n}_e - \dot{n}_h) - N^{\mathrm{QD}}(\dot{w}_h - \dot{w}_e) = 0 \tag{1.42}$$

which can be integrated giving

$$N^{\mathrm{sum}}(n_e - n_h) - N^{\mathrm{QD}}(w_h - w_e) = N^{\mathrm{QD}}(w_e^0 - w_h^0). \tag{1.43}$$

By increasing w_e^0 or w_h^0 and keeping the other at the small value of $10^{-2} D_e kT$, it is possible to model n- or p-doping, respectively. Because the rate equation system treats 2D densities, the doping concentrations $n \approx w_e^0$ and $p \approx w_h^0$ are also given per area. To compare this to 3D doping densities, the areal densities have to be divided by the QW height, which is $h = 4$ nm. Thus, $n = 2 \times 10^{11} \mathrm{cm}^{-2}$ corresponds to $n_{3D} = 5 \times 10^{17} \mathrm{cm}^{-3}$. Figure 1.13c,d shows that changes in initial conditions drastically modify the QD laser turn-on dynamics. For n-doping, the damping of the ROs is increased, whereas the damping is drastically reduced if p-doping is introduced. This behavior can be understood by discussing the steady-state values for the QW carrier densities w_b.

N-doping increases the QW electron density, which then leads to higher in-scattering rates S_e^{in} (Figure 1.3) and, therefore, to higher carrier densities n_e^*. On the contrary, p-doping leads to a higher QW hole concentration and thus to higher occupation of the QD hole levels. Note, however, that the increased QW hole density for p-doped samples also has an effect on the out-scattering rate, as this contains a factor that exponentially decreases with w_e through the detailed balance relation (Eq. (1.13)). The scattering time τ_h for holes decreases with increasing p-doping,

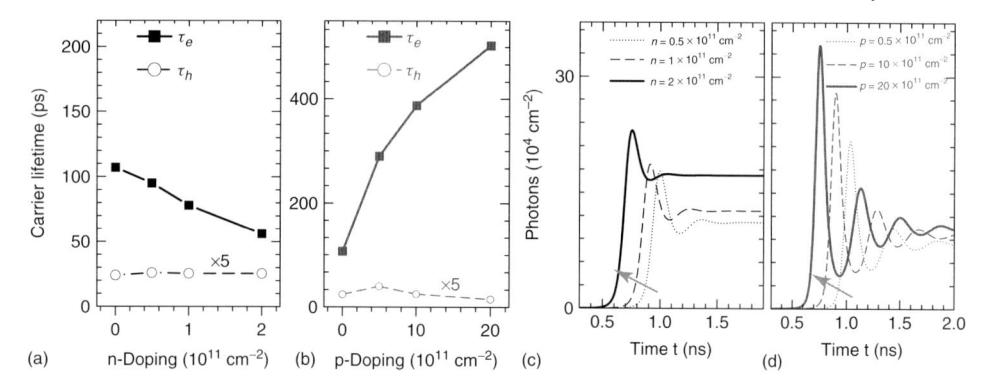

Figure 1.13 (a,b): Carrier lifetimes as a function of (a) n-doping and (b) p-doping density. (b,c): Time series of the photon density during laser turn-on for (c) different n-doping density and (d) different p-doping density. Dotted, dashed, and solid lines correspond to doping of 0.1, 0.2, and 0.4 times the degeneracy concentration $D_{e/h}kT$, respectively. ($D_e kT = 4.7 \times 10^{11}$ cm^{-2} and $D_h kT = 48 \times 10^{11}$ cm^{-2}). Parameters as in Table 1.1; pump current is $J = 2.5 J_{th}$.

while the scattering time for electrons increases as plotted in Figure 1.13b. The ratio between the timescales of both carriers decreases from $\tau_h/\tau_e = 5/100$ for the undoped case to a value of $\tau_h/\tau_e = 3/500$ for a p-doping of $p = 20 \times 10^{11}$ cm^{-2}. Using the analytic expression Eq. (1.37) for the damping rate explains that the lower damping results from the longer lifetime of the small species, whereas the changes in the hole lifetime only marginally affect the damping rate.

Figure 1.14a shows the steady-state characteristics of the QD laser projected onto the (n_h, n_e)-phase space for two different n-doping densities (squares and triangles) and two different p-doping densities (open circles and stars). Figure 1.14b,c shows close-ups for very high p-doping. Going from high p-doping to n-doping, the steady states n_e^* and n_h^* move up along an approximately straight line in (n_h, n_e)-phase space that is given by Eq. (1.23), while the turn-on dynamics becomes more strongly damped and synchronized between electrons and holes. This is different from changing the steady-state values by varying the confinement energy as the increased steady-state values n_e^* (induced by increasing ΔE_e) lead to a desynchronization (separation of timescales) of electrons and holes.

Comparison with Analytic Results

The analytical approximations of Section 1.5 and the obtained predictions about changes in the laser turn-on dynamics are in good agreement with the numerical simulations of a laser with different doping densities discussed in the last section. The increasing n-doping concentration in a QD laser with timescale separation of the carriers ("case D" in Section 1.5) leads to a decrease in the electron lifetime, which was at the same time accompanied by an increased damping. With the analytic formula given in Eq. (1.37), the increased damping can be explained with the decreased lifetime ($S_e^{in} + S_e^{out} = \tau_e^{-1}$ increases). On the other hand, p-doping

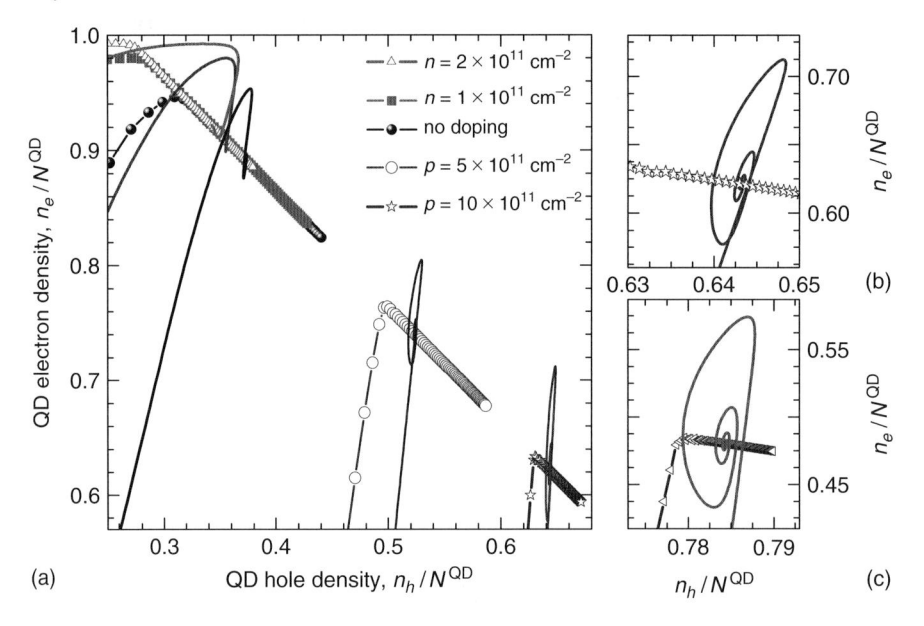

Figure 1.14 Effect of QW doping: steady-state characteristic for a range $J = 0$ to $J/e_0 = 2 \times 10^9$ cm^{-2} (symbols) and turn-on trajectories for $J = 2.5 J_{th}$ (lines) projected onto the (n_h, n_e)-phase space (a) squares, triangles, circles, and stars are doping densities of $n = 1 \times 10^{11}$ cm^{-2}, $n = 2 \times 10^{11}$ cm^{-2}, $p = 5 \times 10^{11}$ cm^{-2}, and $p = 10 \times 10^{11}$ cm^{-2}, respectively; (b) and (c): close-up of (a) for $p = 10 \times 10^{11}$ cm^{-2} and $p = 20 \times 10^{11}$ cm^{-2}, respectively. Parameters as in Table 1.1. (Reprinted from [8].)

of the same device did not yield a higher RO damping. The reason for this counterintuitive result is the separation of timescales of electron and hole lifetimes (which is the case for material with large differences in the effective masses of electrons and holes). The slowest species determines the dynamics and, thus, manipulating its lifetime has a drastic effect on the laser dynamics (Eq. (1.37)). Instead, manipulating the lifetime of the fast species has only a minor effect. The reduced damping for p-doping concentration is based on a reduction of the electron lifetimes, which has its physical origin in the increased rate for mixed electron–hole Coulomb scattering processes because of the excess holes in the reservoir. It confirms that p-doping is beneficial for the modulation response of QD lasers [43]. If a high RO damping rate is a desired property of QD lasers, n-doping should be helpful.

1.7
Model Reduction

One way to simplify the discussed QD laser model is to neglect the density dependence of the scattering rates and to use constant carrier lifetimes; see, for example, [27, 33, 35]. If these lifetimes are chosen properly, they can lead to decent results at a certain point of operation. Nevertheless, the uncertainty in the choice of proper simulation parameters, which is necessary for those models, will lead to a large uncertainty regarding the results. Moreover, effects such as doping or the effect of changing the QD size cannot be studied. There are also approaches that take into account a phenomenological density dependence of the carrier lifetimes; see, for example, [44], but again the problem of choosing the correct parameters remains.

A simpler three-variable rate equation model was formulated by O'Brien et al. [45] and is widely used for QD laser modeling (see [46] for detailed analytic discussions of this model). This model does not distinguish between electrons and holes; it assumes the same dynamics for both species, and it uses in-scattering rates into the QDs that linearly increase with the reservoir carrier density, while the out-scattering rates are constant. Following the analytic results of Section 1.5, a reduction in the full microscopic five-variable model to a model that only assumes one carrier type is possible, but two cases have to be distinguished. If electrons and holes relax on a similar timescale ("case S"), the mean value of electron and hole rates needs to be included; however, for timescale separation between the lifetimes of the species ("case D"), the scattering rate of the slow species will be the important one that determines the dynamics. If this is kept in mind, the (linearly fitted) microscopic in-scattering rates can be used as input parameters for the three-variable rate equation system. Nevertheless, it has to be noted that these parameters need to be adjusted if large variations of the pump current or different doping densities are modeled.

1.8 Comparison to Quantum Well Lasers

If a QD laser model is compared to a QW laser model [47, 48], one striking difference is that the current is not injected directly into the active region, and an additional reservoir has to be included. The relatively slow scattering processes from the carrier reservoir into the QD levels are responsible for the high damping of the turn-on process (Section 1.5) and thus also for the flat modulation response curve of QD lasers if compared to the pronounced peak found for QW lasers. In the limit of large and equal scattering rates for electrons and holes, the QD laser model can be reduced to a QW laser model as shown in [46]. However, the modulation bandwidth (and the cutoff frequency) of QD lasers is also much smaller because of the smaller RO frequency. The reason for this lies in the fact that the threshold currents needed to invert the localized two-level system is much smaller than the current needed to invert a 2D electron gas.

If complex integrated structures, for example, QD lasers subjected to optical feedback [18, 49] or lasers with saturable absorber are discussed (see Chapter 7 [50] and Chapter 8 [51]), the high damping of the turn-on dynamics of the QD laser is one crucial parameter when discussing differences with respect to QW lasers. Another parameter that differs between QD and QW laser devices is the phase–amplitude coupling (linewidth enhancement factor). It comes into play as soon as the phase of the electric field is important, for example, when modeling feedback problems (see Chapter 6 [19]). For a discrete two-level system, the α-factor is zero because of the symmetric gain spectrum. For an inhomogeneously broadened ensemble of QDs inside the QD laser, α still remains small [52] and leads, for example, to a smaller chirp and higher feedback sensitivity of QD lasers (see [18] for a comparison of both laser models with feedback).

1.9
Summary

This chapter reviewed a microscopic rate equation approach that can be used to model the dynamic response of electrically pumped edge emitting QD lasers. Different levels of complexity have been explored. A detailed discussion of the Coulomb scattering rates between localized QD levels and the 2D carrier reservoir in the surrounding QW has underlined the importance of these nonlinear Auger rates for a quantitative modeling of the QD laser device. Two-state lasing properties as well as the effect of additional confined levels on the GS lasing properties have been analyzed. It was shown that temperature, band structure, as well as doping of the carrier reservoir can significantly alter the laser dynamics. Nevertheless, all of these effects can be traced back to the values of the carrier–carrier scattering rates and their nonlinear dependence on the carrier densities in the reservoir. Furthermore, asymptotic analysis has allowed analytic insights into the relations between frequency and damping of the turn-on dynamics and the carrier lifetimes that finally permitted to predict the dynamics of the laser without tedious numeric simulations, and provide insight into the governing physical mechanism.

Acknowledgment

I am grateful to Prof. Eckehard Schöll for giving me the opportunity to do research about QD lasers within his group. My thanks also go to Niels Majer for his excellent work in calculating the microscopic scattering rates and to Thomas Erneux for the great collaboration on the asymptotic analysis of the QD laser model. This work was supported by DFG in the framework of Sfb 787.

References

1. Bimberg, D., Fiol, G., Kuntz, M., Meuer, C., Lämmlin, M., Ledentsov, N.N., and Kovsh, A.R. (2006) High speed nanophotonic devices based on quantum dots. Phys. Stat. Sol. A, **203** (14), 3523–3532. DOI: 10.1002/pssa.200622488.
2. Bimberg, D., Grundmann, M., and Ledentsov, N.N. (1999) *Quantum Dot Heterostructures*, John Wiley & Sons, Inc, New York.
3. Rafailov, E.U., Cataluna, M.A., and Avrutin, E.A. (2011) *Ultrafast Lasers Based on Quantum Dot Structures*, Wiley-VCH Verlag GmbH, Weinheim.
4. Malić, E., Ahn, K.J., Bormann, M.J.P., Hövel, P., Schöll, E., Knorr, A.,

Kuntz, M., and Bimberg, D. (2006) Theory of relaxation oscillations in semiconductor quantum dot lasers. *Appl. Phys. Lett.*, **89**, 101107. DOI: 10.1063/1.2346224.

5. Malić, E., Bormann, M.J.P., Hövel, P., Kuntz, M., Bimberg, D., Knorr, A., and Schöll, E. (2007) Coulomb damped relaxation oscillations in semiconductor quantum dot lasers. *IEEE J. Sel. Top. Quantum Electron.*, **13** (5), 1242–1248. DOI: 10.119/jqstqe.2007.905148.

6. Lüdge, K., Bormann, M.J.P., Malić, E., Hövel, P., Kuntz, M., Bimberg, D., Knorr, A., and Schöll, E. (2008) Turn-on dynamics and modulation response in semiconductor quantum dot lasers. *Phys. Rev. B*, **78** (3), 035316. DOI: 10.1103/physrevb.78.035316.

7. Lüdge, K. and Schöll, E. (2009) Quantum-dot lasers - desynchronized nonlinear dynamics of electrons and holes. *IEEE J. Quantum Electron.*, **45** (11), 1396–1403.

8. Lüdge, K. and Schöll, E. (2010) Nonlinear dynamics of doped semiconductor quantum dot lasers. *Eur. Phys. J. D*, **58**, 167–174. DOI: 10.1140/epjd/e2010-00041-8.

9. Chow, W.W. and Koch, S.W. (1999) *Semiconductor-Laser Fundamentals*, Springer.

10. Su, Y., Carmele, A., Richter, M., Lüdge, K., Schöll, E., Bimberg, D., and Knorr, A. (2011) Theory of single quantum dot lasers: Pauli-blocking enhanced anti-bunching. *Semicond. Sci. Technol.*, **26**, 014015.

11. Chow, W.W. and Koch, S.W. (2005) Theory of semiconductor quantum-dot laser dynamics. *IEEE J. Quantum Electron.*, **41**, 495–505. DOI: 10.1109/jqe.2005.843948.

12. Lingnau, B., Lüdge, K., Schöll, E., and Chow, W.W. (2010) Many-body and nonequilibrium effects on relaxation oscillations in a quantum-dot microcavity laser. *Appl. Phys. Lett.*, **97** (11), 111102. DOI: 10.1063/1.3488004.

13. Gomis-Bresco, J., Dommers, S., Temnov, V.V., Woggon, U., Martinez-Pastor, J., Lämmlin, M., and Bimberg, D. (2009) InGaAs quantum dots coupled to a reservoir of nonequilibrium free carriers. *IEEE J. Quantum Electron.*, **45** (9), 1121–1128.

14. Wegert, M., Majer, N., Lüdge, K., Dommers-Völkel, S., Gomis-Bresco, J., Knorr, A., Woggon, U., and Schöll, E. (2011) Nonlinear gain dynamics of quantum dot optical amplifiers. *Semicond. Sci. Technol.*, **26**, 014008.

15. Majer, N., Lüdge, K., and Schöll, E. (2010) Cascading enables ultrafast gain recovery dynamics of quantum dot semiconductor optical amplifiers. *Phys. Rev. B*, **82**, 235301.

16. Majer, N., Lüdge, K., Gomis-Bresco, J., Dommers-Völkel, S., Woggon, U., and Schöll, E. (2011) Impact of carrier-carrier scattering and carrier heating on pulse train dynamics of quantum dot semiconductor optical amplifiers. *Appl. Phys. Lett.*, **99**, 131102, doi:10.1063/1.3643048.

17. Sciamanna, M. Mode competition driving laser nonlinear dynamics, Chapter 3 this book.

18. Otto, C., Lüdge, K., and Schöll, E. (2010) Modeling quantum dot lasers with optical feedback: sensitivity of bifurcation scenarios. *Phys. Stat. Sol. B*, **247** (4), 829. DOI: 10.1002/pssb.200945434.

19. Otto, C. Lüdge, K., Viktorov, E.A., and Erneux, T. (2011) Quantum dot laser tolerance to optical feedback, Chapter 6 this book.

20. Lorke, M., Nielsen, T.R., Seebeck, J., Gartner, P., and Jahnke, F. (2006) Influence of carrier-carrier and carrier-phonon correlations on optical absorption and gain in quantum-dot systems. *Phys. Rev. B*, **73**, 085324. DOI: 10.1103/physrevb.73.085324.

21. Nielsen, T.R., Gartner, P., and Jahnke, F. (2004) Many-body theory of carrier capture and relaxation in semiconductor quantum-dot lasers. *Phys. Rev. B*, **69**, 235314.

22. Nilsson, H.H., Zhang, J.Z., and Galbraith, I. (2005) Homogeneous broadening in quantum dots due to auger scattering with wetting layer carriers. *Phys. Rev. B*, **72** (20), 205331. DOI: 10.1103/physrevb.72.205331.

23. Schöll, E., Amann, A., Rudolf, M., and Unkelbach, J. (2002) Transverse

spatio-temporal instabilities in the double barrier resonant tunneling diode. *Physica B*, **314**, 113.
24. Schöll, E. (1987) *Nonequilibrium Phase Transitions in Semiconductors*, Springer, Berlin.
25. Lasher, G. and Stern, F. (1964) Spontaneous and stimulated recombination radiation in semiconductors. *Phys. Rev.*, **133** (2A), A553.
26. Lüdge, K., Aust, R., Fiol, G., Stubenrauch, M., Arsenijević, D., Bimberg, D., and Schöll, E. (2010) Large signal response of semiconductor quantum-dot lasers. *IEEE J. Quantum Electron.*, **46** (12), 1755–1762. DOI: 10.1109/jqe.2010.2066959.
27. Markus, A., Chen, J.X., Paranthoen, C., Fiore, A., Platz, C., and Gauthier-Lafaye, O. (2003) Simultaneous two-state lasing in quantum-dot lasers. *Appl. Phys. Lett.*, **82** (12), 1818.
28. Lingnau, B., Lüdge, K., Schöll, E., and Chow, W.W. (2010) in *Dynamic Many-Body and Nonequilibrium Effects in a Quantum Dot Microcavity Laser*, Proceedings of SPIE, vol. 7720, (eds K. Panajotov, M. Sciamanna, A.A. Valle, and R. Michalzik), pp. 77201F-77201F-11. DOI: 10.1117/12.854671.
29. Wolters, J., Dachner, M.R., Malić, E., Richter, M., Woggon, U., and Knorr, A. (2009) Carrier heating in light emitting quantum-dot heterostructures at low injection currents. *Phys. Rev. B*, **80**, 245401.
30. Ji, H.M., Yang, T., Cao, Y.L., Xu, P.F., and Gu, Y.X. (2010) Self-heating effect on the two-state lasing behaviors in 1.3-μm InAs-GaAs quantum-dot lasers. *Jpn. J. Appl. Phys.*, **49**, 072103.
31. Uskov, A.V., Meuer, C., Schmeckebier, H., and Bimberg, D. (2011) Auger capture induced carrier heating in quantum dot lasers and amplifiers. *Appl. Phys. Express*, **4** (2), 022202. DOI: 10.1143/apex.4.022202.
32. Wu, D.C., Su, L.C., Lin, C.Y., Mao, M.H., Wang, J.S., Lin, G., and Chi, J.Y. (2009) Experiments and simulation of spectrally-resolved static and dynamic properties in quantum dot two-state lasing. *Jpn. J. Appl. Phys.*, **48**, 032101.
33. Gioannini, M., Sevega, A., and Montrosset, I. (2006) Simulations of differential gain and linewidth enhancement factor of quantum dot semiconductor lasers. *Opt. Quantum Electron.*, **38**, 381–394. DOI: 10.1007/s11082-006-0038-1.
34. Pikal, J.M., Menoni, C.S., Thiagarajan, P., Robinson, G.Y., and Temkin, H. (2000) Temperature dependence of intrinsic recombination coefficients in 1.3 μm InAsP/InP quantum-well semiconductor lasers. *Appl. Phys. Lett.*, **76** (19), 2659–2661.
35. Rossetti, M., Fiore, A., Sek, G., Zinoni, C., and Li, L. (2009) Modeling the temperature characteristics of InAs/GaAs quantum dot lasers. *J. Appl. Phys.*, **106** (2), 023105. DOI: 10.1063/1.3176499.
36. Kovsh, A.R., Maleev, N.A., Zhukov, A.E., Mikhrin, S.S., Vasil'ev, A.V., Semenova, A., Shernyakov, Y.M., Maximov, M.V., Livshits, D.A., Ustinov, V.M., Ledentsov, N.N., Bimberg, D., and Alferov, Z.I. (2003) InAs/InGaAs/GaAs quantum dot lasers of 1.3 μm range with enhanced optical gain. *J. Crystal Growth*, **251**, 729–736.
37. Ouyang, D., Ledentsov, N.N., Bimberg, D., Kovsh, A.R., Zhukov, A.E., Mikhrin, S.S., and Ustinov, V.M. (2003) High performance narrow stripe quantum-dot lasers with etched waveguide. *Semicond. Sci. Technol.*, **18**, L53–L54.
38. Derickson, D. and Müller, M. (2007) *Digital Communications Test and Measurement: High-Speed Physical Layer Characterization*, Prentice Hall Press, Upper Saddle River, NJ.
39. Erneux, T. and Glorieux, P. (2010) *Laser Dynamics*, Cambridge University Press, UK.
40. Lüdge, K., Schöll, E., Viktorov, E.A., and Erneux, T. (2011) Analytic approach to modulation properties of quantum dot lasers. *J. Appl. Phys.* **109** (9), 103112.
41. Olshansky, R., Hill, P., Lanzisera, V., and Powazinik, W. (1987) Frequency response of 1.3 μm InGaAsP high speed semiconductor lasers. *IEEE J. Quantum Electron.*, **23** (9), 1410–1418.
42. Klotzkin, D. and Bhattacharya, P. (1999) Temperature dependence of dynamic and dc characteristics of quantum-well

43. Shchekin, O.B. and Deppe, D.G. (2002) The role of p-type doping and the density of states on the modulation response of quantum dot lasers. *Appl. Phys. Lett.*, **80** (15), 2758–2760. DOI: 10.1063/1.1469212.
44. Veselinov, K., Grillot, F., Cornet, C., Even, J., Bekiarski, A., Gioannini, M., and Loualiche, S. (2007) Analysis of the double laser emission occurring in 1.55 μm InAs−InP (113)B quantum-dot lasers. *IEEE J. Quantum Electron.*, **43** (9), 810–816. DOI: 10.1109/jqe.2007.902386.
45. O'Brien, D., Hegarty, S.P., Huyet, G., and Uskov, A.V. (2004) Sensitivity of quantum-dot semiconductor lasers to optical feedback. *Opt. Lett.*, **29** (10), 1072–1074.
46. Erneux, T., Viktorov, E.A., and Mandel, P. (2007) Time scales and relaxation dynamics in quantum-dot lasers. *Phys. Rev. A*, **76**, 023819. DOI: 10.1103/physreva.76.023819.
47. Schöll, E., Bimberg, D., Schumacher, H., and Landsberg, P.T. (1984) Kinetics of picosecond pulse generation in semiconductor lasers with bimolecular recombination at high current injection. *IEEE J. Quantum Electron.*, **20**, 394.
48. Schöll, E. (1988) Dynamic theory of picosecond optical pulse shaping by gain-switched semiconductor laser amplifiers. *IEEE J. Quantum Electron.*, **24** (2), 435–442.
49. Otto, C., Globisch, B., Lüdge, K., Schöll, E., and Erneux, T. (2012) Complex dynamics of semiconductor quantum dot lasers subject to delayed optical feedback. *Int. J. Bifurcat. Chaos* (accepted publication).
50. Krauskopf, B. and Walker, J.J. (2011) Bifurcation study of a semiconductor laser with saturable absorber and delayed optical feedback, Chapter 7 this book.
51. Vladimirov, G., Rachinskii, D., and Wolfrum, (2011) Modeling of passively mode-locked semiconductor lasers, Chapter 8 this book.
52. Kim, J. and Chuang, S.L. (2006) Theoretical and experimental study of optical gain, refractive index change, and linewidth enhancement factor of p-doped quantum-dot lasers. *IEEE J. Quantum Electron.*, **42** (9), 942–952. DOI: 10.1109/jqe.2006.880380.

2
Exploiting Noise and Polarization Bistability in Vertical-Cavity Surface-Emitting Lasers for Fast Pulse Generation and Logic Operations

Jordi Zamora-Munt and Cristina Masoller

2.1
Introduction

Nonlinear systems and noise are ubiquitous in nature, and their interplay can result in some unexpected behaviors, which can be very different from those of linear systems. Noise is usually considered a drawback, degrading a system's performance, and a lot of efforts have been devoted to minimize its effect. However, in recent years, a better understanding of the role of noise in nonlinear systems has resulted in the development of novel methods for taking advantage and exploiting noise for controlling these systems. Relevant examples can be found in the work of Prof. Schöll and coworkers [1, 2], who have shown that the regularity of noise-induced motion can be controlled by exploiting the interplay of noise and time-delayed feedback. In coupled stochastic excitable systems, such as neurons, Prof. Schöll and coworkers have also shown that delayed feedback can be employed to control the system's coherence, either to enhance or to decrease synchronization [3–7].

An alternative way to control a stochastic nonlinear system is by modulation of a system's parameter. Here we consider an optical system, specifically, a vertical-cavity surface-emitting laser (VCSEL), and study numerically the influence of modulating the laser current parameter.

A VCSEL is a type of semiconductor laser that has several advantages over conventional, edge-emitting semiconductor lasers (compactness, low cost, fast response, etc.) and is nowadays widely used in photonics, nanotechnology, optical signal processing, and optical networks [8]. As is discussed also by Sciamanna (Chapter 3) and by Shore (Chapter 15), VCSELs often display a complex polarization nonlinear dynamics that, driven solely by noise (due to, e.g., spontaneous emission), can significantly degrade the laser performance. However, by controlling a laser parameter or by fine tuning the strength of an external noisy signal (e.g., via weak incoherent optical injection), the so-called laser consistency or reliability, that is, the ability to encode irregular signals in the laser output in a reproducible manner, can be enhanced, which can in turn lead to stochastic synchronization as discussed by Wieczorek in Chapter 11. These effects, combined with polarization competition, can be exploited for novel applications, such as noise-assisted binary information transmission [9, 10].

A relevant feature of VCSELs is related to the stability of two orthogonal linear polarizations. In VCSELs, polarization bistability and competition often arise because of the laser circular transverse geometry (a schematic representation of the VCSEL geometry is presented in Figure 2.1a). Owing to anisotropies that break the circular symmetry, the output of a VCSEL is linearly polarized along one of two orthogonal directions, referred to as x and y. When the VCSEL begins to lase one linear polarization dominates, and, in many devices, when the injection current is increased above a certain value, it is observed that the emission switches to the orthogonal linear polarization. The polarization switching (PS) usually involves hysteresis, as when the current decreases, the switch back occurs at a lower current value (Figure 2.1b). Stochastic polarization switching (Figure 2.1c) can also occur and has been interpreted in terms of Kramers' hopping in an effective 1D double-well potential [13, 14].

In this chapter, we focus on the interplay of noise, current modulation, and nonlinear dynamics in VCSELs. In the first part of the chapter, we consider a triangular asymmetric current modulation; in the second part, a three-level aperiodic current modulation.

In the first part of this chapter, we review the results presented in [16], where we proposed a way to exploit noise to generate fast optical pulses with a triangular modulating signal that is, on average, below the static threshold (i.e., the threshold for cw operation). Our work was motivated by the experimental and theoretical results of Glorieux and coworkers [15], which employed a Nd 3+:YVO4 diode-pumped laser, and an asymmetric triangular modulation was applied to the power delivered by the pumping diode laser. They showed that a signal with a slow-rising ramp lead to the emission of pulses, even when the laser was operated, on average, below the threshold. The emitted pulses were larger than those emitted when the modulation was a symmetric triangular signal with the same averaged value, see Figure 2.2a,b. In contrast, a signal with a fast-rising ramp and the same averaged value did not lead to pulse emission, and the laser intensity remained at the noise level during all the modulation cycle (Figure 2.2c). In [15], the modulation period was of the order of tens of microseconds, while in [16], we showed that a similar effect could be observed in VCSELs but with much faster modulating signals (of the order of nanoseconds). In the first part of this chapter, we review our previous work and discuss the role of spontaneous emission noise.

In the second part of this chapter, we review the results presented in [17], were we analyzed the response of a polarization-bistable VCSEL to a three-level aperiodic current modulation and showed that it can display the phenomenon of Logic Stochastic Resonance (LSR). The concept of LSR was first introduced by Murali *et al.* [18], who demonstrated that a two-state system with two adjustable thresholds, modeled by a one-dimensional double-well potential, can act as a reliable and flexible logic gate in the presence of an appropriate amount of noise, as shown in Figure 2.3. Since the pioneer work of Murali *et al.* [18], the LSR

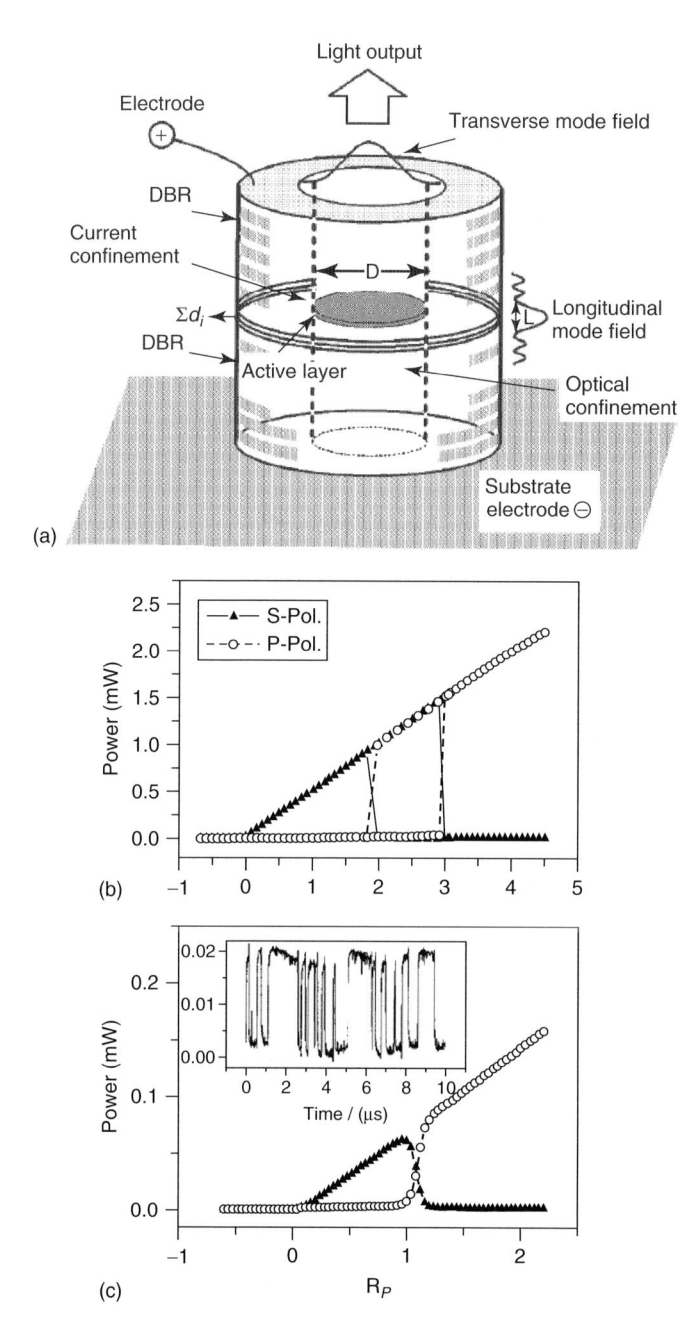

Figure 2.1 (a) Schematic structure of a VCSEL. Adapted from [11]. (b,c) Polarization-resolved optical power as a function of the pump rate for two oxide-confined single-transverse-mode VCSELs. VCSEL A (b) displays polarization hysteresis, while VCSEL C (c) displays stochastic polarization switching. Inset: time series of the optical power of one polarization at the center of the switching region ($R_P = 1.1$). Adapted from [12].

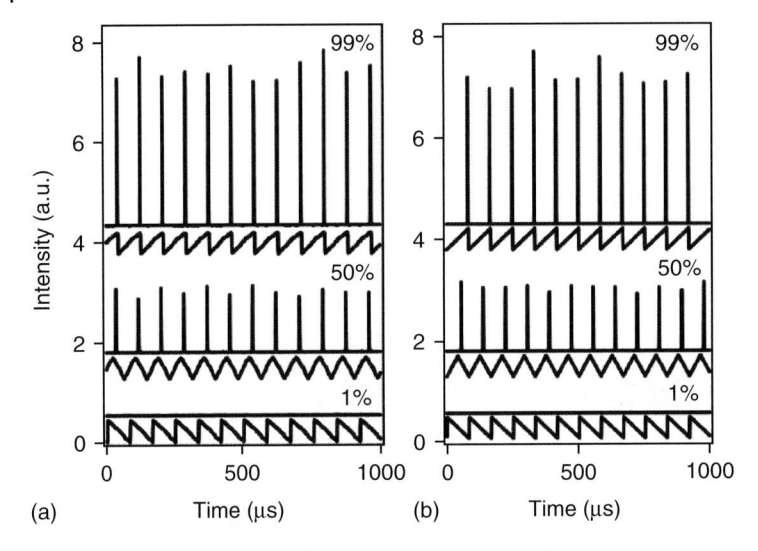

Figure 2.2 Time evolution of the laser intensity (top) and of the pump power (bottom) for an symmetry parameter (which is the ratio between the signal rising time and the modulation period) equal to 1, 50, and 99%. (a) Experiments, (b) Simulations. Adapted from [15].

phenomenon has been numerically and experimentally demonstrated in various nonlinear systems such as electronic circuits, tunneling diodes, and chemical systems [19–25].

In [17], we showed that LSR can also occur in VCSELs, which, under appropriated conditions, can give a reliable logic response to two logic inputs, even in the presence of a significant amount of noise. The two logic inputs are encoded in a three-level aperiodic signal directly applied to the laser bias current. Exploiting polarization bistability, one can consider that the laser response is a logic 1 if one linear polarization is emitted (e.g., x) and a logic 0 if the orthogonal one is emitted (e.g., y). Then, the truth table of the fundamental logical operators AND and OR (and their negations, NAND and NOR) can be reproduced with a probability of a correct response equal to one in a wide range of noise strengths.

The rest of this chapter is organized as follows. In Section 2.2, we present the VCSEL spin-flip model (SFM) used to study both asymmetric triangular current modulation and aperiodic three-level current modulation. In Section 2.3, we discuss the phenomenon of polarization-switching within the framework of the SFM model. In Section 2.4, we demonstrate the generation of fast pulses via asymmetric current modulation, and in Section 2.5, we discuss the influence of spontaneous emission noise. In Section 2.6, we demonstrate numerically that a VCSEL can respond to a three-level aperiodic current modulation as an stochastic logic gate, and in Section 2.7, we analyze the reliability of the VCSEL-based logic gate. Finally, in Section 2.8, we summarize our conclusions.

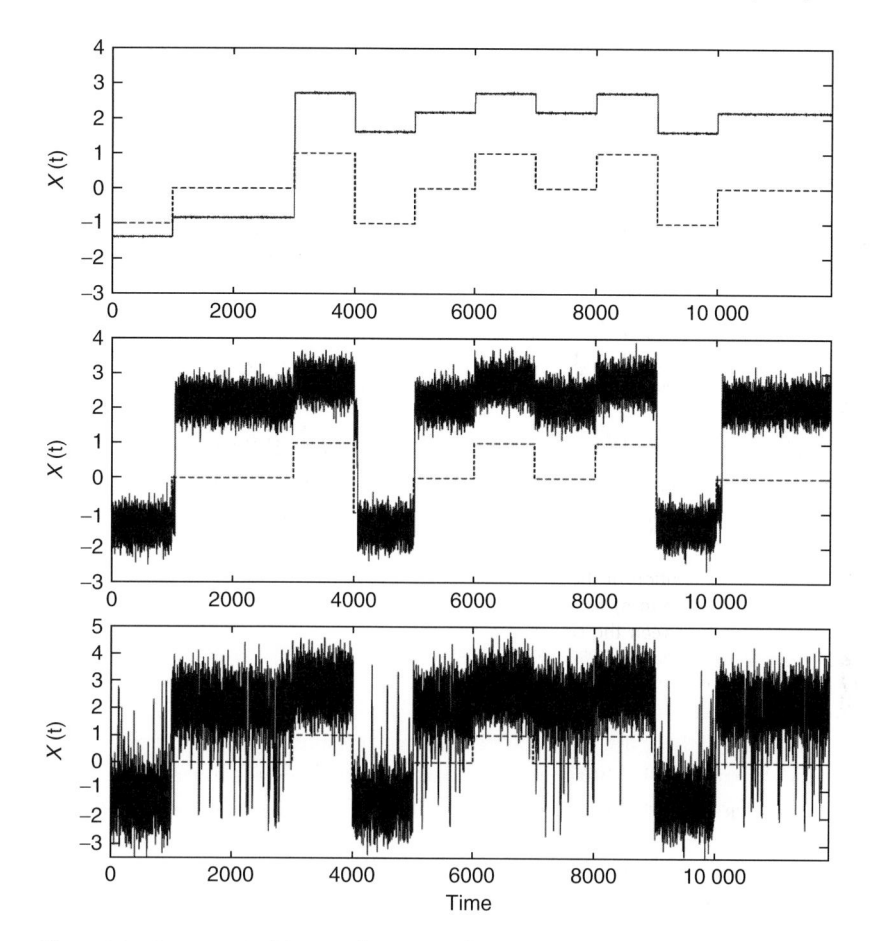

Figure 2.3 Response (solid line) of a two-state piecewise linear system to an input signal (dashed line) that is the sum of randomly switched square pulse trains. In the panels, the additive noise intensity increases from top to bottom, and the input signal is the same in each panel. For an optimal noise intensity (center panel) the system responds to the input signal as a OR/NOR gate. Adapted from [18].

2.2
Spin-Flip Model

The dynamics of a VCSEL can be described by the following set of rate equations for the orthogonal linearly polarized slowly varying complex amplitudes, E_x and E_y, the total carrier density, $N = N_+ + N_-$, and the carrier difference, $n = N_+ - N_-$ (N_+ and N_- being carrier populations with opposite spin) [26, 27]:

$$\frac{dE_x}{dt} = \kappa(1+i\alpha)\left[(N-1)E_x + inE_y\right] + (\gamma_a + i\gamma_p)E_x + \sqrt{\beta_{sp}\gamma_N}\xi_x, \quad (2.1)$$

$$\frac{dE_y}{dt} = \kappa(1+i\alpha)\left[(N-1)E_y - inE_x\right] - (\gamma_a + i\gamma_p)E_y + \sqrt{\beta_{sp}\gamma_N}\xi_y, \quad (2.2)$$

$$\frac{dN}{dt} = \gamma_N\left[J(t) - N(1+|E_x|^2+|E_y|^2) - in(E_y E_x^* - E_x E_y^*)\right], \quad (2.3)$$

$$\frac{dn}{dt} = -\gamma_s n - \gamma_N\left[n(|E_x|^2+|E_y|^2) + iN(E_y E_x^* - E_x E_y^*)\right] \quad (2.4)$$

where κ is the field decay rate, γ_N is the decay rate of the total carrier population, γ_s is the spin-flip rate, α is the linewidth enhancement factor, γ_a and γ_p are linear anisotropies representing dichroism and birefringence, and $J(t)$ is the time-dependent injection current parameter normalized such that the static, cw threshold in the absence of anisotropies is at $J_{th,s} = 1$. Spontaneous emission noise is represented by the terms ξ_x and ξ_y, which are uncorrelated Gaussian white noises with zero mean and unit variance. In the following, the noise strength is defined as $D = \beta_{sp}\gamma_N$, where β_{sp} is the coefficient of spontaneous emission.

The model has two linearly polarized steady-state solutions given by

$$E_x = \mathcal{E}_x e^{i\omega_x t}, \quad E_y = 0, \quad N = \mathcal{N}_x, \quad n = 0, \quad (2.5)$$
$$E_x = 0, \quad E_y = \mathcal{E}_y e^{i\omega_y t}, \quad N = \mathcal{N}_y, \quad n = 0 \quad (2.6)$$

where $\omega_x = \alpha\gamma_a - \gamma_p$, $\omega_y = -\alpha\gamma_a + \gamma_p$, $\mathcal{N}_x = 1 + \gamma_a/\kappa$, $\mathcal{N}_y = 1 - \gamma_a/\kappa$, $\mathcal{E}_x = \sqrt{J/\mathcal{N}_x - 1}$, and $\mathcal{E}_y = \sqrt{J/\mathcal{N}_y - 1}$.

The stability of these linearly polarized steady-state solutions (referred to as x and y modes) depends on the various model parameters. There are parameter regions where there is monostability (where either the x mode or the y mode is stable) and parameter regions where there is bistability (where both x and y modes are stable); see Figure 2.4a, which shows the linear stability of the two polarizations. There are also parameter regions where neither the x mode nor the y mode are stable. Steady-state solutions representing elliptically polarized light (where E_x and E_y have same optical frequency) also exist, as discussed in [27].

Unless otherwise specifically stated, in the rest of this chapter, Eqs. (2.1)–(2.4) are simulated with the following parameters: $\kappa = 300$ ns^{-1}, $\alpha = 3$, $\gamma_N = 1$ ns^{-1}, $\gamma_a = 0.5$ ns^{-1}, $\gamma_p = 50$ rad/ns, $\gamma_s = 50$ ns^{-1}, and $D = 10^{-6}$ ns^{-1}, which are typical for VCSELs.

2.3
Polarization Switching

Within the framework of the SFM, the PS phenomenon induced by the variation of a parameter (typically, the injection current) is interpreted as due to a change of stability of the two linearly polarized steady-state solutions, which can result in a switch from one polarization to the orthogonal one. This is shown in Figure 2.4b, which displays the polarization-resolved intensity–current characteristic for increasing and for decreasing pump current, calculated numerically by integrating

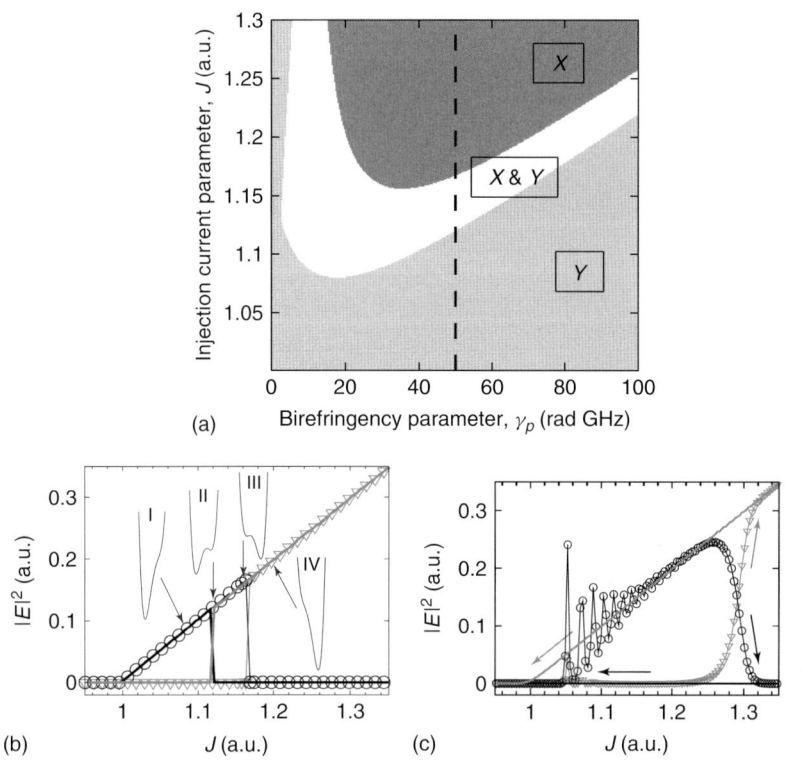

Figure 2.4 (a) Stability of the x and y polarizations in the parameter plane (birefringence, injection current). Other model parameters are $\kappa = 300$ ns^{-1}, $\alpha = 3$, $\gamma_N = 1$ ns^{-1}, $\gamma_a = 0.5$ ns^{-1}, and $\gamma_s = 50$ ns^{-1}. (b) Intensities of the x and y polarizations when the injection current increases and decreases linearly, from $J_i = 0.95$ to $J_f = 1.4$ in 40 μs. x polarization in gray for increasing ($\triangledown$) and decreasing (gray solid line) current; y polarization in black for increasing ($\bigcirc$) and decreasing (black solid line) current. $\gamma_p = 50$ rad ns^{-1}, $D = 10^{-6}$ ns^{-1} and other parameters as in the left panel. The black curves are schematic representations of the effective one-dimensional potential at four pump current values, corresponding to labels I to IV (see Section 2.6 for details). The dash vertical line in the left panel indicates the scan in the right panel. (c) As panel (b), but the current varies from $J_i = 0.95$ to $J_f = 1.4$ in 25 ns.

the model equations with a slow, quasi-adiabatic variation of the injection current (J varies between $J_i = 0.95$ and $J_f = 1.4$, upwards and downwards, in 40 μs).

It can be noticed that the stability scenario presented in Figure 2.4b and the PS points for increasing and decreasing current agree very well with the predictions of the linear stability analysis, displayed in Figure 2.4a.

Near the PS points, noise-induced switching can also occur. It has been shown that, in spite of the potentially complicated polarization dynamics, key features of the PS (such as the distribution of residence times in each polarization state) can be well understood as stochastic hopping in an effective one-dimensional double-well potential [13, 14].

The effective potential associated with the PS scenario is also displayed schematically in Figure 2.4b. In the low current region, labeled I, the laser can emit only the y polarization, which is represented as an effective potential with only one well. For increasing pump, there is a region of pump current values, labeled II, where there is bistability and a small probability of emission of the x polarization. The effective potential is a double-well potential, with a small right well. In this region of pump current values, if the laser emits the x polarization, a weak perturbation or a small amount of noise has a large probability to trigger a PS to the y polarization. On the contrary, if the laser emits the y polarization, there is only a small probability that a fluctuation will trigger a PS. As the pump increases, the switching probabilities vary, and at the right boundary of the bistable region, label III, the most probable polarization is the x polarization. If the laser emits the y polarization, a weak perturbation or a small amount of noise can trigger a switch to the x polarization. In this region, the effective potential is the double-well potential, which has a small left well. Finally, for high pump current (region label IV), the laser emits the x polarization and the effective potential has only one well.

However, this stability scenario changes drastically when the injection current variation is not slow or quasi-adiabatic, as compared to the laser characteristic time scales.

In nonlinear systems, when a control parameter is varied in time and is swept across a bifurcation point, the phenomenon of critical slowing down occurs near the bifurcation point and results in dynamical hysteresis [28]. In semiconductor lasers, critical slowing down has been demonstrated experimentally near the laser threshold and produces a delay in the laser turn-on, as shown in Figure 2.5, which depends on the pump current sweep rate and on the noise strength, among other parameters [28, 29].

The phenomena of critical slowing down and dynamical hysteresis near the PS points can be investigated numerically by simulating the SFM rate equations [30]. Figure 2.4c shows the polarization intensities versus the pump current, when the current varies between $J_i = 0.95$ and $J_f = 1.4$ in 25 ns (note that in Figure 2.4b, the current varied between the same extreme values but in 40 μs). By comparing both figures, one can observe that, with a faster current modulation:

(i) The threshold is delayed to a higher current value, that is, the dynamic lasing threshold is larger than the static one, $J_{s,th} = 1$ in the absence of anisotropies.
(ii) The laser turns on with relaxation oscillations.
(iii) The PS for increasing current is also delayed to a higher current value.
(iv) The PS for decreasing current is also delayed to a lower current value and can even disappear (in Figure 2.4c, one can notice that the x polarization remains on until the laser turns off).
(v) As a consequence of (iii) and (iv), the size of the bistability region increases, as compared to that predicted by both the linear stability analysis and the simulations with quasi-static current variation.

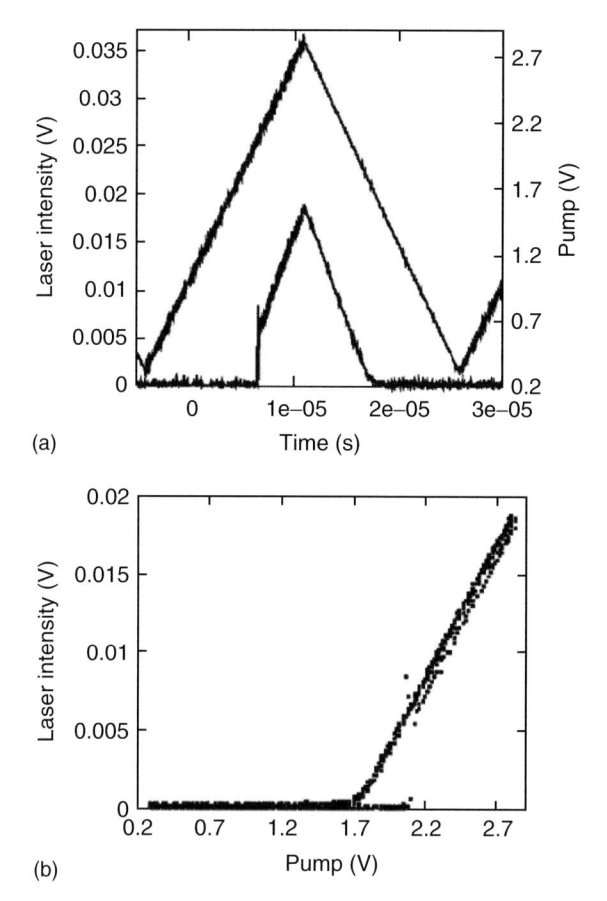

Figure 2.5 (a) Laser intensity and pump voltage as a function of time when a triangle wave of frequency 40 kHz is applied by the function generator. The laser switches on at a voltage V^* larger than the static threshold at $V_{th} = 1.78$, while the turn-off occurs at $V \approx V_{th}$ for decreasing pump. (b) A plot of the laser intensity as a function of pump voltage shows bistability in the interval $V_{th} < V < V^*$. The time traces–plotted with points to better highlight the effect–are slightly separated on the diagonal branch: the lower occurs for increasing pump and the higher for decreasing pump because of the speed at which the laser is driven. Adapted from [29].

To summarize, the quasi-static intensity-current response, as that shown in Figure 2.4a, is a good representation of the polarization response of a VCSEL when the injection current variation is slower than the longest time scale of the laser but fails to describe the laser polarization with faster modulation.

The above described phenomenon, due to critical slowing down near the PS points, has been demonstrated experimentally in directly modulated VCSELs (see Figure 2.6, adapted from [31]).

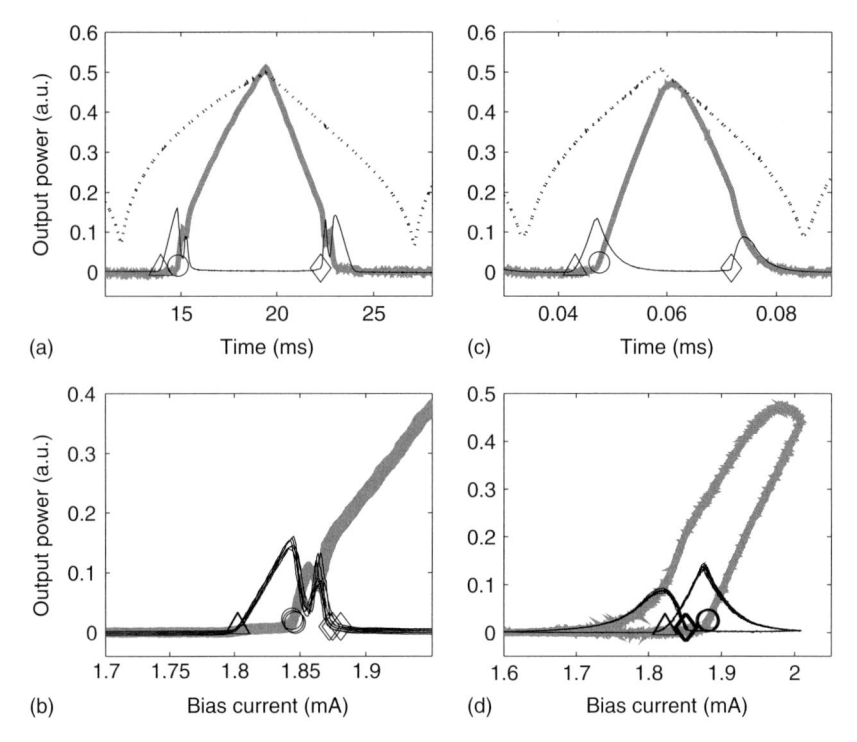

Figure 2.6 Experimentally measured polarization intensities versus time (a), (c) and versus pump current (b), (d) when the current variation is slow and quasi-adiabatic (the pump current is modulated at 60 Hz) (a), (b) and when it is much faster (20 kHz) (c), (d) across the PS points. Adapted from [31].

2.4
Pulse Generation Via Asymmetric Triangular Current Modulation

As discussed in the previous section, when the injection current variation is not quasi-static, the laser turns on with a few relaxation oscillations at the dynamic lasing threshold. As the current sweep becomes faster, the amplitude of those oscillations grows and, eventually, the intensity falls to 0 before the second oscillation occurs. If we now repeat periodically the linear increase and decrease of the pump current parameter with a period short enough, only one pulse per cycle is emitted per modulation cycle, as shown in Figure 2.7.

Therefore, a fast enough triangular current modulation (symmetric or not symmetric) crossing the static cw threshold, $J_{s,th}$, can result in the emission of short pulses of both orthogonal polarizations even when the average current value is below $J_{s,th}$. Because of the presence of noise, which is crucial at threshold, these pulses are irregular, both in amplitude and in timing.

2.4 Pulse Generation Via Asymmetric Triangular Current Modulation | 45

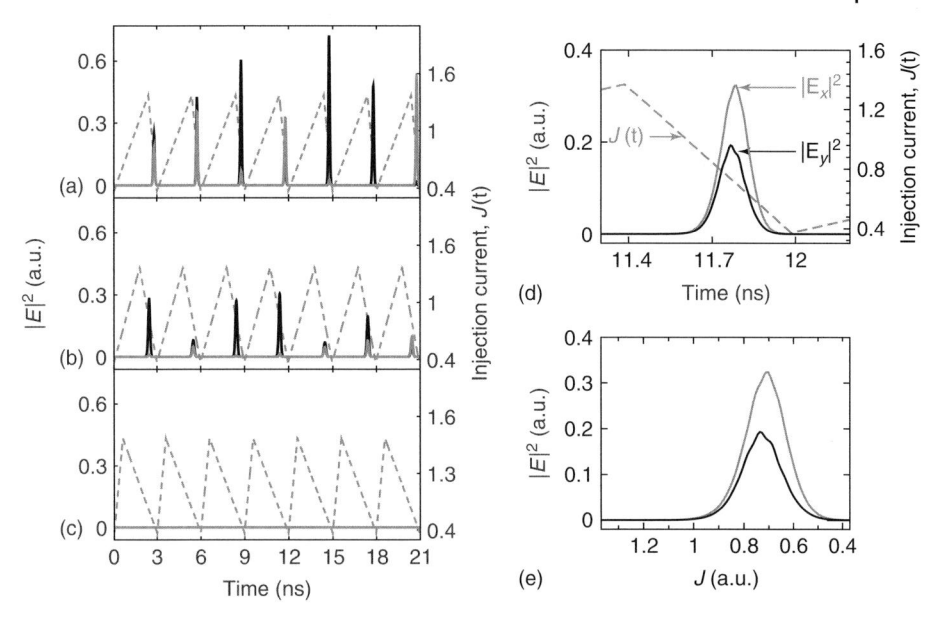

Figure 2.7 (a) Time traces of the intensities of the orthogonal linear polarizations (x, gray solid line; y, black solid line) and the injection current $J(t)$ (dashed line) when the asymmetry parameter is $\alpha_a = 80\%$ (a), 60% (b), and 20% (c). The modulation parameters are $J_0 = 0.37$, $\Delta J = 1$ (which give $J_m = 0.87 < 1$), and the modulation period is $T = 3$ ns. Panel (d) displays a detail of panel (a), and panel (e) displays the same detail as a function of the injection current.

The characteristics of these pulses depend on the shape of the triangular signal modulating the current, $J(t)$, which is defined in terms of four parameters:

- the lowest current value, J_0,
- the modulation amplitude, ΔJ,
- the modulation period, $T = T_1 + T_2$, where T_1 and T_2 are the time intervals during which the current increases or decreases linearly, and
- the asymmetry parameter, which is the ratio between the rising time and the modulation period, $\alpha_a = T_1/T$.

With these definitions, the mean value of the pump current is $J_m = J_0 + \Delta J/2$.

In Figure 2.7, three different shapes of asymmetric current modulation, with the same period and amplitude, are shown (dashed lines): slow rising and fast decreasing ($\alpha_a = 80\%$), almost symmetric ($\alpha_a = 60\%$), fast rising and slow decreasing ($\alpha_a = 20\%$).

The amplitude and the modulation period are chosen such that the laser emits only one sharp pulse per modulation cycle that is triggered at the end of the cycle, and the emission starts when $J(t)$ is still above $J_{s,th}$ as can be seen in Figure 2.7e. This is in good agreement with the observations of [15] and can be interpreted as due to the nonlinear interplay of the photons and the carriers in the VCSEL active region as is discussed later.

In Figure 2.7a–c, the solid lines show the time traces of the intensities of the two linear polarizations, $|E_x|^2$ and $|E_y|^2$. For $\alpha_a = 80\%$, large pulses are emitted (Figure 2.7a). A detail of a pulse in Figure 2.7a is shown in Figure 2.7d. For decreasing α_a, that is, going to a more symmetric modulation, the pulse amplitude gradually decreases (see Figure 2.7b, where $\alpha_a = 60\%$). If we continue decreasing α_a (considering the opposite asymmetry shape, with a fast-rising and slow-decreasing ramp), the pulses become smaller, and eventually, there are no pulses, as the intensities of the two polarizations remain at the noise level (see Figure 2.7c, where $\alpha_a = 20\%$).

Therefore, when the current modulation is asymmetric, there is a clear difference between the two asymmetry shapes: a slow-rising ramp followed by a fast-decreasing one and the opposite situation, a fast-rising ramp followed by a slow-decreasing one.

The effect of the asymmetry of the current modulation on the characteristics of the intensity pulses is presented in Figure 2.8. Figure 2.8a displays the time-averaged intensities of the two polarizations, $\langle|E_x|^2\rangle$ and $\langle|E_y|^2\rangle$, and the time-averaged total intensity, $\langle|E_T|^2\rangle = \langle|E_x|^2 + |E_y|^2\rangle$, versus the asymmetry parameter, α_a. In Figure 2.8b, we display the time-averaged pulse amplitude given by the maximum intensity in a cycle, $\langle A_x\rangle$, $\langle A_y\rangle$, and $\langle A_T\rangle$. When there is more than one pulse per modulation cycle, we calculate the average amplitude of the largest pulse. The averaged amplitudes are one order of magnitude larger than the averaged intensities because the laser emits sharp pulses and is off during most of the modulation cycle. Figure 2.8c displays the dispersion of the amplitude of the pulses, characterized in terms of the standard deviation normalized to the mean amplitude. In the three measures, there is an optimal modulation asymmetry, $\alpha_a \cong 80\%$, for which the averaged intensity and averaged pulse amplitude reach their maximum value, and for this asymmetry, the dispersion of the pulse amplitude is minimum.

The emitted pulses strongly depend on the initial conditions of the cycle, which are given by the dominance of one of the following mechanisms: the spontaneous emission and the radiation left by the previous pulse. When the radiation left by the previous pulse is absorbed by the carriers during the fall part of the cycle, spontaneous emission is the dominant mechanism for triggering the next pulse in the next cycle. On the contrary, when the radiation left has not been completely absorbed, it dominates over spontaneous emission for triggering the next pulse. We interpret our results as in [15], where the authors found that for small asymmetries the spontaneous emission is the dominant mechanism, whereas large asymmetries dominate the radiation left by the previous pulse.

The averaged total amplitude of the pulses is shown in Figure 2.8d as a function of the asymmetry parameter, α_a, and the mean value of the modulation, J_m, for a fixed modulation amplitude, $\Delta J = 1$ (ΔJ is the same as in Figure 2.88a–c; thus, Figure 2.8b is a horizontal scan of Figure 2.8d). We have used J_m instead of J_0 to emphasize that the laser emission occurs with a pump current that is on average below $J_{s,th} = 1$. As the pump current is modulated, it is suitable to define an effective lasing threshold as the averaged pump current above which the laser turns on. The modulation reduces the effective threshold, which depends on the asymmetry, giving the largest threshold reduction and the maximum amplitude for

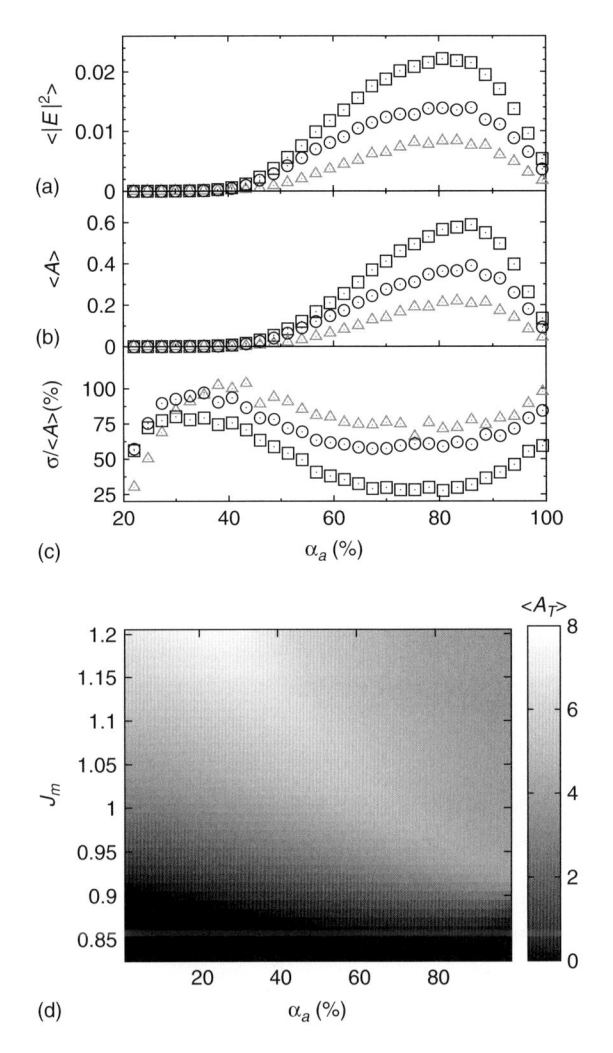

Figure 2.8 (a) Time-averaged intensities, (b) pulse amplitudes, and (c) normalized standard deviation of the pulse amplitude as a function of the asymmetry parameter, α_a. In (a–c) x polarization ($\triangle$), y polarization ($\bigcirc$), and total intensity ($\square$). (d) plot of the average pulse total amplitude $\langle A_T \rangle$ in the parameter plane for the asymmetry parameter, α_a, and the mean pump current, J_m. The modulation amplitude is $\Delta J = 1$ and the period is $T = 3$ ns. J_0 is fixed in figures (a–c), $J_0 = 0.37$, and is varied in figure (d).

an optimal $\alpha_a \sim 80\%$. For increasing J_m, the maximum amplitude moves to lower asymmetries, for which it has a fast-rising ramp followed by a slow-decreasing one.

The effective threshold can be reduced about 20% for large enough modulation amplitude, ΔJ, and small enough J_0. The modulation also enhances the emission of both orthogonal polarizations in a large range of pump currents, which can be understood as an effective stabilization of both polarizations. A discussion on the influence of the modulation parameters can be found in [16].

2.5
Influence of the Noise Strength

Noise plays a key role in the emission of the pulses and the interplay between the pump current modulation, and the noise is expected to produce constructive effects, enhancing the input signal.

In Figure 2.9a–c, we show the time traces of the intensities of the polarizations for a fixed modulation asymmetry, $\alpha_a = 80\%$, and three different noise strengths, D. Large spontaneous emission triggers the pulses at the end of the rising ramp (Figure 2.9a). An optimal noise strength $D \sim 10^{-3}$ ns^{-1} produces pulses with the largest amplitude triggered at the very beginning of the falling ramp (Figure 2.9b). For lower D values, the pulses are emitted at the end of the falling ramp, and their amplitude gradually decreases to zero (Figure 2.9c).

In Figure 2.9d–f, we show the time-averaged intensities, pulse amplitudes, and dispersion of the pulse amplitudes, respectively, as a function of the noise strength, D. While the intensity grows monotonously with the noise strength until it saturates, the amplitude of the pulses shows a maximum at $D \sim 10^{-3}$ ns^{-1}, which is accompanied by the minimum dispersion. This optimal emission for a finite noise strength is a hallmark of stochastic resonance [32] in our system. Here, the effect of the noise over the amplitude of the pulses is much clearer than in the period of the pulses, which occurs almost synchronized with the current modulation.

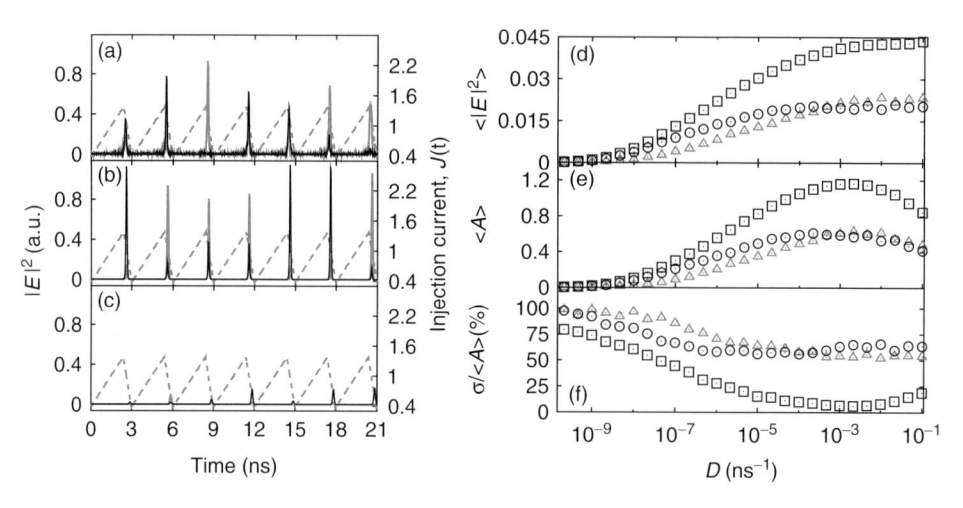

Figure 2.9 Time traces of the intensities for fixed asymmetry parameter $\alpha_a = 80\%$ and three noise strengths (a) $D = 10^{-1}$ ns^{-1}, (b) $D = 10^{-3}$ ns^{-1}, and (c) $D = 10^{-8}$ ns^{-1}. x polarization (gray solid line), y polarization (black solid line), and injection current (dashed line). (d) Time-averaged intensities, (e) pulse amplitudes, and (f) normalized standard deviation of the pulse amplitude as a function of the noise strength. In (d–f), x polarization ($\triangle$), y polarization ($\bigcirc$) and total intensity ($\square$). The modulation parameters are $\Delta J = 1$, $J_0 = 0.37$, $\alpha_a = 80\%$, and $T = 3$ ns.

The appearance of a stochastic resonance highlights the interplay between the pulse triggering mechanisms. As was previously discussed, the radiation left by the previous pulse dominates for large α_a. Thus, by increasing D, we are enhancing the effect of the spontaneous emission in the triggering process. The turn-on occurs earlier, in a pump current modulation cycle, since the noise makes the system easier to reach the lasing state. On the other hand, the earlier the pulse, the lower the radiation left to the next pulse. The maximum amplitude in Figure 2.9e gives the optimal noise intensity for which the radiation left is not too weak and the noise is not too strong.

2.6
Logic Stochastic Resonance in Polarization-Bistable VCSELs

Stochastic resonance phenomena in directly modulated semiconductor lasers have been extensively investigated in the literature [33–37]. Recently, a new kind of stochastic resonance has been demonstrated, named logical stochastic resonance (LSR) [18], which uses the nonlinear response of a bistable system to reproduce logical operations such as the AND and OR operations, under the influence of the right amount of noise. The main idea behind LSR is that the input levels can be chosen such that the probability of the switchings between two logical outputs is controlled by the noise strength [18].

In this section, we discuss the implementation of a VCSEL-based stochastic logical operator using an aperiodic three-level current modulation and the linearly polarized light as the output signal [17]. The case of optical modulation is briefly discussed.

We consider that the pump current parameter, $J(t)$, is the sum of two aperiodic square waves, $J(t) = J_1(t) + J_2(t)$, which encode the two logic inputs. Since the logic inputs can be either 0 or 1, we have four distinct input sets: (0, 0), (0, 1), (1, 0), and (1, 1). Sets (0, 1) and (1, 0) give the same value of J, and thus, the four distinct logic sets reduce to three J values.

Therefore, it is convenient to introduce the following three parameters characterizing the three-level aperiodic current modulation:

- the mean value, J_m,
- the modulation amplitude, ΔJ, and
- the bit time, T, which is the sum of the time interval during which the pump current is constant, T_1, and the time interval T_2 during which there is a fast-increasing or decreasing ramp to the next current value ($T = T_1 + T_2$ with $T_1 \gg T_2$).

Since the four distinct logic sets reduce to three J values, we will use the values J_{II}, J_{III}, and J_{IV} that correspond to the regions indicated in Figure 2.9b. We avoid using J_I because it is close to the static threshold, and with fast modulation, the PS for decreasing current does not occur, as was mentioned before. Then, J_m and ΔJ determine the three current levels as $J_{II} = J_m - \Delta J$, $J_{III} = J_m$, and $J_{IV} = J_m + \Delta J$.

The laser response is determined by the polarization of the emitted light. We chose parameters such that the laser emits either the x or the y polarization (parameter regions where there is polarization coexistence or elliptically polarized light are avoided). The laser response is considered a logical 1 if, for instance, the x polarization is emitted, and a logical 0, if the y polarization is emitted. Which polarization represents a logic 1, and which a logic 0 can depend on the logic operation, as discussed later.

In this way, the polarization emitted at each of the three current levels, encoding the four possible combinations of the two logic inputs, allows to implement the operations OR, AND, NOR, and NAND, according to Table 2.1. One should notice that by detecting one polarization, one obtains a logic response and the negation of that logic response by detecting the orthogonal polarization. In the following, we focus only on the nonnegation operations AND and OR.

Table 2.2 illustrates the encoding scheme. For the OR operation, x represents a logical 1 and y represents a logical 0. If the laser is emitting the y polarization, the current levels J_{III} and J_{IV} [representing the inputs (0,1), (1,0), and (1,1)] will both induce a switch to the x polarization. For the AND operation, the definition of the laser logic response changes: it is a logic 0 if the x polarization is emitted and a logic 1 if the y polarization is emitted. Also, the encoding criterion changes, in the sense that the lower current level J_{II} encodes the input (0, 0) for the OR operation, whereas it encodes the input (1, 1) for the AND operation; the highest current level J_{IV} encodes the input (1, 1) for OR and encodes (0, 0) for AND; the middle level J_{III} encodes (1,0) and (0, 1) for both operations.

Table 2.1 Relationship between the two inputs and the output of the logic operations.

Logic inputs	AND	NAND	OR	NOR
(0,0)	0	1	0	1
(1,0)/(0,1)	0	1	1	0
(1,1)	1	0	1	0

Table 2.2 Encoding scheme: relationship between the logic inputs, the encoding current levels, the output polarization, and the logic output for the AND and OR operations.

Logic inputs	AND: Current	x/y	Logic output	OR: Current	x/y	Logic output
(0,0)	J_{IV}	x	0	J_{II}	y	0
(1,0)/(0,1)	J_{III}	x	0	J_{III}	x	1
(1,1)	J_{II}	y	1	J_{IV}	x	1

There are other ways to associate the four possible logic inputs (0,0), (1,0), (0,1), and (1,1) to three input levels. In the scheme in [18], the levels J_I, J_{II}, J_{III} lead to the operation AND and the levels J_{II}, J_{III}, J_{IV} to the operation OR, in the presence of the right amount of noise.

In the following, we focus on the OR operation implemented with the encoding scheme described in Table 2.2, as the results apply also for the symmetric AND operation, because both operations are implemented with the same three current levels. This scheme allows fast modulation in AND and OR operations by preventing the drawback that at level J_I, the PS disappears for decreasing current, as discussed previously.

Unless otherwise explicitly stated, we use the following parameters for the three-level aperiodic signal: $J_m = 1.3$, $\Delta J = 0.27$, and $T = T_1 + T_2 = 31.5$ ns, with $T_1 = 31$ ns, and $T_2 = 0.5$ ns.

Figure 2.10a–c displays the laser response for the same logic input and three values of the noise strength. The three current levels are such that the laser emits one polarization (x) for two of them, while for the third one, it can switch to the orthogonal polarization (y), in the presence of the right amount of noise. With weak noise, the PS is delayed with respect to the current modulation (Figure 2.10a); with too strong noise, both polarizations are emitted simultaneously within the same bit (Figure 2.10c). Therefore, the operation of the VCSEL as a logic gate depends on the noise strength, in good agreement with [18]. For an intermediate amount of noise (Figure 2.10b), the PS occurs a short time after the beginning of a bit, whereas the noise is not strong enough to stimulate the emission of large intensities on both polarizations.

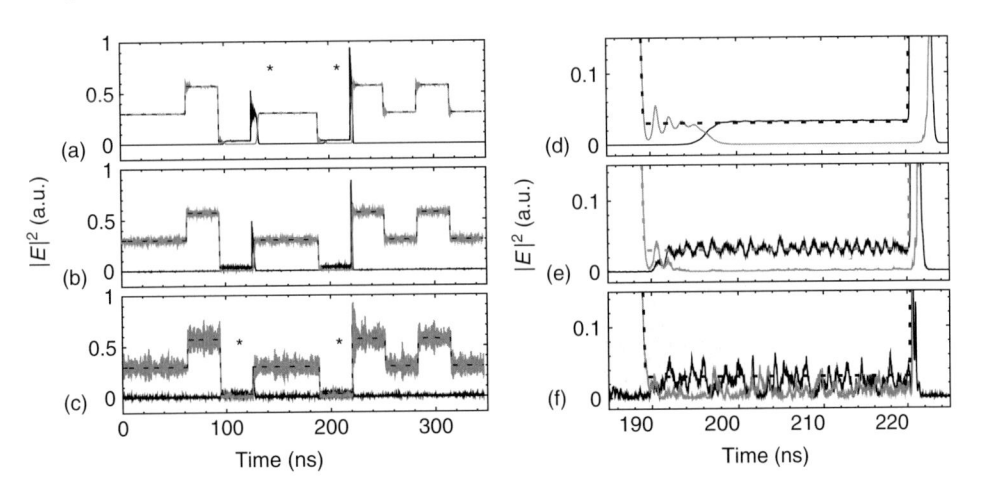

Figure 2.10 (a–c) Laser response under aperiodic three-level current modulation. Time traces of the x polarization (gray), y polarization (black), and the injection current $J - 1$ (dashed) for different noise intensities: (a) $D = 5 \times 10^{-7}$ ns^{-1}, (b) $D = 4 \times 10^{-4}$ ns^{-1} and (c) $D = 6 \times 10^{-3}$ ns^{-1}. The asterisks mark the wrong bits. (d–f) A detail of panels (a–c) to show the main errors in a bit.

2.7
Reliability of the VCSEL-Based Stochastic Logic Gate

To evaluate the reliability of the VCSEL-based stochastic logic gate, we calculate the success probability, that is, the probability to obtain the desired logic output. For the two logic inputs, we generate two random uncorrelated sequences of $N \geq 2^{10}$ bits and compute the success probability, P, as the ratio of the number of correct bits to the total number of bits. We define that a bit is correct, as follows. When x is the "right" output polarization (according to Table 2.2), we count a bit as correct if a given percentage (say, 80%) or more of the emitted power is emitted in the x polarization; if x is the "wrong" polarization, we count a bit as correct if a given percentage (say, 20%) or less of the emitted power is emitted in the x polarization.

Figure 2.11a displays the success probability as a function of the noise strength, for three success criteria: 80–20%, 90–10%, and 70–30%. One can notice that there is a range of noise strengths in which $P = 1$, and this noise range vanishes (increases) when choosing a more restrictive (a more permissive) threshold for the emitted power in the x polarization. Within this noise range, there is optimal noise-activated PSs (the "interwell" dynamics in the double-well potential picture) and optimal sensitivity to spontaneous emission in each polarization (the "intrawell" dynamics in the double-well potential picture). It should be noticed that $P = 1$ occurs for noise strengths D that do not have to be unusually small. On the

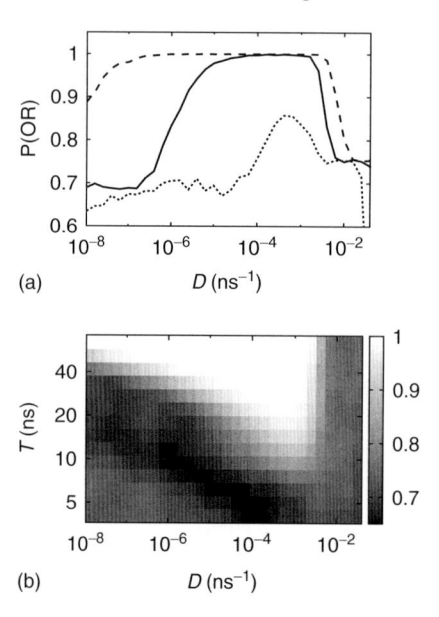

Figure 2.11 (a) Success probability P as a function of the noise strength, D, keeping fixed the bit time $T = 31.5$ ns and using a success criterion of 80–20% (solid line), 90–10% (doted line), and 70–30% (dashed line). (b) Log–log plot of the success probability P as a function of the noise intensity, D, and the bit time, T. Other parameters are as in Figure 2.4.

contrary, they are realistic values for semiconductor lasers, which typically have $D \sim 10^{-4}$.

The success probability depends strongly on the bit time, T, and the noise strength, D, as shown in Figure 2.11b. Short bits (≤ 5 ns) prevent logical operations because of the finite time required for the PS. For increasing T, the success probability grows until it saturates at $P = 1$ for long enough bits, for which the PS time is $\ll T$. The interplay between the duration of the bit and the noise strength can be interpreted as follows. The time needed to escape from a potential well decreases with increasing noise [32]. Then, for weak noise, as D increases, the escape time decreases and the probability of a correct response grows. On the other hand, too strong noise results in spontaneous emission in both polarizations and thus, for large enough noise, the power emitted in the "wrong" polarization grows above the threshold for detecting the response as correct, and thus, above a certain noise level, the success probability decreases monotonously. The dependence of the success probability on the noise strength is due to the interplay of noise-induced escapes (interwell stochastic dynamics) and spontaneous emission noise in the two polarizations (intrawell stochastic dynamics).

The effect of LSR in VCSELs is reliable in a large range of laser parameters and current modulations and is also robust under small feedback strengths [17], which makes the VCSEL logic gate attractive for applications in systems subjected to noisy backgrounds. Our proposed implementation does not require a fine tuning of parameters, and there is a wide range of realistic noise strengths in which the device gives a reliable and correct logic response.

A desired property of the optical logic circuits is that the light should be the input and the output of the system. For that reason, an all-optical logical operator has been considered for our system. A more complex scenario occurs when we use an optical injection as the input modulation. A finite optical injection leads to the bistability of both polarizations, which is required for LSR. However, a complicate route to the PS occurs in the borders of the bistable region [38], and oscillatory or even chaotic dynamics appear. The results obtained for this configuration reveal that, despite the complex dynamics, results similar to those shown in Figure 2.11 can be obtained. A fast (on the order of tens of nanoseconds) and reliable all-optical noise-induced logical operator is possible in VCSELs, but further investigation is required on this topic.

2.8
Conclusions

We have shown that in VCSELs, the interplay of noise, current modulation, and polarization bistability can result in two noise-controlled effects that have potential applications in optical information processing systems. First, we considered a triangular asymmetric current modulation and showed that, for appropriated asymmetry parameters and noise strength, the laser emits large coherent intensity pulses even when the cw value of the bias current is, on average, below the lasing threshold. Second, we have shown that, when the laser current is modulated with a three-level aperiodic signal, in a wide range of noise strengths, the laser responds as a reliable stochastic logic gate. Our results are promising because it has recently been demonstrated experimentally that polarization of bistable VCSELs can be used to build an optical buffer memory [39–41], in which the bit state of the optical signal, "0" or "1", is stored as a lasing linear polarization state (x or y), and it can be transferred from one VCSEL to another that is optically connected in cascade.

Acknowledgment

This research was supported in part by the European Office of Aerospace Research and Development (EOARD) grant FA-8655-10-1-3075, the Spanish Ministerio de Educacion y Ciencia through grant Fis2009-13360-C03-02, and the Agència de Gestió d'Ajuts Universitaris i de Recerca, Generalitat de Catalunya, through grant 2009 SGR 1168 and the ICREA Academia programme.

References

1. Janson, N.B., Balanov, A.G., and Schöll, E. (2004) Delayed feedback as a means of control of noise-induced motion. *Phys. Rev. Lett.*, **93**, 010601.
2. Balanov, A.G., Janson, N.B., and Schöll, E. (2004) Control of noise-induced oscillations by delayed feedback. *Physica D*, **199**, 1.
3. Hauschildt, B., Janson, N.B., Balanov, A., and Schöll, E. (2006) Noise-induced cooperative dynamics and its control in coupled neuron models. *Phys. Rev. E*, **74**, 051906.
4. Prager, T., Lerch, H.-P., Schimansky-Geier, L., and Schöll, E. (2007) Increase of coherence in excitable systems by delayed feedback. *J. Phys. A: Math. Theor.*, **40**, 11045–11055.
5. Flunkert, V. and Schöll, E. (2007) Suppressing noise-induced intensity pulsations in semiconductor lasers by means of time-delayed feedback. *Phys. Rev. E*, **76**, 066202.
6. Dahms, T., Hövel, P., and Schöll, E. (2008) Stabilizing continuous-wave output in semiconductor lasers by time-delayed feedback. *Phys. Rev. E*, **78**, 056213.
7. Schikora, S., Hövel, P., Wünsche, H.-J., Schöll, E., and Henneberger, F. (2006) All-optical noninvasive control of unstable steady states in a semiconductor laser. *Phys. Rev. Lett.*, **97**, 213902.
8. Koyama, F. (2006) Recent advances of VCSEL photonics. *J. Lightwave Technol.*, **24**, 4502–4513.
9. Barbay, S., Giacomelli, G., and Marin, F. (2000) Noise-assisted binary information transmission in vertical cavity surface emitting lasers. *Opt. Lett.*, **25**, 1095–1097.
10. Barbay, S., Giacomelli, G., and Marin, F. (2001) Noise-assisted transmission of binary information: theory and experiment. *Phys. Rev. E*, **63**, 051110.
11. Iga, K. (2000) Surface-emitting laser: its birth and generation of new optoelectronics field. *IEEE J. Sel. Top. Quantum Electron.*, **6**, 1201.
12. Kaiser, J., Degen, C., and Elsässer, W. (2002) Polarization-switching influence on the intensity noise of vertical-cavity surface-emitting lasers. *J. Opt. Soc. Am. B*, **19**, 672.
13. Willemsen, M.B., Khalid, M.U.F., van Exter, M.P., and Woerdman, J.P. (1999) Polarization switching of a vertical-cavity semiconductor laser as a Kramers hopping problem. *Phys. Rev. Lett.*, **82**, 4815–4818.

14. Prati, F., Giacomelli, G., and Marin, F. (1999) Competition between orthogonally polarized transverse modes in vertical-cavity surface-emitting lasers and its influence on intensity noise. *Phys. Rev. A*, **62**, 033810.
15. Preda, C.E., Segard, B., and Glorieux, P. (2006) Weak temporal ratchet effect by asymmetric modulation of a laser. *Opt. Lett.*, **31**, 2347–2349.
16. Zamora-Munt, J. and Masoller, C. (2008) Numerical implementation of a VCSEL-based stochastic logic gate via polarization bistability. *Opt. Express*, **16**, 17848–17853.
17. Zamora-Munt, J. and Masoller, C. (2010) Numerical implementation of a VCSEL-based stochastic logic gate via polarization bistability. *Opt. Express*, **18**, 16418–16429.
18. Murali, K., Shina, S., Ditto, W.L., and Bulsara, A.R. (2009) Reliable logic circuit elements that exploit nonlinearity in the presence of a noise floor. *Phys. Rev. Lett.*, **102**, 104101.
19. Murali, K., Rajamohamed, I., Shina, S., Ditto, W.L., and Bulsara, A.R. (2009) Realization of reliable and flexible logic gates using noisy nonlinear circuits. *Appl. Phys. Lett.*, **95**, 194102.
20. Sinha, S., Cruz, J.M., Buhse, T., and Parmananda, P. (2009) Exploiting the effect of noise on a chemical system to obtain logic gates. *Europhys. Lett.*, **86**, 60003.
21. Worschech, L., Hartmann, F., Kim, T.Y., Hofling, S., Kamp, M., Forchel, A., Ahopelto, J., Neri, I., Dari, A., and Gammaitoni, L. (2010) Universal and reconfigurable logic gates in a compact three-terminal resonant tunneling diode. *Appl. Phys. Lett.*, **96**, 042112.
22. Ditto, W.L., Miliotis, A., Murali, K., Sinha, S., and Spano, M.L. (2010) Chaogates: Morphing logic gates that exploit dynamical patterns. *Chaos*, **20**, 037107.
23. Guerra, D.N., Bulsara, A.R., Ditto, W.L., Sinha, S., Murali, K., and Mohanty, P. (2010) A Noise-Assisted Reprogrammable Nanomechanical Logic Gate. *Nano Lett.*, **10**, 1168–1171.
24. Bulsara, A.R., Dari, A., Ditto, W.L., Murali, K., and Sinha, S. (2010) Logical stochastic resonance. *Chem. Phys.*, **375**, 424–434.
25. Zhang, L., Song, A., and He, J. (2010) Effect of colored noise on logical stochastic resonance in bistable dynamics. *Phys. Rev. E*, **82**, 051106.
26. San Miguel, M., Feng, Q., and Moloney, J.V. (1995) Light-polarization dynamics in surface-emitting semiconductor lasers. *Phys. Rev. A*, **52**, 1728–1739.
27. Martin-Regalado, J., Prati, F., San Miguel, M., and Abraham, N.B. (1997) Polarization properties of vertical-cavity surface- emitting lasers. *IEEE J. Quantum Electron.*, **33**, 765–783.
28. Mandel, P. (1997) *Theoretical Problems in Cavity Nonlinear Optics*, Cambridge University Press, Cambridge, England.
29. Tredicce, J.R., Lippi, G.L., Mandel, P., Charasse, B., Chevalier, A., and Picque, B. (2004) Critical slowing down at a bifurcation. *Am. J. Phys.*, **72**, 799.
30. Masoller, C., Torre, M.S., and Mandel, P. (2006) Influence of the injection current sweep rate on the polarization switching of vertical-cavity surface-emitting lasers. *J. Appl. Phys.*, **99**, 026106.
31. Paul, J., Masoller, C., Hong, Y., Spencer, P.S., and Shore, K.A. (2006) Experimental study of polarization switching of vertical-cavity surface-emitting lasers as a dynamical bifurcation. *Opt. Lett.*, **31**, 748–750.
32. Gammaitoni, L., Hänggi, P., Jung, P., and Marchesoni, F. (1998) Stochastic resonance. *Rev. Mod. Phys.*, **70**, 223–287.
33. Barbay, S., Giacomelli, G., and Marin, F. (2000) Stochastic resonance in vertical cavity surface emitting lasers. *Phys. Rev. E*, **61**, 157.
34. Buldu, J.M., Garcia-Ojalvo, J., Mirasso, C.R., and Torrent, M.C. (2002) Stochastic entrainment of optical power dropouts. *Phys. Rev. E*, **66**, 021106.
35. Nagler, B., Peeters, M., Veretennicoff, I., and Danckaert, J. (2003) Stochastic resonance in vertical-cavity surface-emitting lasers based on a multiple time-scale analysis. *Phys. Rev. E*, **67**, 056112.
36. Arteaga, M.A., Valencia, M., Sciamanna, M., Thienpont, H., Lopez-Amo, M.,

and Panajotov, K. (2007) Experimental evidence of coherence resonance in a time-delayed bistable system. *Phys. Rev. Lett.*, **99**, 023903.

37. Arecchi, F.T. and Meucci, R. (2009) Stochastic and coherence resonance in lasers: homoclinic chaos and polarization bistability. *Eur. Phys. J. B*, **69**, 93.

38. Sciamanna, M. and Panajotov, K. (2005) Two-mode injection locking in vertical-cavity surface-emitting lasers. *Opt. Lett.*, **30**, 2903–2905.

39. Katayama, T., Ooi, T., and Kawaguchi, H. (2009) Experimental demonstration of multi-bit optical buffer memory using 1.55-mu m polarization bistable vertical-cavity surface-emitting lasers. *IEEE J. Quantum Electron.*, **45**, 1495–1504.

40. Kawaguchi, H. (2009) Polarization-bistable vertical-cavity surface-emitting lasers: application for optical bit memory. *Optoelectron. Rev.*, **17**, 265–274.

41. Mori, T., Sato, Y., and Kawaguchi, H. (2009) 10-Gb/s optical buffer memory using a polarization bistable VCSEL. *IEICE Trans. Electron.*, **E92C**, 957–963.

3
Mode Competition Driving Laser Nonlinear Dynamics
Marc Sciamanna

3.1
Introduction

It is known that semiconductor lasers allow for the emission of several lasing modes that compete for the optical gain: longitudinal modes with frequency spacing related to the laser internal cavity length and dispersive properties, transverse modes related to the optical cavity geometry, and polarization modes when the polarization of the emitted light is not pinned by the cavity and gain properties. The coexistence of several laser modes not only affects the laser performances (relative intensity noise, laser linewidth, etc.) but also impacts on the laser dynamics when it is subject to, for example, optical feedback, optical injection, or large current modulation. In such configurations the laser relaxation oscillations (RO) can become undamped and the laser starts behaving like an autonomous nonlinear oscillator. Additional bifurcations – that is, qualitative changes in the dynamics – may occur and destabilize the time-periodic dynamics into either quasiperiodicity or even chaos. When the laser exhibits such a rich and complex set of dynamical behaviors, the multimode emission can lead either to an apparently more regular total laser output or, by contrast, can lead to a more complex laser chaotic dynamics with a higher dimension.

In this chapter, we review recent results showing the influence of mode competition on semiconductor laser nonlinear dynamics. Examples are mostly taken from the author's own contributions. Focus is made on experiments with, whenever it is possible, a comparison with theoretical modeling and simulations. The chapter is organized as follows:

- In Section 3.2, we summarize those configurations where the laser dynamics could be considered as resulting from mode competition and multimode lasing. More specifically, we talk about longitudinal modes, transverse modes, polarization modes of vertical-cavity surface-emitting lasers (VCSELs), and external-cavity modes (ECMs) in compound-cavity lasers.
- In Section 3.3, we review the properties of multimode semiconductor lasers in the presence of a moderately strong optical feedback, such that the laser dynamics is brought into optical chaos.

Nonlinear Laser Dynamics: From Quantum Dots to Cryptography, First Edition. Edited by Kathy Lüdge.
© 2012 Wiley-VCH Verlag GmbH & Co. KGaA. Published 2012 by Wiley-VCH Verlag GmbH & Co. KGaA.

- In Section 3.4, we show how the ECMs of a compound-cavity laser can beat and this beating results in a high-frequency self-pulsating laser output.
- In Section 3.5, we focus on a dynamical regime that is specific to optical feedback from short external cavities and called *regular pulse package* (RPP). We discuss about the impact of polarization mode competition in VCSELs in this dynamical regime, where the laser output fires a regular stream of pulses.
- In Section 3.6, we review the interesting polarization competition arising in VCSEL, and in particular, the polarization dynamics accompanying polarization switching (PS) induced by optical feedback: the optical feedback induces a competition between two orthogonal linearly polarized (LP) laser modes, and together with the induced switching the two laser modes exhibit a time-periodic pulsing dynamics at the period of the time delay in the feedback loop. This example shows the generic properties of two-mode systems with time delay and noise, such as coherence resonance (CR).
- In Section 3.7, we show how the two-polarization-mode lasing of VCSEL influences the dynamics of a laser in the presence of optical injection. We also highlight the interesting injection locking properties of the laser when it emits several high-order transverse modes.
- Finally, in Section 3.8, we discuss recent results on a gain switching dynamics that occurs in quantum dot (QD) semiconductor lasers with optical injection and lasing in both excited and ground energy states.

3.2
Mode Competition in Semiconductor Lasers

There are several configurations where a semiconductor laser does not emit in a single-frequency lasing mode, but instead shows competition and/or simultaneous lasing in nondegenerate laser modes.

1) The laser threshold condition for a semiconductor optical gain medium inserted in a resonating optical cavity (typically a Fabry–Perot cavity obtained by cleavage of the two mirror facets) is deduced from the fact that an optical wave propagating through the laser cavity forms a standing wave between the two mirror facets of the laser. The distance L between the two mirrors determines the period of oscillation of this curve. This standing optical wave resonates only when the cavity length L is an integer number m of half wavelengths existing between the two mirrors. In other words, a node must exist at each end of the cavity. The only way that this can take place is for L to be exactly a whole number multiple of half wavelengths $\lambda/2$. This means that $L = m(\lambda/2)$, where λ is the wavelength of light in the semiconductor matter. As a result of this situation, there can exist many *longitudinal modes* in the cavity of the laser diode, each resonating at its distinct wavelength λ_m. Two adjacent longitudinal laser modes are separated by a wavelength of $\Delta\lambda = \lambda_0^2/2n_g L$ with n_g the so-called group refractive index. The number of longitudinal modes that a semiconductor laser is capable of supporting is a function of the cavity structure, and also because of the gain spectrum: only modes whose optical gains compensate the corresponding

optical losses reach the threshold condition for lasing. Specific cavity laser designs can be suggested, such as distributed feedback lasers (DFB) or distributed Bragg reflector lasers (DBR), to improve the single-mode laser performances: a grating is designed along the laser cavity or in the laser mirrors to suppress the propagation or reduce the reflectivity of undesired longitudinal modes.

2) Since a realistic optical cavity has a finite transverse cross-sectional area, the resonant optical field in the laser cavity cannot be a plane wave. Therefore, there exist certain modes called *transverse modes*, which differ in their transverse distributions of the field in the optical cavity. Since the transverse modes must be sustained by the cavity boundary conditions, it is clear that the transverse modes are modes that reproduce themselves after a round-trip pass in the cavity although they can be attenuated or amplified in amplitude and phase shifted. The transverse modes that exist depend on the optical properties of the gain medium and by any boundary conditions imposed on the wave equation by the optical structures in the medium. The transverse modes are usually described in a rectangular basis by the Hermite–Gaussian functions or are also called TEM$_{m,n}$ modes. Figure 3.1 shows different optical patterns observed in a VCSEL (see hereafter for a more complete overview of VCSEL dynamics) for different injection currents. The conventional transverse electro-magnetic (TEM) solution is the TEM$_{0,0}$ mode with a Gaussian radial intensity profile, but other combinations of higher-order transverse modes can be observed when increasing the optical power.

3) VCSELs exhibit several advantages over the conventional edge-emitting semiconductor lasers: very small threshold current (less than mA), single longitudinal mode emission, circular output beam profile with narrow divergence, on-wafer testing capabilities and the possibility to easily fabricate large bidimensional laser arrays. However VCSELs also exhibit peculiar light polarization properties. In general, VCSELs emit a linear polarization with its linear direction oriented along one of the two preferential crystallographic directions

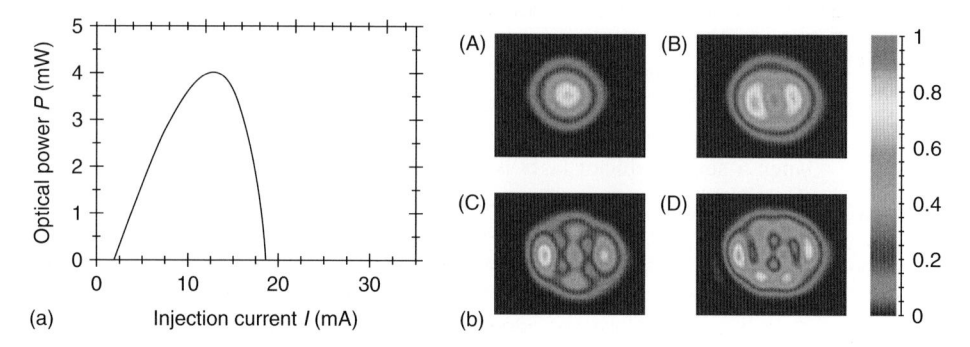

Figure 3.1 (a) Optical power versus injection current for a 6 μm oxide aperture VCSEL, (b) near-field optical patterns at different injection currents: (A) 3.0 mA, (B) 6.2 mA, (C) 14.7 mA, and (D) 18 mA. Taken from C. Degen et al. [1].

[110] and [1−10] for VCSELs grown on substrate oriented in (100). The presence of *two lasing modes with orthogonal linear polarizations* has been attributed to residual polarization anisotropies emerging from the fabrication process. As a result of the residual linear birefringence, the two LP modes exhibit slightly different wavelengths and since the gain curve in semiconductor lasers is wavelength dependent, the two modes exhibit also slightly different gains. In some VCSELs, the light selects at threshold its polarization among the two orthogonal LP modes, and the light polarization remains stable whatever the modifications of the laser operating conditions. However, most often the light polarization is not well defined and may strongly vary depending on the temperature or the injection current. A very common observation is the polarization switching (PS), which occurs when increasing the injection current [2]: the VCSEL starts lasing at threshold in one of the two LP modes, but as the current increases, the VCSEL polarization switches to the orthogonal LP mode. Two-polarization-mode emission has also been reported close to the lasing threshold [3]. As we increase the injection current further, multiple transverse modes start lasing and the multitransverse mode emission can also modify the polarization properties of the emitted light: the first-order transverse mode usually lases with a linear polarization orthogonal to one of the fundamental transverse modes [2].

4) In several and sometimes unintentional situations, a semiconductor laser is subject to optical feedback, that is, when part of the emitted light is reflected and reinjected back to the laser with a given time delay (the time for the light to make a round trip in this extended cavity). A compound cavity is created between the laser output mirror and the external mirror creating an optical feedback. A typical model used to study the effects of optical feedback is the so-called Lang–Kobayashi (LK) model [5]: it consists of a set of two differential equations for the optical field and the carrier inversion, and moreover the differential equation for the field contains a time-delayed field variable to account for the optical feedback. The LK model considers a single longitudinal mode laser with no transverse degrees of freedom. Without optical feedback the laser therefore emits with a single-frequency lasing mode. When looking for the steady-state solutions or lasing modes of this laser system, it appears that besides this free-running lasing mode, additional lasing modes with different frequencies can be created in pairs when increasing the feedback strength [6]. These modes are called *ECMs*. Figure 3.2 shows a typical dynamics observed when a semiconductor laser is subject to a relatively weak optical feedback: the laser hops between different ECMs at irregular time intervals. A frequency versus time analysis allows one to distinguish between the ECMs and their frequency splitting. The frequency separation between ECMs is related to the external cavity length and therefore also to the time delay. As will be shown in the following sections, the possibility for the compound-cavity laser to sustain lasing in several frequency-splitted ECMs allows for the observation of complex new nonlinear dynamics features. ECMs and their stability properties are also discussed in Chapter 6 by Otto *et al.* when analyzing the sensitivity of QD lasers to optical feedback.

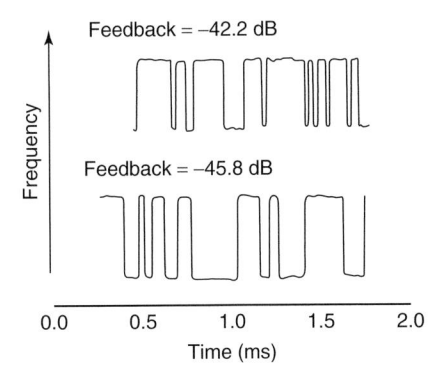

Figure 3.2 Hopping between external-cavity modes in a laser diode with weak optical feedback, and for two different values of the feedback strength. Taken from [4].

3.3
Low-Frequency Fluctuations in Multimode Lasers

Subject to external, delayed, and optical feedback, laser diodes present a large variety of qualitatively different dynamical behaviors. Among them, the low-frequency fluctuation (LFF) regime consists of sudden dropouts in the laser intensity followed by gradual recoveries [7]. The time between power dropouts is a random quantity, but on average is much larger than any of the laser system time-scales (in particular, the period of the RO or the external-cavity round-trip time). A close look into the intensity time trace shows that besides this slow time scale, the laser is also firing sequences of pulses on a much faster, picosecond time scale. A popular interpretation of the LFF phenomenon relies on the LK equations, which assume a single-mode operation of the laser and a weak or moderate amount of external optical feedback. In experimental studies, however, the semiconductor laser is often lasing on several longitudinal models in the case of edge-emitting lasers (EELs), or with two orthogonal polarization modes in the case of VCSELs. The impact of multimode emission on the characteristics of the LFF regime has been investigated by several groups in the recent years. Here is a summary of the main achievements.

1) Experiments on EELs have shown that multimode operation often occurs within the LFF regime, unless the laser is forced to operate in a single longitudinal mode by the use of a grating or a frequency etalon. They found that the mode competition results in either inphase or antiphase pulsating dynamics in the individual modes on a picosecond time scale, together with often synchronous dropouts on the slow time scale of the LFF [8, 9]. As a result the total intensity dynamics (as one would observe if not resolving the individual longitudinal mode dynamics) may be less strongly pulsating than the individual mode dynamics. The conclusion is important because one could observe a qualitatively very different LFF dynamics when looking into the

total intensity or the individual mode dynamics. Theoretical works based on either a multimode extension of the LK equations [10] or a model derived from the Tang–Statz–de Mars equations [11] have been able to reproduce this interesting feature of the multimode LFF dynamics.

2) Often in experiments looking for chaotic LFF dynamics, a frequency selective optical component such as an etalon or a grating is often placed in the external cavity. This device is adequately tuned so that only one longitudinal mode is selected and reinjected into the laser cavity. The other modes are not subject to the optical feedback and are called *free modes*. In this way, the laser is restricted to oscillate essentially in the selected mode. However, experiments have shown that intensity bursts in the free modes occurring simultaneously with dropouts in the mode selected by the feedback [13]. Using a multimode extension of the LK equations, we have reproduced qualitatively similar dynamics – see Figure 3.3 [12]. It appears that the sudden bursts in the free modes is caused by a sudden increase in the carrier density that results from the dropouts in the selected longitudinal mode. The burst lasts until the carrier density decreases to reach again its almost steady state value.

3) LFF has also been found in simulations of VCSELs subject to optical feedback from a distant mirror [14]. As mentioned above, VCSELs exhibit many differences with EELs and, in particular, may exhibit a switching between two orthogonal LP modes. This two-mode feature of VCSELs and the related optical bistability may be responsible for the new dynamical features not seen

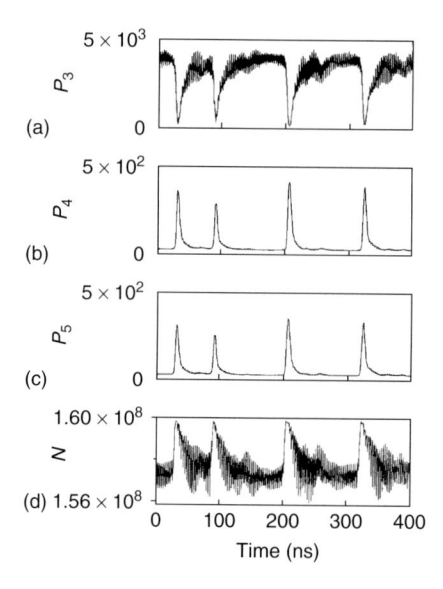

Figure 3.3 Numerical simulation of a multimode laser model extending the LK equations, in the case of an EEL subject to a frequency filtered feedback. Power dropouts characteristic of the LFF dynamics in the selected mode (P_3) are followed by bursts in the free modes (P_4 and P_5). The dynamics can be understood by looking to the time-evolution of the carrier density N [12].

in conventional semiconductor lasers. The LFF dynamics of VCSELs consists of two qualitatively different regimes [15]: one called *type I LFF*, where the two polarization modes exhibit synchronous power dropouts and other called *type II LFF*, where one of the two modes exhibits power dropouts immediately followed by power bursts in the orthogonal mode. As has been demonstrated theoretically, the important parameters deciding on the type of LFF dynamics are the linear birefringence and dichroism inherent to the epitaxial growth of the VCSEL cavity. Figure 3.4 shows the transition between the two types of LFF dynamics, when varying the strength of the VCSEL cavity linear anisotropies. Type I and type II LFF also lead to different dynamics in the total intensity on the time scale of the fast pulsing underlying LFF [16], in a similar way than explained above for multimode EELs. Experiments have later confirmed these features of LFF in VCSELs [17].

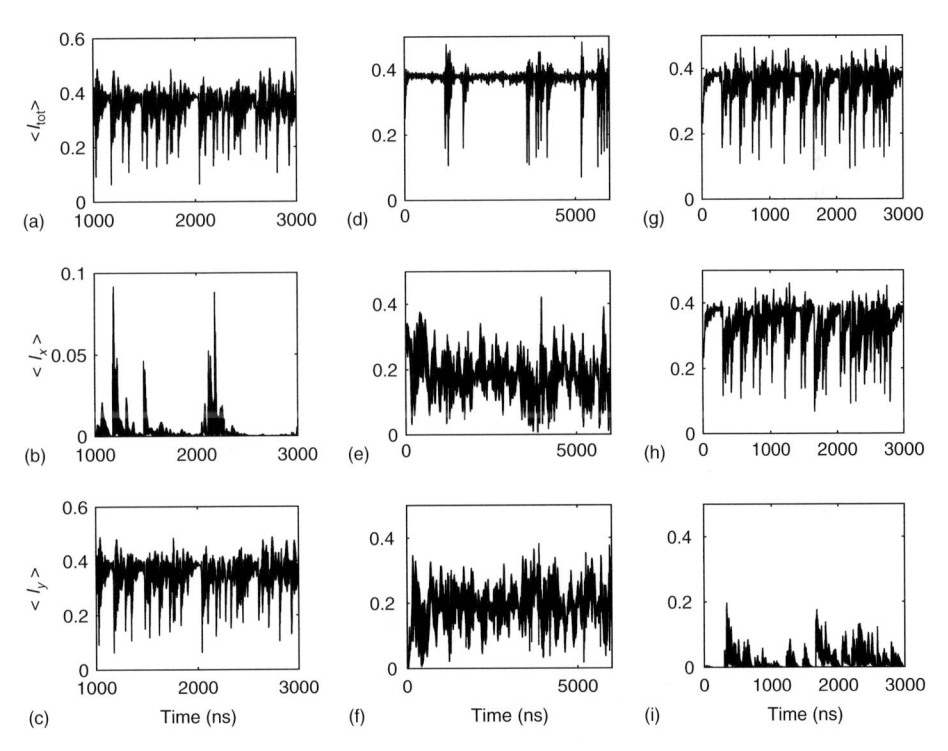

Figure 3.4 Simulated LFF dynamics in VCSEL for different parameters. From top to bottom: time traces of total, x-LP, and y-LP mode intensities. The left panel shows LFF dynamics in the total intensity and in one of the LP mode, while the other LP mode exhibits power bursts simultaneously to the other mode power dropouts. The right panel illustrates a similar case. In the middle panel, however, the two modes fluctuate in anticorrelated dynamics resulting in less frequent dropouts of the total intensity. Taken from [15].

3.4
External-Cavity Mode Beating and Bifurcation Bridges

The LFF dynamics disappears for short external cavity (EC) (typically less than 1 cm) as the frequency of the EC $1/\tau_{ext}$ becomes much larger than the frequency of the laser RO. However, Tager and Petermann [18], have shown that new dynamical regimes, other than LFF, may appear at short EC and correspond to fast harmonic oscillations of the laser intensity, at a frequency close to the EC frequency. Following their definition, EC is called *short* if the EC round-trip time τ_{ext} is such that $\omega_{RO}\tau_{ext} = O(1)$, where ω_{RO} is the RO angular frequency. Whether the laser diode operates in the short or the long EC regime can therefore be roughly classified in comparison with the frequency of the RO in the free-running laser (i.e., the laser without optical feedback). Thanks to mathematical continuation techniques (see also Chapter 7 by Krauskopf and Walker), these high-frequency intensity oscillations have been identified by Pieroux et al. [19] as the result of a Hopf bifurcation bridge between two steady-state solutions of the compound-cavity problem, that is, between two ECMs. The Hopf bifurcation bridge means that one ECM solution exhibits a Hopf bifurcation and the emerging branch of time-periodic solution connects to another Hopf bifurcation located on another ECM solution. ECMs are therefore connected by bridges of time-periodic solutions. From a physics point of view, these bridges correspond to a beating between the two interacting ECMs. The beating yields high-frequency oscillations of the laser intensity, as reported by Tager and Petermann [18].

The possibility for the laser diode to sustain lasing in two frequency-separated ECMs and the consecutive beating between these modes has given new interesting ways to generate all-optically signal at high frequency (theoretically tens of gigahertz or even more). However, a drawback seems to rely on the fact that the high-frequency self-pulsation results from a beating between a stable ECM and an unstable ECM (also called an *antimode*). More specifically the previously mentioned bridge between two Hopf bifurcations is unstable and the time-periodic beating solution destabilizes with a torus bifurcation to quasiperiodic or even chaotic dynamics. If one would think of an application, this would mean also that significant effort must be made to precisely control the feedback strength and the feedback phase in an experiment. This situation is illustrated in Figure 3.5a. However, interestingly when decreasing the laser linewidth enhancement factor (α) the bifurcation bridge creating the beating solution can be fully stable, see Figure 3.5b–d. The reason for this is that the connection of Hopf bifurcation points now happens between two stable ECMs. Not only a decreasing α leads to stabilization of ECM beating but also it doubles the beating frequency, all other parameters being fixed [20]. Laser diodes with small α have become today of great interest, in the context of the development of QD and quantum cascade semiconductor laser diodes (Chapters 1 and 4).

Similar self-pulsating dynamics resulting from ECM beating have been also reported in VCSELs with polarization rotating optical feedback [21] and have been attributed to Hopf bifurcation bridges between orthogonally polarized ECMs [22]. Experimental evidence of the beating mechanism between ECMs (and the resulting

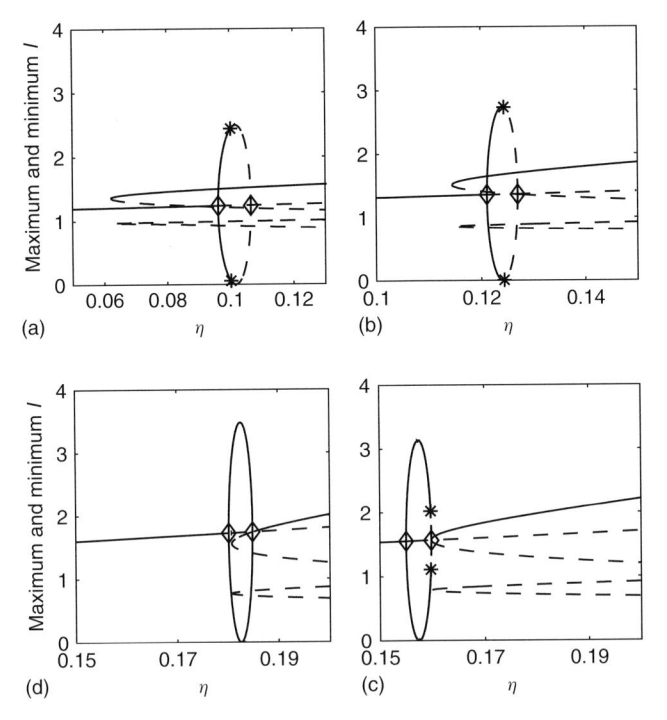

Figure 3.5 Numerical bifurcation diagrams showing the extrema of the laser intensity I as a function of the optical feedback rate η, for different values of the linewidth enhancement factor: (a) $\alpha = 4$, (b) $\alpha = 2$, (c) $\alpha = 1.25$, and (d) $\alpha = 1$. Full and broken lines correspond to stable and unstable solutions, respectively. All figures show a closed branch of two ECM solutions connecting two Hopf bifurcation points (symbol ◊). This branch changes stability at a torus bifurcation point (symbol ∗). The torus bifurcation point progressively moves to the right Hopf bifurcation point as α progressively decreases. In (d) the closed branch of two ECM solutions is stable. Taken from [20].

high-frequency intensity oscillations) has been given in two different systems: laser diode with T-shaped EC (double feedback) [23] and two-section semiconductor laser with an integrated passive cavity [24].

3.5
Multimode Dynamics in Lasers with Short External Cavity

The LFF dynamics typically occurs when a laser is subject to a moderately strong optical feedback from a typically quite long EC (tens of centimeters). When the EC delay time decreases and becomes comparable or smaller than the RO time scale of the laser dynamics, it is said that the EC is short – as detailed in Section 3.3. Besides the ECM beating, another peculiar dynamics that has been recently observed experimentally is called the *RPP dynamics* [25]: the laser fires pulses at each EC round-trip time, but the amplitude of the pulses is modulated by a

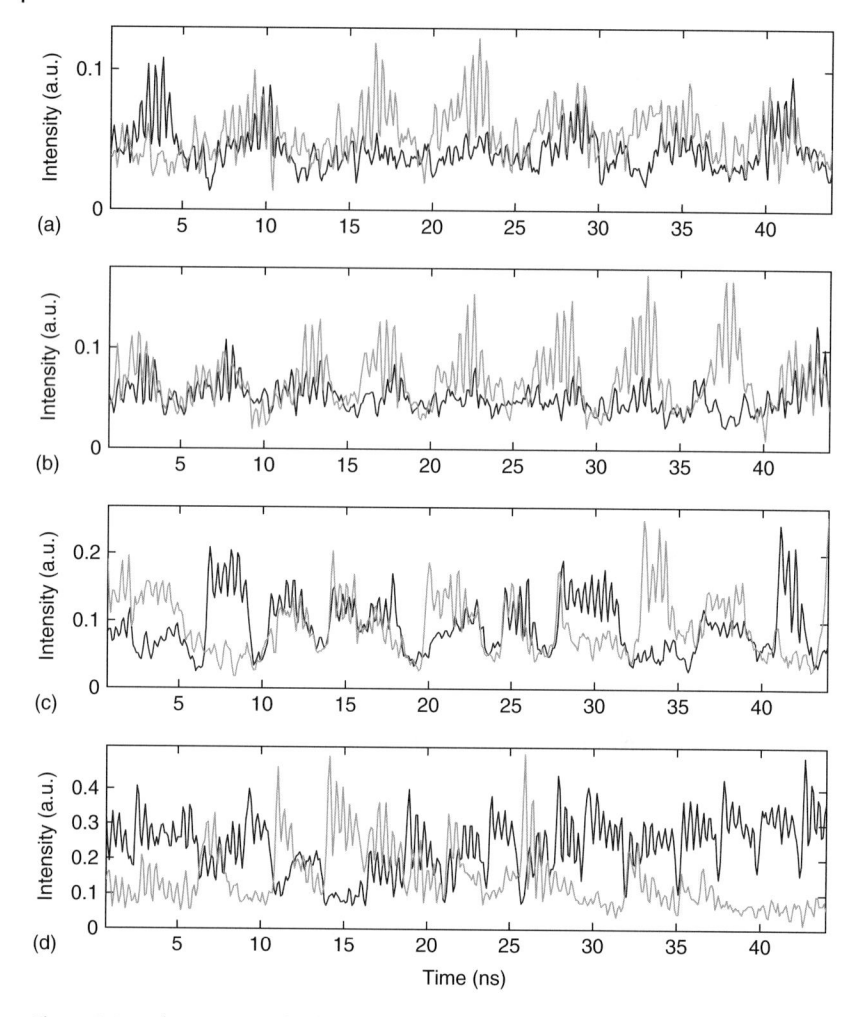

Figure 3.6 Polarization-resolved dynamics of a VCSEL in the pulse package regime at different increasing values of the injection current. In black (gray), we plot the x(y)-LP mode of the VCSEL. The EC length is around 6.5 cm. It shows how the mode intensities tend to exhibit regular pulse package dynamics (case b) and progressively more irregular pulsating dynamics as the injection current increases and approaches a VCSEL polarization switching point. Taken from [28].

much slower envelope. The pulses are grouped by packages that repeat periodically with the period of the slow envelope. A cascade of bifurcations on the ECMs leads to a stable and robust quasiperiodic attractor at large values of the feedback strength [26]. A progressive sweep of the EC length from short to long also allows observing a transition from RPP type of dynamics to a LFF type of dynamics [27].

In Figure 3.6 shows an example of RPP dynamics, but here observed in a multimode laser, that is, a VCSEL lasing simultaneously in two orthogonal polarization modes [28]. In black (gray), we plot the $x(y)$-LP mode intensity of the VCSEL. At the lowest value of the injection current, $J = 3.2$ mA (Figure 3.6a), the amplitude of the pulse peaks is still small and the shape of the single pulse package (PP) envelope is not very regular. However, the envelope of the packages can be clearly identified, which indicates that the PPs in the two LP modes are almost periodic. The PP dynamics in the two LP modes can be much better recognized at $J = 3.4$ mA (Figure 3.6b), however, it can be seen that the polarization-resolved PP dynamics is not as regular as for the total intensity. The reason for this is that we observe polarization mode competition, underlying the PP dynamics, reducing the regularity of the PP dynamics in each polarization mode. This mechanism becomes more relevant at a higher injection current, approaching the solitary VCSEL polarization switching (PS) point. Accordingly, we find a gradual loss of the regularity of the PP dynamics as J is increased from 3.2 to 3.8 mA. A closer look at the dynamics presented in Figure 3.6 reveals that in some cases the PP dynamics temporarily takes place in one of the LP modes only, whereas the second mode is almost turned off. In other cases the PP dynamics take place in the two LP modes simultaneously. We refer to the first case of dynamics, in which the pulses are emitted in one LP mode only, as *type I PP dynamics*. The second case of dynamics, in which the PP dynamics take place in the two LP modes simultaneously, is called *type II PP dynamics*.

3.6
Polarization Mode Hopping in VCSEL with Time Delay

In the previous sections, we have already introduced the interesting features of VCSEL polarization dynamics. The fact that such a laser easily emits two modes with slightly different frequencies and almost similar gain/loss ratio makes it very appealing for the study of mode competition in configurations where the laser is brought into complex nonlinear dynamical scenarios. We have already mentioned the inphase or antiphase polarization dynamics occurring either in chaotic LFF dynamics or in the RPP dynamics, which are characteristic of optical feedback bifurcations. Here, we shall focus more on the bistable properties arising from the (polarization) mode competition of VCSELs and how they interplay with time delay. We shall introduce our prototype experiment for the study of the interplay between bistable mode switching and time delay: a VCSEL subject to a *weak optical feedback* and *noise*. The experimental demonstration of CR is then a remarkable example of how much noise can influence mode competition and switching. Other interesting features that relate to polarization bistability and noise can be found in Chapter 2 by Zamora-Munt and Masoller.

3.6.1
Polarization Switching Induced by Optical Feedback

It is well known that a weak optical feedback modifies the threshold gain of a semiconductor laser as a function of the constructive or destructive interference conditions between the emitted and the reflected lights. As mentioned above, in the VCSEL configuration, the two orthogonal LP modes have slightly different optical frequencies and also slightly different threshold gains. The optical feedback effect would therefore modify the threshold gain difference between the two LP modes, by adding a modulation term of the form $\kappa[\cos(\omega_x \tau) - \cos(\omega_y \tau)]$, where $\omega_{x,y}$ are the optical frequencies of the LP modes, κ is the feedback strength (amount of light reinjected back into the cavity relative to the amount of emitted light) and τ is the delay time in the EC. For a large enough feedback strength the threshold gain difference between LP modes may therefore change its sign and if the mechanism determining PS is related to the change of net gain between modes, then this also means that PS would occur. A progressive sweep of the EC length and therefore of the delay time would induce several PSs for specific values of the delay time. Similarly, if the injection current is increased progressively then the optical frequencies of the LP modes may exhibit a significant red shift that finally modifies the argument of the cosine functions and may lead to successive PSs for specific values of the injection current.

We have made the corresponding experiment, as reported in [29]. We use a proton-implanted 850 nm VCSEL with a threshold current of 6 mA (in the solitary case). The solitary VCSEL emits light in the fundamental transverse mode with a stable linear polarization along the horizontal direction (x) for currents up to 2.25 times the threshold current. The vertical LP mode (y) is strongly suppressed. The frequency splitting between the two LP modes is measured to be about 8 GHz. The VCSEL is then subject to an optical feedback from a distant semitransparent mirror. The EC length is 20.2 cm (delay time equal to 1.3 ns). Figure 3.7 shows the

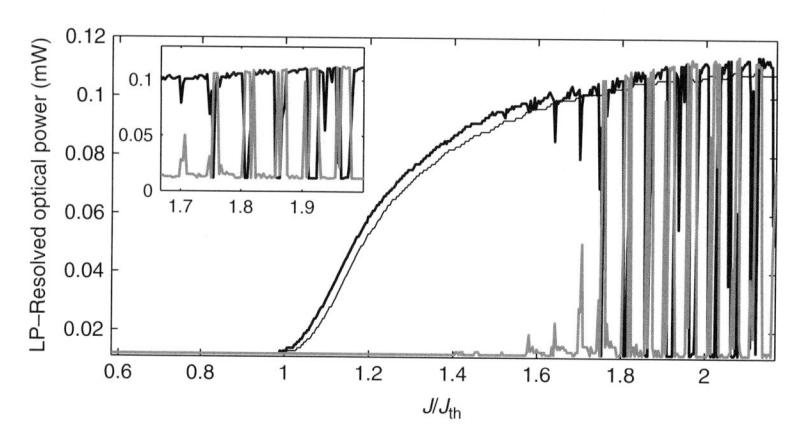

Figure 3.7 Experimental L–I curve of a VCSEL subject to a weak optical feedback (thick lines), to be compared with the case without optical feedback thin line). The x-LP (y-LP) mode intensity is plotted in black (gray). The inset shows enlargement to better illustrate the channeled L–I curve with multiple polarization switching points. Taken from [29].

L–I curve of the solitary VCSEL (thin line) together with the L–I curves resulting from the weak optical feedback (large black and gray lines).

The first effect of the optical feedback on the L–I VCSEL characteristics is to reduce the threshold condition. In our case, the feedback strength is such that the threshold current is reduced by about 2%. The solitary VCSEL exhibits lasing only in the x-LP mode, but even this weak amount of optical feedback induces a dramatic effect on the L–I curve. The polarization-resolved L–I curve with optical feedback exhibits a so-called channeled behavior, that is, it shows multiple PSs at periodically separated values of the injection current (see the inset).

3.6.2
Polarization Mode Hopping with Time-Delay Dynamics

When a VCSEL exhibits a PS as a function of the injection current, it is typically observed that the polarization state is not well defined around the switching point: a continuous transition between the two LP modes occurs as we increase the injection current from a value slightly below the switching point to a value slightly above the switching point. If we set the injection current close to the PS point, we observe that the light randomly alternates between emission of the x-LP mode and emission of the y-LP mode, a situation called *polarization mode hopping* [30]. From the statistical analysis of the mode-hopping dynamics, it appears that the dwell time in one of the two LP modes follows a Kramers law, with a probability distribution function exponentially decaying with time. The mean dwell time depends on the injection current and spontaneous emission level.

In our optical feedback experiment also, a polarization mode hopping is observed if we fix the injection current at one PS point [29]. The laser system then randomly dwells in the x- or the y-LP mode, on a slow time scale; see Figure 3.8a. The two LP modes are anticorrelated at the time scale of the slow mode hopping. This behavior resembles that of the mode-hopping solitary VCSEL. However, a careful observation shows that superimposed on the slow polarization mode hopping, a fast oscillatory behavior appears, at the frequency of the EC (750 MHz), that is, the inverse of the optical feedback time delay. These fast oscillations more clearly appear during an attempt to or a successful polarization switch. Figure 3.8b shows an example of fast oscillations in the two LP modes during a PS. It shows that the LP modes are anticorrelated at the time scale of the EC frequency. These oscillations therefore vanish at the time trace of the total intensity; see Figure 3.8c.

To further compare the statistical properties of this optical feedback induced mode hopping with those of the mode hopping in solitary VCSELs, we measure the residence times for the x- and y-LP modes (also called *dwell times*) from the stored oscilloscope time traces [31]. In a symmetric mode-hopping regime, these two dwell times exhibit the same distribution and consequently the same statistical properties. The resulting experimental distribution of the residence time (RTD) in one LP mode is then shown in Figure 3.9. The RTD is distinctly different from the one reported in the experiments on solitary mode hopping VCSELs [30]. Instead of an exponentially decaying behavior for all residence times, the RTD of

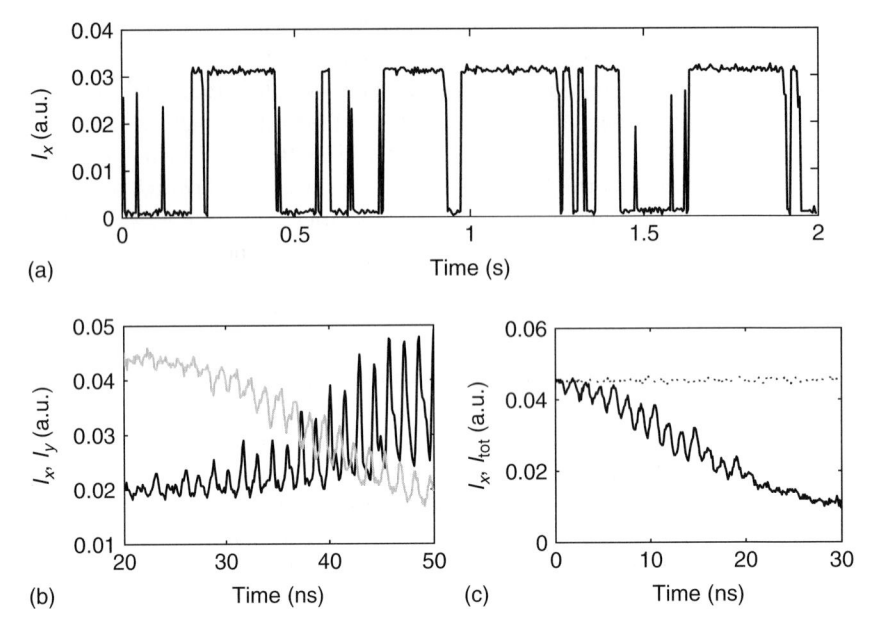

Figure 3.8 (a) Typical time trace of mode hopping in the LP mode intensity of a VCSEL close to a polarization switching induced by weak optical feedback, (b) is an enlargement of (a) showing the two LP mode intensities when the light polarization hops between the two orthogonal directions. (c) shows a time trace of the LP mode intensity (solid curve) together with the one of the total intensity (dotted curve). Taken from [29].

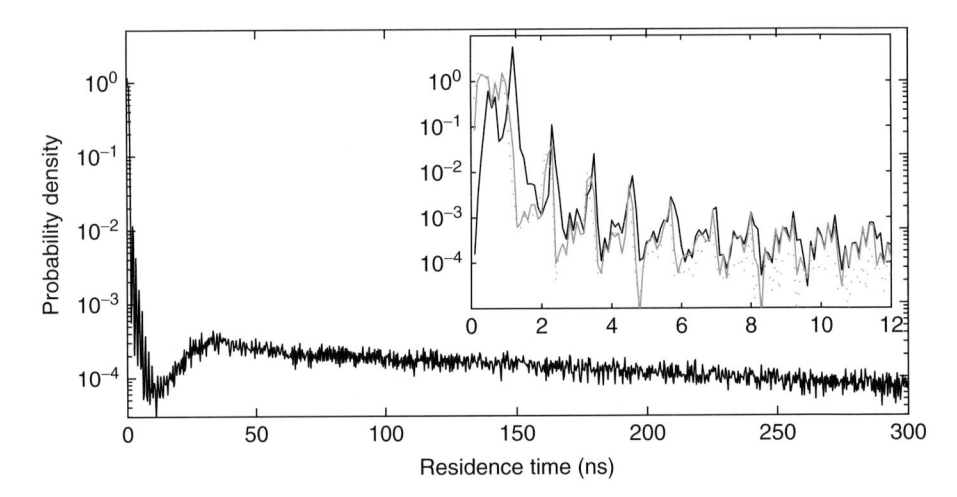

Figure 3.9 Residence time distribution (RTD) of the polarization mode hopping induced by optical feedback. The inset shows an enlargement of the RTD and in black line is shown the so-called joint residence time distribution. Taken from [31].

delayed mode-hopping VCSEL exhibits a discontinuity for small residence times. The slow mode hopping is responsible for the exponential decay at large residence times. This exponential decaying RTD is typical for the polarization mode hopping in solitary VCSELs. However, the fingerprints of the time-delayed feedback, that is, the fast intensity oscillations at the EC frequency that complement the slow mode hopping, are responsible for RTD at smaller residence times. As these fast oscillations always appear during a PS (or an attempt to switch), the probability for measuring residence times up to the EC round-trip time is quite large. Our experimental results confirm theoretical predictions made by Masoller on a simpler system, that is, the presence of a time-delayed feedback in a bistable system yields an increased probability to find events at times smaller than the delay time [32]. Moreover, the RTD exhibits an increased probability at each multiple of the delay time. Indeed, because of the stochastic nature of the mode hopping and the corresponding noise intensity fluctuations, some rapid oscillations do not cross the detection threshold. As a result, we observe an increased probability to detect the fast oscillations with one, two, and more cycles missed, which corresponds to an oscillatory behavior in the RTD (see the inset of Figure 3.9).

The inset also shows the distribution of what we define as a joint residence time in black line, that is, the time interval needed for the system to visit the two LP modes consecutively. Interestingly, while the statistics for each LP state shows a discontinuity close to the delay time, the statistics of the joint residence time distribution (JRTD) shows a clear maximum at the delay time. The time the system spends in visiting the two modes consecutively tends to follow the regularity imposed by the EC and its associated round-trip time, independently on whether the system spends on average more time in one mode or in the other.

A theoretical study based on the two-mode rate equations reproduces qualitatively well the optical feedback induced PS, mode hopping, and the corresponding RTD. The numerical simulations for different values of the spontaneous emission rate confirm that the noise plays an important role in the slow mode-hopping dynamics, the mean dwell time tends to decrease as the noise level increases [29]. The polarization dynamics under investigation is therefore an interesting example of a bistable dynamics controlled by noise and influenced by the optical feedback delay time. In the following, we make use of this new interesting polarization dynamics to demonstrate experimentally a more general concept: the existence of CR in a bistable system with time delay, that is, the fact that adding an optimal amount of noise to the system dynamics may finally bring the system into a pulsating dynamics with optimal regularity. As discussed in [31], the CR in our VCSEL configuration is better observed when the time delay is of the same order of magnitude than the mean dwell time in the LP modes (typically several tens of nanosecond).

3.6.3
Coherence Resonance in a Bistable System with Time Delay

It is commonly accepted now that the noise can play constructive role in nonlinear dynamical systems. After the seminal paper of Benzi *et al.* [33] the phenomenon of stochastic resonance, namely, the fact that adding noise can better synchronize dynamical system to an external periodic signal, has attracted a lot of interest (for a review, see [34]). It has later been realized that noise can enhance regular dynamics in nonlinear systems even in the absence of external signal, when an internal time scale is present in the system [35]. This phenomenon was initially considered as stochastic resonance in autonomous system and later named CR [36]. CR has been first predicted for excitable dynamical systems, that is, systems that emit quasiregular pulses as a result of an excitation threshold and with a refractory period. It has been demonstrated experimentally in several systems including semiconductor lasers subject to optical feedback and driven into chaotic excitable dynamics [37]. Theoretical works on different models furthermore reported that not only excitable but also bistable or multistable systems driven by noise can exhibit CR [38], as also confirmed experimentally in bistable chaotic electronic Chua circuits [39]. Recently, CR has been predicted in another class of systems, which exhibit bistability together with time delay [40]. Time delay, bistability, and noise are important ingredients in a large variety of systems in physics, biology, and chemistry.

In the following, we summarize an experimental demonstration of CR in a bistable time-delayed system, namely, a VCSEL subject to optical feedback [41].

Figure 3.10 shows a typical time trace of the LP mode intensity when the injection current is set close to a PS point (induced by optical feedback). The output power in each LP mode is made of successive pulses emitted with a repetition rate given by the long EC delay time (27 ns) that complement a random switching dynamics between the two LP mode states. 1 and 2 are the defined residence times, and JRT is the joint residence time as defined above. The system dynamics is not symmetric here, since the system spends more time in one mode than the other. However, as mentioned above, the joint residence time is always very close to the value of the delay time, irrespective of the mode-hopping symmetry.

We then add noise to the injection current of our VCSEL and analyze the effect of the noise level in the distribution of the residence times and, in particular, of the joint residence time, since this directly reflects the optical feedback induced mode-hopping dynamics. Figure 3.11 plots the experimental results for three different noise levels.

If the noise is weak, the system needs a lot of time to consecutively visit the two stable states and the peak of the JRTD at the EC round-trip time (27 ns) is very small (Figure 3.11a). As the noise strength is increased the peak at 27 ns dramatically increases reaching its maximum (Figure 3.11b). For higher noise intensities, more and more fast PSs occur and the background masks the peak structure (Figure 3.11c). The right panel (Figure 3.11d) confirms the existence of an optimal noise level for which the JRTD exhibits a maximum peak at the delay time. By plotting the evolution of the area of the first peak of the JRTD once, the background is subtracted as a function of the noise intensity, we observe a maximum value for a noise intensity close to -120 dBm Hz^{-1}. This constitutes a clear evidence of CR in our system. Different indicators have complemented our observations. In particular, we have observed the RF (resonance frequency) spectrum of the laser output for different noise levels. A peak appears close to the

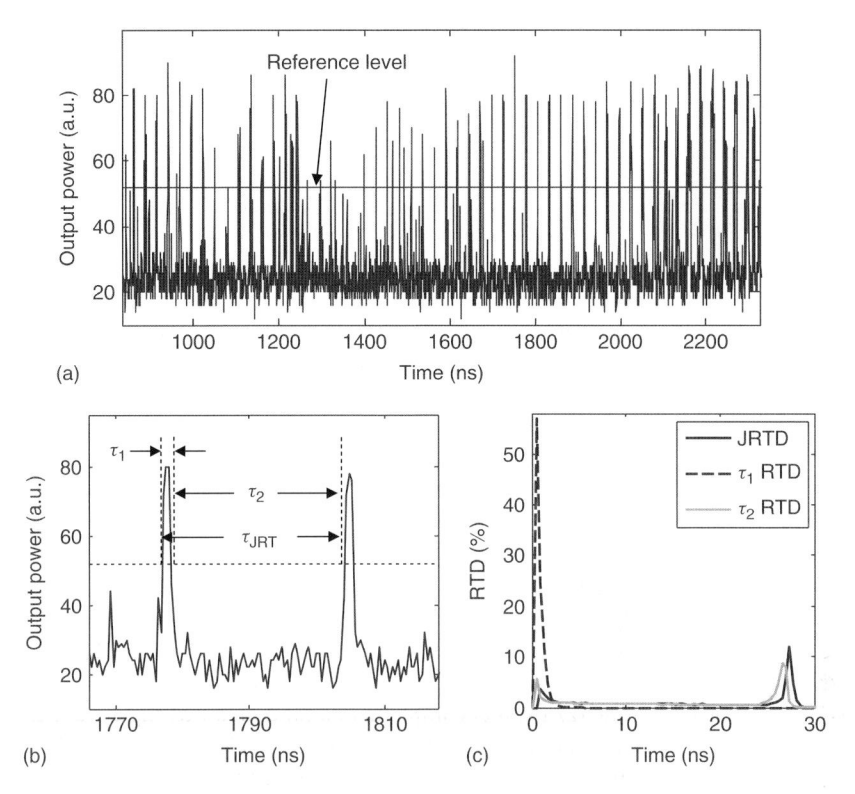

Figure 3.10 (a) Experimental time trace of the VCSEL LP mode intensity when it exhibits a polarization mode hopping induced by optical feedback (delay time 27 ns). Also shown is the reference level used for the following statistical analysis. In (b) is shown an enlargement with the definitions of the different dwell times and of the joint residence time.

long EC frequency and the peak height relative to the RF spectrum noise floor reaches a maximum for a given noise level (similar to the noise level that brings the maximum JRTD value at the delay time).

3.7
Polarization Injection Locking Properties of VCSELs

As has been demonstrated in the previous sections, VCSELs and their unique two-mode dynamics can show nonlinear dynamics not seen in conventional single-mode EELs. We have so far illustrated cases where the laser is subject to either a strong optical feedback (LFF or RPP dynamics) or a weak optical feedback (PS and bistable mode hopping). In this section, we show that the two-polarization-mode properties of VCSELs also impact on the bifurcation scenarios leading to chaos in the presence of an external optical injection.

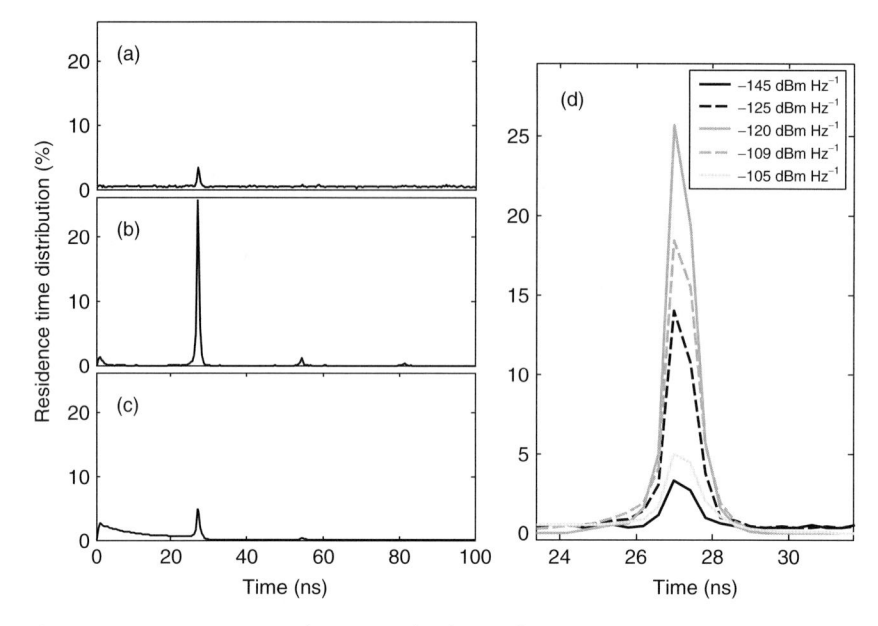

Figure 3.11 (a)–(c) Joint residence time distribution for three increasing noise levels, showing an optimal regularity at the delay time for an optimal noise level (case b). (d) is an enlargement of the JRTD around time values corresponding to the delay time, and for increasing values of noise added to the injection current. Taken from [41].

3.7.1
Optical Injection Dynamics

Optical injection is an important case of additional degree of freedom that can easily destabilize a semiconductor laser. A laser called *slave laser* (SL) is injected with light from an external laser source (master laser, ML). The two lasers emit in approximately the same wavelength range and it is assumed that an optical isolator prevents reciprocal coupling between the two lasers. Depending on the strength of the injected signal, the SL can either change its frequency of operation to that of the ML, and thus, lock to the ML frequency, or it may also engage in a more complicated dynamics in response to the external signal. The injection locking was known more than 30 years ago in several types of oscillators [42]. Its application for semiconductor lasers is of great interest. Indeed the injection locking of semiconductor lasers was shown to significantly improve the coherence properties of the emitted signal, leading, for example, to a reduction of the mode hopping and mode partition noise, a reduction of the laser linewidth and of the frequency chirping, and an enhancement of the modulation bandwidth.

We can summarize the dynamics of a laser with optical injection with three regimes of operation (for a review, see [43]):

- Stable locking region in which the SL is frequency locked to the ML, which corresponds to the "steady state" of the laser system. For locking to occur, two conditions must be satisfied: (i) the detuning between the frequency of the injected field and the RF of the laser diode should not be too large and (ii) the injected power should be large enough.
- Nonlocking region in which the SL does not manage to lock to the frequency of the ML, but the nonlinear interaction between the wave component at the slave frequency and the wave component at the master frequency being amplified by the SL active medium may give rise to interesting wave-mixing effects.
- Destabilized locking region in which the stationary locked state is destabilized to a more complex dynamics such as a time-periodic dynamics, a period doubling (PD) regime, quasiperiodic, or even chaotic behaviors.

While numerous theoretical and experimental papers have dealt with the optical injection-induced instabilities and dynamics in single-mode EEL, only few contributions have so far addressed the question of the influence of multimode laser emission. The question of multimode optical injection dynamics has been of interest recently in the context of two-mode laser systems:

1) In a pioneering experiment in 1993, Pan *et al.* have studied a first configuration, where a VCSEL is subject to optical injection. The VCSEL emits in a single x-LP mode and is injected with light polarized along the orthogonal direction (y) [44]. This configuration is called *orthogonal optical injection*. For sufficiently large injection strength, the VCSEL switches its polarization to that of the injected light, and may exhibit an injection locking depending on the frequency detuning between the two lasers. The injection power required for PS depends on the frequency detuning and moreover the PS occurs with a large bistability region. In the following, we shall summarize several additional features in the same experimental configuration: first, the PS induced by optical injection is accompanied by severe laser instabilities, and second that the PS phenomenon interplays with the bifurcations typically observed in optical injection problems and also makes possible the observation of new bifurcation mechanisms.

2) Recent investigations have concerned a so-called two-color laser device, that is, a device that can lase simultaneously on two different modes with possibly quite a large (terahertz) frequency spacing [45–47]. One of the two modes is subject to optical injection and still the laser exhibits a large variety of two-mode inphase and antiphase dynamics including limit cycle, quasiperiodicity and chaos [48]. The large frequency spacing between the lasing modes makes it possible to investigate the properties of optically injected multimode lasers far beyond the approximation of the single-mode laser equations. Among the interesting features, it is worth mentioning a bistability between a one-mode injection-locked state and a two-mode equilibrium state, that is, bistability between two steady states. As it is the case for VCSELs, such a bistability can be applied to all-optical signal processing [49]. This point is further discussed in Chapter 10 by Amann.

3.7.2
Polarization and Transverse Mode Switching and Locking: Experiment

Details on the performed experiments can be found in [50, 51]. An oxide-confined quantum well VCSEL emitting around 845 nm is used as a SL. Optical injection is achieved from an EC tunable ML. The solitary VCSEL threshold is about 1 mA and the VCSEL switches its polarization from horizontal to vertical LP mode at 4.60 mA and backwards at 2.25 mA, forming a large hysteresis region. The PS is from the low to the high frequency LP mode. For currents above 5 mA, first-order transverse mode appears. In our experiment, the solitary VCSEL is biased at 2.105 mA, so that it emits only in the fundamental transverse mode with horizontal polarization. The injected LP light is set to be vertical.

In order to represent the richness of the polarization dynamics in VCSEL with orthogonal optical injection, we show in Figure 3.12 the boundaries of qualitatively different dynamics in the plane of the injection parameters (the detuning and the injected power). The injected power has been normalized to the output power of the VCSEL $P_{out} = 1.28$ mW at the bias current of 2.105 mA and has been taken in logarithmic scale $\log(P_{inj}/P_{out})$. We have defined the frequency detuning as the frequency of the ML minus the frequency of the SL. For each value of the frequency detuning, we perform a sweeping along the horizontal axis, that is, increasing and then decreasing the injected power. The horizontal axis is limited in the positive part by the maximum output power of the ML. The maximum negative detuning corresponds to the largest detuning for which we observe injection locking with the maximum injected power. The polarization switch on and switch off points

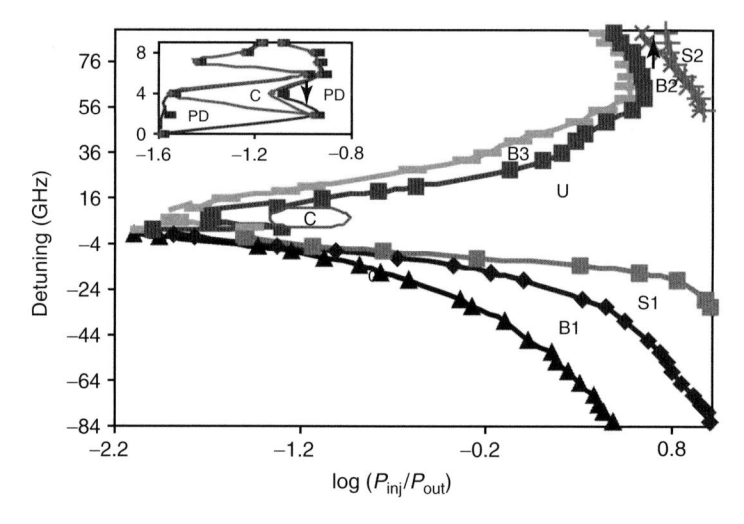

Figure 3.12 Experimental mapping of bifurcations to qualitatively different polarization dynamics in a VCSEL subject to orthogonal optical injection. Labels and symbols are explained in the text. Taken from [50].

for increasing (decreasing) the injected power are represented by the lines with diamonds and squares (lines with dots and triangles) in Figure 3.12. In the regions S1 and S2, the frequency of VCSEL emission is locked to the ML. However, in the case of S2, it is the first-order transverse mode and not the fundamental transverse mode that locks to the master laser, the fundamental transverse mode being then suppressed when crossing the line with crosses. The unlocking of the first-order transverse mode happens at smaller values of P_{inj}, describing bistable region B2 between the fundamental and the first-order transverse mode both with the same polarization.

We observe two polarization bistable regions in a regime of fundamental mode emission, which correspond to two different ways of PS. The first one is with frequency locking (B1). The second polarization bistable region (B3) is without frequency locking. The two bistable regions are connected at a detuning of 2 GHz, which coincides with the birefringence frequency splitting between the two VCSEL LP modes. This means that when the ML is biased at the frequency of the VCSEL vertical mode (the suppressed mode), a dramatic change of dynamics occurs: from PS with injection locking to PS without locking. For larger positive or negative detunings, the switching power is larger, and moreover the switching power is larger for a negative than for a positive detuning value. It is worthy to notice that the widths of the injection locking regions S1 and S2 and of the bistability region B1 increases with the detuning. By contrast, the width of the bistability region B3 remains approximately constant when changing the frequency detuning. This bistable region B3 is also strongly influenced by the locking of the first-order LP mode (S2), its borderline turning backwards at a detuning of 50 GHz.

The mapping of dynamical states shows that richer nonlinear dynamics including PD route to chaos and even reverse PD from the chaotic zone are found for detunings in the range of 2–10 GHz. Cascade of complex dynamics involving chaotic instabilities is presented in Figure 3.13 corresponding to a detuning of 2 GHz. As the injection strength increases, the injection-locked steady state (Figure 3.13a) undergoes a Hopf bifurcation to a limit cycle at the RO frequency (Figure 3.13b). For larger injected power, harmonics of the RO frequency are even observed (Figure 3.13c). As the injection power is increased further, the injected VCSEL undergoes a PD dynamics (Figure 3.13d), leading to a chaotic dynamics (Figure 3.13e). Chaotic instabilities correspond to the presence of a large pedestal in the VCSEL spectra and involve both vertical and horizontal LP modes. The case (Figure 3.13f) shows that, if the injected power is increased further, the chaotic regime is exited with a reverse PD cascade, leading to a limit cycle dynamics (Figure 3.13g). For still larger injection strength the limit cycle dynamics may even undergo PD again, as shown in the case (Figure 3.13h).

In the following, we analyze more systematically the competition between transverse modes and injection locking phenomena that occur when the detuning is positive and large, close to the frequency separation between fundamental and first-order transverse modes (around 150 GHz in our case) [52].

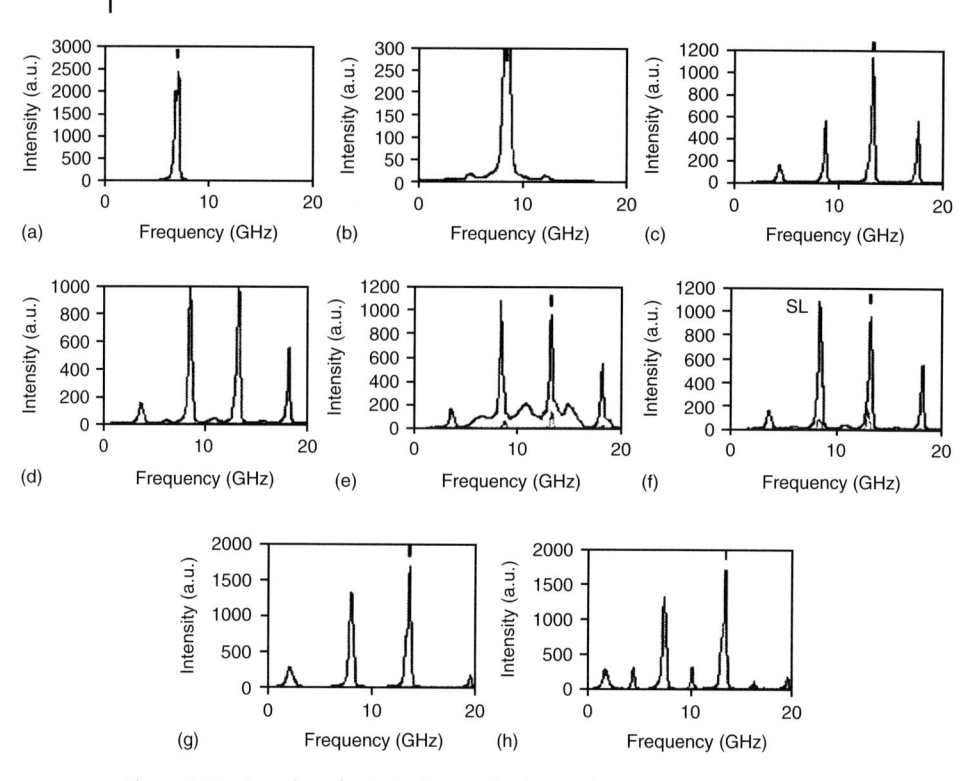

Figure 3.13 Samples of polarization-resolved optical spectra showing a period doubling cascade of bifurcations to chaos (e), following by a reverse period doubling cascade, as the injection strength increases for a fixed detuning (24 GHz). Taken from [50].

Figure 3.14 shows the mapping of the VCSEL subject to optical injection for a very large positive detuning range, that is, from 2 to 180 GHz. For a fixed detuning value, polarization-resolved dynamics as well as transverse mode competition are analyzed when the injection strength is scanned. If the injection strength is increased, and depending on the frequency detuning, different switching scenarios are resolved. A switching mechanism that involves the VCSEL fundamental orthogonal transverse modes, that is, from the horizontal (x-LP) to the vertical (y-LP) mode, is observed for the whole frequency detuning range. The corresponding boundary is labeled by black triangles. This boundary exhibits two minima for the switching power. A first minimum is located at a detuning of 2 GHz for which PS is achieved at 7.1 µW. A second minimum for the switching power is found for a detuning of 150 GHz and an injection power of 623.9 µW. It is worth mentioning that the second minimum is at much larger power than the one for a detuning of 2 GHz. We analyze in more detail the transverse mode competition behavior for detunings ranging from 61 to 120 GHz. With increasing the injection power we

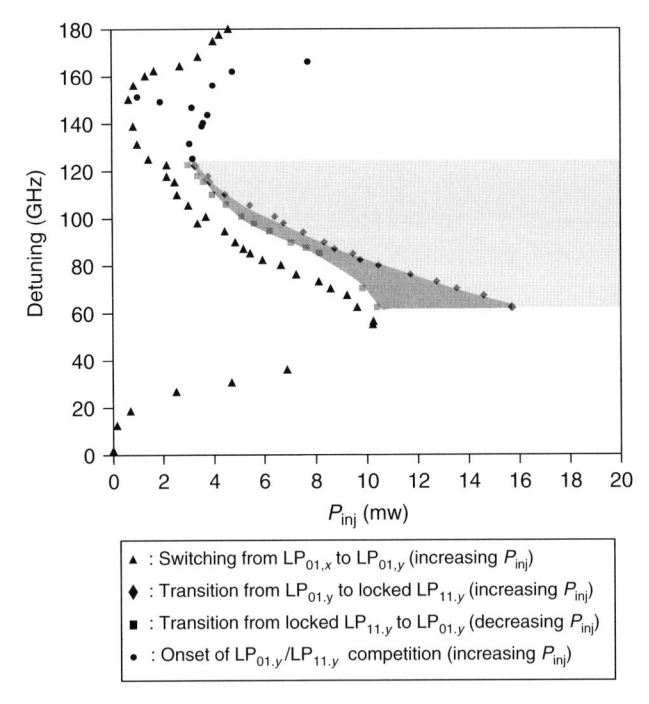

Figure 3.14 Experimental mapping of transverse mode switching and locking for a positive and large detuning (of the order of the frequency splitting between fundamental and first-order transverse modes). Taken from [52].

Legend:
- ▲ : Switching from $LP_{01,x}$ to $LP_{01,y}$ (increasing P_{inj})
- ◆ : Transition from $LP_{01,y}$ to locked $LP_{11,y}$ (increasing P_{inj})
- ■ : Transition from locked $LP_{11,y}$ to $LP_{01,y}$ (decreasing P_{inj})
- ● : Onset of $LP_{01,y}/LP_{11,y}$ competition (increasing P_{inj})

first observe PS between the fundamental gaussian mode polarized along x, $LP_{01,x}$, and the fundamental gaussian mode polarized along y, $LP_{01,y}$. These PS points are denoted by black triangles. When increasing further the injection power, we observe injection locking of the first-order transverse mode polarized along y, $LP_{11,y}$ mode – its frequency locks to the one of the ML, together with suppression of the fundamental transverse mode $LP_{01,y}$. The corresponding injection locking boundary is denoted by black diamonds. As the injection strength is increased, the VCSEL is initially frequency pushed but still emits a horizontal LP mode. For a further increase in the injection strength, switching from horizontal x-LP to vertical y-LP fundamental mode is achieved. By still increasing the injection strength, an abrupt injection locking of the first-order transverse mode to ML with suppression of the fundamental mode occurs. Bistability is observed if the injection power is decreased after injection locking of the mode is achieved, that is, the VCSEL unlocks for an injection strength smaller than the one necessary to induce the locking regime (see the boundary labeled with light gray squares in Figure 3.14). The width of the bistable region associated to the locking of the $LP_{11,y}$ mode decreases as we increase the detuning as indicated by the zone with a dark gray shadding.

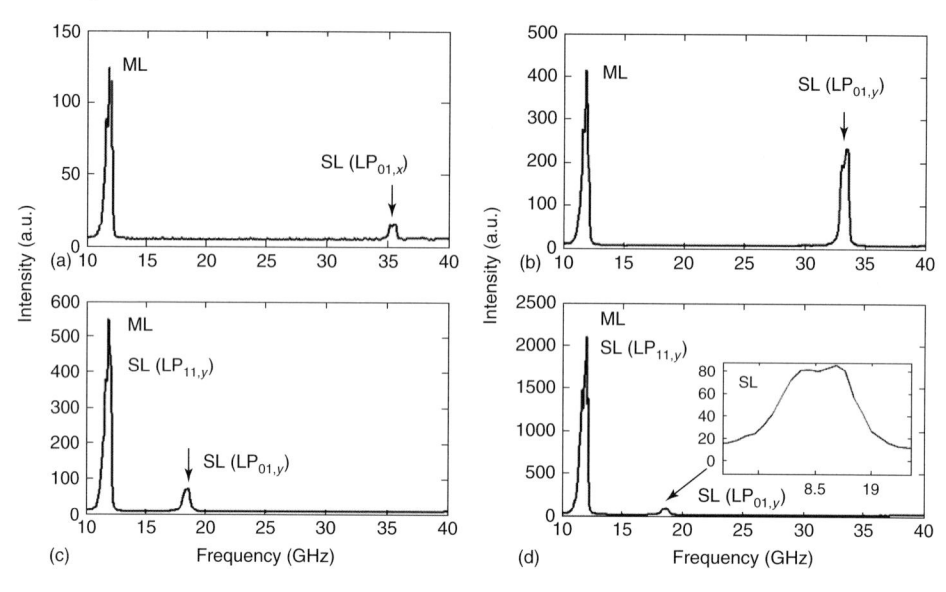

Figure 3.15 Example of polarization and transverse mode competition appearing as the injection strength increases from (a) to (d), for a fixed detuning equal to 125 GHz. The polarization switching from (a) to (b) is progressive and the first-order transverse mode in (c) competes with the fundamental transverse mode with the same polarization (c)–(d). Taken from [52].

For frequency detunings larger than 120 GHz, injection locking of the $LP_{11,y}$ mode accompanied by suppression of the fundamental transverse mode $LP_{01,y}$ is not observed anymore. Figure 3.15 represents the situation for which the VCSEL is under optical injection but the injection strength is not sufficient to induce PS (Figure 3.15a). By increasing the injection level PS from x-LP to y-LP fundamental mode is achieved (Figure 3.15b). A further increase in the injection strength leads to a strong competition between the $LP_{01,y}$ and $LP_{11,y}$ modes. The onset of such a mode competition is shown on the mapping in Figure 3.14 by black circles, which correspond to the observation of a progressive decrease of the intensity at the SL frequency and a relatively strong increase of power at the ML frequency (Figure 3.15c). Again, at a much stronger injection, a weak increase of the intensity at the SL frequency, that is, a recovery of the y-LP fundamental mode, has been observed; see Figure 3.15d and the inset. As shown in Figure 3.14, the transverse mode competition appears at much lower injection power for a detuning of 150 GHz, which corresponds to the second minimum of the switching power. For larger positive detunings up to around 165 GHz, the mode competition is still resolved but at progressively increasing injection levels. Above this detuning range and as we increase the injection power, PS between the fundamental modes is still observed, but afterwards the VCSEL keeps emitting an unlocked y-LP fundamental mode.

3.7.3
Bifurcation Picture of a Two-Mode Laser

The experimental results were detailed before clearly pointing out the important role played by both polarization and transverse mode competition in VCSELs on the laser dynamics. On the basis of a theoretical model for VCSEL, it is possible to gain insight into the bifurcation scenarios leading to nonlinear dynamics. A theoretical approach for our VCSEL orthogonal optical injection configuration can be obtained from the analysis of a set of rate equations. The PS mechanism in VCSELs can be modeled using, for example, the spin-flip approach, as done in [53–56], but we have obtained similar results using a two-mode model with gain compression terms and not including spin-flip relaxation mechanisms [57].

It is possible to reproduce theoretically several of our experimental results, as summarized hereafter:

- An increase of optical injection strength leads to PS with bistability. The range of injection strength corresponding to the bistability region increases as the frequency detuning increases, the minimum being obtained when the detuning is close to the VCSEL frequency splitting between LP modes.
- Injection locking accompanies the PS mostly in the negative frequency detuning side, whereas for positive detunings PS is typically accompanied by a PD route to chaos and a transition to time-periodic unlocked dynamics.
- For large positive frequency detunings, a strong competition can occur between transverse modes, which may lead to injection locking of a first-order transverse mode with suppression of the VCSEL fundamental transverse mode.

To bring new light into the bifurcation picture, we have made use of the continuation techniques to follow the bifurcations of steady states and time-periodic solutions in the plane of the injection parameters (frequency detuning versus injection strength). Figure 3.16 shows a typical bifurcation mapping that we obtained theoretically from the analysis of a set of equations for single transverse mode VCSEL [55, 56].

Qualitative changes in the VCSEL dynamics are detected and followed using the continuation package AUTO 97 [58]. Different bifurcation curves are plotted: a saddle-node (SN), two Hopf (H1 and H2), and a torus (TR). The supercritical and subcritical parts of each bifurcation curve are represented in black and gray, respectively. When increasing the injection strength, the VCSEL switches its polarization to that of the injected field. When decreasing the injection strength, the VCSEL switches back to its free-running polarization but for a smaller injection strength. These "PS off" (x-LP mode off) and "PS on" (x-LP mode on) points are shown with circles and squares, respectively. The PS curves interplay with the bifurcation curves. SN and H1 are bifurcations on a stationary injection-locked state and have also been reported in the case of optically injected EEL. In the conventional case of EEL, the locking region is then delimited by the codimension two point G where SN and H1 intersect. In our VCSEL system, the locking region is delimited

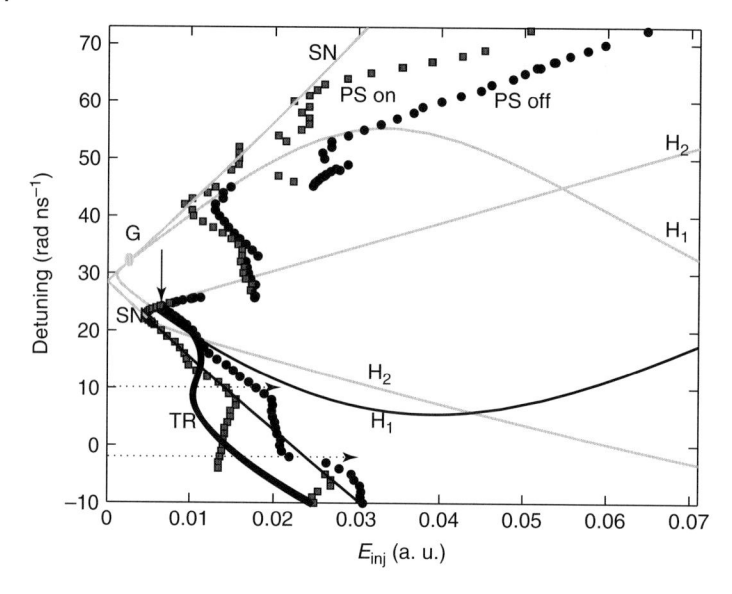

Figure 3.16 Theoretical mapping of the bifurcation boundaries of VCSEL with orthogonal optical injection in the plane of the injection parameters (detuning versus injection strength). Labels are detailed in the text. Together with the bifurcation lines are shown the polarization switching points, which therefore interplay with bifurcations to nonlinear dynamics [55, 56].

by SN, H1 but also by a new bifurcation H2. The maximum detuning leading to injection locking therefore stays well below the codimension-two saddle-node-Hopf point G. Apart from its effect on the locking, H2 also affects the PS mechanism. The supercritical part of H2 coincides with "PS on" points. Moreover, the smallest injection strength needed to achieve PS is located on H2 and corresponds to a dramatic change in the PS curve (see the solid vertical arrow in Figure 3.16). As a result, the PS curve exhibits a snakelike shape with local minima of the injected power required for switching. The observed shape agrees qualitatively with our experimental results. Additional bifurcations on the time-periodic solutions (not shown) lead to a PD route to chaos as observed experimentally and are located close to these local minima in the PS curve.

The torus bifurcation Tr gives rise to a time-periodic dynamics at the RO frequency in the noninjected mode (x) and to a wave-mixing dynamics in the injected mode (y), which have been also found in experiment [55].

Interestingly, the model allows for another type of injection locking solution, where the VCSEL locks its two orthogonal LP modes to the injected field [53, 54] (not shown here). This two-mode injection-locked solution is observed when the detuning is negative and such that the ML frequency is close to the frequency of the noninjected mode of the solitary VCSEL. The two-mode injection-locked solution is born from SN bifurcation and destabilizes through a Hopf bifurcation.

3.8
Dynamics of a Two-Mode Quantum Dot Laser with Optical Injection

In the previous section, we have unveiled several dynamical features, which result from polarization and/or transverse mode competition in VCSELs with optical injection. Another example of two-mode laser has been discussed and is related to a two-color laser with terahertz frequency spacing between modes with similar power levels. Here, we discuss about another example of two-mode laser in the presence of optical injection, namely, a *QD semiconductor laser*.

Self-assembled QD lasers and amplifiers have attracted much interest in recent years. For example, they have shown a significantly reduced sensitivity to optical feedback, resulting from a small linewidth enhancement factor and a strong RO damping rate [60, 61]. These advantages make them appealing for high-frequency direct modulation and isolator-free laser operation. The three-dimensional quantum confinement of a QD gives rise to discrete energy levels for both electrons and holes. GS emission, resulting from the recombination of a GS electron hole pair, generally occurs at low injection currents. The finite number of QDs within the active region and the discrete energy structure of QDs can lead, however, to saturation of the GS already at moderate currents. As a result, the occupation of the excited states (ES) grows with the current and the laser can start to lase from these states too. Simultaneous emission from both states has been demonstrated for a solitary QD laser in [62]. Figure 3.17 shows an example of such simultaneous lasing in GS and ES observed experimentally in 1.55 µm. InAs/InP QD laser diodes when increasing the pump power (normalized with respect to the GS pump power threshold P_{th}). The two-mode lasing of QDs has been studied also in the context

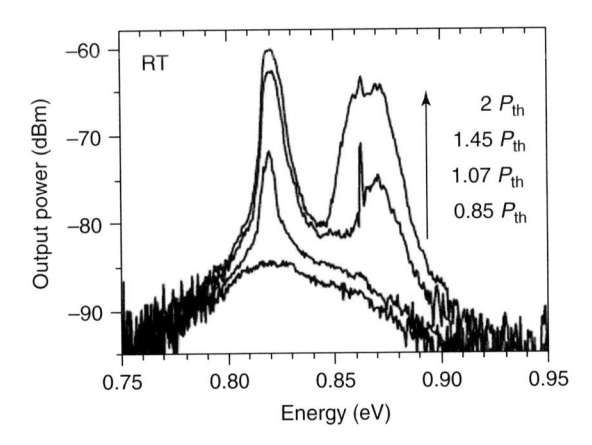

Figure 3.17 Room temperature emission spectra of a QD broad area laser with six QD stacked layers under a pulsed pumping excitation. When increasing the pump power (normalized with respect to the GS pump power threshold P_{th}), we observe the simultaneous lasing of ground state (GS) (energy centered on 0.82 eV) and excited state (energy centered on 0.87 eV). Taken from [59].

of optical feedback [63] and for dual-wavelength mode-locking [64]. Chapter 1 by Lüdge also details the physical modeling of QD devices including their two-mode lasing properties.

When subject to optical feedback, it has been experimentally demonstrated that both ES and GS modes can be excited simultaneously and that the two-mode emission influences the properties of, for example, LFFs dynamics [63]. As discussed in Section 3.2, the LFF is made of power dropouts in the laser intensity occurring at irregular time intervals. Experiments on QD two-mode lasers have shown that for in the LFF dynamics the ES emits bursts of pulses simultaneously to power dropouts in the GS dynamics. As a result the total intensity dynamics is almost stationary. This result complements the previous investigations of LFF on multilongitudinal mode EELs and two-polarization-mode VCSELs, and adds to the conclusion saying that mode competition may strongly influence the chaotic features of the total laser intensity dynamics.

Let us now discuss more specifically the case of optical injection. We have studied a model for QD semiconductor laser that allows for lasing in both ES and GS [65]. This theoretical work hence complements recent bifurcation analysis of a single-mode QD laser with optical injection [66–68]. Owing to the large frequency spacing between GS and ES (terahertz range), it is expected that the injected light interacts directly only with the GS mode. In the model, carriers from the wetting layer (WL) are first being captured into the ES and then relax to the GS. Both capture and relaxation times depend on the fixed parameter corresponding to the empty destination state, that is, ES for the capture process and GS for the relaxation process, and on the actual occupation of the destination state. In the presence of optical injection, we observe the following:

1) As it was also observed for two-mode VCSELs (see the previous section), the two-mode QD laser system provides a new injection locking possibility, where the GS is injection locked and coexists with an unlocked ES dynamics. The locking of the GS depends on both the frequency detuning between master and slave (QD) laser and on the injected power. But inside the locking region there exists therefore a locked solution with two-mode steady-state dynamics. The transition between the two regions is through a so-called transcritical bifurcation, which marks the threshold for the lasing onset of the ES mode.

2) In those parameter regions, where the GS dynamics is unlocked, the oscillations in the GS intensity time trace – and therefore the corresponding oscillations in the GS occupation caused by optical injection – cause a modulation of the relaxation time and, consecutively to oscillations in the occupation of the ES. Such a modulation creates a gain switching mechanism that leads to the emission of very short, picosecond pulses, from the ES. As shown in Figure 3.18, the two-mode dynamics can be very complex and depend on the optical injection parameters. Time traces of the GS and ES intensities are plotted for different increasing detuning values and for a fixed injection strength. In all cases the GS dynamics is a regularly modulated output, and the

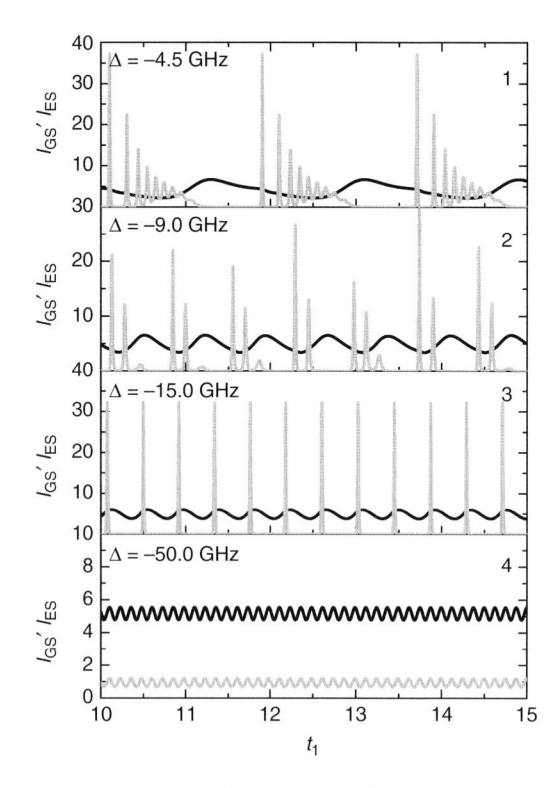

Figure 3.18 Two-mode quantum dot laser with optical injection. Time trace of GS and ES intensities for four increasing negative detuning values Δ. Panels 1–4 show qualitatively different behaviors of the ES dynamics, but in all cases the modulation of the GS intensity leads to an in-phase modulation of the ES intensity [65].

ES dynamics is related to this modulation of the GS intensity and occupation. In panel 1 is shown a case where the ES dynamics is made of packages of pulses where inside each package the laser exhibits ROs. The period between PPs corresponds to the modulation period in the GS intensity time trace. In panel 2 the ES has a more chaotic dynamics, but for a slightly larger negative detuning, in panel 3, the ES exhibits a very regular pulsating output. Finally, for a still larger detuning the ES dynamics is a weak modulation in phase with the time-evolution of the GS intensity; see panel 4.

3.9
Conclusions

To summarize, we have illustrated different examples of semiconductor laser nonlinear dynamics that can be attributed to mode competition. In Section 3.3 the competition between longitudinal modes in an EEL diode results in different correlation properties of the modal dynamics in the presence of a moderately strong optical feedback (in a dynamical regime called LFF dynamics). In Section 3.4, the beating between ECMs leads to the generation of fast all-optical self-pulsation, beyond several tens of gigahertz. Section 3.5 shows the importance of multimode laser dynamics in the case of an optical feedback from a short EC (time delay smaller than the laser RO period). The short cavity experiment is realized with a VCSEL emitting in two polarization modes. In Section 3.6, a new dynamics is observed in a time-delayed laser system, which results from a bistable mode hopping in VCSELs. The hopping occurs between orthogonal LP modes and is driven by noise. The addition of an external noise in this two-mode lasing system brings the system into so-called CR: the mode-hopping dynamics gets an optimal regularity at the time scale of the time delay. In Section 3.7, the mode competition in VCSEL manifests itself in the context of optical injection-induced nonlinear dynamics. We report on in depth experiments and numerical results specifically devoted to the role of (two) mode competition in an optically injected laser system. Finally, Section 3.8 illustrates some recent modeling results on the competition between excited and GS dynamics in a QD laser diode with optical injection. Fast pulsing dynamics are observed and motivate additional experiments.

Several other examples of multimode nonlinear laser dynamics are discussed in the following chapters. Not discussed in this book are also the cases of polarization and transverse mode chaos in a VCSEL with large current modulation [69, 70], and recent observations of dynamics with elliptical polarization in a QD VCSEL [71].

Acknowledgments

The author would like to thank the different coworkers, who contributed to the reported results: Krassimir Panajotov, Hugo Thienpont, Ignace Gatare, Mikel Arizaleta, Andrzej Tabaka, Lukasz Olejniczak, Angel Valle, Cristina Masoller, Neal B. Abraham, Thomas Erneux, Michel Nizette, Athanasios Gavrielides, Fabien Rogister, Ingo Fischer, and Wolfgang Elsaesser. The author thanks Conseil Régional de Lorraine, Fondation Supélec, Institut Carnot C3S, and European Action COST MP0702 for their supports.

References

1. Degen, C., Elsaber, W. and Fischer, I. (1999) Transverse modes in oxide confined VCSELs: Influence of pump profile, spatial hole burning, and thermal effects. *Opt. Express*, **5** (3), 38–47.
2. Chang-Hasnain, C.J., Harbison, J.P., Hasnain, G., von Lehmen, A.C., Florez, L.T., and Stoffel, N.G. (1991) Dynamic, polarization, and transverse mode characteristics of vertical-cavity surface-emitting lasers. *IEEE J. Quantum Electron.*, **27** (6), 1402–1409.
3. Choquette, K.D. and Leibenguth, R.E. (1994) Control of vertical-cavity laser polarization with anisotropic transverse cavity geometries. *IEEE Photon. Technol. Lett.*, **6** (1), 40–42.
4. Sivaprakasam, S., Saha, R., Lakshmi, P.A., and Singh, R. (1996)

Mode hopping in external-cavity diode lasers. *Opt. Lett.*, **21** (6), 411–413.

5. Lang, R. and Kobayashi, K. (1980) External optical feedback effects on semiconductor injection laser properties. *IEEE J. Quantum Electron.*, **16** (3), 347–355.
6. Petermann, K. (1995) External optical feedback phenomena in semiconductor lasers. *IEEE J. Select. Top. Quantum. Electron.*, **1** (2), 480–489.
7. Risch, C. and Voumard, C. (1977) Self-pulsation in the output intensity and spectrum of GaAs-AlGaAs cw diode lasers coupled to a frequency-selective external optical cavity. *J. Appl. Phys.*, **48** (5), 2083–2085.
8. Sukow, D.W., Heil, T., Fischer, I., Gavrielides, A., Hohl-AbiChedid, A., and Elsäßer, W. (1999) Picosecond intensity statistics of semiconductor lasers operating in the low-frequency fluctuation regime. *Phys. Rev. A*, **60** (1), 667–673.
9. Huyet, G., White, J.K., Kent, A.J., Hegarty, S.P., Moloney, J.V., and McInerney, J.G. (1999) Dynamics of a semiconductor laser with optical feedback. *Phys. Rev. A*, **60** (2), 1534–1537.
10. Rogister, F., Mégret, P., Deparis, O., and Blondel, M. (2000) Coexistence of in-phase and out-of-phase dynamics in a multimode external-cavity laser diode operating in the low-frequency fluctuations regime. *Phys. Rev. A*, **62** (6), 061803.
11. Viktorov, E.A. and Mandel, P. (2000) Low frequency fluctuations in a multimode semiconductor laser with optical feedback. *Phys. Rev. Lett.*, **85** (15), 3157–3160.
12. Rogister, F., Sciamanna, M., Deparis, O., Mégret, P., and Blondel, M. (2001) Low-frequency fluctuation regime in a multimode semiconductor laser subject to a mode-selective optical feedback. *Phys. Rev. A*, **65** (1), 015602.
13. Giudici, M., Giuggioli, L., Green, C., and Tredicce, J.R. (1999) Dynamical behavior of semiconductor lasers with frequency selective optical feedback. *Chaos Solitons Fractals*, **10** (4–5), 811–818.
14. Masoller, C. and Abraham, N.B. (1999) Low-frequency fluctuations in vertical-cavity surface-emitting semiconductor lasers with optical feedback. *Phys. Rev. A*, **59** (4), 3021–3031.
15. Sciamanna, M., Masoller, C., Abraham, N.B., Rogister, F., Mégret, P., and Blondel, M. (2003) Different regimes of low-frequency fluctuations in vertical-cavity surface-emitting lasers. *J. Opt. Soc. Am. B*, **20** (1), 37–44.
16. Sciamanna, M., Masoller, C., Rogister, F., Mégret, P., Abraham, N.B., and Blondel, M. (2003) Fast pulsing dynamics of a vertical-cavity surface-emitting laser operating in the low-frequency fluctuation regime. *Phys. Rev. A*, **68** (1), 015805.
17. Sondermann, M., Bohnet, H., and Ackemann, T. (2003) Low-frequency fluctuations and polarization dynamics in vertical-cavity surface-emitting lasers with isotropic feedback. *Phys. Rev. A*, **67** (2), 021802.
18. Tager, A.A. and Petermann, K. (1994) High-frequency oscillations and self-mode locking in short external-cavity laser diodes. *IEEE J. Quantum Electron.*, **30** (7), 1553–1561.
19. Pieroux, D., Erneux, T., Haegeman, B., Engelborghs, K., and Roose, D. (2001) Bridges of periodic solutions and tori in semiconductor lasers subject to delay. *Phys. Rev. Lett.*, **87** (19), 193901.
20. Erneux, T., Gavrielides, A., and Sciamanna, M. (2002) Stable microwave oscillations due to external-cavity-mode beating in laser diodes subject to optical feedback. *Phys. Rev. A*, **66** (3), 033809.
21. Li, H., Hohl, A., Gavrielides, A., Hou, H., and Choquette, K.D. (1998) Stable polarization self-modulation in vertical-cavity surface-emitting lasers. *Appl. Phys. Lett.*, **72** (19), 2355–2357.
22. Sciamanna, M., Erneux, T., Rogister, F., Deparis, O., Mégret, P., and Blondel, M. (2002) Bifurcation bridges between external-cavity modes lead to polarization self-modulation in vertical-cavity surface-emitting lasers. *Phys. Rev. A*, **65** (4), 041801.
23. Sukow, D., Hegg, M.C., Wright, J.L., and Gavrielides, A. (2002) Mixed external cavity mode dynamics in a semiconductor laser. *Opt. Lett.*, **27** (10), 827–829.

24. Wunsche, H.J., Brox, O., Radziunas, M., and Henneberger, F. (2001) Excitability of a semiconductor laser by a two-mode homoclinic bifurcation. *Phys. Rev. Lett.*, **88** (2), 023901.
25. Heil, T., Fischer, I., Elsäßer, W., and Gavrielides, A. (2001) Dynamics of semiconductor lasers subject to delayed optical feedback: the short cavity regime. *Phys. Rev. Lett.*, **87** (24), 243901.
26. Tabaka, A., Panajotov, K., Veretennicoff, I., and Sciamanna, M. (2004) Bifurcation study of regular pulse packages in laser diodes subject to optical feedback. *Phys. Rev. E*, **70** (3), 036211.
27. Sciamanna, M., Tabaka, A., Thienpont, H., and Panajotov, K. (2005) Intensity behavior underlying pulse packages in semiconductor lasers that are subject to optical feedback. *J. Opt. Soc. Am. B*, **22** (4), 777–785.
28. Tabaka, A., Peil, M., Sciamanna, M., Fischer, I., Elsäßer, W., Thienpont, H., Veretennicoff, I., and Panajotov, K. (2006) Dynamics of vertical-cavity surface-emitting lasers in the short external cavity regime: Pulse packages and polarization mode competition. *Phys. Rev. A*, **73** (1), 013810.
29. Sciamanna, M., Panajotov, K., Thienpont, H., Veretennicoff, I., Mégret, P., and Blondel, M. (2003) Optical feedback induces polarization mode hopping in vertical-cavity surface-emitting lasers. *Opt. Lett.*, **28** (17), 1543–1545.
30. Willemsen, M.B., Khalid, M.U.F., van Exter, M.P., and Woerdman, J.P. (1999) Polarization switching of a vertical-cavity semiconductor laser as a kramers hopping problem. *Phys. Rev. Lett.*, **82** (24), 4815–4818.
31. Panajotov, K., Sciamanna, M., Tabaka, A., Mégret, P., Blondel, M., Giacomelli, G., Marin, F., Thienpont, H., and Veretennicoff, I. (2004) Residence time distribution and coherence resonance of optical-feedback-induced polarization mode hopping in vertical-cavity surface-emitting lasers. *Phys. Rev. A*, **69** (1), 011801.
32. Masoller, C. (2003) Distribution of residence times of time-delayed bistable systems driven by noise. *Phys. Rev. Lett.*, **90** (2), 020601.
33. Benzi, R., Sutera, A., and Vulpiani, A. (1981) The mechanism of stochastic resonance. *J. Phys.*, **A14** 453–457.
34. Gammaitoni, L., Hänggi, P., Jung, P., and Marchesoni, F. (1998) Stochastic resonance. *Rev. Mod. Phys.*, **70** (1), 223–287.
35. Gang, H., Ditzinger, T., Ning, C.Z., and Haken, H. (1993) Stochastic resonance without external periodic force. *Phys. Rev. Lett.*, **71** (6), 807–810.
36. Pikovsky, A.S. and Kurths, J. (1997) Coherence resonance in a noise-driven excitable system. *Phys. Rev. Lett.*, **78** (5), 775–778.
37. Giacomelli, G., Giudici, M., Balle, S., and Tredicce, J.R. (2000) Experimental evidence of coherence resonance in an optical system. *Phys. Rev. Lett.*, **84** (15), 3298–3301.
38. Lindner, B. and Schimansky-Geier, L. (2000) Coherence and stochastic resonance in a two-state system. *Phys. Rev. E*, **61** (6), 6103–6110.
39. Palenzuela, C., Toral, R., Mirasso, C.R., Calvo, O., and Gunton, J.D. (2001) Coherence resonance in chaotic systems. *Europhys. Lett.*, **56** (3), 347–353.
40. Tsimring, L.S. and Pikovsky, A. (2001) Noise-induced dynamics in bistable systems with delay. *Phys. Rev. Lett.*, **87** (25), 250602.
41. Arizaleta Arteaga, M., Valencia, M., Sciamanna, M., Thienpont, H., López-Amo, M., and Panajotov, K. (2007) Experimental evidence of coherence resonance in a time-delayed bistable system. *Phys. Rev. Lett.*, **99** (2), 023903.
42. Stover, H.L. and Steier, W.H. (1966) Locking of laser oscillators by light injection. *Appl. Phys. Lett.*, **8** (4), 91–93.
43. Wieczorek, S., Krauskopf, B., Simpson, T., and Lenstra, D. (2005) The dynamical complexity of optically injected semiconductor lasers. *Phys. Rep.*, **416** (1–2), 1–128.
44. Pan, Z.G., Jiang, S., Dagenais, M., Morgan, R.A., Kojima, K., Asom, M.T., and Leibenguth, R.E. (1993) Optical injection induced polarization bistability

in vertical-cavity surface-emitting lasers. *Appl. Phys. Lett.*, **63** (22), 2999–3001.
45. Wang, C.L. and Pan, C.L. (1995) Tunable multiterahertz beat signal generation from a two-wavelength laser-diode array. *Opt. Lett.*, **20** (11), 1292–1294.
46. Matus, M., Kolesik, M., Moloney, J.V., Hofmann, M., and Koch, S.W. (2004) Dynamics of two-color laser systems with spectrally filtered feedback. *J. Opt. Soc. Am. B*, **21** (10), 1758–1771.
47. O'Brien, S., Osborne, S., Buckley, K., Fehse, R., Amann, A., O'Reilly, E.P., Barry, L.P., Anandarajah, P., Patchell, J., and O'Gorman, J. (2006) Inverse scattering approach to multiwavelength Fabry-Pérot laser design. *Phys. Rev. A*, **74** (6), 063814.
48. Osborne, S., Amann, A., Buckley, K., Ryan, G., Hegarty, S.P., Huyet, G., and O'Brien, S. (2009) Antiphase dynamics in a multimode semiconductor laser with optical injection. *Phys. Rev. A*, **79** (2), 023834.
49. Osborne, S., Buckley, K., Amann, A., and O'Brien, S. (2009) All-optical memory based on the injection locking bistability of a two-color laser diode. *Opt. Express*, **17** (8), 6293–6300.
50. Gatare, I., Sciamanna, M., Buesa, J., Thienpont, H., and Panajotov, K. (2006) Nonlinear dynamics accompanying polarization switching in vertical-cavity surface-emitting lasers with orthogonal optical injection. *Appl. Phys. Lett.*, **88** (10), 101 106/1–101 106/3.
51. Gatare, I., Sciamanna, M., Buesa, J., Thienpont, H., and Panajotov, K. (2006) Mapping of the dynamics induced by orthogonal optical injection in vertical-cavity surface-emitting lasers. *IEEE J. Quantum Electron.*, **42** (2), 198–207.
52. Valle, A., Gatare, I., Panajotov, K., and Sciamanna, M. (2007) Transverse mode switching and locking in vertical-cavity surface-emitting lasers subject to orthogonal optical injection. *IEEE J. Quantum Electron.*, **43** (4), 322–333.
53. Sciamanna, M. and Panajotov, K. (2005) Two-mode injection locking in vertical-cavity surface-emitting lasers. *Opt. Lett.*, **30** (21), 2903–2905.
54. Sciamanna, M. and Panajotov, K. (2006) Route to polarization switching induced by optical injection in vertical-cavity surface-emitting lasers. *Phys. Rev. A*, **73** (2), 023811.
55. Gatare, I., Sciamanna, M., Nizette, M., and Panajotov, K. (2007) Bifurcation to polarization switching and locking in vertical-cavity surface-emitting lasers with optical injection. *Phys. Rev. A*, **76** (3), 031803.
56. Gatare, I., Sciamanna, M., Nizette, Thienpont, H., and Panajotov, K. (2009) Mapping of two-polarization-mode dynamics in vertical-cavity surface-emitting lasers with optical injection. *Phys. Rev. E*, **80** (2), 026218.
57. Nizette, M., Sciamanna, M., Gatare, Thienpont, H., and Panajotov, K. (2009) Dynamics of vertical-cavity surface-emitting lasers with optical injection: a two-mode model approach. *J. Opt. Soc. Am. B*, **26** (8), 1603–1613.
58. Doedel, E., Fairgrieve, T., Sandstede, B., Champneys, A., Kuznetsov, Y., and Wang, X. (1997) AUTO97: continuation and bifurcation software for ordinary differential equations, (Software available at http://www.indy.cs.concordia.ca/auto/main.html).
59. Veselinov, K., Grillot, F., Gioannini, M., Montrosset, I., Homeyer, E., Piron, R., Even, J., Bekiarski, A., and Loualiche, S. (2008) Lasing spectra of 1.55 μm inas/inp quantum dot lasers: theoretical analysis and comparison with the experiments. *Opt. Quantum. Electron.*, **40** (2), 227–237.
60. O'Brien, D., Hegarty, S.P., Huyet, G., and Uskov, A.V. (2004) Sensitivity of quantum-dot semiconductor lasers tooptical feedback. *Opt. Lett.*, **29** (10), 1072–1074.
61. Otto, C., Lüdge, K., and Schöll, E. (2010) Modeling quantum dot lasers with optical feedback: sensitivity of bifurcation scenarios. *Phys. Status Solidi b*, **247** 829–845.
62. Markus, A., Chen, J.X., Paranthoën, C., Fiore, A., Platz, C., and Gauthier-Lafaye, O. (2003) Simultaneous two-state lasing in quantum-dot lasers. *Appl. Phys. Lett.*, **82** (12), 1818–1820.

63. Viktorov, E.A., Mandel, P., O'Driscoll, I., Carroll, O., Huyet, G., Houlihan, J., and Tanguy, Y. (2006) Low-frequency fluctuations in two-state quantum dot lasers. *Opt. Lett.*, **31** (15), 2302–2304.
64. Cataluna, M.A., Nikitichev, D.I., Mikroulis, S., Simos, H., Simos, C., Mesaritakis, C., Syvridis, D., Krestnikov, I., Livshits, D., and Rafailov, E.U. (2010) Dual-wavelength mode-locked quantum-dot laser, via ground and excited state transitions: experimental and theoretical investigation. *Opt. Express*, **18** (12), 12832–12838.
65. Olejniczak, L., Panajotov, K., Wieczorek, S., Thienpont, H., and Sciamanna, M. (2010) Intrinsic gain switching in optically injected quantum dot laser lasing simultaneously from the ground and excited state. *J. Opt. Soc. Am. B*, **27** (11), 2416–2423.
66. Goulding, D., Hegarty, S.P., Rasskazov, O., Melnik, S., Hartnett, M., Greene, G., McInerney, J.G., Rachinskii, D., and Huyet, G. (2007) Excitability in a quantum dot semiconductor laser with optical injection. *Phys. Rev. Lett.*, **98** (15), 153903.
67. Olejniczak, L., Panajotov, K., Thienpont, H., and Sciamanna, M. (2010) Self-pulsations and excitability in optically injected quantum-dot lasers: Impact of the excited states and spontaneous emission noise. *Phys. Rev. A*, **82** (2), 023807.
68. Erneux, T., Viktorov, E.A., Kelleher, B., Goulding, D., Hegarty, S.P., and Huyet, G. (2010) Optically injected quantum-dot lasers. *Opt. Lett.*, **35** (7), 937–939.
69. Sciamanna, M., Valle, A., Mégret, P., Blondel, M., and Panajotov, K. (2003) Nonlinear polarization dynamics in directly modulated vertical-cavity surface-emitting lasers. *Phys. Rev. E*, **68** (1), 016207.
70. Valle, A., Sciamanna, M., and Panajotov, K. (2007) Nonlinear dynamics of the polarization of multitransverse mode vertical-cavity surface-emitting lasers under current modulation. *Phys. Rev. E*, **76** (4), 046206.
71. Olejniczak, L., Sciamanna, M., Thienpont, H., Panajotov, K., Mutig, A., Hopfer, F., and Bimberg, D. (2009) Polarization switching in quantum dot vertical-cavity surface-emitting lasers. *IEEE Photon. Technol. Lett.*, **21** (14), 1008–1010.

4
Quantum Cascade Laser: An Emerging Technology
Andreas Wacker

Since the realization of semiconductor heterostructures in the early 1970s, the idea of using optical transitions between subbands for the amplification of light [1] has been a long-standing goal. Although many groups worked actively on this topic, see, for example, [2], the first lasing structure was established finally in 1994 at AT&T Bell Laboratories [3]. As discussed in detail below, one of the central ideas was the repetition of several identical active regions, where light amplification takes place, and thus, the new device was called Quantum Cascade Laser (QCL). The first designs emitted radiation in the infrared (IR) range of the optical spectrum with wavelengths around 3.4–24 µm [4, 5] (corresponding to frequencies of 88 and 12.5 THz). Here, the optical phonon frequency (7–12 THz for typical (In/Ga/Al)As semiconductors) sets a lower bound for such IR-QCLs because of the presence of the Reststrahlen band, where no radiation propagates. A detailed compilation of experimental and technological issues of the first generation of QCLs is given in [6].

In 2002, two major breakthroughs occurred: based on the two-phonon resonance design, continuous wave (cw) operation at room temperature [7] could be achieved for IR-QCLs, which sets the stage for commercial applications. Quite astonishing was the realization of QCLs, which operate below the Reststrahlen band [8] in the terahertz (THz) region of the optical spectrum. These devices are referred to as THz-QCLs and constitute a promising technology with a wide variety of possible applications [9, 10]. Presently, QCL structures have been established in a wide range of the optical spectrum, covering almost two orders of magnitude from 1.2 THz (250 µm) [11] to 114 THz (2.63 µm) [12]. Even frequencies below 1 THz can be obtained in a strong magnetic field [13]. Thus, the QCL concept is extremely versatile and is now considered to be *ready for take-off* [14] in a large variety of technological applications covering environmental science, process control, medical diagnostics, and chemical physics [15]. Following the applications, aspects of nonlinear dynamics are currently becoming the focus of research [16].

In this chapter, the physics behind this successful device concept has been outlined. Furthermore, a detailed overview of different design strategies and theoretical concepts has been presented.

Nonlinear Laser Dynamics: From Quantum Dots to Cryptography, First Edition. Edited by Kathy Lüdge.
© 2012 Wiley-VCH Verlag GmbH & Co. KGaA. Published 2012 by Wiley-VCH Verlag GmbH & Co. KGaA.

4.1
The Essence of QCLs

As outlined below, the defining concept of a QCL is based on the following ideas:

1) The use of the *electronic subbands* in semiconductor heterostructures as *upper* (up) and *lower* (low) laser levels, see Figure 4.1.
2) Electric pumping by a bias along the growth direction of the heterostructure.
3) The periodic repetition of active elements (as in Figure 4.1) enhances light amplification and allows to cover the entire waveguide.

Figure 4.2 illustrates the concept of a standard design of an IR-QCL [17]. Here, each period consists of essentially two parts: (i) an *active region*, which is designed to contain a few subbands, namely, the upper laser level (red), the lower laser level (blue), and the extraction level (green), with appropriate energies for the laser transition and efficient extraction (ii) an *injector region*, which removes electrons from the lowest laser level or the extraction level in one period and guides them into the upper laser of the next period at the operating bias. Typically, this region is considered as a kind of superlattice, with high transmission in the miniband region and low transmission in the minigap [18], thus stopping electrons from tunneling out of the upper laser level into continuum states.

4.1.1
Semiconductor Heterostructures

Epitaxial growth techniques allow for the realization of semiconductor heterostructures, where layers of different materials (with similar lattice constant) alternate. This provides a spatial variation of the conduction band edge in the growth direction

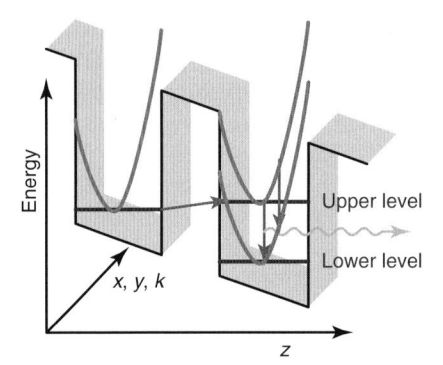

Figure 4.1 Sketch of the lasing transition in a semiconductor heterostructure as suggested by Kazarinov and Suris in 1971 [1]. The black line indicates the conduction band edge of the semiconductor heterostructure in the presence of an electric field. Here, the z-axis is the growth direction, and the structure is translational invariant in the (x, y) plane. The horizontal lines indicate the energy of the quantized levels $\varphi_n(z)$ in the quantum wells. Adding the plane wave behavior $e^{i(k_x x + k_y y)}$, one obtains the dispersion of the subbands indicated by the parabolae.

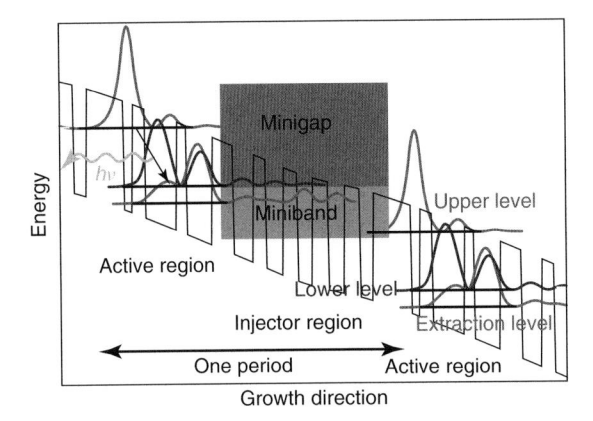

Figure 4.2 Conduction band profile for a conventional IR-QCL [17] together with the squares of the wavefunctions of the most important states. The lasing transition occurs between discrete levels in the active region, while the carriers are guided from the extraction level through the injector region to the upper laser level in the next period.

(here the z direction), as shown in Figure 4.1. Similar to the textbook problem in introductory quantum mechanics, this provides bound states $\varphi_n(z)$ with level energies E_n in the region of the semiconductor with the lower conduction band edge (called well), and these states are used as upper and lower laser levels. In addition, there is a plane wave behavior $e^{i(k_x x + k_y y)}$ in the perpendicular direction with energy $E(k) = \hbar^2(k_x^2 + k_y^2)/2m_{\text{eff}}$, as indicated by the parabolae to be added to E_n. The use of heterostructures has two strong advantages compared to other systems. First, the level energies $E_{\text{up/low}}$ can be widely tuned by the choice of semiconductor material and layer thickness, which allows realization of devices lasing in an enormous range of frequencies. Second, the states in the growth plane follow approximately the same parabolic dispersion with respect to k_x, k_y. Thus, the transition frequency does not vary much with k, and the operation is less sensitive to the electron temperature in contrast to interband quantum-well lasers.[1]

As all k values contribute with the same transition frequency, the amplification of the optical field directly depends on the densities n_{up} and n_{low} (in units 1 cm^{-2}) for the upper and lower level, respectively. Fermi's golden rule with an effective broadening γ provides the modal gain (in units 1 cm^{-1}) for an optical field of angular frequency ω with electric field component in z direction

$$g(\omega) = \frac{e^2 |z_{\text{up,low}}|^2 \omega (n_{\text{up}} - n_{\text{low}})}{2 L_z c \epsilon_0 \sqrt{\epsilon_r}} \frac{\gamma}{(E_{\text{up}} - E_{\text{low}} - \hbar\omega)^2 + \gamma^2/4}, \qquad (4.1)$$

1) Actually, there is some deviation due to nonparabolicity, which is, however, usually neglected. On the other hand, nonparabolicity actually allows for inversion-less lasers [19, 20], which is not further discussed here.

where $z_{up,low}$ is the dipole matrix element between the lasers, and L_z is the effective width of the waveguide. A derivation is given in Appendix 4.7. Thus, gain is proportional to the inversion $n_{up} - n_{low}$. Furthermore, the matrix element $z_{up,low}$ should also be optimized for improving gain.

Although all operating QCL structures are based on layered structures, there have been a few attempts to confine the lateral structure using quantum dots [21, 22]. This reduced dimensionality may provide better performance because of reduction of final states in scattering processes [23, 24].

4.1.2
Electric Pumping

If an electric field is applied along the growth direction of the structure, the conduction band edge gets tilted as can be seen in Figures 4.1 and 4.2. At the designed field of operation, the precise layer structure guides the electrons into the upper laser level via combinations of tunneling and scattering processes. The level from which the upper laser level is fed is typically referred to as *injector level* (in). While the further propagation at the energy E_{up} is essentially blocked by a gap in the energy spectrum of the heterostructure, efficient pathways are provided for the emptying of the lower level into an *extraction level* (ex). For an optimal design, there is thus a current channel in $\to$ up $\overset{weak}{\to}$ low $\to$ ex $\to \ldots \to$ in (in next period). Commonly one describes this by rate equations for the electron densities

$$\dot{n}_{up} = \frac{J}{e} - \frac{n_{up}}{\tau_{up,ex}} - \frac{n_{up}}{\tau_{up,low}} \quad \text{and} \quad \dot{n}_{low} = \frac{n_{up}}{\tau_{up,low}} - \frac{n_{low}}{\tau_{low,ex}}. \tag{4.2}$$

Here, J is the current density feeding the upper lasing level, and $\tau_{n,m}$ are the scattering times between levels n and m. (The extraction level may actually stand for a set of levels all involved in the transport.) In the stationary state, one obtains

$$n_{up} - n_{low} = \frac{J}{e} \frac{\tau_{up,ex}(\tau_{up,low} - \tau_{low,ex})}{\tau_{up,low} + \tau_{up,ex}}. \tag{4.3}$$

Thus, inversion (i.e., $n_{up} > n_{low}$) requires that the upper laser level has a long lifetime $\tau_{up,low} > \tau_{low,ex}$. Furthermore, the difference $n_{up} - n_{low}$ increases linearly with the current. In real structures, this is not entirely the case because of parasitic current paths not reaching the upper laser level.

4.1.3
Cascading

The intersubband transitions between states with inverted populations provide local amplifications of the optical field. This field is confined in a waveguide of lateral dimensions L_z, which is typically about half a wavelength in the hosting material[2], that is, $L_z \gtrsim \lambda/(2\sqrt{\epsilon_r})$, with the vacuum wavelength $\lambda = 2\pi c/\omega$ and

---
2) Metal–metal waveguides can actually be thinner.

the refractive index $\sqrt{\epsilon_r}$. Waveguide losses are typically of the order of several 10/cm, and thus laser operation can hardly be achieved by a single intersubband transition. Thus, one needs several transitions, which can be conveniently achieved by a cascade of identical structures with length d (called *period*) filling the entire waveguide. This provides in the optimal case a factor L_z/d to the modal gain. A common scale of a single period is about $d = 50$ nm. In order to have amplification over the entire thickness of the waveguide, one requires 30 periods in an IR-QCL [17] (for a vacuum wavelength of 10 µm and an refractive index of 3). Owing to the larger wavelength, THz-QCLs require a significantly larger number of periods, where 200 is a typical value.

The periodic repetition of identical elements requires a homogeneous electric field inside the structure. As it is well known from the Gunn diode and other nonlinear elements, a homogeneous electric field becomes unstable if there is a local negative differential conductance (NDC) [25]. In this case, domains with different electric fields occur, which has also been observed in QCL structures [26]. In order to match the resonance condition in each period, this has to be avoided, and thus a *positive differential conductance* (PDC) for each period is required. A straightforward way to achieve this is the incorporation of a *tunneling resonance* TL → TR in the currently path, where the right level has a slightly higher energy than the left level for the operating bias as shown in Figure 4.3. In order to be effective, the tunneling transition TL→TR requires

$$n_{TL} \gtrsim n_{TR}, \tag{4.4}$$

as otherwise it has no essential effect on the current flow.

Another central issue is the electron distribution. As the electrons gain energy with respect to the bottom of the conduction band while moving along the field, the electron distribution is heated up. This energy has to be removed, as otherwise the electrons will eventually reach states above the conduction band edge of the barriers, which are rather delocalized. Electrons can lose energy by transferring it either to the radiation field, which can play some role in very effective lasers,

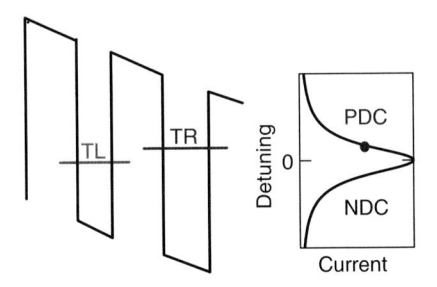

Figure 4.3 Sketch of a tunnel transition between a left (TL) and a right (TR) level. The current shows a distinct peak if the levels are aligned. (See Section 10.2 of [27] for details.) With increasing bias, the right level moves down, so that the displayed level alignment corresponds to a situation with positive differential conductance.

or to the phonons in the lattice. Here, the polar scattering at optical phonons is dominating for III/V semiconductors. Thus the time for transfer through a period must be significantly smaller than the optical phonon scattering time, which is typically of the order of 1 ps.

4.2
Different Designs

Based on the concepts addressed in Section 4.1 a variety of QCL designs has been realized. Their key features and essential differences shall be reviewed in this section.

4.2.1
Optical Transition and Lifetime of the Upper State

The gain transition is the heart of any QCL design. Equations (4.1) and (4.3) indicate, that it is advantageous to have a large dipole matrix element $z_{up,low}$ and a long scattering time $\tau_{up,low}$. Both quantities essentially depend on the overlap between the upper and the lower laser state. As $z_{up,low}$ increases and $\tau_{up,low}$ decreases with overlap, a compromise between both needs has to be found. QCL designs with a high spatial overlap are called *vertical* [28], whereas the opposite case is referred to as *diagonal*. The vertical design allowed for pulsed room temperature operation for IR-QCLs [29] and is dominating the development of IR-QCLs since then. Indeed comparing matrix elements for optical and phonon transitions indicates that vertical transitions are advantageous for lower transition energies [30]. On the other hand, recent results indicate that for THz-QCLs more diagonal designs can be of advantage [31–33].

In addition, different designs have been developed, where the upper and lower laser levels belong to an ensemble of levels. The first example is the interminiband laser [34], where the active region consists of a short superlattice and the lasing transitions occur between the different minibands, which was also the design of the first THz-QCL [8]. A similar concept is the bound to continuum design [35], where only the lower laser level is part of a continuum of states which shall facilitate the extraction process. However, other extraction designs such as the double phonon resonance for IR-QCLs and the resonant phonon extraction for THz-QCLs (Section 4.2.2) have dominated the further development. Recently, there is a renewed interest in continuum designs due to the increased tunability of the lasing wavelength by bias [36].

4.2.2
Effective Extraction from the Lower Laser Level

The first QCL designs were based on three levels in the active region (see also Figure 4.2), where the extraction level was located about one optical phonon energy

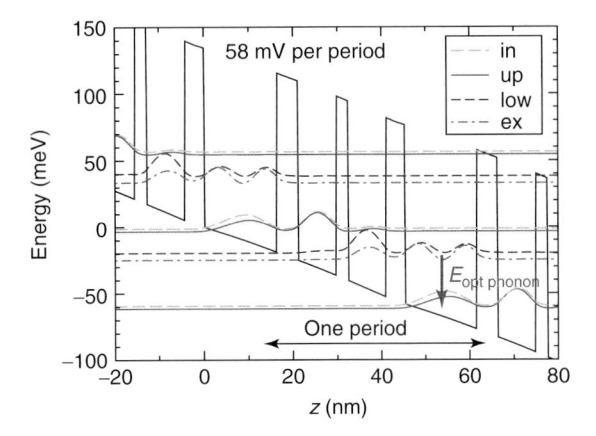

Figure 4.4 Band diagram of the QCL from 32 with a resonant phonon depopulation design and a diagonal laser transition. Note that only four levels are involved in the main operation. The dashed and dash-dotted lines refer to levels in adjacent periods of the cascade.

below the lower laser level. For this energy separation, intersubband phonon scattering is particular strong [37], leading to a short scattering time $\tau_{\text{low,ex}}$ in Eq. (4.3). As the optical phonon energy is not much larger than the thermal energy at room temperature, the lower laser level is only emptied to a certain degree, as long as the carriers are in the extraction level or further levels aligned with it. This phenomenon is often referred to as thermal backfilling. It can be avoided by adding a forth level located about one optical phonon energy below the extraction level, so that extraction occurs via two subsequent phonon emission processes, called *double phonon resonance* [38], which allowed for the realization of continuous wave (cw) room temperature operation [7]. For THz-QCLs, the energy difference between the upper and the lower laser level is small compared to the optical phonon energy. Thus, the scattering times $\tau_{\text{up,ex}}$ and $\tau_{\text{low,ex}}$ become similar for vertical designs. This problem is solved by emptying the lower laser level by resonant tunneling to an auxiliary level, which is subsequently emptied by optical phonon scattering. This *resonant phonon depopulation* scheme [39] has been proven very successful for the improvement of THz-QCLs, which quickly reached operation temperatures above the temperature of liquid nitrogen [40]. Figure 4.4 shows the band diagram for an optimized design [32] lasing up to 186 K at 3.9 THz.

4.2.3
Injection

Except for interminiband designs, most QCL structures have a well-defined upper laser level, which has to be fed by the driving current, as assumed in Eq. (4.2). In most designs, carriers are injected into the upper laser level by resonant tunneling from the injector level (or possibly several such states as indicated by the miniband in Figure 4.2) called *tunneling injection*. Here, the width of the injector barrier is of high relevance, as a wide barrier prevents from effective filling, while a thin barrier broadens the laser level [41, 42]. This tunneling transition also ensures the PDC of the device, as required for a stable electric field configuration along the cascade (Section 4.1.3), where we can identify the levels TL = in and TR = up. This tunneling injection design has, however, two shortcomings. (i) As the tunneling resonance should be an effective source of PDC, the injector level must exhibit an occupation at least comparable to the upper laser level, see Eq. (4.4). This restricts the possible inversion for a given total carrier density. (ii) Tunneling from the injector to the lower laser level constitutes a second resonance which is a further source of NDC and thus of particular concern for low lasing frequencies, when it mixes with the tunneling resonance into the upper laser level. A possible solution to these issues is the development of *scattering injection* designs [43, 44], where the upper laser level is fed from the injector level by a scattering process, that is, these levels are not aligned. Such structures were recently shown to exhibit improved temperature performance in the THz region [45–47]. In order to guarantee PDC and the stability of the electric field distribution along the sample, scattering injection designs should have a further tunnel resonance TL→TR in the current path. Frequently, this is a further transition between the extraction level and the injector of the subsequent period. Alternatively, one can use this tunneling transition to depopulate the lower laser level. In these *extraction controlled* designs, the lower laser level is the carrier reservoir until the design field is reached, where the lower laser level is efficiently emptied by a tunnel transition and subsequent scattering events, as suggested in [47]; see also Figure 4.5b.

4.3
Reducing the Number of Levels Involved

The first QCL designs contained a relatively large number of semiconductor layers per period and involved of the order of 10 levels in the current path through each period (see, e.g., Figure 4.2). A four-level design was later realized for IR-QCLs [49], showing that this complexity is not essential. However, the performance of this structure was not comparable with other designs. This is most likely due to the need to dissipate a large amount of energy in each period to avoid the continuous heating of the electron distribution while propagating along the cascade.

The bias per period under operating conditions is given by the energy of the lasing transition plus the energy difference due to additional scattering processes in the current path (at least one or two optical phonons). Typically, only a smaller fraction of this energy gained by the carriers in each period is emitted by radiation, as scattering transitions between the upper laser level and other levels can never be entirely avoided. Thus, a large fraction of this energy has to be transferred to lattice vibrations, mostly by optical phonon scattering processes. Thus, for IR-QCLs, each electron should be able to emit several optical phonons while transversing one

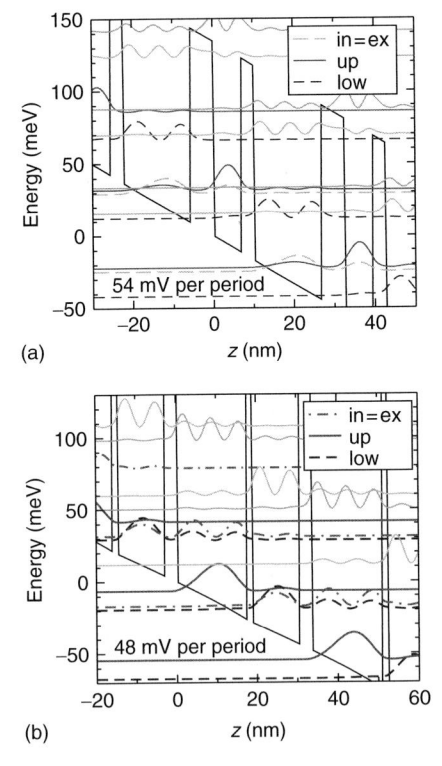

Figure 4.5 Band diagrams of the two different three-level designs using tunneling injection (a) and scattering injection (b). The designs are taken from [48] and [47], respectively. Note that each period contains only two wells in both cases.

period, which is facilitated by the presence of many levels involved. This explains why efficient designs show such a complexity of levels.

For THz-QCLs, however, the bias per period required is only slightly larger than the optical phonon frequency, and thus, less energy has to be dissipated per period. Thus, few-level designs are more likely to work and have the essential advantage of a shorter period, allowing for more active regions in the waveguide and consequently a larger modal gain. For example, the very successful structure in Figure 4.4 includes only four levels in the current path, which is given by the sequence in→ up → low→ ex → in (in next period).

Actually, a further simplification with only three remaining levels is possible if the extraction and injection levels are identical. This corresponds to the current path in/ex→ up → low → in/ex (in next period). This allows for two different designs, either with tunneling injection (Figure 4.5a) or with scattering injection (Figure 4.5b). Note that both designs include a tunneling transition in the current path. For the tunneling injection design, this is TL = in/ex and TR = up, while for scattering injection TL = low and TR = in/ex holds, which is actually an extraction controlled design. The design of Figure 4.5a is a working device [48], while Figure 4.5b is proposed to have a high operating temperature, but this has not been realized yet. A similar three-level design is presented in [50], which shows lasing both to tunnel and scattering injection.

Such three-level designs constitute the minimal version for QCLs based on population inversion. A hypothetical QCL design with two levels per period requires the current flow up → low→ up (in next period). Now the alignment condition requires that the tunneling transition is due to TL = low and TR = up. In order to achieve gain, generally $n_{up} > n_{low}$ is required.[3] This, however, does not match Eq. (4.4) for a two-level design, which is therefore not expected to work.

4.4
Modeling

The crucial part of any QCL is the achievement of inversion between subbands by resonant tunneling. In order to limit heating of the electronic distribution, phonon scattering is an important ingredient for the operation as well. On the other hand, phonon scattering as well as other scattering processes provide undesirable nonoptical transitions between the laser levels and broaden the tunneling transitions. This shows that the operation of any QCL constitutes an intricate interplay between quantum effects (tunneling and subbands) and scattering.

Typically the operation of QCLs is modeled by rate equations [54] between the levels of the active regions, while transitions through the injector are taken into account by phenomenological tunneling rates. The transition rates are evaluated microscopically within Fermi's golden rule for phonon scattering [55, 56] and partially for electron–electron scattering [57, 58]. In addition, confined phonon modes [59] and hot phonon effects [60] have been studied. While rate equations take into account only the electron density of subbands, Monte-Carlo simulations of the Boltzmann equation [61, 62] allow for a study of nonequilibrium distributions within the subbands. Here the importance of electron–electron scattering is debated. Reference [63] shows that the impact of electron–electron scattering is strong if no elastic scattering mechanism is taken into account. In contrast [64], elastic impurity scattering gives stronger effects than electron–electron scattering. If one includes the injector states in such a simulation, one obtains a self-consistent simulation of the entire structure [30, 63, 65] within the semiclassical carrier dynamics.

These *semiclassical models* for the carrier dynamics relate on various assumptions:

- Quantum mechanical correlations between different states are negligible.
- Broadening effects are of minor importance for the energetic selectivity of the transitions.
- Different scattering processes can be summed neglecting correlation effects.

3) Alternatively, dispersive gain [51–53] is possible, which will not be addressed here.

Two different approaches exist to include such quantum effects: density matrices and nonequilibrium Green's functions (NEGF).

Density matrices include the correlations ρ_{nm} between different quantum states $n \neq m$. These are of particular importance for the tunneling through the injection barrier, where their neglect provides the wrong result that the peak tunnel current does not drop with the barrier width [66, 67]. In a more phenomenological way, this can be done on the level of densities [68–70], which is very cost-effective. Taking into account the k-resolution, the equations for the density matrix $\rho_{nm}(k_x, k_y)$ become much more involved [63, 67, 71]. Here, it is a well-known problem that unphysical negative occupations may occur in frequently used approximation schemes; see [67] for a thorough discussion. A possible solution by using further approximations is outlined in [72] on the level of densities.

Alternatively, *nonequilibrium Green's functions* allow for the energetic resolution of coherences, which avoids any problems with negative occupations and is also of particular relevance for the gain spectrum, where corresponding density matrix calculations are particularly cumbersome [73]. Simulation schemes based on NEGF have been developed for periodic structures [74–76] based on the simplification for k-independent self-energies (but keeping the full energy dependence). This k-dependence is included in the schemes of [77] and [78], where the latter treats only a single period with an injection from a thermalized contact. A particular strength of the NEGF technique is the detailed description of the gain spectrum [79], including the important effect that the decay rates of the different levels do not just add up to the width of the gain transition [80]. A detailed overview on different gain features is given in [81].

As an example for the capabilities of NEGF simulations, some results for the three-level tunneling injection design of [48] are presented in the following. Figure 4.6a shows the calculated current using the nominal sample parameters and a roughness with an average height of 1 Å and lateral correlation length of 10 nm. The calculated currents in the peak and the lower plateau agree reasonably well with the data of [48]. As the NEGF approach of [76, 80] systematically includes all possible couplings with the alternating field without referring to the rotating wave approximation, the gain spectrum can be calculated over the entire frequency range as shown in Figure 4.6b. Note that the gain for $\omega \to 0$ correctly approaches the value

$$g(\omega = 0) = -\frac{1}{c\sqrt{\epsilon_r \epsilon_0}} \frac{dj}{dF}, \quad (4.5)$$

where dj/dF is just the slope in the dc current-field relationship at the operation points marked in Figure 4.6a.

Figure 4.7 displays the distribution of electrons before and after the main current peak around $Fd \approx 57$ mV together with the injector/extraction level (in/ex), the upper laser level (up), and the lower laser level (low). One can clearly detect the population inversion between the upper and the lower laser level. The alignment of the levels in and up causes the current peak, and consequently, the conductance is negative at 60 mV per period, where the upper laser level is below the injector

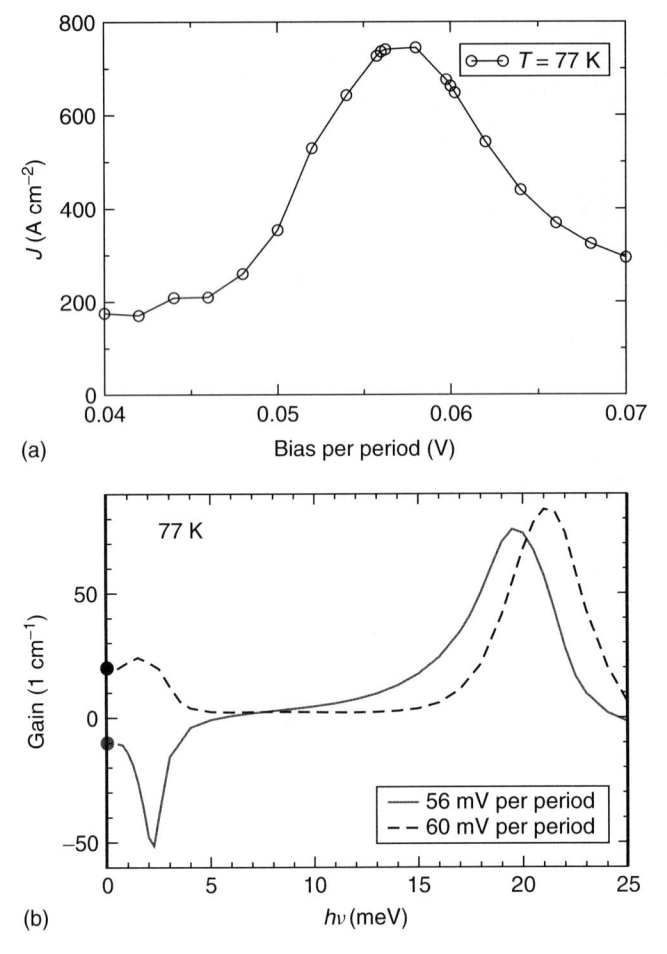

Figure 4.6 (a) Calculated current versus voltage drop per period for the design of [48]. (b) Gain at two different biases. The dots show the gain obtained by $g = -\frac{1}{c\sqrt{\epsilon_r\epsilon_0}} \frac{dj}{dF}$ from the dc conductance at the respective operation point.

level, compare Figure 4.3. This shift has only minor consequences on the main gain peak around 20 meV, as shown in Figure 4.6. However, the gain/absorption is strongly altered in the low-frequency region around 2 meV, where it turns from absorption at 56 mV to gain at 60 mV, associated with NDC in the static current field relationship. The second scenario resembles the standard Bloch gain in superlattices [82]. Furthermore, note that the occupation of the injector and upper laser level are comparable; thus, less than half of the carriers can occupy the upper laser level in the PDC region, which is a shortcoming of the injection design addressed in Section 4.2.3.

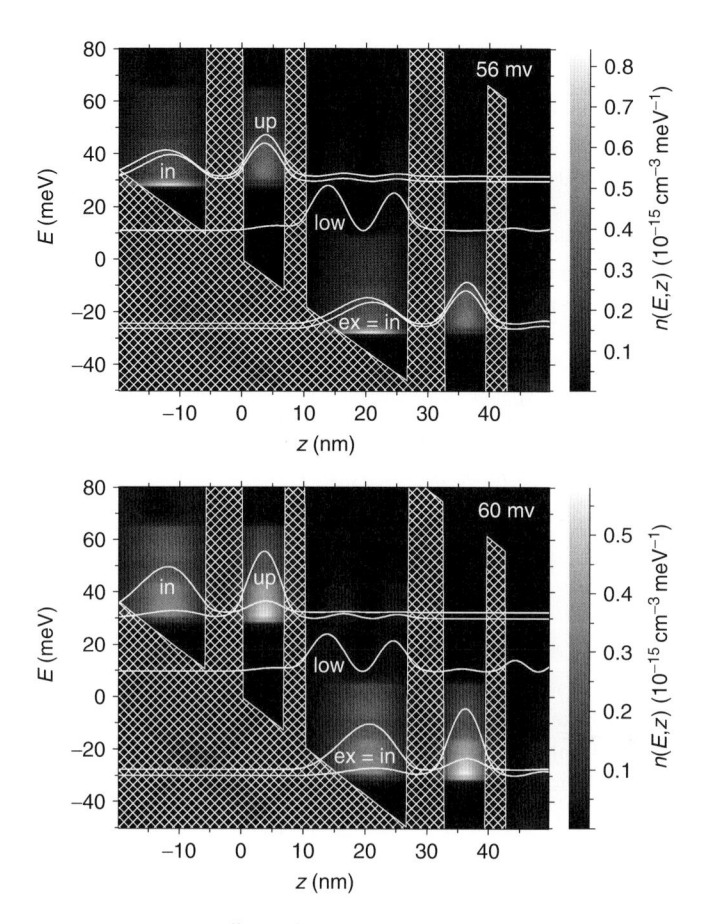

Figure 4.7 Energetically resolved electron density for the design of [48] as obtained from lesser Green's function for 77 K at two biases.

It is important to note that both density matrix theory [83] and Green's function [76] clearly show that the actual motion of carriers is due to coherences, that is nondiagonal elements of density matrix ρ_{nm}. However, these coherences are implicitly included in the semiclassical rate equation in a subtle way; for details see [84]. Thus, in many cases, the semiclassical approach provides similar currents as quantum kinetic calculations, which lead to an earlier conclusion that coherent effects are not of relevance [63].

4.5
Outlook

Specific tailoring of semiconductor heterostructures has allowed for the realization of QCLs covering almost two decades of the optical spectrum. In the IR range, they successfully operate at room temperature and are currently becoming the method of choice for spectroscopic applications. Current research focuses on increased output power and efficiency, where it recently became possible to transfer 50% of electrical power into lasing light [85, 86]. Another issue is the improvement of the tunability in a single device in order to cover several molecular transitions [36].

In contrast, the effective use of THz-QCLs is still hindered by the low operation temperature. Since the empirical limit $k_B T \lesssim \hbar\omega$ [9] has now been overcome [46], there is renewed hope for a further improvement of the maximum operation temperature, allowing for device operation with simple Peltier cooling or even directly at room temperature.

A further interesting issue is the optical field and its coupling to internal and external resonators [87]. In this context, phase locking between different transverse modes due to the optical nonlinearity was observed [88]. The authors state that *quantum cascade lasers are a unique laboratory for studying nonlinear dynamics because of their high intracavity intensity, strong intersubband optical nonlinearity, and an unusual combination of relaxation times,* suggesting a rich field for further studies of the general features outlined in Part II of this book.

Acknowledgments

My deepest thanks go to Eckehard Schöll who taught me the principles of semiconductor transport and nonlinear dynamics. Equally important, he guided my development to become an independent researcher. He was always a reliable help when times were difficult. On the occasion of his birthday I wish him all the best for the future!

I want to thank S.-C. Lee, M. Pereira, F. Banit, C. Weber, and R. Nelander for their essential contributions to the development of the theoretical concepts used here.

4.6
Appendix: Derivation of Eq. (4.1)

Here, the gain resulting from a single intersubband transition shall be calculated. We consider an electromagnetic field propagating in y direction with the electric field component

$$\mathbf{F}(\mathbf{r}, t) = F_0 \mathbf{e}_z \cos\left(\frac{\sqrt{\epsilon_r}\omega}{c} y - \omega t\right).$$

Then Fermi's golden rule provides the transition rate between the upper and lower laser level

$$R_{\text{up}\to\text{low}}(\mathbf{k}) = R_{\text{low}\to\text{up}}(\mathbf{k}) = \frac{2\pi}{\hbar} \left|\frac{eF_0 z_{\text{up,low}}(\mathbf{k})}{2}\right|^2 \delta(E_{\text{up}} - E_{\text{low}} - \hbar\omega)$$

where the spontaneous emission is neglected. Here, the matrix element is given by

$$z_{up,low}(\mathbf{k}) = \frac{1}{A}\int d^3 r\, \varphi_{up}^*(z) e^{-i(k_x x+k_y y)} z \varphi_{low}(z) e^{i(k_x x+k_y y)}$$

$$= \int dz\, \varphi_{up}^*(z) z \varphi_{low}(z) = z_{up,low}$$

where A is the normalization area. Summing over all k values and taking into account the occupations $f_n(k)$ provides the total number of photons coherently emitted per time via stimulated processes.

$$\dot{N}_{photon} = \frac{2\pi}{\hbar}\left|\frac{eF_0 z_{up,low}}{2}\right|^2 \underbrace{\delta(E_{up}-E_{low}-\hbar\omega)}_{\to \frac{1}{2\pi}\frac{\gamma}{(E_{up}-E_{low}-\hbar\omega)^2+\gamma^2/4}} 2(\text{for spin})\underbrace{\sum_k [f_{up}(k)-f_{low}(k)]}_{=A(n_{up}-n_{low})} \quad (4.6)$$

where the δ-function was replaced by a Lorentzian taking into account lifetime broadening in a phenomenological way. These photons add the power $P_{gain} = \dot{N}_{photon}\hbar\omega$ to the electromagnetic field.

The Poynting vector provides the average energy flux density

$$I(y) = \frac{1}{2}F_0^2 \sqrt{\epsilon_r \epsilon_0} c. \quad (4.7)$$

We consider a cube of dimensions $L_x L_y L_z$ in which the electromagnetic wave propagates. Owing to the power transferred to the electromagnetic field, the flux through the facets differs as $[I(L_y) - I(0)]L_x L_z = P_{gain}$, which is just the continuity equation. Now the gain is given by the relative change of intensity per length resulting from the stimulated transitions. Thus we have

$$g = \frac{I(L_y) - I(0)}{I(0)L_y} = \frac{P_{gain}}{I(0)AL_z} \quad (4.8)$$

where we used $A = L_x L_y$ as the area of the heterostructure layer. Inserting $P_{gain} = \dot{N}_{photon}\hbar\omega$ and applying Eqs. (4.6) and (4.7) provides Eq. (4.1), which is the desired result.

References

1. Kazarinov, R.F., and Suris, R.A. (1971) Possibility of the amplification of electromagnetic waves in a semiconductor with a superlattice. *Sov. Phys. Semicond.*, **5**, 707.
2. Yee, W.M., Shore, K.A., and Schöll, E. (1993) Carrier transport and intersubband population inversion in coupled quantum wells. *Appl. Phys. Lett.*, **63 (8)**, 1089.
3. Faist, J., Capasso, F., Sivco, D.L., Sirtori, C., Hutchinson, A.L., and Cho, A.Y. (1994) Quantum cascade laser. *Science*, **264**, 553.
4. Faist, J., Capasso, F., Sivco, D.L., Hutchinson, A.L., Chu, S.N.G., and Cho, A.Y. (1998) Short wavelength ($\lambda \sim 3.4\,\mu m$) quantum cascade laser based on strained compensated In-GaAs/AlInAs. *Appl. Phys. Lett.*, **72 (6)**, 680.

5. Colombelli, R., Capasso, F., Gmachl, C., Hutchinson, A.L., Sivco, D.L., Tredicucci, A., Wanke, M.C., Sergent, A.M., and Cho, A.Y. (2001) Far-infrared surface-plasmon quantum-cascade lasers at 21.5 μm and 24 μm wavelengths. *Appl. Phys. Lett.*, **78** (18), 2620.
6. Gmachl, C., Capasso, F., Sivco, D.L., and Cho, A.Y. (2001) Recent progress in quantum cascade lasers and applications. *Rep. Prog. Phys.*, **64**, 1533.
7. Beck, M., Hofstetter, D., Aellen, T., Faist, J., Oesterle, U., Ilegems, M., Gini, E., and Melchior, H. (2002) Continuous wave operation of a mid-infrared semiconductor laser at room temperature. *Science*, **295**, 301.
8. Köhler, R., Tredicucci, A., Beltram, F., Beere, H.E., Linfield, E.H., Davies, A.G., Ritchie, D.A., Iotti, R.C., and Rossi, F. (2002) Terahertz heterostructure laser. *Nature*, **417**, 156.
9. Williams, B.S. (2007) Terahertz quantum-cascade lasers. *Nat. Phot.*, **1**, 517.
10. Lee, M. and Wanke, M.C. (2007) Applied physics: searching for a solid-state terahertz technology. *Science*, **316**, 64.
11. Walther, C., Fischer, M., Scalari, G., Terazzi, R., Hoyler, N., and Faist, J. (2007) Quantum cascade lasers operating from 1.2 to 1.6 THz. *Appl. Phys. Lett.*, **91** (13), 131122.
12. Cathabard, O., Teissier, R., Devenson, J., Moreno, J.C., and Baranov, A.N. (2010) Quantum cascade lasers emitting near 2.6 μm. *Appl. Phys. Lett.*, **96** (14), 141110.
13. Scalari, G., Walther, C., Fischer, M., Terazzi, R., Beere, H., Ritchie, D., and Faist, J. (2009) THz and sub-THz quantum cascade lasers. *Laser Photon. Rev.*, **3**, 45.
14. Müller, A. and Faist, J. (2010) The quantum cascade laser: ready for take-off. *Nat. Photon.*, **4**, 291.
15. Curl, R.F., Capasso, F., Gmachl, C., Kosterev, A.A., McManus, B., Lewicki, R., Pusharsky, M., Wysocki, G., and Tittel, F.K. (2010) Quantum cascade lasers in chemical physics. *Chem. Phys. Lett.*, **487**, 1.
16. Wojcik, A.K., Yu, N., Diehl, L., Capasso, F., and Belyanin, A. (2010) Nonlinear dynamics of coupled transverse modes in quantum cascade lasers. *J. Mod. Opt.*, **57**, 1892.
17. Sirtori, C., Kruck, P., Barbieri, S., Collot, P., Nagle, J., Beck, M., Faist, J., and Oesterle, U. (1998) GaAs/AlGaAs quantum cascade lasers. *Appl. Phys. Lett.*, **73**, 3486.
18. Pacher, C., Rauch, C., Strasser, G., Gornik, E., Elsholz, F., Wacker, A., Kießlich, G., and Schöll, E. (2001) Antireflection coating for miniband transport and fabry-perot resonances in GaAs/AlGaAs superlattices. *Appl. Phys. Lett.*, **79** (10), 1486.
19. Faist, J., Capasso, F., Sirtori, C., Sivco, D.L., Hutchinson, A.L., Hybertsen, M.S., and Cho, A.Y. (1996) Quantum cascade lasers without intersubband population inversion. *Phys. Rev. Lett.*, **76** (3), 411.
20. Pereira, M.F. Jr. (2008) Intervalence transverse-electric mode terahertz lasing without population inversion. *Phys. Rev. B*, **78** (24), 245305.
21. Anders, S., Rebohle, L., Schrey, F.F., Schrenk, W., Unterrainer, K., and Strasser, G. (2003) Electroluminescence of a quantum dot cascade structure. *Appl. Phys. Lett.*, **82**, 3862.
22. Ulbrich, N., Bauer, J., Scarpa, G., Boy, R., Schuh, D., Abstreiter, G., Schmult, S., and Wegscheider, W. (2003) Midinfrared intraband electroluminescence from AlInAs quantum dots. *Appl. Phys. Lett.*, **83**, 1530.
23. Apalkov, V.M. and Chakraborty, T. (2003) Influence of dimensionality on the emission spectra of nanostructures. *Appl. Phys. Lett.*, **83**, 3671.
24. Vukmirovic, N., Indjin, D., Ikonic, Z., and Harrison, P. (2008) Quantum-dot cascades. *IEEE Photon. Tech. Lett.*, **20**, 129.
25. Shaw, M.P., Mitin, V.V., Schöll, E., and Grubin, H.L. (1992) *The Physics of Instabilities in Solid State Electron Devices*, Plenum Press, New York.
26. Lu, S.L., Schrottke, L., Teitsworth, S.W., Hey, R., and Grahn, H.T. (2006) Formation of electric-field domains in GaAs - $Al_xGa_{1-x}As$ quantum cascade laser structures. *Phys. Rev. B*, **73** (3), 033311.

27. Schöll, E. (ed.) (1998) *Theory of Transport Properties of Semiconductor Nanostructures*, Electronic Materials Series, vol. 4, Chapman and Hall, London.
28. Faist, J., Capasso, F., Sirtori, C., Sivco, D.L., Hutchinson, A.L., and Cho, A.Y. (1995) Continuous wave operation of a vertical transition quantum cascade laser above $T = 80$ K. *Appl. Phys. Lett.*, **67** (21), 3057. Erratum in (1996) 68, 2024.
29. Faist, J., Capasso, F., Sirtori, C., Sivco, D.L., Hutchinson, A.L., and Cho, A.Y. (1996) Room temperature mid-infrared quantum cascade lasers. *Electron. Lett.*, **32**, 560.
30. Donovan, K., Harrison, P., and Kelsall, R.W. (1999) Comparison of the quantum efficiencies of interwell and intrawell radiative transitions in quantum cascade lasers. *Appl. Phys. Lett.*, **75**, 1999.
31. Belkin, M., Wang, Q.J., Pflugl, C., Belyanin, A., Khanna, S., Davies, A., Linfield, E., and Capasso, F. (2009) High-temperature operation of terahertz quantum cascade laser sources. *Sel. Top. Quantum Electron. IEEE J. Sel. Top. Quantum Electr.*, **15** (3), 952.
32. Kumar, S., Hu, Q., and Reno, J.L. (2009) 186 K operation of terahertz quantum-cascade lasers based on a diagonal design. *Appl. Phys. Lett.*, **94** (13), 131105.
33. Mátyás, A., Belkin, M.A., Lugli, P., and Jirauschek, C. (2010) Temperature performance analysis of terahertz quantum cascade lasers: vertical versus diagonal designs. *Appl. Phys. Lett.*, **96** (20), 201110.
34. Scamarcio, G., Capasso, F., Faist, J., Sirtori, C., Sivco, D.L., Hutchinson, A.L., and Cho, A.Y. (1997) Tunable interminiband infrared emission in superlattice electron transport. *Appl. Phys. Lett.*, **70**, 1796.
35. Faist, J., Beck, M., Aellen, T., and Gini, E. (2001) Quantum-cascade lasers based on a bound-to-continuum transition. *Appl. Phys. Lett.*, **78** (2), 147.
36. Yao, Y., Wang, X., Fan, J.Y., and Gmachl, C.F. (2010) High performance continuum-to-continuum quantum cascade lasers with a broad gain bandwidth of over $400 \, \text{cm}^{-1}$. *Appl. Phys. Lett.*, **97** (8), 081115.
37. Ferreira, R. and Bastard, G. (1989) Evaluation of some scattering times for electrons in unbiased and biased single- and multiple-quantum-well structures. *Phys. Rev. B*, **40** (2), 1074.
38. Hofstetter, D., Beck, M., Aellen, T., and Faist, J. (2001) High-temperature operation of distributed feedback quantum-cascade lasers at 5.3 μm. *Appl. Phys. Lett.*, **78**, 396.
39. Williams, B.S., Kumar, S., Callebaut, H., Hu, Q., and Reno, J.L. (2003) Terahertz quantum-cascade laser operating up to 137 K. *Appl. Phys. Lett.*, **83** (25), 5142.
40. Kumar, S., Williams, B.S., Kohen, S., Hu, Q., and Reno, J.L. (2004) Continuous-wave operation of terahertz quantum cascade lasers above liquid-nitrogen temperature. *Appl. Phys. Lett.*, **84**, 2494.
41. Sirtori, C., Capasso, F., Faist, J., Hutchinson, A., Sivco, D.L., and Cho, A.Y. (1998) Resonant tunneling in quantum cascade lasers. *IEEE J. Quantum Electron.*, **34**, 1772.
42. Khurgin, J.B., Dikmelik, Y., Liu, P.Q., Hoffman, A.J., Escarra, M.D., Franz, K.J., and Gmachl, C.F. (2009) Role of interface roughness in the transport and lasing characteristics of quantum-cascade lasers. *Appl. Phys. Lett.*, **94** (9), 091101.
43. Yamanishi, M., Fujita, K., Edamura, T., and Kan, H. (2008) Indirect pump scheme for quantum cascade lasers: dynamics of electron-transport and very high T_0-values. *Opt. Express*, **16** (25), 20748–20758.
44. Yasuda, H., Kubis, T., Vogl, P., Sekine, N., Hosako, I., and Hirakawa, K. (2009) Nonequilibrium green's function calculation for four-level scheme terahertz quantum cascade lasers. *Appl. Phys. Lett.*, **94** (15), 151109.
45. Kumar, S., Chan, C.W.I., Hu, Q., and Reno, J.L. (2010) Operation of a 1.8-THz quantum-cascade laser above 160 K, *Conference on Lasers and Electro-Optics (CLEO) and the Quantum Electronics and Laser Science Conference (QELS)*, Optical Society of America, Washington, DC. CThU1.

46. Kumar, S., Chan, C.W.I., Hu, Q., and Reno, J.L. (2011) A 1.8-THz quantum cascade laser operating significantly above the temperature of $\hbar\omega/k_B$. *Nat. Phys.*, **7**, 166.
47. Wacker, A. (2010) Extraction-controlled quantum cascade lasers. *Appl. Phys. Lett.*, **97** (8), 081105.
48. Kumar, S., Chan, C.W.I., Hu, Q., and Reno, J.L. (2009) Two-well terahertz quantum-cascade laser with direct intrawell-phonon depopulation. *Appl. Phys. Lett.*, **95**, 141110.
49. Ulbrich, N., Scarpa, G., Böhm, G., Abstreiter, G., and Amann, M. (2002) Intersubband staircase laser. *Appl. Phys. Lett.*, **80**, 4312.
50. Scalari, G., Amanti, M.I., Walther, C., Terazzi, R., Beck, M., and Faist, J. (2010) Broadband THz lasing from a photon-phonon quantum cascade structure. *Opt. Express*, **18**, 8043.
51. Terazzi, R., Gresch, T., Giovannini, M., Hoyler, N., Sekine, N., and Faist, J. (2007) Bloch gain in quantum cascade lasers. *Nat. Phys.*, **3**, 329.
52. Revin, D.G., Soulby, M.R., Cockburn, J.W., Yang, Q., Manz, C., and Wagner, J. (2008) Dispersive gain and loss in midinfrared quantum cascade laser. *Appl. Phys. Lett.*, **92** (8), 081110.
53. Wacker, A. (2007) Lasers: coexistence of gain and absorption. *Nat. Phys.*, **3**, 298.
54. Capasso, F., Faist, J., and Sirtori, C. (1996) Mesoscopic phenomena in semiconductor nanostructures by quantum design. *J. Math. Phys.*, **37**, 4775.
55. Paulavičius, D., Mitin, V., and Stroscio, M.A. (1998) Hot-optical-phonon effects on electron relaxation in an AlGaAs/GaAs quantum cascade laser. *J. Appl. Phys.*, **84**, 3459.
56. Slivken, S., Litvinov, V.I., Razeghi, M., and Meyer, J.R. (1999) Relaxation kinetics in quantum cascade lasers. *J. Appl. Phys.*, **85**, 665.
57. Hyldgaard, P. and Wilkins, J.W. (1996) Electron-electron scattering in far-infrared quantum cascade lasers. *Phys. Rev. B*, **53**, 6889.
58. Harrison, P. (1999) The nature of the electron distribution functions in quantum cascade lasers. *Appl. Phys. Lett.*, **75**, 2800.
59. Becker, C., Sirtori, C., Page, H., Robertson, A., Ortiz, V., and Marcadet, X. (2002) Influence of confined phonon modes on the thermal behavior of AlAs/GaAs quantum cascade structures. *Phys. Rev. B*, **65**, 085305.
60. Compagnone, F., Di Carlo, A., and Lugli, P. (2002) Monte Carlo simulation of electron dynamics in superlattice quantum cascade lasers. *Appl. Phys. Lett.*, **80**, 920.
61. Tortora, S., Compagnone, F., Di Carlo, A., Lugli, P., Pellegrini, M.T., Troccoli, M., and Scamarcio, G. (1999) Theoretical study and simulation of electron dynamics in quantum cascade lasers. *Phys. B*, **272**, 219.
62. Iotti, R.C. and Rossi, F. (2001) Carrier thermalization versus phonon-assisted relaxation in quantum cascade lasers: A monte carlo approach. *Appl. Phys. Lett.*, **78**, 2902.
63. Iotti, R.C. and Rossi, F. (2001) Nature of charge transport in quantum-cascade lasers. *Phys. Rev. Lett.*, **87**, 146603.
64. Callebaut, H., Kumar, S., Williams, B.S., Hu, Q., and Reno, J.L. (2004) Importance of electron-impurity scattering for electron transport in terahertz quantum-cascade lasers. *Appl. Phys. Lett.*, **84**, 645.
65. Indjin, D., Harrison, P., Kelsall, R.W., and Ikonic, Z. (2002) Self-consistent scattering theory of transport and output characteristics of quantum cascade lasers. *J. Appl. Phys.*, **91**, 9019.
66. Callebaut, H. and Hu, Q. (2005) Importance of coherence for electron transport in terahertz quantum cascade lasers. *J. Appl. Phys.*, **98**, 104505.
67. Weber, C., Wacker, A., and Knorr, A. (2009) Density-matrix theory of the optical dynamics and transport in quantum cascade structures: the role of coherence. *Phys. Rev. B*, **79**, 165322.
68. Kumar, S. and Hu, Q. (2009) Coherence of resonant-tunneling transport in terahertz quantum-cascade lasers. *Phys. Rev. B*, **80** (24), 245316.
69. Dupont, E., Fathololoumi, S., and Liu, H.C. (2010) Simplified density-matrix model applied to three-well terahertz quantum cascade lasers. *Phys. Rev. B*, **81** (20), 205311.

70. Terazzi, R. and Faist, J. (2010) A density matrix model of transport and radiation in quantum cascade lasers. *New J. Phys.*, **12** (3), 033045.
71. Waldmueller, I., Chow, W.W., Young, E.W., and Wanke, M.C. (2006) Nonequilibrium many-body theory of intersubband lasers. *IEEE J. Quantum Electron.*, **42** (3), 292.
72. Gordon, A. and Majer, D. (2009) Coherent transport in semiconductor heterostructures: a phenomenological approach. *Phys. Rev. B*, **80** (19), 195317.
73. Willenberg, H., Döhler, G.H., and Faist, J. (2003) Intersubband gain in a Bloch oscillator and quantum cascade laser. *Phys. Rev. B*, **67**, 085315.
74. Wacker, A. and Jauho, A.P. (1998) Quantum transport: the link between standard approaches in superlattices. *Phys. Rev. Lett.*, **80**, 369.
75. Lee, S.C. and Wacker, A. (2002) Nonequilibrium Green's function theory for transport and gain properties of quantum cascade structures. *Phys. Rev. B*, **66**, 245314.
76. Lee, S.C., Banit, F., Woerner, M., and Wacker, A. (2006) Quantum mechanical wavepacket transport in quantum cascade laser structures. *Phys. Rev. B*, **73**, 245320.
77. Schmielau, T. and Pereira, M. (2009) Nonequilibrium many body theory for quantum transport in terahertz quantum cascade lasers. *Appl. Phys. Lett.*, **95**, 231111.
78. Kubis, T., Yeh, C., Vogl, P., Benz, A., Fasching, G., and Deutsch, C. (2009) Theory of nonequilibrium quantum transport and energy dissipation in terahertz quantum cascade lasers. *Phys. Rev. B*, **79**, 195323.
79. Nelander, R. and Wacker, A. (2008) Temperature dependence of the gain profile for terahertz quantum cascade lasers. *Appl. Phys. Lett.*, **92**, 081102.
80. Banit, F., Lee, S.C., Knorr, A., and Wacker, A. (2005) Self-consistent theory of the gain linewidth for quantum cascade lasers. *Appl. Phys. Lett.*, **86**, 041108.
81. Wacker, A., Nelander, R., and Weber, C. (2009) Simulation of gain in quantum cascade lasers. *Proc SPIE*, **7230**, 72301A.
82. Wacker, A. (2002) Gain in quantum cascade lasers and superlattices: A quantum transport theory. *Phys. Rev. B*, **66**, 085326.
83. Iotti, R.C., Ciancio, E., and Rossi, F. (2005) Quantum transport theory for semiconductor nanostructures: A density matrix formulation. *Phys. Rev. B*, **72**, 125347.
84. Wacker, A. (2008) Coherence and spatial resolution of transport in quantum cascade lasers. *Phys. Stat. Sol. C*, **5**, 215.
85. Liu, P.Q., Hoffman, A.J., Escarra, M.D., Franz, K.J., Khurgin, J.B., Dikmelik, Y., Wang, X., Fan, J., and Gmachl, C.F. (2010) Highly power-efficient quantum cascade lasers. *Nat. Photon.*, **4** (2), 95.
86. Bai, Y., Slivken, S., Kuboya, S., Darvsih, S.R., and Razeghi, M. (2010) Quantum cascade lasers that emit more light than heat. *Nat. Photon.*, **4** (2), 99.
87. Walther, C., Scalari, G., Amanti, M., Beck, M., and Faist, J. (2010) Microcavity laser oscillating in a circuit-based resonator. *Science*, **327**, 1495.
88. Yu, N., Diehl, L., Cubukcu, E., Bour, D., Corzine, S., Höfler, G., Wojcik, A.K., Crozier, K.B., Belyanin, A., and Capasso, F. (2009) Coherent coupling of multiple transverse modes in quantum cascade lasers. *Phys. Rev. Lett.*, **102** (1), 013901.

5
Controlling Charge Domain Dynamics in Superlattices
Mark T. Greenaway, Alexander G. Balanov, and T. Mark Fromhold

In this chapter, we show how an applied magnetic field can transform the structure and THz dynamics of charge domains in a biased semiconductor superlattice. It has been shown that the electrical current through a superlattice can be modulated by using an applied bias voltage and a tilted magnetic field to switch on and off stochastic web patterns, which thread the electron phase space and act as a network of conduction channels through which the electrons can propagate in real space [1–12]. When the web is switched on, the electrons undergo chaotic diffusive motion along its filaments, thereby producing a sharp increase in the measured and calculated current flow. Delocalization of the electron paths produces a series of strong resonant peaks in the electron drift velocity versus electric field curves. Self-consistent, static calculations of the current in the superlattice agree well with the experimental data and reveal strong resonant features originating from the sudden delocalization of the stochastic single-electron paths. In this chapter, we use the drift velocity characteristics to make dynamic self-consistent calculations of the self-sustained current oscillations generated in the superlattice. We find that the extra resonant features in the drift velocity–field curve dramatically affect the collective electron behavior by inducing multiple propagating charge domains and high-frequency current oscillations, whose amplitude and frequency are greatly increased by the tilted field.

In semiconductor physics, chaotic electron transport has been explored using a variety of two-dimensional billiard structures [13–22], antidot arrays [13, 14, 23–25], superlattices [26–29], and resonant tunneling diodes containing a wide quantum well enclosed by two tunnel barriers [13, 30–49].

Despite the diversity of experimental studies of quantum chaos in semiconductor nanostructures, they all involve systems in which the transition to chaos occurs by the gradual and progressive destruction of stable orbits in response to an increasing perturbation. This gradual onset of chaos occurs for all systems used in previous quantum chaos experiments, which obey the KAM theorem [1, 3, 13].

However, by connecting a series of quantum wells together to form a superlattice, it is possible to create a much rarer type of "weak" chaos – studied by Zaslavsky and coworkers [50–57] for driven harmonic oscillator systems that do not obey the KAM theorem – which is characterized by abrupt delocalization of the classical paths. The theory of such "non-KAM" chaos is of great interest because of the diverse applications in, for example, plasma physics, tokamak fusion, turbulent fluid dynamics, ion traps, quasicrystals, and ultracold atoms in optical lattices [9, 50–65]. However, it has proven difficult to realize and explore the rich phase space structure of a driven harmonic oscillator in experiments. In the next section, we show that non-KAM chaos can have a fundamental effect not only on the single-electron dynamics in superlattices but also on the collective dynamics.

The collective dynamics of electrons in superlattices, and the resulting fields and charge densities, have been studied extensively and a number of interesting phenomena have been observed and studied [1–12, 26–29, 66–83]. For a complete review of these models, refer to [67, 82, 84]. In [3], the authors used a model to investigate the effect of a tilted magnetic field, which calculated the *static* (unstable) solution for the field and charge profiles and found good correspondence with the experimental $I(V)$ curves, confirming the non-KAM-chaos-induced electron dynamics. In this chapter, we consider a *dynamic* model of charge and field domains for superlattices in tilted magnetic fields, based closely on that developed by E. Schöll and coworkers [67, 84, 85], and predict that the magnetic field not only changes the shape of the $I(V)$ curve but also significantly modifies the charge dynamics.

5.1
Model of Charge Domain Dynamics

To investigate the collective behavior of the electrons, the current continuity and Poisson equations were solved self-consistently throughout the device by adapting a model used previously to describe interwell transitions in superlattices [66, 67, 78] to the case of miniband transport. This allows us to investigate how the electron density, $n(x, t)$, and electric field, $F(x, t)$, in the superlattice vary spatially and temporally.

Figure 5.1a shows a schematic diagram of the GaAs/AlAs/InAs superlattice used in recent experiments [3, 7]. Together, 14 unit cells, each of width $d = 8.3$ nm, form the superlattice region of length L, which is enclosed by GaAs ohmic emitter and collector contacts, of length l, on the left- and right-hand edges, respectively. Electrons are confined to the first miniband with kinetic energy versus wave number, k_x, dispersion relation $E(k_x) = \Delta[1 - \cos(k_x d)]/2$, where the miniband width $\Delta = 19.1$ meV [3]. Semiclassical miniband transport corresponds to modeling the superlattice as a continuum (depicted in Figure 5.1b), where electrons move freely (with the GaAs effective mass $m^* = 0.067 m_e$) in the y and z directions but have dispersion $E(k_x)$, along the superlattice axis.

In our calculations, the superlattice region of the device was discretized into $N = 480$ layers with width $\Delta x = L/N = 0.24$ nm, which, after analysis of the

5.1 Model of Charge Domain Dynamics

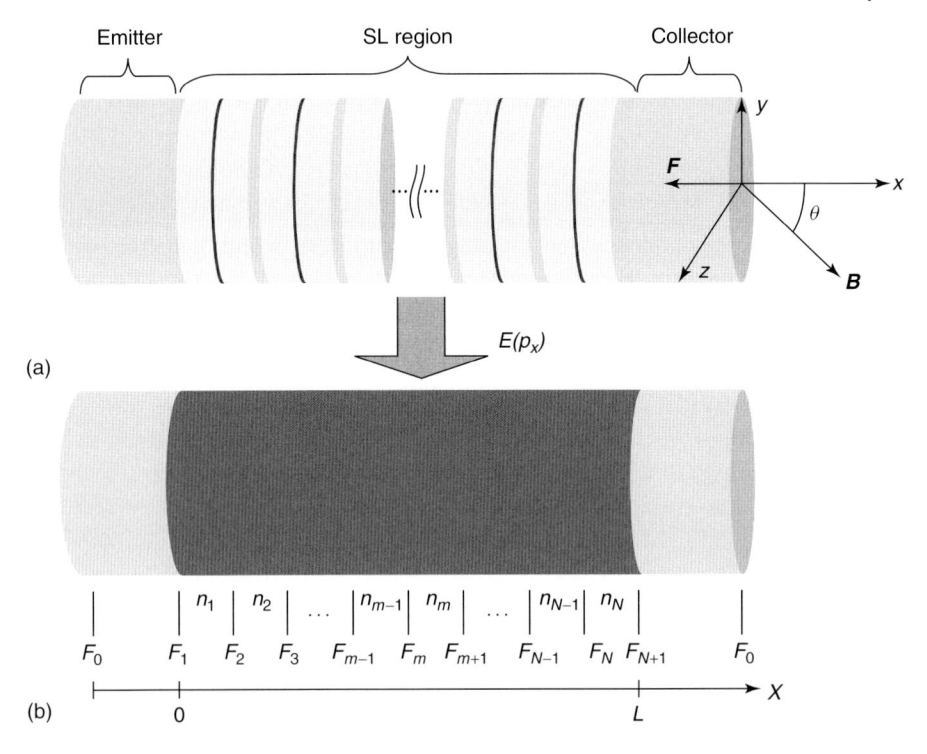

Figure 5.1 Schematic diagram showing how the superlattice structure (a) can be described by a continuum (b), where electron transport is defined by the miniband dispersion relation $E(p_x)$. The continuum is discretized into N layers where in the mth layer the electron density is n_m. The electric fields at the left- and right-hand edges of the layer are F_m and F_{m+1}, respectively, and F_0 is the field at the left- and right-hand edges of the device. The coordinate axes show the orientation of the electric and magnetic field vectors F and B, respectively.

results, was shown to be small enough to approximate a continuum[1]. The volume electron density of electrons in the mth layer is n_m and the field values at the left- and right-hand edge of the layer (see the vertical lines in Figure 5.1b) are F_m and F_{m+1}, respectively.

The dynamical equation that describes the evolution of the charge density in each layer is given by the following current continuity equation:

$$e\Delta x \frac{dn_m}{dt} = J_{m-1} - J_m \qquad m = 1, \ldots, N \qquad (5.1)$$

where $e > 0$ is the electron charge and J_m is the current density (C s^{-1} m^{-2}) of electrons moving from the mth into the $m+1^{th}$ layer. This equation is integrated numerically using a fourth-order Runge–Kutta scheme [86]. Note that in this

---

1) The model is considered to approximate a continuum when the field in the device varies on a spatial scale $\ll \Delta x$.

analysis, diffusion of electrons is neglected, as in previous models [3, 66], since for low temperature systems with bias applied, its effect is small compared to electron drift. J_m depends on the local drift velocity of electrons in layer m, v_d^m, and is given by

$$J_m = en_m v_d^m, \quad m = 1, \ldots, N. \tag{5.2}$$

Thus, J_m depends directly on the single-electron orbits and, therefore, we can expect that the features of the single-electron transport will have an effect on the collective electron dynamics[2]. The Esaki–Tsu approach was used to calculate the drift velocity for a field corresponding to the average field in layer m, $\overline{F_m}$, so that

$$v_d^m = v_d\left(\overline{F_m}, \mathbf{B}, \theta\right) \tag{5.3}$$

where

$$v_d\left(\overline{F_m}, B, \theta\right) = \sum \frac{\delta}{\tau} \int_0^\infty v_x\left(\overline{F_m}, \mathbf{B}, \theta, t\right) e^{-t/\tau} dt. \tag{5.4}$$

In this equation, the velocity of the electron, v_x, is found by calculating semiclassical trajectories for a miniband electron with an applied magnetic field, $\mathbf{B}$, tilted at an angle θ to the superlattice x-axis (Figure 5.1a), and an electric field equal to $\overline{F_m}$ [1, 3, 10]. The effective electron-scattering time is equal to

$$\tau = \tau_i \sqrt{\frac{\tau_e}{\tau_i + \tau_e}} = 250 \, \text{fs} \tag{5.5}$$

and is determined from the elastic (interface roughness) scattering time $\tau_e = 29$ fs and the inelastic (phonon) scattering time $\tau_i = 2.1$ ps [3, 7]. Since we are using this expression for τ, the drift velocity is modified by the coefficient, $\delta = \tau/\tau_i$ [71].

In these simulations, we average the drift velocity over ~2500 initial energies determined by the average temperature, T, of the electrons. For a lattice temperature of 4 K, we consider a range of initial electron energies up to 10 meV. This energy consists of a thermal component $k_B T \sim 0.4$ meV (where k_B is the Boltzmann's constant) and also a contribution due to voltage heating, that is, kinetic energy imparted by the electric field, $\approx \Delta_{SL}/2 \sim 10$ meV. The initial conditions were linearly spaced over a "sphere" of initial momenta in p_x, q_y, and p_z with a radius equal to the defined maximum energy of the system. This method of defining the thermal distribution was explored in [12] and has shown good correspondence with experimental results [3, 12].

When the drift–diffusion model is considered, in the case of a weakly coupled superlattice, the electron density and electric fields are discrete variables specified within, and at the edges of, each well. However, in a strongly coupled superlattice where electron dynamics is governed by miniband transport, the model of charge

---

[2] Note that the scattering time, τ, is less than the characteristic time scale of the domain dynamics, allowing us to assign a local drift velocity.

transport must approximate a continuum[3]. In this model, it is assumed, therefore, that the field an electron experiences at a particular point must be an average of the field values surrounding it. In particular, to create a smooth field profile, the field was averaged over a distance of one superlattice period[4]. Therefore, we use the average field at point x, $\overline{F(x)}$, which is defined as

$$\overline{F(x)} = \frac{1}{d}\int_{-d/2}^{d/2} F(x)dx. \tag{5.6}$$

After discretization, this average field is given by

$$\overline{F_m} = \frac{1}{N_{FA}} \sum_{m-N_{FA}/2}^{m+N_{FA}/2} F_m \tag{5.7}$$

where $N_{FA} = d/\Delta x$ is the number of discretization layers in one quantum well.

In each layer, F_m obeys the discretized version of Poisson's equation

$$F_{m+1} = \frac{e\Delta x}{\varepsilon_0 \varepsilon_r}(n_m - n_D) + F_m \quad m = 1,\ldots,N \tag{5.8}$$

where $\varepsilon_0 = 8.85 \times 10^{-12}$ F m^{-1} and $\varepsilon_r = 12.5$ are, respectively, the absolute and relative permittivities, and $n_D = 3 \times 10^{22}$ m^{-3} is the n-type doping density in the superlattice layers.

To properly simulate the charge domain dynamics of a system, it is important to consider the boundary conditions of the system, that is, the physical properties of the contact regions. There have been many theoretical works that investigate the effect of the boundary conditions on the dynamics of the charge in the superlattice [74, 75]. The choice of the boundary conditions has been shown to induce stationary charge domains, and also moving charge domains that give rise to both periodic [87] and chaotic currents [75]. In this investigation, however, we wish to compare these numerical results with experimental data. Therefore, the contact regions are modeled using a realistic picture of the contact doping profiles in order to obtain good correspondence between theory and experiment.

Ohmic boundary conditions [67] are used to determine the current injected into the superlattice region from the emitter, which is

$$J_0 = \sigma F_1 \tag{5.9}$$

where σ (Ω^{-1} m^{-1}) is the electrical conductivity given by Fromhold et al. [3] and Hardwick [12],

$$\sigma = \frac{n_0 e^2 \tau_c}{m^*} \tag{5.10}$$

where n_0 and τ_c are, respectively, the doping density and electron-scattering time in the contact regions. The voltage applied to the system is the global constraint. It is

---

3) Equivalent to models describing charge in Gunn diodes [84].

4) Which has physical sense if we consider the wavefunction of the electron to have its maximum across a single quantum well.

determined from the sum of the potential dropped across each discrete superlattice layer and across the contact regions. In fact, most of the field is dropped across the contact regions and any external resistance (e.g., measuring equipment) that is in series with the superlattice, so these regions must be considered rigorously. In the contact, just to the left of the superlattice layers, we assume that there is a charge accumulation layer. Applying Gauss's law to this layer we find that

$$F_1 - F_0 = \frac{en_L}{\varepsilon_0 \varepsilon_r} \tag{5.11}$$

where n_L is the areal density of electrons in the accumulation layer, which is modeled as a delta function sheet of negative charge at distance $l - s$ from the left-hand edge of the device, and F_0 is the field at the left-hand edge of the device. The electric field, F_{N+1}, at the right-hand edge of the superlattice region is screened by a depletion layer of length q and electron density n_0, which ensures that the fields at the left- and right-hand edge of the *device* are equal, meaning that

$$F_0 = F_{N+1} - \frac{en_0 q}{\varepsilon_0 \varepsilon_r}. \tag{5.12}$$

The potential drop across the collector contact is found by spatial integration of the electric field

$$V_C = \int_0^q F(x) dx = \int_0^q \left(F_{N+1} - \frac{en_0 x}{\epsilon_0 \epsilon_r} \right) dx.$$

Therefore, the voltage dropped across the depletion region is

$$V_C = q F_{N+1} - \frac{en_0 q^2}{2\varepsilon_0 \varepsilon_r}. \tag{5.13}$$

The voltage drop across the entire device, V, can be found by assuming that the field in the remaining sections of the superlattice is constant across each layer so that

$$V = F_0(l - s) + F_0(l - q) + F_1 s + V_C$$
$$+ \frac{\Delta x}{2} \sum_{m=1}^{N} (F_m + F_{m+1}) + \sigma F_0 A R_{ext} \tag{5.14}$$

where R_{ext} is the resistance that describes the physical contacts to the device and the remaining circuit of the experimental system, and A is the cross-sectional area of the device. In this system of equations, the applied voltage V is the global constraint that determines the dynamics of each n_m and F_m.

We define the global current density, $J(t)$, in the layers of the superlattice region as [67]

$$J(t) = \frac{1}{(N+1)} \sum_{m=0}^{N} J_m. \tag{5.15}$$

The corresponding current is then

$$I(t) = J(t) A. \tag{5.16}$$

The system of equations (Eqs. (5.1)–(5.16)) can now be solved self-consistently (ensuring that the voltage dropped across the device is constant and equal to the applied voltage) to obtain $n(x, t)$, $F(x, t)$ and, therefore, $I(t)$.

It should be noted here that there are three assumptions made for this model:

- For a given voltage across the device, the electric field is constant in the ohmic contacts and is equal to F_0, which corresponds to the device remaining overall neutral.
- The width of the superlattice sections, Δx, is small enough that the changes in electric field and charge density in the sections are negligible.
- Rigorously, the proportion of ionized donors will depend exponentially on the local electric field [3]. However, if this is applied for collective dynamics in a magnetic field, the model is numerically unstable. Therefore, here we assume that all donors are ionized.

5.2 Results

In this section, we show the electron dynamics for the GaAs/AlAs/InAs superlattice with lattice period $d = 8.3$ nm and miniband width $\Delta = 19.1$ meV, as described in the previous section. We consider a semiclassical formulation for electrons in the first miniband of the superlattice, to obtain the electron velocity along the x-axis, v_x, in the equation for drift velocity (5.4). The drift velocity was calculated by averaging over 2500 electron trajectories corresponding to a lattice temperature of 4 K (see previous section). The experimentally obtained (and verified [3, 12]) parameters for this superlattice are summarized in the following table.

Parameter	Symbol	Value
Mean doping density of layers	n_D	3×10^{22} m^{-3}
Effective scattering time	τ	250 fs
Drift velocity correction	δ	1/8.5
External resistance	R_{ext}	17 Ω
Contact length	l	500 Å
Contact doping density	n_0	1.0×10^{23} m^{-3}
Contact scattering time	τ_c	90 fs
Position of accumulation layer	s	150 Å
Diameter of superlattice messa	D_{SL}	25 μm

These parameters are used in the system of equations derived in Section 5.1 to calculate the charge domain dynamics for superlattice. In the simulations, we solved equations (5.1)–(5.16) self-consistently, solving the current continuity equation, Eq. (5.1), using the fourth-order Runge–Kutta method [86]. Initially, the density of electrons in the layers is equal to the doping density ($n_m(t=0) = n_D$) and the fields in the layers are given a nominal value to avoid divisions by 0 ($F_m(t=0) = 1 \times 10^3$ V m^{-1}). In this analysis, we only consider the case when $B = 15$ T. The results are presented in the following sections.

5.2.1
Drift Velocity Characteristics for $\theta = 0°, 25°,$ and $40°$

The formulation of the drift model, and Eq. (5.2), imply that the structure and dynamics of the charge domains in the superlattice will strongly depend on the drift velocity–field characteristics. Therefore, in Figure 5.2 we show how the drift velocity varies with $r = \omega_B/\omega_\parallel \propto F$, where $\omega_B = eFd/\hbar$ is the Bloch oscillation frequency and $\omega_\parallel = e\mathbf{B}\cos\theta/m^*$ is the cyclotron frequency corresponding to the x component of B.

When $\theta = 0°$ (lower curve in Figure 5.2), the cyclotron motion in the $y - z$ plane is separable from the Bloch motion along the x-axis. Consequently, we see only one maximum in the drift velocity curve (dashed line labeled ET) when $r = 1/\omega_\parallel \tau$. Beyond the "Esaki–Tsu" peak, drift velocity is suppressed as electrons are able to perform more Bloch oscillations [67, 69, 88].

When $\theta \neq 0$, there is strong coupling between the Bloch and cyclotron motion, which drives the electron trajectories chaotic [1, 3]. When r is irrational, the electron orbits remain localized. However, when r is an integer, electrons map out intricate "stochastic-web" and their orbits become unbounded. This abrupt delocalization of the orbit creates sharp resonant peaks in the electron's drift velocity–field characteristic. Therefore, when $\theta = 25°$ (middle curve in Figure 5.2), in addition to the Esaki–Tsu peak, there is a $r = 1$ resonance (dashed line in Figure 5.2 labeled 1.0) and by increasing θ to $40°$ (top curve in Figure 5.2) we also find that the $r = 2$ resonance becomes apparent (dashed line in Figure 5.2 labeled 2.0). For r

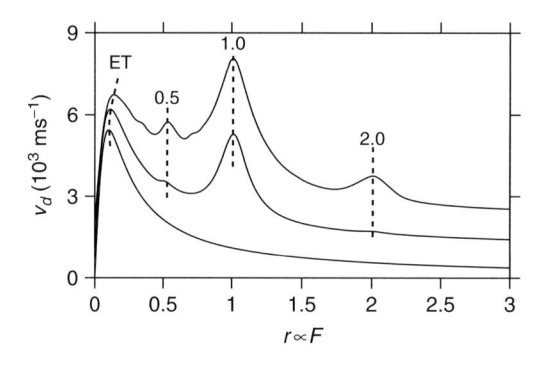

Figure 5.2 v_d versus $r \propto F$ curves calculated for $\mathbf{B} = 15$ T with (from bottom to top) $\theta = 0, 25°$ and $40°$. For clarity, curves are offset vertically by 10^3 ms^{-1}. The dashed colored lines show the positions of the Esaki-Tsu (ET) peak, the $r = 0.5$ peak, the $r = 1$ peak and the $r = 2$ peak.

values that are rational but not integer, the electron orbits are finite, but exhibit some resonant extension along x. This causes the small additional peaks visible at $r = 0.5$ in the middle and top curves of Figure 5.2.

In the following sections, it will be shown that, via the resonant features in the drift velocity curves, the single-electron chaotic trajectories can drastically alter the collective electron dynamics.

5.2.2
Current–Voltage Characteristics for $\theta = 0°, 25°,$ and $40°$

In this section, we consider the $I(V)$ curves for the superlattice. Generally, following the initial transient behavior, the current–time characteristics reach a constant value or perform self-sustained oscillations between the minimum and maximum current, I_{min} and I_{max}, respectively. The behavior of the current oscillations depends strongly on V, **B**, and θ. Figure 5.3 shows the current–voltage characteristics for values of $\theta = 0°, 25°,$ and $40°$ when $\mathbf{B} = 15$ T, offset for clarity. The plot shows that for all values of θ, the current is single valued at low V and then at some critical voltage, V_c, the current becomes double valued (the upper line showing I_{max} and the lower line I_{min}), denoted by the shaded region in the plot. At low voltages, the current–voltage curve is approximately linear and has a stationary solution, showing characteristics similar to bulk semiconductor material. This ohmic part of the $I(V)$ curve corresponds to the approximately linear region in the drift velocity characteristic (Figure 5.2). At the critical value, V_c (the value of which depends on **B** and θ), the stationary state loses its stability via Hopf bifurcation and starts to oscillate between I_{max} and I_{min}, corresponding to the nonlinear region of the drift velocity characteristic.

With further increase in V, the size of the oscillations $I_a = I_{max} - I_{min}$ and, consequently, their power grows for all θ (in the range of voltages presented here). However, as clearly shown in Figure 5.3, the presence of a tilted magnetic field

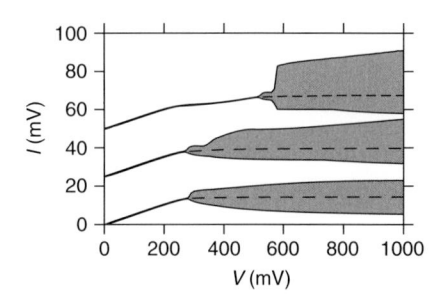

Figure 5.3 $I(V)$ characteristics calculated for (from bottom to top) $\theta = 0°, 25°,$ and $40°$. For clarity, curves are vertically offset by 15 mA. Current oscillations occur within the shaded regions, whose upper (lower) bounds are $I_{max}(V)$ [$I_{min}(V)$]. Dashed curves show I values corresponding to the unstable steady state solution of Eq. (5.1).

($\theta \neq 0$) can also increase the amplitude of the current oscillations, for example, in the voltage range $0 - 1$ V $I_a \approx 18$ mA when $\theta = 0°$ compared to ≈ 24 mA for $\theta = 25°$, and ≈ 33 mA for $\theta = 40°$.

The dashed curves in Figure 5.3 show the static solution to the dynamical equations obtained by setting Eq. (5.1) to 0. Such static solutions have been shown previously to closely correspond to experimental DC $I(V)$ measurements [3]. For low voltages (in the stationary regime), the full time-dependent solution of the equations of motion correspond exactly to the static solutions. When $\theta = 40°$, features corresponding to the $r = 1$ resonance are visible between 200 mV and 400 mV. In the current oscillation regime, the static current effectively bisects the extremal values, I_{min} and I_{max}, of the dynamical solution, showing that even when we enter the oscillating regime the two solutions are still broadly consistent with each other, also suggesting good correspondence with experimental results.

5.2.3
$I(t)$ Curves for $\theta = 0°$, $25°$, and $40°$

In this section, the $I(t)$ curves and how they vary with V, $\mathbf{B}$, and θ are considered in detail. We consider θ values of $0°$, $25°$, and $40°$ because the drift velocity curves have a simple form with, respectively, only 1, 2, and 3 resonant features (Figure 5.2). The traces shown in the left-hand column of Figure 5.4 are the $I(t)$ characteristics

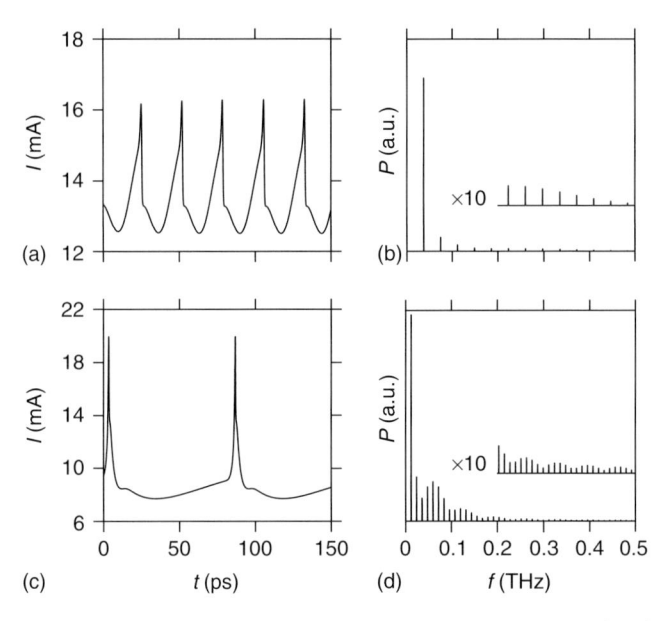

Figure 5.4 $I(t)$ oscillations calculated when $\theta = 0°$ for (a) $V = 290$ mV and (c) $V = 490$ mV. The corresponding Fourier power spectra of $I(t)$ are given in (b) and (d) for $V = 290$ mV and $V = 490$ mV, respectively, with a common vertical scale in arbitrary units. Inset within the Fourier power spectra are ×10 magnifications of the spectra for $f > 0.2$ THz.

for $\theta = 0°$ when $V = 290$ mV and 490 mV. Figure 5.4a is for $V = 290$ mV, a value close to the critical voltage $\approx V_c$. The $I(t)$ curve exhibits periodic oscillations, whose frequency ~ 37 GHz corresponds to the single dominant peak in the Fourier power spectrum shown in Figure 5.4b. The sparsity of the power spectrum confirms the periodicity of the current oscillations at the Hopf bifurcation. Increasing V to 490 mV has little qualitative effect on the shape of the current oscillations (Figure 5.4c). However, the fundamental frequency of the oscillations falls to ~ 12 GHz. In addition, the peaks in $I(t)$ sharpen and also increase in amplitude. These effects combine to strengthen the higher frequency harmonics in the Fourier power spectrum (Figure 5.4d).

The plots in Figure 5.5 show the $I(t)$ curves and comparative Fourier power spectra when $\theta = 25°$ for different V values. The $I(t)$ curve and frequency spectrum calculated for $\theta = 25°$ and $V = 290$ mV $\approx V_c$ (Figure 5.5a,b) are very similar to those

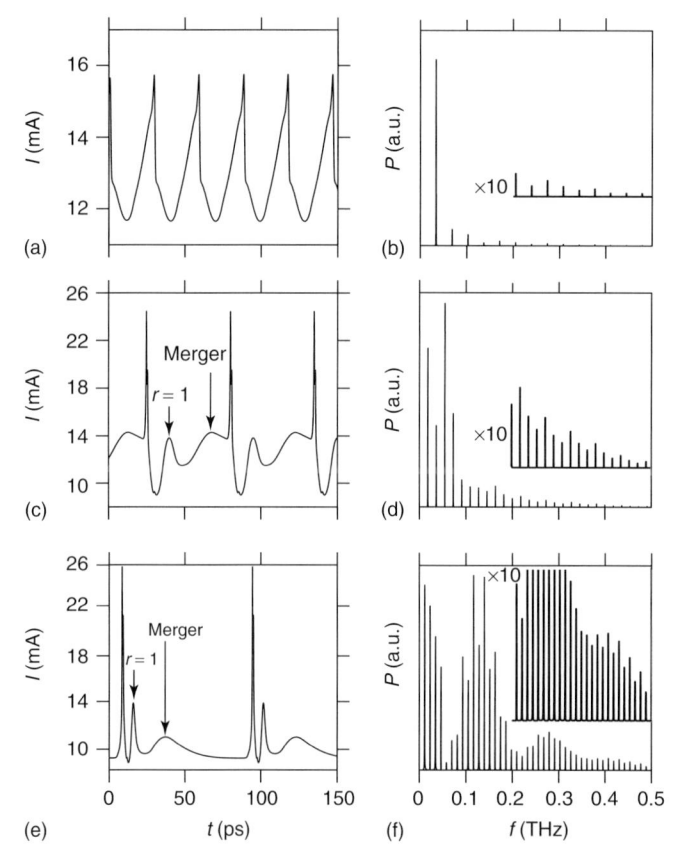

Figure 5.5 $I(t)$ oscillations calculated when $\theta = 25°$ for (a) $V = 290$ mV, (c) $V = 490$ mV, and (e) $V = 690$ mV. The corresponding Fourier power spectra of $I(t)$ are given in (b), (d), and (f) for $V = 290$ mV, 490 mV, and 690 mV, respectively, with a common vertical scale in arbitrary units. Inset within the Fourier power spectra are ×10 magnifications of the spectra for $f > 0.2$ THz.

for $\theta = 0°$ (Figure 5.4a,b). However, we find in Figure 5.5c that when V increases to 490 mV, the current–time characteristics for $\theta = 25°$ differ markedly from those for $\theta = 0°$ (Figure 5.5c). Comparison of the figures reveals that, in particular, tilting B almost doubles the fundamental frequency (compare Figures 5.4d and 5.5d) and also introduces new features in $I(t)$ (arrowed in Figure 5.5c), which are absent when $\theta = 0°$. These features originate from the $r = 1$ resonance in the drift velocity curve, as explained in the next section. Compared with the case when $\theta = 0°$, the arrowed features in the $I(t)$ curve strongly enhance the high-frequency components in the Fourier power spectrum: see Figure 5.5d, which reveals a dominant third harmonic at 54 GHz and also strengthened harmonics for frequencies > 0.2 THz. Increasing V further to 690 mV (Figure 5.5e) induces stronger resonant features in the $I(t)$ curve and also strengthens the high-frequency components of the spectra (Figure 5.5f).

Increasing θ to 40°, we find that when $V = 540$ mV $\approx V_c$ the shape of the $I(t)$ curve (Figure 5.6a) and the Fourier power spectrum (Figure 5.6b), which is dominated by the fundamental peak, do not significantly alter from the case when $\theta = 0°$ and 25°. However, the frequency of the fundamental peak does increase substantially to 56 GHz compared to 37 GHz and 34 GHz when $V \approx V_c$ for $\theta = 0°$ and $\theta = 25°$, respectively. In contrast, increasing the voltage to $V = 610$ mV when $\theta = 40°$ (Figure 5.6c) induces complex $I(t)$ fluctuations that are both stronger and richer than for comparable voltages at $\theta = 0°$ (see Figure 5.4c) and $\theta = 25°$ (see, Figure 5.5c). Consequently, the high-frequency peaks in the Fourier power spectrum (Figure 5.6d) are further enhanced, with the fifth harmonic at 92 GHz being the strongest. Increasing V to 740 mV (Figure 5.6e), we find a similar $I(t)$ plot to when $V = 610$ mV. However, in the Fourier power spectrum (Figure 5.6f), the fundamental frequency is decreased compared to when $V = 610$ mV, consistent with other θ considered in this section. Note that there is also no significant enhancement of the high-frequency components seen after increasing V from 490 mV to 690 mV when $\theta = 25°$ (compare Figure 5.5d and f). This suggests that the mechanism of frequency enhancement is complex. The general dependence of θ and V on frequency will be considered in Section 5.2.6.

5.2.4
Charge Dynamics for $\theta = 0°, 25°,$ and $40°$

To understand how $I(t)$ varies with V and θ, in this section we consider how these parameters affect the underlying spatiotemporal electron charge dynamics. The charge dynamics of the system depend strongly on the drift velocity characteristics of electrons in the system (Figure 5.2).

The gray-scale plot in Figure 5.7a shows n_m calculated versus t and x for $\theta = 0°$ and $V = 290$ mV $\approx V_c$. The plot in Figure 5.7b shows the corresponding three-dimensional visualization of the natural log of the charge density. These figures show that for any given x, the local charge density oscillates periodically as a function of t. This is due to the negative differential velocity in the corresponding $v_d(F)$ curve (see lower plot in Figure 5.2) as we now explain.

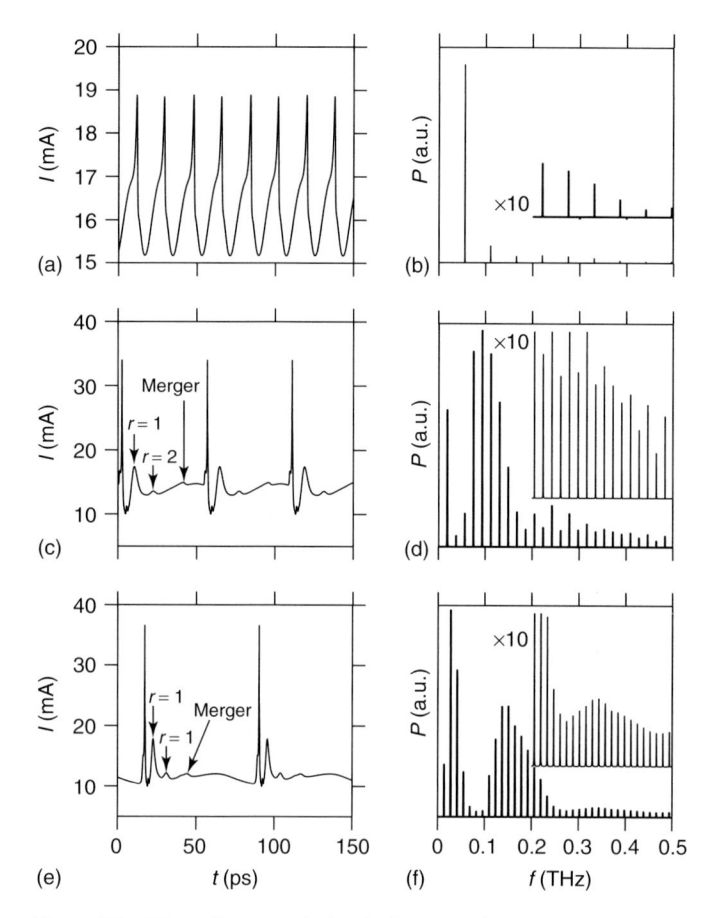

Figure 5.6 $I(t)$ oscillations calculated when $\theta = 40°$ for (a) $V = 540$ mV, (c) $V = 610$ mV, and (e) $V = 740$ mV. The corresponding Fourier power spectra of $I(t)$ are given in (b), (d), and (f) for $V = 540$ mV, 610 mV, and 740 mV, respectively, with a common vertical scale in arbitrary units. Inset within the Fourier power spectra are ×10 magnifications of the spectra for $f > 0.2$ THz.

To demonstrate the influence of the negative differential velocity region, the dashed curve in Figure 5.7a shows the (t, x) locus along which F is fixed at the value corresponding to the ET peak in the lower curve of Figure 5.2. As x passes beyond this locus, the electrons, because of the negative differential velocity, slow, thereby increasing the local values of both n_m and F_m. This further decreases v_d and, thereby, increases n_m, making the electrons accumulate in a charge domain (shown light gray in Figure 5.7a and as a large peak in the $n(x, t)$ surface in Figure 5.7b). Note that the condition for forming the charge domain requires electrons to be in the negative differential velocity regime throughout a sufficiently extended region of the superlattice and also a large enough injection current to "seed" a charge domain [66, 67, 74, 89].

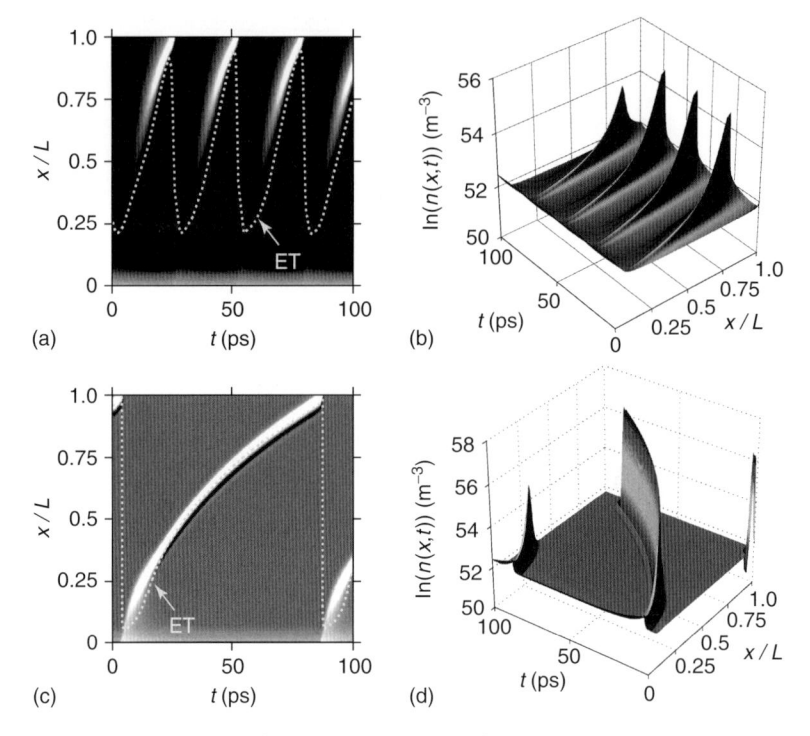

Figure 5.7 Gray scale (left-hand column) and surface plots (right-hand column) of $n_m(t, x)$ calculated for $\theta = 0°$. (a,b) are for $V = 290$ mV and (c,d) are for $V = 490$ mV. In (a) and (c), charge densities $> 10^{23}$ m^{-3} appear white. Dashed curves are loci of constant F values corresponding to the Esaki–Tsu (ET) drift velocity peak (see Figure 5.2).

As the domain propagates through the superlattice, it strengthens as the difference in drift velocity between the low- and high-field regions increases. Also, because of the resulting local increase in field in the high-field region and the corresponding reduction in drift velocity, the domain slows (both effects are clearly shown in Figure 5.7a,b).

When the domain approaches the collector ($x = L$), the high-field region immediately beyond the domain narrows. Consequently, to keep V constant, the electric field in that region increases, which requires the charge within the domain to also increase. When this domain reaches the collector contact, it therefore produces a sharp peak in $I(t)$ (Figure 5.4). Immediately, a new charge domain forms near the emitter, because of an increase in field at the left-hand edge of the superlattice, and the propagation process repeats, thereby producing self-sustained $I(t)$ oscillations [67, 90, 91].

For larger V, (see Figure 5.7c,d when $V = 490$ mV) similar domain dynamics occur. But now there is a higher mean field in the layers of the superlattice. Consequently, the Esaki–Tsu locus (dashed line in Figure 5.7c) is closer to the emitter compared to when $V = 290$ mV. As a result, the charge domain forms closer to the emitter and, as t increases, traverses the entire superlattice. Increasing V also increases the difference between the field values in the high- and low-field regions. This requires more charge to accumulate in the domain, which raises the amplitude of the $I(t)$ oscillations and, since v_d falls at higher fields, decreases the frequency of the oscillations (see $I(t)$ curve in Figure 5.4c).

When $V = 290\,\text{mV} \approx V_c$, increasing θ from $0°$ (Figure 5.7a) to $25°$ (Figure 5.8a) produces little qualitative change in the charge domain dynamics. This is because

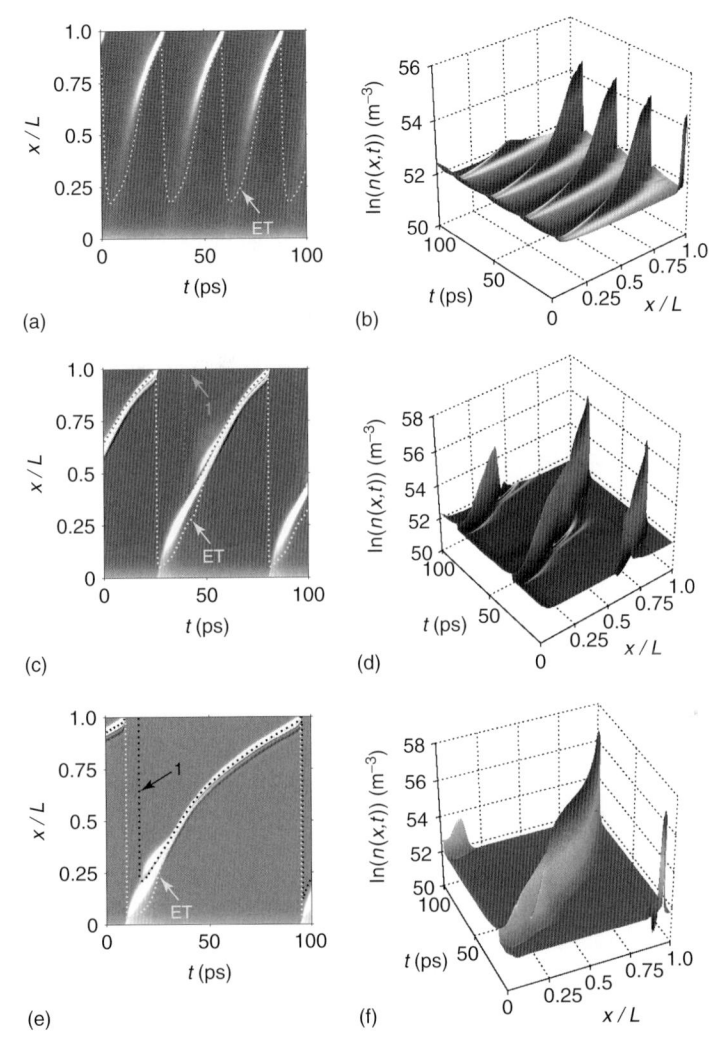

Figure 5.8 Gray scale (left-hand column) and surface plots (right-hand column) of $n_m(t,x)$ calculated for $\theta = 25°$. In (a,b) $V = 290$ mV, (c,d) $V = 490$ mV, and (e,f) $V = 690$ mV. Charge densities $> 10^{23}$ m^{-3} appear white in (a–c). Dashed, labeled, curves are loci of constant F values corresponding, respectively, to the Esaki-Tsu (ET) and $r = 1$ drift velocity peaks (see Figure 5.2).

V is low enough to ensure that $r < 1$ throughout the superlattice: a regime where the $v_d(r)$ curves for $\theta = 25°$ and $\theta = 0°$ have similar shapes (compare the bottom two curves in Figure 5.2).

This picture changes qualitatively when V and, consequently, some F_m values become high enough to ensure that, locally, $r \geq 1$. Figure 5.8c,d illustrates this for $V = 490$ mV and $\theta = 25°$. The dashed, labeled curves in Figure 5.8c show the (t, x) loci along which F equals the values corresponding, respectively, to the leftmost Esaki–Tsu and $r = 1$ v_d peaks in Figure 5.2. When $t \approx 25$ ps, negative differential velocity associated with the Esaki–Tsu peak creates a high-density charge domain for x just beyond the ET locus, as in the low-voltage case. However, when t increases to ≈ 50 ps, a second charge accumulation region appears above the $r = 1$ locus. This domain originates from the negative differential velocity region just beyond the $r = 1$ drift velocity peak. Its appearance produces an additional peak, labeled "$r = 1$," in the $I(t)$ trace in Figure 5.5c. The two charge domains merge when $t \approx 65$ ps, thereby inducing an additional peak in $I(t)$, labeled "Merger" in Figure 5.5c. After merger, the charge within the single domain is almost twice that for $\theta = 0°$. In addition, the presence of the large $r = 1$ v_d peak increases the mean electron drift velocity compared with $\theta = 0°$, thereby also raising the domain propagation speed. These two factors increase both the frequency and amplitude of the $I(t)$ oscillations (compare Figures 5.4c and 5.5c). The appearance of the extra "Merger" peak in $I(t)$ further strengthens the high-frequency harmonics in the Fourier power spectrum (compare Figures 5.4d and 5.5d).

Increasing V to 690 mV produces higher fields in the superlattice, which reduces the spatial distance between the (t, x) loci corresponding to the Esaki–Tsu and $r = 1$ drift velocity peaks (Figure 5.8e). Therefore, the charge domain associated with the $r = 1$ peak is generated closer to the emitter (when $x/L \approx 0.25$) and thus closer, both spatially and temporally, to the generation of the "Esaki–Tsu" domain. In addition, the charge domain grows much quicker than for lower voltages (compare Figure 5.8d and 5.8f). The combination of these factors causes the $r = 1$ and Merger peaks in the $I(t)$ trace for $V = 690$ mV (see arrowed peaks in Figure 5.5e) to occur at lower t and also be much sharper than when $V = 490$ mV (see arrowed peaks in Figure 5.5c). In turn, this makes the high-frequency components in the Fourier power spectra stronger for $V = 690$ mV than for 490 mV (compare Figure 5.5d and 5.5f).

Note that higher voltages do not necessarily achieve higher frequency Fourier components. In fact, as the system approaches $V \approx 1$ V, the effect of the $r = 1$ v_d peak on the charge density profile, and thus the $I(t)$ curve, is negated because the $r = 1$ domain occurs closer to the Esaki–Tsu domain.

Increasing θ to $40°$ further enriches the charge domain patterns (see Figure 5.9). Now the $r = 0.5, 1$, and 2 drift velocity resonances are stronger (see upper curve in Figure 5.2) and occur for smaller F, meaning that their effect on the domain dynamics is apparent even for V very close to V_c. Figure 5.9a is the charge density plot when $V = 540$ mV $\approx V_c$ for $\theta = 40°$. It reveals charge domains near the dashed, labeled, loci along which F coincides, with the Esaki–Tsu, $r = 0.5$, and $r = 1$ v_d peaks. It is interesting to note that the $r = 0.5$ locus effectively splits the

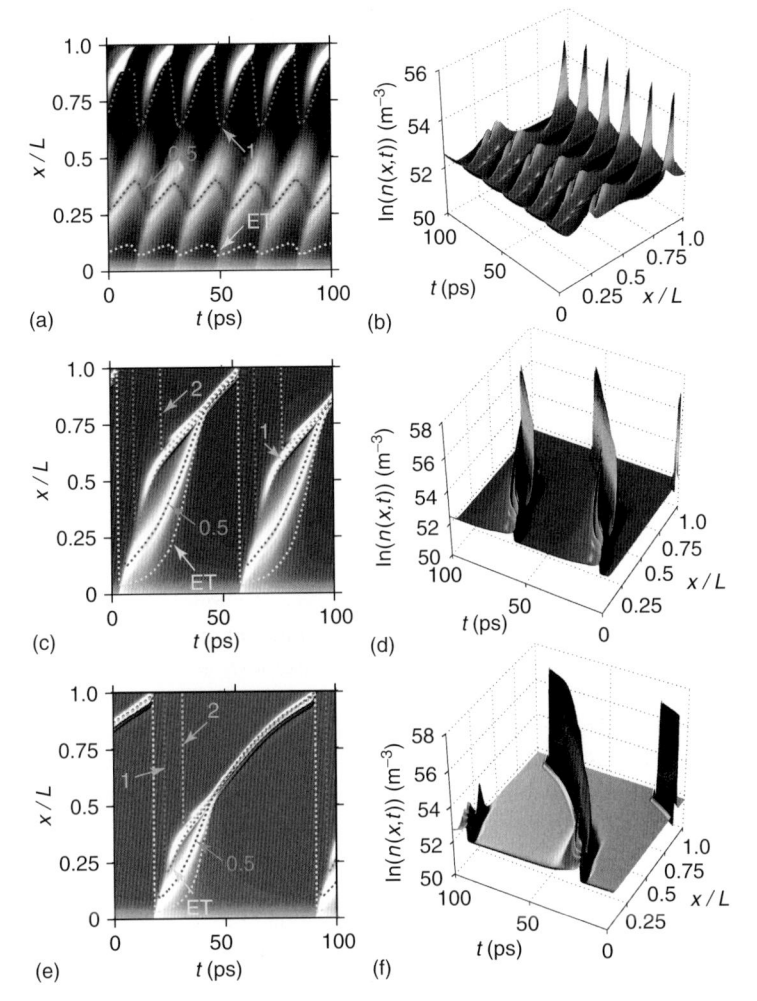

Figure 5.9 Gray scale (left-hand column) and surface plots (right-hand column) of $n_m(t, x)$ calculated for $\theta = 40°$. In (a,b) $V = 540$ mV, (c,d) $V = 610$ mV, and (e,f) $V = 740$ mV. Charge densities $> 10^{23}$ m^{-3} appear white in (a–c). Dashed, labeled, curves are loci of constant F values corresponding to the Esaki-Tsu (ET) and $r = 0.5$, $r = 1$, and $r = 2$ drift velocity peaks (see Figure 5.2).

domain induced by the Esaki–Tsu negative differential velocity (see $r = 0.5$ curve in Figure 5.9a), which clearly demonstrates the effect of the multiple drift velocity peaks on the charge domain dynamics. The shape of these domains is clearly shown in the three-dimensional plot in Figure 5.9b.

In the left-hand region of the superlattice, between the left-hand quantum well and the charge domain, the local field lies in the region of negative differential velocity, where $r < 1$ in the v_d curve. As x increases, the local field increases and enters the region of positive differential velocity. This stops the domain progressing further through the superlattice. For $x > 2L/3$, the local fields lie within the $r > 1$ region of negative differential velocity and a second domain is formed. The coexistence, and in-phase oscillation, of these multiple domains doubles both the amplitude and frequency of the $I(t)$ oscillations (see Figure 5.6a) in the $V \approx V_c$ regime compared to when $\theta = 0°$ (see Figure 5.4a) and $\theta = 25°$ (see Figure 5.5a).

When V reaches 610 mV (Figure 5.9c,d), a new domain associated with the $r = 2$ resonance (see locus) appears. When $t = 40$ ps, the domains with $r = 0.5, 1,$ and 2 resonant peaks all merge. At this voltage, the various domains produce multiple peaks in $I(t)$, as shown in Figure 5.6c where the labels mark peaks arising from the formation of the ET, $r = 0.5, 1,$ and 2 domains and their eventual merger, resulting in strong high-frequency components in the power spectrum (Figure 5.6d).

Increasing the voltage to 740 mV (Figure 5.9e,f), causes the domains to bunch. This "blurs" the shape of the $I(t)$ curve (Figure 5.6e), thereby reducing the high-frequency components in the power spectrum (Figure 5.6f).

5.2.5
Stability and Power of $I(t)$ Oscillations for $0° < \theta < 90°$

In this section, we consider how the stability, strength, and frequency of the current oscillations change over a wide range of θ and V.

Figure 5.10 is a gray scale map showing the variation of $I_a = I_{max} - I_{min}$ for a range of V up to 1V and for $0° < \theta < 90°$. The plot effectively maps the boundary between low V regimes (left-hand white area), where the current does not oscillate, and high V regimes where oscillations do occur. The scale gives a measure of the power of the current oscillations at high V. It is clear that for all θ there are no current oscillations ($I_a = 0$) for $V \lesssim 280$ mV, corresponding with the position of the Esaki–Tsu peak when $\theta = 0°$ (Figure 5.2). Generally, as θ increases, the critical voltage above which the system has current oscillations ($I_a > 0$), V_c, increases. At first glance, this may seem surprising since, as we increase θ, the resonances in $v_d(F)$ shift to lower F, suggesting that we might expect V_c to decrease with increasing θ. However, altering θ also changes the strength of the resonant features, which makes V_c depend in a complicated way on the shape of the $v_d(F)$ curve.

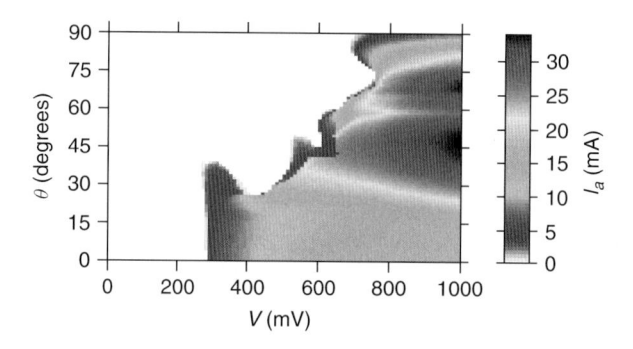

Figure 5.10 Gray scale map of $I_a = I_{max} - I_{min}$ (scale right) calculated for $\mathbf{B} = 15$ T.

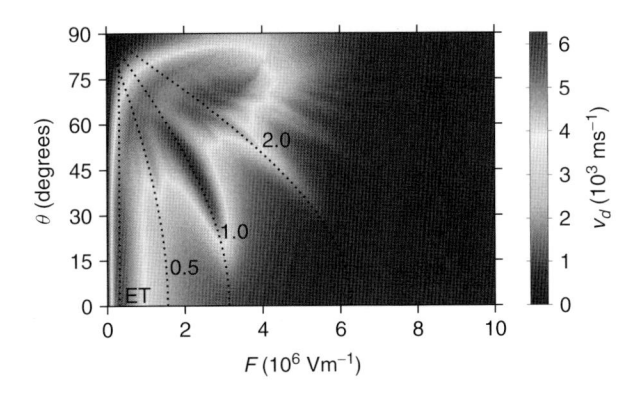

Figure 5.11 Gray scale map of $v_d(F,\theta)$ (scale right) calculated for $\mathbf{B} = 15\,\text{T}$.

To understand the cause of the complex variation of V_c with θ, it is useful to recap how the drift velocity varies with F and θ. Figure 5.11 is a gray scale map showing the variation of $v_d(F,\theta)$. The resonant features due to the Esaki–Tsu peak when $\omega_B \tau = 1$ (dotted line labeled ET), and the features due to the Bloch and cyclotron resonances $r = 0.5$ (dotted line labeled $r = 0.5$), 1 (labeled $r = 1.0$), and 2 (labeled $r = 2.0$) peaks are clearly visible. The resonant features become more pronounced as θ is increased to $45°$ because the coupling between the cyclotron and Bloch oscillations strengthens. Increasing θ beyond $45°$ weakens the resonant peaks as the cyclotron and Bloch oscillations decouple. In addition, the resonances all occur at lower field values, resulting in "bunching" of the drift velocity peaks, clear for $60° < \theta < 80°$. When $\theta > 80°$, the Esaki–Tsu peak again dominates as the resonance effects diminish. Now, however, we find that the Esaki–Tsu peak occurs at a higher F value as θ approaches $90°$. This shift occurs because the magnetic field deflects the electron away from the x-axis toward the z-axis, thus reducing the drift velocity and increasing the electric field required to obtain the maximum drift velocity [92].

The form of the $v_d(F,\theta)$ map shown in Figure 5.11 enables us to explain the variation of V_c with θ. For values of $\theta \lesssim 27°$, Figure 5.11 reveals that the Esaki–Tsu peak is dominant in v_d and, consequently, V_c is constant (Figure 5.10), with current oscillations being induced by the region of negative differential velocity immediately following the Esaki–Tsu peak. When $40° \gtrsim \theta \gtrsim 60°$, the $r = 1$ peak is dominant in Figure 5.11, and, correspondingly, V_c jumps to a higher voltage $\approx 600\,\text{mV}$ corresponding to the region of high negative differential velocity following the $r = 1$ peak. For $60° \lesssim \theta \lesssim 75°$, Figure 5.11 shows that the region of negative differential velocity occurs at increasingly high F as θ increases and, correspondingly, V_c also increases. For $75° \gtrsim \theta \gtrsim 85°$, the amplitude of the higher order resonant peaks in the drift velocity diminish and, correspondingly, V_c decreases. When $\theta \gtrsim 85°$, the Esaki–Tsu peak dominates and, since at higher θ it occurs at higher F, we find a slight increase in V_c.

The surface plot of I_a in Figure 5.10 also gives an estimate of the power of the current oscillations. When $\theta \lesssim 20°$, we find that the current oscillations are relatively weak, although increasing the voltage from V_c to 1 V raises the amplitude of the oscillations by $\approx$15 mA (from $\approx$5 mA to 18 mA when $\theta = 0°$). The increase in the oscillatory amplitude can again be attributed to the shape of the drift velocity curve. Increasing the voltage across the device enhances the field difference between the low- and high-field regions in the superlattice and, hence, the difference in the drift velocity between the high- and low-field regions. This allows more charge to be injected into the charge domain, thereby enhancing the current oscillations. At higher voltages, the low- and high-field regions (corresponding to high- and low-velocity regions) are both larger, which allows the domain to form and grow more quickly and thereby increase the amount of charge in the domain.

As θ increases so that the $r = 1$ resonance strengthens in $v_d(F)$, there is a gradual increase in the amplitude of the current oscillations until $\theta = 45°$, since increasing the peak drift velocity in $v_d(F)$ also increases the associated negative differential velocity (Figure 5.11). The enhanced velocity injects more charge into the domains and the larger negative differential velocity allows the domain to form quicker and thus "collect" more charge. In addition, there are new domain filaments formed by the extra features in the drift velocity curve, which carry more charge through the superlattice. These effects combine to increase the amplitude of the current oscillations.

Increasing θ beyond 45° leads to the appearance of a second region of enhanced current oscillations in Figure 5.10 when $\theta \approx 70°$, where a large number of resonant domains are induced in the superlattice. Further increasing θ to 90° generally decreases I_a until $\theta = 90°$, when I_a has qualitatively the same form as when $\theta = 0°$.

5.2.6
Frequency of I(t) for 0° < θ < 90°

In Sections 5.2.3 and 5.2.4, it was shown that by inducing extra charge domain filaments it was possible to significantly increase the power and frequency of the current oscillations, especially when $\theta = 40°$ and $V = 610$ mV, where the dominant component in the frequency spectrum was the fifth harmonic with a frequency of 92 GHz. To explore this further, Figure 5.12 shows a gray scale map

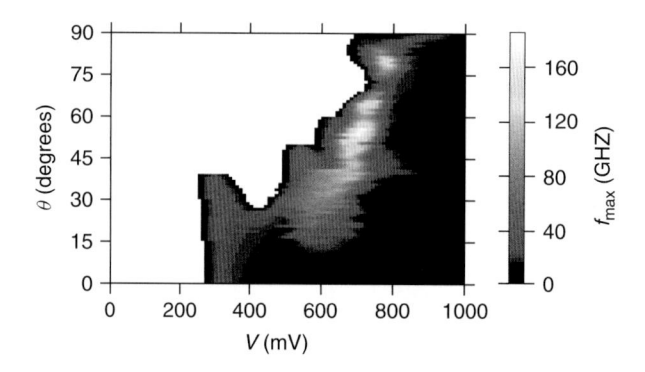

Figure 5.12 Gray scale map of $f_{max}(V, \theta)$ (scale right) calculated for $B = 15$ T.

of the dominant harmonic in the $I(t)$ frequency spectrum, f_{max}, versus θ and V. The figure reveals a startling increase in the characteristic frequency of the system when a tilted magnetic field is applied to the superlattice. Generally for low $\theta \lesssim 5°$ $f_{max} \approx 25$ GHz when $V \approx V_c$, but decreases with increasing $V > V_c$. This is because increasing V raises the electric field in the superlattice, thereby reducing the drift velocity and, hence, the frequency of the oscillations in $I(t)$ (Figure 5.4).

However, as soon as a resonant peak appears in the drift velocity curve when $\theta \approx 15°$ (Figure 5.11), there is an immediate increase in the frequency of the current oscillations to approximately 40 GHz (Figure 5.12) as extra charge domains are induced in the superlattice. Again, increasing the voltage causes a decrease in the frequency of the $I(t)$ oscillations because the fields in the superlattice increase, thus diminishing the effect of the resonant features. The optimum voltage for the maximum frequency output occurs when, throughout the superlattice, the electric fields encompass all resonances in $v_d(F)$ so that all the charge domains contribute to the features in the $I(t)$ curve.

As θ increases, the power of the higher harmonics in the frequency spectra increases because of the creation of extra charge domains until, at $\theta \approx 57°$, the frequency of the highest powered peak is, astonishingly, at ≈ 180 GHz: an order of magnitude increase in the frequency of the dominant peak observed when $\theta = 0°$.

Recent experimental results [93, 94] show successfully that by using backward wave oscillators, it is possible to take advantage of the higher harmonics of frequency generators such as superlattices. Therefore, to quantify the overall power of the high-frequency components, we calculate $P_{int} = \langle P(f) \rangle$, where $\langle . \rangle$ denotes integration over $f > 0.2$ THz. Figure 5.13 shows a gray scale map of P_{int} in the $V - \theta$ plane. For $V < V_c$ (left of the dashed curve in Figure 5.13), $P_{int} = 0$ because there are no charge domain oscillations (see also Figure 5.10).

As for the case of f_{max}, P_{int} generally increases with increasing θ as extra charge domains are formed in the superlattice. We find there is a maximum in P_{int} (dark gray region to right of dashed line) when $V \approx 800$ mV and $\theta \approx 70°$. In this regime, P_{int} is an order of magnitude higher than for $\theta = 0°$ because of the formation of multiple propagating charge domains.

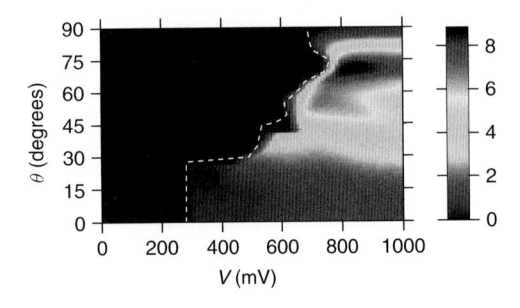

Figure 5.13 Gray scale map of $P_{int}(V, \theta)$ (scale right in arbitrary units) calculated for $B = 15$ T. Dashed curve: variation of V_c with θ.

5.3
Conclusion

In this chapter, we considered the effect of a tilted magnetic field on the dynamics and structure of charge domains in biased superlattices and formulated a modified drift-diffusion model. Simulations of the collective electron dynamics revealed that the extra negative differential drift velocity regions, caused by the resonant peaks, induce multiple charge domains. These extra domains increase both the amplitude and frequency of the current oscillations – both effects that should be experimentally observable.

Recently there has been some work on the demonstration of stable THz gain in superlattices using a modulated bias [11] and also with a tilted magnetic field applied [95]. Thus far, these studies have focused on the homogeneous field case. It would be interesting to study the gain of superlattices in the nonhomogeneous field case to confirm the existence of the high-frequency components in the current spectra. Also, our results suggest that it is possible to control the form and *collective* dynamics of charge domains in superlattices by using *single-electron* miniband transport to tailor $v_d(F)$. Multiple v_d maxima can also be created in other ways, for example, by applying an AC electric field [96], which could be studied in the context of this model.

Acknowledgment

The authors are grateful to Prof. Eckehard Schöll for useful discussions.

References

1. Fromhold, T.M., Krokhin, A.A., Tench, C.R., Bujkiewicz, S., Wilkinson, P.B., Sheard, F.W., and Eaves, L. (2001) Effects of stochastic webs on chaotic electron transport in semiconductor superlattices. *Phys. Rev. Lett.*, **87** (4), 046803. DOI: 10.1103/PhysRevLett.87.046803.
2. Kuraguchi, M., Ohmichi, E., Osada, T., and Shiraki, Y. (2002) Relationship between stark-cyclotron resonance and angular dependent magnetoresistance oscillations. *Physica E*, **12**, 264.
3. Fromhold, T.M., Patanè, A., Bujkievicz, S., Wilkinson, P.B., Fowler, D., Sherwood, D., Stapleton, S.P., Krokhin, A.A., Eaves, L., Henini, M., Sankeshwar, N.S., and Sheard, F.W. (2004) Chaotic electron diffusion through stochastic web enhances current flow in supperlattices. *Nature*, **428**, 726.
4. Stapleton, S.P., Bujkiewicz, S., Fromhold, T.M., Wilkinson, P.B., Patane, A., Eaves, L., Krokhin, A.A., Henini, M., Sankeshwar, N.S., and Sheard, F.W. (2004) Use of stochastic web patterns to control electron transport in semiconductor superlattices. *Physica D*, **199** (1–2), 166–172, DOI: 10.1016/j.physd.2004.08.011. (Workshop on Trends in Pattern Formation, Dresden, Aug 25-Sep 19, 2003).
5. Hardwick, D.P.A., Naylor, S.L., Bujkiewicz, S., Fromhold, T.M., Fowler, D., Patané, A., Eaves, L., Krokhin, A.A., Wilkinson, P.B., Henini, M., and Sheard, F.W. (2006) Effect of inter-miniband tunneling on current resonances due to the formation of

stochastic conduction networks in superlattices. *Physica E*, **32**, 285.
6. Kosevich, Y.A., Hummel, A.B., Roskos, H.G., and Köhler, K. (2006) Ultrafast fiske effect in semiconductor superlattices. *Phys. Rev. Lett.*, **96** (13), 137403. DOI: 10.1103/PhysRevLett.96.137403. http://link.aps.org/abstract/PRL/v96/e137403.
7. Fowler, D., Hardwick, D.P.A., Patané, A., Greenaway, M.T., Balanov, A.G., Fromhold, T.M., Eaves, L., Henini, M., Kozlova, N., Freudenberger, J., and Mori, N. (2007) Magnetic-field-induced miniband conduction in semiconductor superlattices. *Phys. Rev. B*, **76** (24), 245303. DOI: 10.1103/PhysRevB.76.245303. http://link.aps.org/abstract/PRB/v76/e245303.
8. Balanov, A.G., Fowler, D., Patané, A., Eaves, L., and Fromhold, T.M. (2008) Bifurcations and chaos in semiconductor superlattices with a tilted magnetic field. *Phys. Rev. E*, **77** (2), 026209. DOI: 10.1103/PhysRevE.77.026209. http://link.aps.org/abstract/PRE/v77/e026209.
9. Soskin, S.M., McClintock, P.V.E., Fromhold, T.M., Khovanov, I.A., and Mannella, R. (2010) Stochastic webs and quantum transport in superlattices: an introductory review. *Contemp. Phys.*, **51** (3), 233–248. DOI: 10.1080/00107510903539179.
10. Greenaway, M.T., Balanov, A.G., Schöll, E., and Fromhold, T.M. (2009) Controlling and enhancing terahertz collective electron dynamics in superlattices by chaos-assisted miniband transport. *Phys. Rev. B*, **80** (20), 205318. DOI: 10.1103/PhysRevB.80.205318. http://link.aps.org/abstract/PRB/v80/e205318.
11. Hyart, T., Alexeeva, N.V., Mattas, J., and Alekseev, K.N. (2009) Terahertz bloch oscillator with a modulated bias. *Phys. Rev. Lett.*, **102** (14), 140405. DOI: 10.1103/PhysRevLett.102.140405. http://link.aps.org/abstract/PRL/v102/e140405.
12. Hardwick, D.P.A. (2007) Quantum and semiclassical calculations of electron transport through a stochastic system. PhD thesis. School of Physics and Astronomy, University of Nottingham.
13. Stöckmann, H.J. (1999) *Quantum Chaos: An Introduction*, Cambridge University Press, Cambridge.
14. Nakamura, K. and Harayama, T. (2003) *Quantum Chaos and Quantum Dots*, Oxford University Press, Oxford.
15. Marcus, C., Rimberg, A., Westervelt, R., Hopkins, P., and Gossard, A. (1992) *Phys. Rev. Lett.*, **69**, 506.
16. Chang, A., Baranger, H., Pfeiffer, L., and West, K. (1994) *Phys. Rev. Lett.*, **73**, 2111.
17. Folk, J., Patel, S., Godijn, S., Huibers, A., Cronenwett, S., Marcus, C., Campman, K., and Gossard, A. (1996) *Phys. Rev. Lett.*, **76**
18. Ketzmerick, R. (1996) *Phys. Rev. B*, **54**, 10841.
19. Sachrajda, A., Ketzmerick, R., Gould, C., Feng, Y., Kelly, P., Delage, A., and Wasilewski, Z. (1998) *Phys. Rev. Lett.*, **80**, 1948.
20. Bird, J., Akis, R., Ferry, D., Vasileska, D., Cooper, J., Aoyagi, Y., and Sugano, T. (1999) *Phys. Rev. Lett.*, **82**, 4691.
21. Micolich, A., et al. (2001) *Phys. Rev. Lett.*, **87**, 036802.
22. Marlow, C., et al. (2006) *Phys. Rev. B*, **19**, 195318.
23. Weiss, D., Roukes, M., Menschig, A., Grambow, P., von Klitzing, K., and Weimann, G. (1991) *Phys. Rev. Lett.*, **66**, 2790.
24. Fleischmann, R., Geisel, T., and Ketzmerick, R. (1992) *Phys. Rev. Lett.*, **68**, 1367.
25. Weiss, D., Richter, K., Menschig, A., Bergmann, R., Schweizer, H., von Klitzing, K., and Weimann, G. (1993) *Phys. Rev. Lett.*, **70**, 4118.
26. Kastrup, J., Grahn, H., Ploog, K., Prengel, F., Wacker, A., and Schöll, E. (1994) *Appl. Phys. Lett.*, **65**, 1808.
27. Zhang, Y., Kastrup, J., Klann, R., Ploog, K.H., and Grahn, H.T. (1996) Synchronization and chaos induced by resonant tunneling in GaAs/AlAs superlattices. *Phys. Rev. Lett.*, **77** (14), 3001–3004. DOI: 10.1103/PhysRevLett.77.3001.
28. Alekseev, K.N., Berman, G.P., Campbell, D.K., Cannon, E.H., and Cargo, M.C.

29. Luo, K., Grahn, H., Ploog, K., and Bonilla, L. (1998) *Phys. Rev. Lett.*, **81**, 1290.
30. Fromhold, T., Eaves, L., Sheard, F., Leadbeater, M., Foster, T., and Main, P. (1994) *Phys. Rev. Lett.*, **72**, 2608.
31. Fromhold, T., Wilkinson, P., Sheard, F., Eaves, L., Miao, J., and Edwards, G. (1995) *Phys. Rev. Lett.*, **75**, 1142.
32. Fromhold, T., Fogarty, A., Eaves, L., Sheard, F., Henini, M., Foster, T., Main, P., and Hill, G. (1995) *Phys. Rev. B*, **51**, 18029.
33. Shepelyansky, D. and Stone, A. (1995) *Phys. Rev. Lett.*, **74**, 2098.
34. Müller, G., Boebinger, G., Mathur, H., Pfeiffer, L., and West, K. (1995) *Phys. Rev. Lett.*, **75**, 2875.
35. Wilkinson, P., Fromhold, T., Eaves, L., Sheard, F., Miura, N., and Takamasu, T. (1996) *Nature*, **380**, 608.
36. Monteiro, T. and Dando, P. (1996) *Phys. Rev. E*, **53**, 3369.
37. Fromhold, T., Wilkinson, P., Sheard, F., and Eaves, L. (1997) *Phys. Rev. Lett.*, **78**, 2865.
38. Fromhold, T., Wilkinson, P., Eaves, L., Sheard, F., Main, P., Henini, M., Carter, M., Miura, N., and Takamasu, T. (1997) *Chaos Solitons Fractals*, **8**, 1381–1411.
39. Monteiro, T., Delande, D., Fisher, A., and Boebinger, G. (1997) *Phys. Rev. B*, **56**, 3913.
40. Monteiro, T., Delande, D., and Connerade, J. (1997) *Nature*, **387**, 863.
41. Narimanov, E., Stone, A., and Boebinger, G. (1998) *Phys. Rev. Lett.*, **80**, 4024.
42. Narimanov, E. and Stone, A. (1998) *Phys. Rev. B*, **57**, 9807.
43. Narimanov, E. and Stone, A. (1998) *Phys. Rev. Lett.*, **80**, 49.
44. Saraga, D. and Monteiro, T. (1998) *Phys. Rev. Lett.*, **81**, 5796.
45. Saraga, D., Monteiro, T., and Rouben, D. (1998) *Phys. Rev. E*, **58**, 2701.
46. Saraga, D. and Monteiro, T. (1998) *Phys. Rev. E*, **57**, 5252.
47. Bogomolny, E. and Rouben, D. (1998) *Europhys. Lett.*, **43**, 111.
48. Bogomolny, E. and Rouben, D. (1999) *Eur. Phys. J. B*, **9**, 695.
49. Fromhold, T., Wilkinson, P., Hayden, R., Eaves, L., Sheard, F., Miura, N., and Henini, M. (2002) *Phys. Rev. B*, **65**, 155312.
50. Sagdeev, R.Z., Usikov, D.A., and Zaslavsky, G.M. (1988) *Nonlinear Physics*, Harwood Academic Publishers.
51. Vasiliev, A.A., Zaslavsky, G.M., Natenzon, M.Y., Neishtadt, A.I., Petrovichev, B.A., Sagdeev, R.Z., and Chernikov, A.A. (1988) Attractors and stochastic attractors of motion in a magnetic-field. *Zh. Eksp. Teor. Fiz.*, **94** (10), 170–187.
52. Beloshapkin, V.V., Chernikov, A.A., Natenzon, M.Y., Petrovichev, B.A., Sagdeev, R.Z., and Zaslavsky, G.M. (1989) Chaotic streamlines in pre-turbulent states. *Nature*, **337** (6203), 133–137.
53. Zaslavsky, G.M. (1991) *Weak Chaos and Quasi-Regular Patterns*, Cambridge University Press.
54. Shlesinger, M.F., Zaslavsky, G.M., and Klafter, J. (1993) Strange kinetics. *Nature*, **363**, 31.
55. Zaslavsky, G. (2004) Hamiltonian chaos and fractional dynamics.
56. Luo, A. (2004) *Appl. Mech. Rev.*, **57**, 161.
57. Kamenev, D.I. and Berman, G.P. (2000) *Quantum Chaos: a Harmonic Oscillator in Monochromatic Wave*, Rinton, Princeton, NJ.
58. Karney, C.F.F. and Bers, A. (1977) Stochastic ion heating by a perpendicularly propagating electrostatic wave. *Phys. Rev. Lett.*, **39** (9), 550–554.
59. Chia, P.K., Schmitz, L., and Conn, R.W. (1996) Stochastic ion behavior in subharmonic and superharmonic electrostatic waves. *Phys. Plasmas*, **3** (5), 1545–1568.
60. Gardiner, S.A., Cirac, J.I., and Zoller, P. (1997) Quantum chaos in an ion trap: the delta-kicked harmonic oscillator. *Phys. Rev. Lett.*, **79** (24), 4790–4793. DOI: 10.1103/PhysRevLett.79.4790.
61. Demikhovskii, V.Y., Kamenev, D.I., and Luna-Acosta, G.A. (1999) Quantum weak chaos in a degenerate system. *Phys. Rev. E*, **59** (1), 294–302. DOI: 10.1103/PhysRevE.59.294.

62. Demikhovskii, V.Y., Izrailev, F.M., and Malyshev, A.I. (2002) Manifestation of arnol'd diffusion in quantum systems. *Phys. Rev. Lett.*, **88** (15), 154101. DOI: 10.1103/PhysRevLett.88.154101.
63. Hensinger, W.K., Haffner, H., Browaeys, A., Heckenberg, N.R., Helmerson, K., Mckenzie, C., Milburn, G.J., Phillips, W.D., Rolston, S.L., Rubinsztein-Dunlop, H., and Upcroft, B. (2001) Dynamical tunnelling of ultracold atoms. *Nature*, **412** (6842), 52–55. http://dx.doi.org/10.1038/412052a0.
64. Steck, D.A., Oskay, W.H., and Raizen, M.G. (2001) Observation of chaos-assisted tunneling between islands of stability. *Science*, **293** (5528), 274–278. DOI: 10.1126/science.1061569. http://dx.doi.org/10.1126/science.1061569.
65. Scott, R.G., Bujkiewicz, S., Fromhold, T.M., Wilkinson, P.B., and Sheard, F.W. (2002) Effects of chaotic energy-band transport on the quantized states of ultracold sodium atoms in an optical lattice with a tilted harmonic trap. *Phys. Rev. A*, **66** (2), 023 407. DOI: 10.1103/PhysRevA.66.023407.
66. Hizanidis, J., Balanov, A., Amann, A., and Schöll, E. (2006) Noise-induced front motion: Signature of a global bifurcation. *Phys. Rev. Lett.*, **96** (24), 244104. DOI: 10.1103/PhysRevLett.96.244104. URL http://link.aps.org/abstract/PRL/v96/e244104.
67. Wacker, A. (2002) Semiconductor superlattices: a model system for nonlinear transport. *Phys. Rep.*, **357**, 1.
68. Soskin, S.M., Khovanov, I.A., Mannella, R., and McClintock, P.V.E. (2009) *Noise and Fluctuations: 20th International Conference on Noise and Fluctuations (ICNF-2009)*, vol. 1129 AIP, Melville, New York, pp. 17–20.
69. Esaki, L. and Tsu, R. (1970) Superlattice and negative differential conductivity in semiconductors. *IBM J. Res. Dev.*, **14**, 61.
70. Shik, A. (1975) *Sov. Phys. Semicond.*, **8**, 1195.
71. Ignatov, A.A., Dodin, E.P., and Shashkin, I.V. (1991) Transient response theory of semiconductor superlattices: connection with bloch oscillations. *Mod. Phys. Lett. B*, **5** (16), 1087–1094.
72. Canali, L., Lazzarino, M., Sorba, L., and Beltram, F. (1996) Stark-cyclotron resonance in a semiconductor superlattice. *Phys. Rev. Lett.*, **76** (19), 3618–3621. DOI: 10.1103/PhysRevLett.76.3618.
73. Schomburg, E., Grenzer, J., Hofbeck, K., Blomeier, T., Winnerl, S., Brandl, S., Ignatov, A., Renk, K., Pavel'ev, D., Koschurinov, Y., Ustinov, V., Zhukov, A., Kovsch, A., Ivanov, S., and Kop'ev, P. (1998) Millimeter wave generation with a quasi planar superlattice electronic device. *Solid-State Electron.*, **42**, 1495.
74. Schöll, E. (2001) *Nonlinear Spatio-temporal Dynamics and Chaos in Semiconductors*, Cambridge University Press.
75. Amann, A., Schlesner, J., Wacker, A., and Schöll, E. (2002) Chaotic front dynamics in semiconductor superlattices. *Phys. Rev. B*, **65** (19), 193 313. DOI: 10.1103/PhysRevB.65.193313.
76. Alekseev, K. and Kusmartsev, F. (2002) Pendulum limit, chaos and phase-locking in the dynamics of ac-driven semiconductor superlattice. *Phys. Lett. A*, **305**, 281.
77. Patané, A., Sherwood, D., Eaves, L., Fromhold, T.M., Henini, M., Main, P.C., and Hill, G. (2002) *Appl. Phys. Lett.*, **81**, 661.
78. Bonilla, L.L. (2002) Theory of nonlinear charge transport, wave propagation, and self-oscillations in semiconductor superlattices. *J. Phys.: Condens. Matter*, **14**, R341.
79. Shimada, Y., Hirakawa, K., Odnoblioudov, M., and Chao, K.A. (2003) Terahertz conductivity and possible bloch gain in semiconductor superlattices. *Phys. Rev. Lett.*, **90** (4), 046 806. DOI: 10.1103/PhysRevLett.90.046806.
80. Savvidis, P.G., Kolasa, B., Lee, G., and Allen, S.J. (2004) Resonant crossover of terahertz loss to the gain of a bloch oscillating *inas/alsb* superlattice. *Phys. Rev. Lett.*, **92** (19), 196 802, DOI: 10.1103/PhysRevLett.92.196802.
81. Raspopin, A.S., Zharov, A.A., and Cui, H.L. (2005) Spectrum of electromagnetic

excitations in a dc-biased semiconductor superlattice. *J. Appl. Phys.*, **98** (10), 103517. DOI: 10.1063/1.2135413. http://link.aip.org/link/?JAP/98/103517/1.

82. Bonilla, L.L. and Grahn, H.T. (2005) Non-linear dynamics of semiconductor superlattices. *Rep. Prog. Phys.*, **68**, 577.

83. Greenaway, M.T., Balanov, A.G., Fowler, D., Kent, A.J., and Fromhold, T.M. (2010) Using acoustic waves to induce high-frequency current oscillations in superlattices. *Phys. Rev. B*, **81** (23), 235–313. DOI: 10.1103/PhysRevB.81.235313.

84. Shaw, M.P., Mitin, V.V., Schöll, E., and Grubin, H.L. (1992) *The Physics of Instabilities in Solid State Electron Devices*, Plenum Press, New York.

85. Prengel, F., Wacker, A., and Schöll, E. (1994) Simple model for multistability and domain formation in semiconductor superlattices. *Phys. Rev. B*, **50** (3), 1705–1712. DOI: 10.1103/PhysRevB.50.1705.

86. Press, W.H., Teukolsky, S.A., Vetterling, W.T., and Flannery, B.P. (1996) *Numerical recipes in C*, 2nd edn, Cambridge University Press.

87. Schwarz, G. and Schöll, E. (1996) Field Domains in Semiconductor Superlattices. *Phys. Stat. Sol. B*, **194**, 351.

88. Tsu, R. and Döhler, G. (1975) Hopping conduction in a "superlattice". *Phys. Rev. B*, **12**, 680.

89. Schöll, E. (1987) *Nonequilibrium Phase Transitions in Semiconductors*, Springer, Berlin.

90. Kastrup, J., Klann, R., Grahn, H.T., Ploog, K., Bonilla, L.L., Galán, J., Kindelan, M., Moscoso, M., and Merlin, R. (1995) Self-oscillations of domains in doped GaAs-AlAs superlattices. *Phys. Rev. B*, **52** (19), 13-761–13-764. DOI: 10.1103/PhysRevB.52.13761.

91. Amann, A., Peters, K., Parlitz, U., Wacker, A., and Schöll, E. (2003) Hybrid model for chaotic front dynamics: From semiconductors to water tanks. *Phys. Rev. Lett.*, **91** (6), 066 601. DOI: 10.1103/PhysRevLett.91.066601.

92. David, S. (2003) Effect of stochastic webs on electron transport in semiconductor superlattices. PhD thesis. University of Nottingham.

93. Endres, C.P., Müller, H.S.P., Brünken, S., Paveliev, D.G., Giesen, T.F., Schlemmer, S., and Lewen, F. (2006) High resolution rotation-inversion spectroscopy on doubly deuterated ammonia, ND2H, up to 2.6 THz. *J. Mol. Struct.*, **795** (1–3), 242–255. DOI: 10.1016/j.molstruc.2006.03.035. http://www.sciencedirect.com/science/article/B6TGS-4JWFMTK-2/.

94. Endres, C.P., Lewen, F., Giesen, T.F., Schlemmer, S., Paveliev, D.G., Koschurinov, Y.I., Ustinov, V.M., and Zhucov, A.E. (2007) Application of superlattice multipliers for high-resolution terahertz spectroscopy. *Rev. Sci. Instrum.*, **78** (4), 043106. DOI: 10.1063/1.2722401. http://link.aip.org/link/?RSI/78/043106/1.

95. Hyart, T., Mattas, J., and Alekseev, K.N. (2009) Model of the influence of an external magnetic field on the gain of terahertz radiation from semiconductor superlattices. *Phys. Rev. Lett.*, **103** (11), 117401. DOI: 10.1103/PhysRevLett.103.117401. http://link.aps.org/abstract/PRL/v103/e117401.

96. Hyart, T., Shorokhov, A.V., and Alekseev, K.N. (2007) Theory of parametric amplification in superlattices. *Phys. Rev. Lett.*, **98** (22), 220404. DOI: 10.1103/PhysRevLett.98.220404. http://link.aps.org/abstract/PRL/v98/e220404.

**Part II
Coupled Laser Device**

Nonlinear Laser Dynamics: From Quantum Dots to Cryptography, First Edition. Edited by Kathy Lüdge.
© 2012 Wiley-VCH Verlag GmbH & Co. KGaA. Published 2012 by Wiley-VCH Verlag GmbH & Co. KGaA.

6
Quantum Dot Laser Tolerance to Optical Feedback

Christian Otto, Kathy Lüdge, Evgeniy Viktorov, and Thomas Erneux

6.1
Introduction

In optical fiber networks, the semiconductor laser source may be perturbed by unavoidable optical feedback from fiber pigtails or fiber connectors unless expensive optical isolators are used. Analytical expressions for the stable operation of laser diodes are highly desirable and have been a constant preoccupation of researchers in the field [1]. Mork et al. [2] investigated the Lang and Kobayashi equations describing a quantum well (QW) semiconductor laser subject to delayed optical feedback and derived an approximation of the stability boundary in terms of the feedback parameter k. $k^2 \equiv P_r/P_i$ is defined as the ratio of reflected power P_r and emitted power P_i. Mathematically, this stability boundary corresponds to the lowest possible value of the first Hopf bifurcation of an external cavity mode. The external cavity modes (ECMs) are the basic solutions of a laser subject to optical feedback from a distant mirror. In the weak feedback limit, there exists only one mode that is determined by the feedback phase $C = \omega_0 \tau_{ec}$, in first approximation (ω_0 is the angular frequency of the solitary laser and τ_{ec} is the round-trip time). The stability condition derived by Mork et al. [2] is given by

$$k < k_c \equiv \frac{\Gamma^{QW}}{\sqrt{1+\alpha^2}} \qquad (6.1)$$

where α is the linewidth enhancement factor and Γ^{QW} is defined as the damping rate of the relaxation oscillations (ROs) multiplied by the photon lifetime τ_p, so that k_c is dimensionless. Because of a possible confusion with a different definition of the damping rate used by Mork et al. [2], we derive the expressions of the ROs frequency ω^{QW} and damping rate Γ^{QW} from their rate equations in Appendix A. As noted by Mork et al. [2], Eq. (6.1) was previously suggested by Helms and Petermann [3] as a simple analytical criterion for tolerance with respect to optical feedback. Helms and Petermann [3] evaluate the validity of Eq. (6.1) by analyzing numerically the stability of the minimum linewidth ECM. They noted that this approximation gives a good description of the critical feedback level as long as the linewidth enhancement factor α is significantly larger than unity. They then

---

Nonlinear Laser Dynamics: From Quantum Dots to Cryptography, First Edition. Edited by Kathy Lüdge.
© 2012 Wiley-VCH Verlag GmbH & Co. KGaA. Published 2012 by Wiley-VCH Verlag GmbH & Co. KGaA.

proposed an empirical law given by

$$k_c = \Gamma^{QW} \frac{\sqrt{1+\alpha^2}}{\alpha^2}. \tag{6.2}$$

Both Eqs. (6.1) and (6.2) are used in current experimental studies of quantum dot (QD) lasers subject to optical feedback. Specifically, Eq. (6.1) is used in [4], and Eq. (6.2) is used in [5–7]. Note that the minimum linewidth mode appears as the first ECM in the weak feedback limit for the feedback phase $C = -\arctan(\alpha)$. For this mode, the minimum value of the feedback strength of the first Hopf bifurcation, which marks the critical feedback strength below that the laser is guaranteed to be stable, is given by Levine et al. [8]

$$k_c = \Gamma^{QW} \frac{\sqrt{1+\alpha^2}}{\alpha^2 - 1}. \tag{6.3}$$

The approximation of the first Hopf bifurcation in terms of an arbitrary phase and thus for arbitrary ECMs is derived in [9]. Substituting the expression for the frequency of the minimum linewidth mode $\Delta \simeq C = -\arctan(\alpha)$ then leads to Eq. (6.3). The denominator in Eq. (6.3) is different from the denominator of Eq. (6.2), which explains the numerically observed singularity as $\alpha \to 1^+$ [3]. The expression (6.1) follows from analytic considerations of the first Hopf bifurcation at a feedback phase $C = \pi + \arctan(\alpha)$, which provides the lowest possible value of k_c. The inequality in Eq. (6.1) is based on a series of approximations ($k \ll 1$, $\omega^{QW} \tau_{ec}/\tau_p \gg 1$, $\alpha > 1$), which may or may not be appropriate. Asymptotic techniques were later used to determine systematic approximations for a variety of cases (pump parameter close to threshold, short external cavity) [9]. In this approach, all small or large dimensionless parameters appearing in the rate equations are scaled with respect to a unique parameter γ, defined as the ratio of the photon and carrier lifetimes ($\gamma \equiv \tau_p/\tau_s \sim 10^{-2}-10^{-3}$) [10]. Different scalings lead to different limits. We shall use the same strategy for two different rate equation models that are currently used in order to determine useful information on the dynamics of QD lasers. As we shall demonstrate, the stability condition can still be formulated by Eq. (6.1) but with different expressions for the damping rate Γ.

Both models with one carrier type and electron–hole models have been successfully used to describe turn-on experiments [11–14], gain recovery dynamics [15–17], optical injection [18] and optical feedback [19, 20]. The rate equation models with only one carrier type assume the same scattering rates for electrons and holes. They allow the derivation of simple analytical expressions that are useful when examining experimental data. Electron–hole rate equation models take into account the fact that the thermal redistribution occurs on different time scales for electrons and holes. These models aim to bridge the gap between a microscopic description and the simpler rate equation models but are too complicated for direct analysis.

In QD semiconductor devices, the carriers are first injected into a two dimensional carrier reservoir, that is, a QW, before being captured by a dot. The minimal way to describe this process is to formulate three rate equations for the electrical field in the cavity, the carrier density in the reservoir, and the occupation probability of a dot [21, 22]. These equations were analyzed using the asymptotic limit $\gamma \to 0$ in [23], and we shall apply the same analysis for the laser subject to optical feedback. Our main result is described in Section 6.2. The electron-hole rate equation model that we consider next involves five independent variables for the charge carrier densities in the QD, the charge carrier densities in the reservoir, and the photon density in the cavity, and it involves microscopically calculated scattering rates that are strongly nonlinear functions of the carrier densities in the reservoir [11, 12, 24, 25]. (See Chapter 1 for a review on the microscopic modeling). We recently showed that these equations can be simplified by taking advantage of the limit $\gamma \to 0$ [26]. Although coefficients of the reduced equations need to be computed numerically, distinct scaling laws can be extracted for the RO frequency and RO damping rate. We plan to use the same analysis here for the case of a laser subject to optical feedback [19]. The main results are summarized in Section 6.3. The asymptotic studies of the two problems are long and tedious. For clarity, the detailed computations are relegated to Appendixes B and C, and in the following, we only concentrate on the final results.

6.2
QD Laser Model with One Carrier Type

The rate equations for a QD laser subject to optical feedback formulated by O'Brien et al. [27] consist of three equations for the amplitude of the normalized laser field in the cavity E, the occupation probability ρ of a QD in the laser, and the number n of carriers in the reservoir per QD. The dimensionless equations are derived in Appendix B and are of the form

$$E' = \frac{1}{2}\left[-1 + g(2\rho - 1)\right](1 + i\alpha)E + ke^{-iC}E(t' - \tau), \tag{6.4}$$

$$\rho' = \gamma\left[Bn(1 - \rho) - \rho - (2\rho - 1)|E|^2\right], \tag{6.5}$$

$$n' = \gamma\left[J - n - 2Bn(1 - \rho)\right] \tag{6.6}$$

where prime means differentiation with respect to the dimensionless time $t' = t/\tau_p$. The factor 2 in Eq. (6.6) accounts for the twofold spin degeneracy in the quantum dot energy levels. A similar factor 2 is included in the definition of the differential gain factor g in Eq. (6.4) [28]. The parameter $\gamma \equiv \tau_p/\tau_s$ is the ratio of the photon lifetime and the carrier recombination time. The relaxation rates of ρ and n are assumed equal for mathematical simplicity. J is the electrically injected pump current per dot, and it is the control parameter. The nonlinear term $Bn(1 - \rho)$ describes the carrier exchange rate between the reservoir and the dots. $B \equiv \tau_s/\tau_{cap} \sim 10^2-10^3$ is the dimensionless capture rate, and $1 - \rho$ is the Pauli blocking factor. The three parameters B, γ, and $g - 1$ control the time-dependent response of the solitary QD laser. The last term in Eq. (6.4) represents the contribution of the delayed optical feedback. k and τ are the dimensionless feedback rate and round-trip time laser-mirror-laser, respectively, and C is the feedback phase.

As for the conventional laser, our objective is to determine the minimal value of the feedback rate below which a stable operation can be guaranteed. We shall consider γ as our order parameter because it does not appear in the expressions of the steady states and scale B and $g-1$ with respect to γ. Several cases are possible, depending on their respective scalings. The physically most interesting case considers the relation $B(1-\rho) = O(\gamma^{-1/2})$ [23], which basically assumes the carrier capture process into the QDs to be much faster than the radiative recombination time of the carriers in the QDs. The first Hopf bifurcation point k^H is determined by Eq. (6.69) (see Appendix B for the asymptotic analysis), and its lowest possible value and thus the lower bound for the critical feedback strength is given by the same expression as Eq. (6.1) but with a different dimensionless damping rate named Γ_2^{QD}:

$$\Gamma_2^{QD} \equiv \frac{\gamma}{2I^* + B_1^2}\left[2I^*\frac{1+I^*}{1-g^{-1}} + \frac{B_1^2}{2}(1+2I^*)\right] \tag{6.7}$$

with $B_1 \equiv \gamma^{1/2}B(1-g^{-1})$, and the steady state intensity of the solitary laser I^* (Appendix B). In the limit $\gamma \to 0$, I^* is given by

$$I^* \equiv \frac{g}{2}(J - J_{\text{th}}) \tag{6.8}$$

where $J_{\text{th}} \equiv 1 + g^{-1}$ is the threshold current in this limit. The expression for the RO frequency in units of τ_p is $\omega^{QD} \equiv \sqrt{2\gamma I^*}$ and is the same as the one for the QW laser (ω^{QW} is given by Eq. (6.37)). If the damping rate given in Eq. (6.7) is explored in the limits $B_1^2 \to \infty$ (fast capture) or $I^* \to 0$ (close to threshold), the value decreases and approaches the much lower RO damping rate of QW lasers

$$\Gamma^{QW} \equiv \frac{\gamma(1+2I^*)}{2} \tag{6.9}$$

(see Appendix A, Eq. (6.38)), thus in this limits $\Gamma_2^{QD} \to \Gamma^{QW}$.

However, if $B_1^2 = O(1)$ and/or g is close to 1, Γ_2^{QD} is much larger than Γ^{QW}. This can be demonstrated by rewriting Γ_2^{QD} as

$$\Gamma_2^{QD} = \Gamma^{QW} + \frac{\gamma I^*}{2I^* + B_1^2}\frac{g+1+2I^*}{g-1} \tag{6.10}$$

where the correction term clearly indicates the effect of $g-1$ if g is close to 1.

6.3
Electron-Hole Model for QD Laser

The microscopically based rate equation model for a solitary QD laser that separately treats electron and hole dynamics has been formulated and further investigated in [12, 14, 24] (see Chapter 1 for a review). Supplemented by the optical feedback term [19] and formulated with dimensionless quantities [20], it describes the evolution of the occupation probability of the confined QD levels, ρ_e and ρ_h, the number of carriers in the reservoir per QD, W_e and W_h, (e, h stand for electrons and holes, respectively), and the normalized slowly varying amplitude of the laser field inside the cavity $E = \sqrt{I}\exp(i\phi)$ with the normalized intensity I and the phase ϕ.

$$E' = \frac{1}{2}\Big[-1 + g(\rho_e + \rho_h - 1)\Big](1 + i\alpha)E + ke^{-iC}E(t' - \tau), \tag{6.11}$$

$$\rho_e' = \gamma\Big[s_e^{in}(1 - \rho_e) + s_e^{out}\rho_e - (\rho_e + \rho_h - 1)|E|^2 - \rho_e\rho_h\Big], \tag{6.12}$$

$$\rho_h' = \gamma\Big[s_h^{in}(1 - \rho_h) + s_h^{out}\rho_h - (\rho_e + \rho_h - 1)|E|^2 - \rho_e\rho_h\Big], \tag{6.13}$$

$$W_e' = \gamma\Big[J + (s_e^{in} + s_e^{out})\rho_e - s_e^{in} - cW_eW_h\Big], \tag{6.14}$$

$$W_h' = \gamma\Big[J + (s_h^{in} + s_h^{out})\rho_h - s_h^{in} - cW_eW_h\Big]. \tag{6.15}$$

In the above equations, prime means differentiation with respect to the dimensionless time $t' = t/\tau_p$ (with the photon lifetime τ_p). As before, k, C, and τ are the dimensionless feedback rate, the feedback phase, and the external round-trip time, respectively, and g is the linear gain parameter. The parameter c accounts for spontaneous and nonradiative losses in the reservoir, and J is the dimensionless electrically injected pump current per QD. Further s_e^{in}, s_e^{out}, s_h^{in}, s_h^{out} represent dimensionless scattering rates that are rescaled by $s_{e,h}^{in,out} = W^{-1}S_{e,h}^{in,out}$, with W^{-1} being the carrier lifetime because of radiative recombination inside a QD that corresponds to τ_s in the QW and in the QD model with one carrier type. They are computed numerically from a microscopic theory of carrier–carrier scattering events between QD and reservoir [12, 24]. The scattering times $\tau_e \equiv (S_e^{in} + S_e^{out})^{-1}$ and $\tau_h \equiv (S_h^{in} + S_h^{out})^{-1}$ are our reference time scales.

By rescaling time with respect to the RO frequency, which in turn scales like $\gamma^{1/2}$ as $\gamma = \tau_p/\tau_s \to 0$, reformulating the above equations in terms of deviations from the steady state and taking advantage of the small value of γ, we showed in [26] that the rate equations can be reduced to four equations that are given in Appendix C.

As we shall demonstrate, valuable information can be extracted from these equations on the basis of simple scaling assumptions. Three cases were explored in [26], which we now review.

6.3.1
Similar Scattering Times τ_e and τ_h

At first, one case that assumes the scattering times of both carrier types to be on the same timescale will be discussed. Further, this case assumes $s_e^{in} + s_e^{out}$ and $s_h^{in} + s_h^{out}$ to be $O(1)$ quantities compared to $\gamma^{1/2}$. We find that the expression for the critical feedback strength k_c is the same as Eq. (6.1) but with a different damping rate given by

$$\Gamma^S \equiv \frac{\gamma}{2}\left[\frac{s_e^{in} + s_e^{out}}{2} + 2I^* + \rho_h^* + \rho_e^* + \frac{s_h^{in} + s_h^{out}}{2}\right] \tag{6.16}$$

where I^*, ρ_e^*, and ρ_h^* are the dimensionless steady state values for the solitary laser of the light intensity, the electron, and the hole occupation probability in the QDs, respectively, that need to be computed numerically. In [26], we noted that

$\rho_h^* + \rho_e^* = 1 + g^{-1}$, where $g = O(1)$ is the dimensionless gain coefficient. Eq. (6.16) then simplifies as

$$\Gamma^S = \frac{\gamma}{2}\left[\frac{s_e^{in} + s_e^{out}}{2} + 2I^* + 1 + g^{-1} + \frac{s_h^{in} + s_h^{out}}{2}\right]. \tag{6.17}$$

Eq. (6.17) can be reformulated as

$$\Gamma^S = \Gamma^{QW} + \frac{1}{2}\left[\frac{s_e^{in} + s_e^{out}}{2} + \frac{s_h^{in} + s_h^{out}}{2}\right] \tag{6.18}$$

where

$$\Gamma^{QW} \equiv \frac{\gamma}{2}\left[1 + g^{-1} + 2I^*\right] \tag{6.19}$$

has the same format as Eq. (6.9) and can be considered as the contribution of the conventional QW laser.

6.3.2
Different Scattering Times τ_e and τ_h

The microscopic calculations predict very large scattering rates for the holes [12] because of their much larger effective mass. Consequently, this section aims to discuss the effect of holes if they are much faster than the electrons. For the asymptotic analysis, we introduce the dimensionless parameter a as a measure for the hole scattering rates

$$a \equiv \sqrt{\frac{\gamma}{2I^*}}(s_h^{in} + s_h^{out}) \tag{6.20}$$

where I^* is assumed to be an $O(1)$ quantity.

6.3.3
Small Scattering Lifetime of the Holes $a = O(1)$

To this end, we assume that $s_e^{in} + s_e^{out} = O(1)$, while $s_h^{in} + s_h^{out} = O(\gamma^{-1/2})$, which then implies from Eq. (6.20) that $a = O(1)$. Note that this is different from Section 6.3.1, where the scaling $a = O(\gamma^{1/2})$ was discussed. The leading order equation for the growth rate is the same as for the solitary laser [26] and does not contain any contribution of the feedback. In other words, the amplitude of the feedback k is too weak ($k = O(\gamma)$). We would need to consider the case $k = O(\gamma^{1/2})$ in order to find the feedback parameter in the leading equation for the growth rate, but this problem has not been solved analytically yet.

6.3.4
Very Small Scattering Lifetime of the Holes $a = O(\gamma^{-1/2})$

For the case in which the hole scattering time is on the order of pico seconds, another scaling has to be introduced. Thus, for this case, we assume that

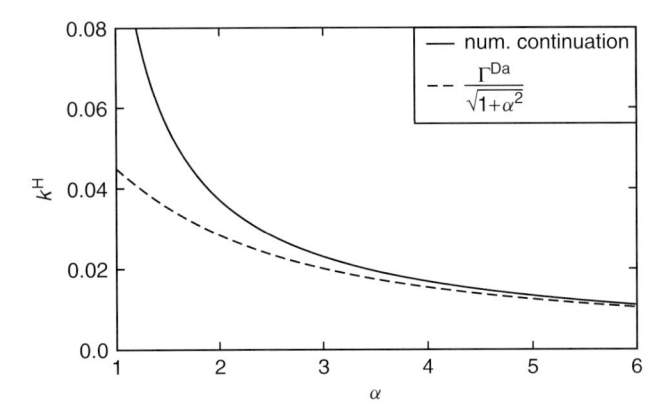

Figure 6.1 Solid line shows the first Hopf bifurcation point $k = k^H$ as a function of α as obtained numerically from the original rate equations using a continuation method [20] ($C = \pi + \arctan(\alpha)$, $\tau = 80$). The broken line represents its analytical approximation given by Eq. (6.1). As α decreases toward zero, k^H increases, and the analytical approximation that assumes $k \ll 1$ begins to fail.

$a = O(\gamma^{-1/2})$, while $s_e^{in} + s_e^{out} = O(1)$. Compared to the case of similar scattering times, the RO frequency is slightly reduced by a factor of $1/\sqrt{1/2}$. The expression for the critical feedback strength is the same as Eq. (6.1) but with a different dimensionless damping rate given by

$$\Gamma^{Da} \equiv \frac{\gamma}{2}\left[\frac{I^*}{\gamma}\frac{1}{s_h^{in} + s_h^{out}} + s_e^{in} + s_e^{out} + I^* + \rho_h^*\right]. \tag{6.21}$$

In Figure 6.1, we compare numerical and analytical predictions for a laser subject to a long external cavity. The numerical determination of the Hopf bifurcation point k^H has been obtained by using a continuation technique (DDE-Biftool [29]) applied to the original electron–hole rate equation model [19, 26] and not from the reduced Eqs. (6.87)–(6.90). Details on the numerical simulations and parameter values are documented in [20]. The broken line in Figure 1 represents the analytical approximation given by Eq. (6.1). As α decreases toward zero, k^H increases and the analytical approximation that assumes $k \ll 1$ begins to fail, while a very good agreement with the analytic results is found for larger α.

6.4 Summary

The expression for the critical feedback strength from Eq. (6.1) provides a sufficient condition for stable operation of a QW laser subject to optical feedback. The critical feedback rate above which pulsating instabilities are observed is determined as a function of the linewidth enhancement factor α and the damping rate of the ROs. Its simplicity has encouraged experimental studies of QD lasers subject to optical feedback. It is shown that Eq. (6.1) is also a good approximation for QD lasers provided that their much larger damping rate of the relaxation oscillations is considered. The damping rate is generally obtained by fitting data. In this review, we examine two different rate equations models for QD lasers and derive the stability condition with analytical expressions for the damping rate. These expressions allow us to anticipate the effect of specific parameters, for example, the carrier scattering rates and the differential gain coefficient, and design lasers with a larger tolerance to optical feedback.

Acknowledgment

C. Otto is grateful for support of the research training group GRK 1558. The work by K. Lüdge was supported by the DFG in the framework of SFB 787. The work by E. Viktorov and T. Erneux was supported by the Fond National de la Recherche Scientifique (Belgium). The research by T. Erneux was also supported by the Air Force Office of Scientific Research (AFOSR) grant FA8655-09-1-3068.

6.5
Appendix A: Rate Equations for Quantum Well Lasers

The rate equations for a solitary QW laser used by Mork et al. [2] are given by

$$\frac{d\mathcal{E}}{dt} = \frac{1}{2}\left[G_N(N-N_0) - \frac{1}{\tau_p}\right]\mathcal{E}, \tag{6.22}$$

$$\frac{dN}{dt} = J - \frac{N}{\tau_s} - G_N(N-N_0)\mathcal{E}^2. \tag{6.23}$$

Here, $\mathcal{E}$ is the amplitude of the electrical field, and N is the carrier density. The linear gain coefficient is denoted by G_N, N_0 is the transparency density of carriers, J is the pumping current and τ_p and τ_s are the photon and carrier lifetimes, respectively. The nonzero intensity steady state is

$$N^* = N_0 + \frac{1}{G_N \tau_p}, \tag{6.24}$$

$$\mathcal{E}^{*2} = \frac{1}{G_N(N-N_0)}\left(J - \frac{N}{\tau_s}\right). \tag{6.25}$$

From the linearized equations, we then determine the characteristic equation for the growth rate λ

$$\lambda^2 + \left(\frac{1}{\tau_s} + G_N \mathcal{E}^{*2}\right)\lambda + \frac{1}{\tau_p}G_N \mathcal{E}^{*2} = 0. \tag{6.26}$$

In order to properly define the relaxation oscillation frequency and its damping rate, we take advantage of the fact that $\tau_p \ll \tau_s$. The roots of the quadratic equation then take the form

$$\lambda = -\Gamma_{RO}^{QW} \pm i\omega_{RO}^{QW} \tag{6.27}$$

where

$$\omega_{RO}^{QW} \equiv \sqrt{G_N \frac{1}{\tau_p}\mathcal{E}^{*2} - \frac{1}{4}(\frac{1}{\tau_s} + G_N\mathcal{E}^{*2})^2} \simeq \sqrt{G_N \frac{1}{\tau_p}\mathcal{E}^{*2}}, \qquad (6.28)$$

$$\Gamma_{RO}^{QW} \equiv \frac{1}{2}\left(\frac{1}{\tau_s} + G_N\mathcal{E}^{*2}\right) = \frac{1}{2}\left(\frac{1}{\tau_s} + \tau_p\omega_{RO}^2\right) \qquad (6.29)$$

are defined as the RO frequency and RO damping rate of the solitary laser, respectively. They are the quantities that are measured experimentally. Mork et al. [2] introduced the RO damping rate as "τ_R^{-1}", which equals $2\Gamma_{RO}^{QW}$ but could wrongly be interpreted as Γ_{RO}^{QW}.

In order to determine asymptotic approximations, we need to reformulate the rate equations in dimensionless form. The simplest way is to measure time in units of the photon lifetime by introducing

$$t' \equiv t/\tau_p. \qquad (6.30)$$

Furthermore, introducing the new dimensionless dependent variables E and Z defined by

$$E \equiv \sqrt{\frac{G_N\tau_s}{2}}\mathcal{E} \text{ and } Z \equiv \frac{1}{2}\left[G_N(N-N_0)\tau_p - 1\right] \qquad (6.31)$$

allows to reduce the number of parameters. Inserting Eqs. (6.30) and (6.31) into Eqs. (6.22) and (6.23), we find

$$\frac{dE}{dt'} = ZE, \qquad (6.32)$$

$$\frac{dZ}{dt'} = \gamma\left[P - Z - (1+2Z)E^2\right] \qquad (6.33)$$

where γ and P are defined by

$$\gamma \equiv \frac{\tau_p}{\tau_s}, \quad P \equiv \frac{G_N\tau_p\tau_s}{2}(J - J_{th}), \text{ with } J_{th} \equiv \frac{N_0}{\tau_s} + \frac{1}{G_N\tau_p\tau_s}. \qquad (6.34)$$

The nonzero intensity steady state is

$$Z^* = 0 \text{ and } I^* = E^{*2} = P, \qquad (6.35)$$

and the characteristic equation for the growth rate σ is given by

$$\sigma^2 + \gamma(1+2I^*)\sigma + 2I^*\gamma = 0. \qquad (6.36)$$

Provided γ is sufficiently small, the roots of Eq. (6.36) are complex-conjugated. The dimensionless RO frequency and RO damping rate (in units of time t') are then defined from the imaginary and real part of these roots. We obtain

$$\omega^{QW} \equiv \sqrt{2\gamma I^* - \frac{\gamma^2}{4}(1+2I^*)^2} \simeq \sqrt{2\gamma I^*} \text{ as } \gamma \to 0 \quad \text{and} \qquad (6.37)$$

$$\Gamma^{QW} \equiv \frac{\gamma(1+2I^*)}{2}. \qquad (6.38)$$

In our analysis of the QD rate equations, we use the same dimensionless time $t' \equiv t/\tau_p$ and reformulate the dynamic equations so that the same γ multiply the right-hand side of the carrier equations.

6.6
Appendix B: Asymptotic Analysis for a QD Laser Model with One Carrier Type

The equations examined by O'Brien et al. [27] are the following three equations for the amplitude of the laser field in the cavity, $\mathcal{E}$; the number of carriers in the reservoir per dot, N; and the occupation probability of the dots in the laser ρ

$$\frac{d\mathcal{E}}{dt} = \frac{1}{2}\left[-\frac{1}{\tau_p} + g_0\theta(2\rho - 1)\right]\mathcal{E} + \frac{i\delta\omega}{2}\mathcal{E} + \frac{\eta}{2}\mathcal{E}(t - \tau_{ec}), \qquad (6.39)$$

$$\frac{d\rho}{dt} = -\frac{\rho}{\tau_s} - g_0(2\rho - 1)|\mathcal{E}|^2 + \widetilde{F}(N, \rho), \qquad (6.40)$$

$$\frac{dN}{dt} = -\frac{N}{\tau_s} + \widetilde{J} - 2N^{QD}\widetilde{F}(N, \rho). \qquad (6.41)$$

For the definition of the various parameters, see [27]. The capture rate is described by the term $\widetilde{F} = \widetilde{C}N^2(1 - \rho)$ in [27] and is proportional to the number of carriers present as well as the probability to find a dot. As in Reference [23], we shall consider $\widetilde{F} = \widetilde{B}N(1 - \rho)$ instead of $\widetilde{F} = \widetilde{C}N^2(1 - \rho)$. Here, the carrier phonon and the Auger carrier capture rates are denoted by $\widetilde{B}$ and $\widetilde{C}$, respectively. We define $\delta\omega = \alpha/\tau_p$, where α is the linewidth enhancement factor. N^{QD} is the two dimensional density of quantum dots. In our analysis, we introduce the α factor in the traditional way, that is, by the term $(1 + i\alpha)$ multiplying the full square bracket in Equation (6.39). Moreover, we take into account the feedback phase $C = \omega_0\tau_{ec}$, where ω_0 is the angular frequency of the solitary laser ($C = 0 \mod 2\pi$ in [27]). Our starting equations are then given by

$$\frac{d\mathcal{E}}{dt} = \frac{1}{2}\left[-\frac{1}{\tau_p} + g_0\theta(2\rho - 1)\right](1 + i\alpha)\mathcal{E} + \frac{\eta}{2}e^{-iC}\mathcal{E}(t - \tau_{ec}), \qquad (6.42)$$

$$\frac{d\rho}{dt} = -\frac{\rho}{\tau_s} - g_0(2\rho - 1)|\mathcal{E}|^2 + \widetilde{B}N(1 - \rho), \qquad (6.43)$$

$$\frac{dN}{dt} = -\frac{N}{\tau_s} + \widetilde{J} - 2N^{QD}\widetilde{B}N(1 - \rho). \qquad (6.44)$$

By introducing a dimensionless time t', the number of carriers in the reservoir per QD n, and a normalized field E, according to

$$t' \equiv t/\tau_p, \qquad n \equiv N/N^{QD}, \qquad E \equiv \sqrt{g_0\tau_s}\mathcal{E}, \qquad (6.45)$$

the Eqs. (6.42)–(6.44) simplify as

$$\frac{dE}{dt'} = \frac{1}{2}[-1 + g(2\rho - 1)](1 + i\alpha)E + ke^{-iC}E(t' - \tau), \qquad (6.46)$$

$$\frac{d\rho}{dt'} = \gamma\left[-\rho - (2\rho - 1)|E|^2 + Bn(1 - \rho)\right], \qquad (6.47)$$

$$\frac{dn}{dt'} = \gamma\left[-n + J - 2Bn(1 - \rho)\right] \qquad (6.48)$$

where

$$\gamma \equiv \frac{\tau_p}{\tau_s}, \quad g \equiv g_0\theta\tau_p, \quad k \equiv \frac{\eta}{2}\tau_p, \quad \tau \equiv \tau_{ec}/\tau_p, \quad J \equiv \frac{\widetilde{J}\tau_s}{N^{QD}}, \quad \text{and } B \equiv \widetilde{B}N^{QD}\tau_s. \qquad (6.49)$$

6.6 Appendix B: Asymptotic Analysis for a QD Laser Model with One Carrier Type

If we consider the rate equations (6.46)–(6.48), in terms of the normalized intensity I and the phase ϕ of the field $E = \sqrt{I}\exp(i\phi)$, the equations can be rewritten as

$$I' = [-1 + g(2\rho - 1)] I + 2k\sqrt{I(t'-\tau)I(t')}\cos(-C + \phi(t'-\tau) - \phi), \quad (6.50)$$

$$\phi' = \frac{1}{2}[-1 + g(2\rho - 1)]\alpha + k\sqrt{\frac{I(t'-\tau)}{I(t')}}\sin(-C + \phi(t'-\tau) - \phi), \quad (6.51)$$

$$\rho' = \gamma \left[Bn(1-\rho) - \rho - (2\rho - 1)I \right], \quad (6.52)$$

$$n' = \gamma \left[J - n - 2Bn(1-\rho) \right]. \quad (6.53)$$

6.6.1
External Cavity Modes

The basic solutions of the feedback problem are the external cavity modes (ECMs). They are defined as solutions with constant field intensity and carrier numbers, that is, $I = I_s$, $\rho = \rho_s$, $n = n_s$, and a phase of the field that varies linearly in time given by $\phi = \phi_s \equiv -C\frac{t'}{\tau} + \Delta_s \frac{t'}{\tau}$ with the ECM frequency Δ_s. For simplicity of notation, the index s is omitted in the following.

From Eqs. (6.50)–(6.53), the basic solutions satisfy the following conditions:

$$\frac{1}{2}[-1 + g(2\rho - 1)] = -k\cos(\Delta),$$

$$\Delta = C - k\tau \left[\alpha \cos(\Delta) + \sin(\Delta) \right],$$

$$n = \frac{J}{1 + 2B(1-\rho)},$$

$$I = \frac{Bn(1-\rho) - \rho}{2\rho - 1}.$$

Solving for ρ, we obtain

$$\rho = \frac{1}{2}(1 + g^{-1}) - \frac{k}{g}\cos(\Delta), \quad (6.54)$$

$$n = \frac{J}{1 + 2B(1-\rho)}, \quad (6.55)$$

$$I = \frac{B(1-\rho)}{1 + 2B(1-\rho)} \frac{J - J_{th}}{2\rho - 1}, \quad (6.56)$$

where

$$J_{th} \equiv \frac{\rho(1 + 2B(1-\rho))}{B(1-\rho)}. \quad (6.57)$$

We note the following relations, which will be useful when we eliminate n from the coefficients of the characteristic equation:

$$Bn + 1 + 2I = \frac{1+I}{1-\rho},$$

$$\left[\begin{array}{c} (1 + 2B(1-\rho))(Bn + 1 + 2I) \\ -2B^2 n(1-\rho) \end{array} \right] = \frac{1+I}{1-\rho} + 2B(1-\rho)(1+2I).$$

6.6.2
Stability

From the linearized equations, we determine the following condition for the growth rate σ:

$$\begin{bmatrix} \begin{pmatrix} k\cos(\Delta)F \\ -\sigma \end{pmatrix} & k\sqrt{I}\sin(\Delta)F & g\sqrt{I} & 0 \\ \frac{-k}{\sqrt{I}}F\sin(\Delta) & \begin{pmatrix} \cos(\Delta)F \\ -\sigma \end{pmatrix} & g\alpha & 0 \\ -2\gamma(2\rho-1)\sqrt{I} & 0 & -\gamma\begin{pmatrix} Bn+1 \\ +2I \\ -\sigma \end{pmatrix} & \gamma B(1-\rho) \\ 0 & 0 & 2\gamma Bn & -\gamma\begin{pmatrix} 1 \\ +2B(1-\rho) \end{pmatrix} \\ & & & -\sigma \end{bmatrix} = 0 \quad (6.58)$$

where

$$F \equiv \exp(-\sigma\tau) - 1. \quad (6.59)$$

Expanding the determinant (Eq. (6.58)), we find the following characteristic equation for the growth rate σ

$$0 = \sigma^4 + \sigma^3 \left[\gamma\left(1 + 2B(1-\rho) + \frac{1+I}{1-\rho}\right) - 2k\cos(\Delta)F \right]$$

$$+ \sigma^2 \begin{bmatrix} 2\gamma(2\rho-1)gI + \gamma^2\left(\frac{1+I}{1-\rho} + 2B(1-\rho)(1+2I)\right) + k^2F^2 \\ -\gamma 2k\cos(\Delta)F\left(2B(1-\rho) + \frac{2+I-\rho}{1-\rho}\right) \end{bmatrix}$$

$$+ \sigma \begin{bmatrix} 2\gamma(2\rho-1)gI\left[\gamma\left(1 + 2B(1-\rho)\right) + k(\alpha\sin(\Delta) - \cos(\Delta))F\right] \\ -\gamma^2 2k\cos(\Delta)F\left(\frac{1+I}{1-\rho} + 2B(1-\rho)(1+2I)\right) \\ +\gamma k^2 F^2 \left(2B(1-\rho) + \frac{2+I-\rho}{1-\rho}\right) \end{bmatrix}$$

$$+ \begin{bmatrix} \gamma^2 k^2 F^2 \left(\frac{1+I}{1-\rho} + 2B(1-\rho)(1+2I)\right) \\ +2k\gamma^2(2\rho-1)gI\left(1 + 2B(1-\rho)\right)(\alpha\sin(\Delta) - \cos\Delta))F \end{bmatrix}. \quad (6.60)$$

We next investigate two cases that depend on the size of parameter B.

6.6.3
$B(1 - \rho) = O(1)$

We solve Eq. (6.60) by seeking a solution of the form

$$\sigma = \gamma^{1/2}\sigma_0 + \gamma\sigma_1 + \cdots,$$
$$k = \gamma k_1 + \cdots. \quad (6.61)$$

From the Eqs. (6.54)–(6.56) we note the following scaling laws:

$$I = I^* + O(\gamma), \ \rho = \rho^* + O(\gamma), \text{ and } \Delta = \Delta_0 + O(\gamma) \quad (6.62)$$

6.6 Appendix B: Asymptotic Analysis for a QD Laser Model with One Carrier Type

where

$$I^* = \frac{B(1-\rho^*)}{(1+2B(1-\rho^*))} \frac{(J-J_{th,0})}{(2\rho^*-1)} = \frac{g}{2} \frac{B(1-g^{-1})}{1+B(1-g^{-1})}(J-J_{th,0}),$$

$$\rho^* = \frac{1}{2}(1+g^{-1}), \quad \text{and} \quad \Delta_0 = C. \tag{6.63}$$

Here, I^* and ρ^* denote intensity and occupation probability of the QDs for the solitary laser, respectively, and the threshold current of the solitary laser is given by Erneux et al. [23]

$$J_{th,0} \equiv \frac{\rho^*(1+2B(1-\rho^*))}{(1-\rho^*)B} = \frac{1+B(1-g^{-1})}{B(1-g^{-1})}(1+g^{-1}).$$

We find from Eq. (6.60) the following sequence of problems

$$O(\gamma^2) : 0 = \sigma_0^4 + \sigma_0^2 2I^*, \tag{6.64}$$

$$O(\gamma^{5/2}) : 0 = \left(4\sigma_0^2 + 4I^*\right)\sigma_0\sigma_1$$

$$+ \sigma_0^3 \left[1 + 2B(1-\rho^*) + \frac{1+I^*}{1-\rho^*} - 2k_1\cos(\Delta_0)F_0\right]$$

$$+ 2I^*\sigma_0 \left[k_1(\alpha\sin(\Delta_0) - \cos(\Delta_0))F_0\right.$$

$$+ 1 + 2B(1-\rho^*)\right] \tag{6.65}$$

where

$$F_0 \equiv \exp(-\gamma^{1/2}\sigma_0\tau) - 1. \tag{6.66}$$

From Eq. (6.64), we determine σ_0 as

$$\sigma_0 = i\sqrt{2I^*}$$

and from Eq. (6.65) with

$$\omega^{QD} \equiv \sqrt{2\gamma I^*}, \tag{6.67}$$

we find σ_1 as

$$\sigma_1 = -\Gamma + \frac{k_1}{2}(\alpha\sin(\Delta_0) + \cos(\Delta_0))\left[(\cos(\omega^{QD}\tau) - 1) - i\sin(\omega^{QD}\tau)\right]$$

where

$$\Gamma \equiv \frac{1+I^*}{1-g^{-1}}.$$

The real part of σ_1 then is

$$\text{Re}(\sigma_1) = -\Gamma - k_1(\alpha\sin(\Delta_0) + \cos(\Delta_0))\sin^2(\frac{\omega^{QD}\tau}{2}), \tag{6.68}$$

which implies stability for all values of k_1 if $(\alpha\sin(\Delta_0) + \cos(\Delta_0)) > 0$ or provided that

$$k_1 < k_1^H \equiv \frac{-\Gamma}{(\alpha\sin(\Delta_0) + \cos(\Delta_0))\sin^2\left(\frac{\omega^{QD}\tau}{2}\right)}$$

$$= \frac{-2}{(1-\cos(\omega^{QD}\tau))(\cos(\Delta_0 - \arctan(\alpha))} \frac{\Gamma}{\sqrt{1+\alpha^2}} \tag{6.69}$$

if $\alpha \sin(\Delta_0) + \cos(\Delta_0) < 0$. From Eq. (6.69), we see that the lowest possible value for k_1^H is for

$$\Delta_0 = C = \pi + \arctan(\alpha) \text{ and } \omega^{QD}\tau = \pi \pmod{2\pi}.$$

It is given by

$$k_{1c} \equiv \frac{\Gamma}{\sqrt{1+\alpha^2}}. \tag{6.70}$$

In terms of the original parameters, the stability condition in Eq. (6.70) implies that

$$k < k_c \equiv \frac{\Gamma_1^{QD}}{\sqrt{1+\alpha^2}} \tag{6.71}$$

where $\Gamma_1^{QD} \equiv \gamma \Gamma$, or equivalently,

$$\Gamma_1^{QD} = \gamma \frac{1 + R_0^2}{1 - g^{-1}}. \tag{6.72}$$

6.6.4
$B(1 - \rho) = O(\gamma^{-1/2})$

Taking into account that $B(1 - \rho) = O(\gamma^{-1/2})$, we introduce a $O(1)$ quantity B_1 as $B_1 \equiv \gamma^{1/2} 2B(1 - \rho^*)$. With the scaling of $\rho = \rho^* + O(\gamma)$ (Eq. (6.62)), we may expand J_{th} from Eq. (6.57) and I from Eq. (6.56) in powers of $\gamma^{1/2}$, which yields

$$J_{th} = 2\rho^* + \frac{1}{B(1-\rho^*)}\rho^* + O(\gamma) = 2\rho^* + \gamma^{1/2} 2\rho^* B_1^{-1} + O(\gamma),$$
$$I = I^* + \gamma^{1/2} I_1 + O(\gamma) \tag{6.73}$$

where we have defined the steady state intensity of the solitary laser I^* in the limit $\gamma \to 0$ and its first order correction I_1

$$I^* = \frac{1}{2}\frac{1}{2\rho^* - 1}(J - 2\rho^*) = \frac{g}{2}(J - (1 + g^{-1})), \tag{6.74}$$

$$I_1 = -\frac{g}{2} B_1^{-1} J. \tag{6.75}$$

Inserting Eq. (6.61) and Eq. (6.73) into the characteristic Eq. (6.60), we find the following problems for σ_0 and σ_1

$$O(\gamma^2): \quad \sigma_0^4 + \sigma_0^3 B_1 + \sigma_0^2 2I^* + \sigma_0 2I^* B_1 = 0, \tag{6.76}$$

$$O(\gamma^{5/2}): \quad \left[4\sigma_0^3 + 3\sigma_0^2 B_1 + 2\sigma_0 2I^* + 2I^* B_1\right]\sigma_1$$
$$= -(\sigma_0^2 + \sigma_0 B_1)2I_1$$
$$- \sigma_0^3 \left[1 + 2\frac{1 + I^*}{1 - g^{-1}} - 2k_1 \cos(\Delta_0) F_0\right]$$
$$- \sigma_0^2 \left[B_1(1 + 2I^*) - 2k_1 \cos(\Delta_0) F_0 B_1\right]$$
$$- \sigma_0 2I^* \left[1 + k_1(\alpha \sin(\Delta_0) - \cos(\Delta_0))F_0\right]$$
$$- \left[2I^* B_1 k_1(\alpha \sin(\Delta_0) - \cos(\Delta_0))F_0\right] \tag{6.77}$$

where F_0 is defined by (6.66). Equation (6.76) admits the solution

$$\sigma_0^2 = -2I^* \tag{6.78}$$

and from Eq. (6.77) with Eqs. (6.78) and (6.67), we find

$$\sigma_1 = \frac{2\sigma_0 I_1}{4I^*} + \frac{k_1(\alpha \sin(\Delta_0) + \cos(\Delta_0))\left[(\cos(\omega^{QD}\tau) - 1) - i\sin(\omega^{QD}\tau)\right]}{2}$$

$$- \frac{1}{4I^*(\sigma_0 + B_1)}\left[\sigma_0 4I^* \frac{1+I^*}{1-g^{-1}} + 2I^* B_1(1+2I^*)\right]. \tag{6.79}$$

Equation (6.79) implies that

$$\mathrm{Re}(\sigma_1) = -\Gamma + \frac{k_1(\alpha \sin(\Delta_0) + \cos(\Delta_0))(\cos(\omega^{QD}\tau) - 1)}{2} \tag{6.80}$$

where

$$\Gamma \equiv \frac{1}{2I^* + B_1^2}\left[2I^* \frac{1+I^*}{1-g^{-1}} + \frac{B_1^2}{2}(1+2I^*)\right] \tag{6.81}$$

is the damping rate of the solitary laser [9]. Our stability conditions are now similar to those of Eqs. (6.71)–(6.72) with Γ_2^{QD} replacing Γ_1^{QD}, where

$$\Gamma_2^{QD} \equiv \gamma\Gamma = \frac{\gamma}{2I^* + \gamma B^2(1-g^{-1})}\left[2I^* \frac{1+I^*}{1-g^{-1}} + \frac{\gamma B^2(1-g^{-1})}{2}(1+2I^*)\right]. \tag{6.82}$$

6.7
Appendix C: Asymptotic Analysis for a QD Laser Model with Two Carrier Types

The microscopically based electron-hole rate equation model describes the evolution of the charge carrier densities in the QD (n_e and n_h), the carrier densities in the reservoir (w_e and w_h) (e, h stand for electrons and holes, respectively), and the photon density n_{ph}. See Chapter 1 for the equations with dimensions, while the dimensionless form is given in Eqs. (6.11)–(6.15). To make the equations dimensionless, we introduced the dimensionless time $t' \equiv t/\tau_p$ (with $\tau_p^{-1} = 2\kappa$) as well as the dimensionless variables

$$I \equiv n_{\mathrm{ph}}A, \quad \rho_{e/h} \equiv N_{e/h}/N^{QD}, \quad W_{e/h} \equiv w_{e/h}/N^{\mathrm{sum}} \tag{6.83}$$

and the dimensionless parameters with

$$g \equiv \frac{\Gamma W A N^{QD}}{2\kappa}, \quad \gamma \equiv \frac{W}{2\kappa}, \quad k \equiv \frac{1}{2\kappa}\frac{K}{\tau_{\mathrm{in}}}, \quad \tau \equiv 2\kappa\tau_{ec}, \tag{6.84}$$

$$s_{e/h}^{\mathrm{in/out}} \equiv \frac{1}{W} S_{e/h}^{\mathrm{in/out}}, \quad c \equiv \frac{BN^{\mathrm{sum}}}{W}, \quad J \equiv \frac{j}{e_0 N^{\mathrm{sum}} W}. \tag{6.85}$$

The resulting Eqs. (6.11)–(6.15) for the solitary laser ($k = 0$) are singular in the limit $\gamma \to 0$, because the leading order equations do not admit physical solutions. The basic idea to remove this singularity is to scale time with respect to the RO frequency [10], which in turn scales like $\gamma^{1/2}$ in the limit $\gamma \to 0$. We showed in [26] that by taking advantage of the small value of $\gamma \to 0$ the five rate equations without feedback can be reduced to three equations for the deviations of the intensity and the QD populations from their steady state values. The reason basically is that in these coordinates the dynamical variables for the reservoir populations follow passively the QD variables and can thus be neglected in first approximation.

Supplemented by the optical feedback term [19], the dynamical equations consist of four equations for the deviation of the intensity from its steady state, y; the phase of the electrical field ϕ; and the deviations $u_{e/h}$ of the QD occupation probabilities from their steady state values. Specifically, the new dynamic variables y, u_e, and u_h are defined via

$$I = I^*(1+y) \quad \text{and} \quad \rho_{e/h} = \rho_{e/h}^* + \sqrt{\gamma} \omega g^{-1} u_{e/h} \tag{6.86}$$

where the superscript $*$ denotes the steady state values of the solitary laser.

The new set of rate equations is given by

$$y' = (u_e + u_h)(1+y)$$
$$+ 2\varepsilon \zeta \sqrt{(1+y)(1+y(s-s_c))} \cos(-C + \phi(s-s_c) - \phi), \tag{6.87}$$

$$\phi' = \alpha \frac{1}{2}(u_e + u_h)$$
$$+ \varepsilon \zeta \sqrt{\frac{1+y(s-s_c)}{1+y}} \sin(-C + \phi(s-s_c) - \phi), \tag{6.88}$$

$$u'_e = -\frac{1}{2}y - \varepsilon(s_e^{in} + s_e^{out})u_e$$
$$- \varepsilon(u_e + u_h)I^* - \varepsilon(u_e \rho_h^* + \rho_e^* u_h) + O(\gamma), \tag{6.89}$$

$$u'_h = -\frac{1}{2}y - au_h$$
$$- \varepsilon(u_e + u_h)I^* - \varepsilon(u_e \rho_h^* + \rho_e^* u_h) + O(\gamma) \tag{6.90}$$

where prime means differentiation with respect to the dimensionless time s with

$$s \equiv \omega t' = \omega t / \tau_p \quad \text{and} \quad \omega \equiv \sqrt{2\gamma I^*} \tag{6.91}$$

is proportional to the RO frequency of the solitary laser. Equation (6.91) is identical to ω^{QW} given by Eq. (6.37) and ω^{QD} given by Eq. (6.67). I^*, ρ_e^*, ρ_h^* are dimensionless steady state values of the solitary laser that need to be computed numerically. The new feedback amplitude $\zeta = O(1)$, the delay s_c, the small parameter ε, and a are defined by

$$\zeta \equiv \frac{k}{\gamma}, \quad s_c \equiv \omega \tau, \quad \varepsilon \equiv \sqrt{\frac{\gamma}{2I^*}}, \quad \text{and} \tag{6.92}$$

$$a \equiv \varepsilon(s_h^{in} + s_h^{out}). \tag{6.93}$$

The dimensionless scattering rates that also need to be computed numerically are denoted by s_e^{in}, s_e^{out}, s_h^{in}, and s_h^{out}. As we shall now demonstrate, valuable information can be extracted from these equations on the basis of simple scaling assumptions.

6.7.1
External Cavity Modes

The basic solutions are the external cavity modes (ECMs). Analog to Section 6.6.1, they are defined as solutions with constant deviations from the steady state values of intensity and carrier occupation probabilities, i.e. $y = y_s$, $u_e = u_{e,s}$, $u_h = u_{h,s}$, and a phase that changes linearly in time

$$\phi = \phi_s \equiv -C\frac{s}{s_c} + \Delta_s \frac{s}{s_c} \qquad (6.94)$$

with the ECM frequency Δ_s. For simplicity of notation the index s is omitted in the following.

From Eqs. (6.87) and (6.88), we find that Δ satisfies the following transcendental equation

$$\Delta = C - \varepsilon \zeta s_c \left(\alpha \cos(\Delta) + \sin(\Delta)\right), \qquad (6.95)$$

which implies that $\Delta \simeq C$ as $\varepsilon \to 0$, that is, Δ is independent of the feedback amplitude ζ, in first approximation. For the subsequent asymptotics we write

$$\Delta = \Delta_0 + O(\epsilon) \qquad (6.96)$$

with $\Delta_0 = C$. From Eq. (6.87), we also note that

$$u_e + u_h = -2\varepsilon\zeta \cos(\Delta), \qquad (6.97)$$

which indicates that both u_e and u_h are $O(\varepsilon)$ small. From Eq. (6.89), we then find that y is $O(\varepsilon^2)$ small. These scaling laws for u_e, u_h, and y are useful when we reorganize the coefficients of the characteristic equation in powers of ε. Three cases were explored in [26] which we now examine.

6.7.2
Stability

From the linearized equations, we determine the following condition for the growth rate μ:

$$\begin{vmatrix} \begin{pmatrix} -\varepsilon\zeta \cos(\Delta)F \\ -\mu \end{pmatrix} & \begin{pmatrix} -2\varepsilon\zeta(1+\gamma) \\ \times \sin(\Delta)F \end{pmatrix} & 1+\gamma & 1+\gamma \\ \left(\varepsilon\zeta \frac{\sin(\Delta)}{2(1+\gamma)}F\right) & \begin{pmatrix} -\varepsilon\zeta \cos(\Delta)F \\ -\mu \end{pmatrix} & \frac{\alpha}{2} & \frac{\alpha}{2} \\ -\frac{1}{2} & 0 & \begin{pmatrix} -\varepsilon\begin{pmatrix} s_e^{in} + s_e^{out} \\ +I^* + \rho_h^* \end{pmatrix} \\ -\mu \end{pmatrix} & -\varepsilon\left(I^* + \rho_e^*\right) \\ -\frac{1}{2} & 0 & -\varepsilon\left(I^* + \rho_h^*\right) & \begin{pmatrix} -a \\ -\varepsilon(I^* + \rho_e^*) \\ -\mu \end{pmatrix} \end{vmatrix} = 0$$

$$(6.98)$$

where

$$F \equiv 1 - e^{-\mu s_c}. \tag{6.99}$$

Expanding the determinant, we obtain

$$\mu^4 + \mu^3 \left[(s_e^{in} + s_e^{out} + 2I^* + \rho_h^* + \rho_e^*)\varepsilon + a + 2\varepsilon\zeta\cos(\Delta)F\right]$$

$$-\mu^2 \left[-(1+\gamma) - \begin{bmatrix} \varepsilon\left(s_e^{in} + s_e^{out} + I^* + \rho_h^*\right)\left(a + \varepsilon(I^* + \rho_e^*)\right) \\ -\varepsilon^2\left(I^* + \rho_h^*\right)\left(I^* + \rho_e^*\right) + \varepsilon^2\zeta^2 F^2 \\ +2\varepsilon\zeta\cos(\Delta)F\left(a + \varepsilon(2I^* + \rho_e^* + \rho_h^* + s_e^{in} + s_e^{out})\right) \end{bmatrix}\right]$$

$$+\mu \begin{bmatrix} \varepsilon^2\zeta^2 F^2 \left(a + \varepsilon(2I^* + \rho_e^* + s_e^{in} + s_e^{out} + \rho_h^*)\right) \\ +2\varepsilon\zeta\cos(\Delta)F \left[\varepsilon\begin{pmatrix} s_e^{in} + s_e^{out} \\ +I^* + \rho_h^* \end{pmatrix}\begin{pmatrix} a \\ +\varepsilon(I^* + \rho_e^*) \\ -\varepsilon^2(I^* + \rho_h^*)(I^* + \rho_e^*) \end{pmatrix}\right] \\ -\varepsilon\zeta(1+\gamma)\sin(\Delta)F\alpha \\ +2\varepsilon\zeta\cos(\Delta)F\frac{1}{2}(1+\gamma) + \frac{1}{2}(1+\gamma)\left[a + (s_e^{in} + s_e^{out})\varepsilon\right] \end{bmatrix}$$

$$+\varepsilon^2\zeta^2 F^2 \left[\varepsilon\begin{pmatrix} s_e^{in} + s_e^{out} \\ +I^* + \rho_h^* \end{pmatrix}\begin{pmatrix} a + \varepsilon I^* \\ +\varepsilon\rho_e^* \end{pmatrix} - \varepsilon^2\left(I^* + \rho_h^*\right)\left(I^* + \rho_e^*\right)\right]$$
$$-2\varepsilon\zeta(1+\gamma)\sin(\Delta)F\frac{\alpha}{4}\left[a + (s_e^{in} + s_e^{out})\varepsilon\right]$$
$$+\varepsilon\zeta\cos(\Delta)F\frac{1}{2}(1+\gamma)\left[a + (s_e^{in} + s_e^{out})\varepsilon\right].$$

$$\tag{6.100}$$

6.7.3
Similar Carrier Lifetimes τ_e and τ_h with $a = O(\varepsilon)$

In this subsection, we assume that both $s_e^{in} + s_e^{out}$ and $s_h^{in} + s_h^{out}$ are $O(1)$ quantities. We seek a solution of the form

$$\mu = \mu_0 + \varepsilon\mu_1 + \ldots \tag{6.101}$$

and assume that $s_h^{in} + s_h^{out} = O(1)$. Inserting Eqs. (6.93), (6.96) and (6.101) into Eq. (6.100), we obtain the following sequence of problems for μ_0 and μ_1

$$O(1): \mu_0^4 + \mu_0^2 = 0, \tag{6.102}$$

$$O(\varepsilon): 4\mu_0^3\mu_1 + 2\mu_0\mu_1$$
$$+\mu_0^3\left[s_e^{in} + s_e^{out} + 2I^* + \rho_h^* + \rho_e^* + s_h^{in} + s_h^{out} + 2\zeta\cos(\Delta_0)F_0\right]$$
$$+\mu_0\left[\begin{matrix} -\zeta\sin(\Delta_0)F_0\alpha \\ +\zeta\cos(\Delta_0)F_0 + \frac{1}{2}(s_h^{in} + s_h^{out} + s_e^{in} + s_e^{out}) \end{matrix}\right] = 0, \tag{6.103}$$

where we have introduced

$$F_0 \equiv 1 - e^{-\mu_0 s_c}. \tag{6.104}$$

The solution of Eq. (6.102) is

$$\mu_0^2 = -1$$

and from Eq. (6.103), we then obtain

$$\mu_1 = -\Gamma - \frac{1}{2}\zeta F_0(\cos(\Delta_0) + \sin(\Delta_0)\alpha) \tag{6.105}$$

where

$$\Gamma \equiv \frac{1}{2}\left[\frac{s_e^{in} + s_e^{out}}{2} + 2I^* + \rho_h^* + \rho_e^* + \frac{s_h^{in} + s_h^{out}}{2}\right] \tag{6.106}$$

is the damping rate of the solitary laser [26]. Using Eq. (6.99) and $\lambda_0 = i$, Eq. (6.105) then implies that

$$\text{Re}(\mu_1) = -\Gamma - \frac{1}{2}\zeta(1 - \cos(s_c))(\cos(\Delta_0) + \sin(\Delta_0)\alpha) \tag{6.107}$$

$$= -\Gamma - \zeta \sin^2\left(\frac{s_c}{2}\right)(\cos(\Delta_0) + \sin(\Delta_0)\alpha).$$

The stability condition now is

$$\zeta < \zeta_H \equiv -\frac{\Gamma}{\sin^2(\frac{s_c}{2})(\cos(\Delta_0) + \sin(\Delta_0)\alpha)}, \tag{6.108}$$

if $\alpha \sin(\Delta_0) + \cos(\Delta_0) < 0$. The lowest possible value for ζ_H is for

$$\Delta_0 = C = \pi + \arctan(\alpha) \text{ and } s_c = \pi \pmod{2\pi}.$$

It is given by

$$\zeta_c \equiv \frac{\Gamma}{\sqrt{1+\alpha^2}}. \tag{6.109}$$

In terms of the original parameters, the stability condition is the same as for Eqs. (6.71)–(6.72), with Γ^S replacing Γ_1^{QD}, where

$$\Gamma^S \equiv \frac{\gamma}{2}\left[\frac{s_e^{in} + s_e^{out}}{2} + 2I^* + \rho_h^* + \rho_e^* + \frac{s_h^{in} + s_h^{out}}{2}\right]. \tag{6.110}$$

6.7.4
Different Carrier Lifetimes τ_e and τ_h with $a = O(1)$

Assuming that $s_e^{in} + s_e^{out} = O(1)$ and $s_h^{in} + s_h^{out} = O(\varepsilon^{-1})$ (or equivalently $= O(1)$), we now find from (6.100) that μ_0 satisfies

$$O(1): \mu_0^4 + \mu_0^3 a + \mu_0^2 + \mu_0 \frac{a}{2} = 0, \tag{6.111}$$

which is analyzed in [26]. We note that ζ does not appear in Eq. (6.111) meaning that the feedback is too weak ($k = O(\gamma)$) to have an effect in this case.

6.7.5
Very Small Scattering Lifetime of the Holes with $a = O(\varepsilon^{-1})$

In this subsection we assume that $s_e^{in} + s_e^{out} = O(1)$ and $s_h^{in} + s_h^{out} = O(\varepsilon^{-2})$ (or equivalently $a = O(\varepsilon^{-1})$). We introduce a $O(1)$ quantity $a_1 \equiv \varepsilon^2(s_h^{in} + s_h^{out})$, such that

$$a = \frac{a_1}{\varepsilon}. \tag{6.112}$$

Inserting Eq. (6.112) in Eq. (6.100), we now find the following problems for μ_0 and μ_1

$$O(\varepsilon^{-1}) : a_1\mu_0^3 + \mu_0\frac{a_1}{2} = 0, \tag{6.113}$$

$$O(1) : 3a_1\mu_0^2\mu_1 + \mu_1\frac{a_1}{2} + \mu_0^4$$
$$- \lambda_0^2\left[-1 - \left(s_e^{in} + s_e^{out} + I^* + \rho_h^*\right)a_1 - 2\zeta\cos(\Delta_0)F_0 a_1\right]$$
$$- 2\zeta\sin(\Delta_0)F_0\frac{\alpha}{4}a_1 + \zeta\cos(\Delta_0)F_0\frac{1}{2}a_1 = 0 \tag{6.114}$$

where F_0 is defined in Eq. (6.104). The solution of Eq. (6.113) is

$$\mu_0^2 = -\frac{1}{2}$$

and from Eq. (6.114), we then obtain

$$\mu_1 = -\Gamma - \frac{\zeta}{2}F_0(\cos(\Delta_0) + \sin(\Delta_0)\alpha) \tag{6.115}$$

where

$$\Gamma \equiv \frac{1}{2}\left[\frac{1}{2a_1} + s_e^{in} + s_e^{out} + I^* + \rho_h^*\right] \tag{6.116}$$

is the damping rate of the ROs for the solitary laser [26]. Using Eq. (6.99) and $\mu_0 = i/\sqrt{2}$, Eq. (6.115) implies

$$\text{Re}(\mu_1) = -\Gamma - \frac{1}{2}\zeta(1 - \cos(s_c))(\cos(\Delta_0) + \sin(\Delta_0)\alpha)$$
$$= -\Gamma - \zeta\sin^2(\frac{s_c}{2\sqrt{2}})(\cos(\Delta_0) + \sin(\Delta_0)\alpha). \tag{6.117}$$

The stability conditions are the same as for Eqs. (6.71)–(6.72), with Γ^{Da} replacing Γ_1^{QD} where

$$\Gamma^{Da} \equiv \frac{\gamma}{2}\left[\frac{1}{2a_1} + s_e^{in} + s_e^{out} + I^* + \rho_h^*\right]. \tag{6.118}$$

References

1. Kane, D.M. and Shore, K.A. (eds) (2005) *Unlocking Dynamical Diversity: Optical Feedback Effects on Semiconductor Lasers*, Wiley-VCH Verlag GmbH, Weinheim.

2. Mørk, J., Tromborg, B., and Mark, J. (1992) Chaos in semiconductor lasers with optical feedback-Theory and experiment. *IEEE J. Quantum Electron.*, **28**, 93.

3. Helms, J. and Petermann, K. (1990) A simple analytic expression for the stable operation range of laser diodes with optical feedback. *IEEE J. Quantum. Electron.*, **26** (5), 833.
4. Gioannini, M., Thé, G.A.P., and Montrosset, I. (2008) Multi-population rate equation simulation of quantum dot semiconductor lasers with feedback. NUSOD '08, 8th International Conference on Numerical Simulation of Optoelectronic Devices. September 1–5, 2008, Nottingham (UK), pp. 101–1021. DOI: 10.1109/nusod.2008.4668262.
5. Azouigui, S., Dagens, B., Lelarge, F., Provost, J.G., Make, D., Le Gouezigou, O., Accard, A., Martinez, A., Merghem, K., Grillot, F., Dehaese, O., Piron, R., Loualiche, S., Zou, Q., and Ramdane, A. (2009) Optical feedback tolerance of quantum-dot- and quantum-dash-based semiconductor lasers operating at 1.55 μm. *IEEE J. Sel. Top. Quantum Electron.*, **15** (3), 764–773. DOI: 10.1109/jstqe.2009.2013870.
6. Grillot, F., Naderi, N.A., Pochet, M., Lin, C.Y., and Lester, L.F. (2008) Variation of the feedback sensitivity in a 1.55 μm InAs/InP quantum-dash fabry-perot semiconductor laser. *Appl. Phys. Lett.*, **93** (19), 191108. DOI: 10.1063/1.2998397.
7. O'Brien, D., Hegarty, S.P., Huyet, G., McInerney, J.G., Kettler, T., Lämmlin, M., Bimberg, D., Ustinov, V., Zhukov, A.E., Mikhrin, S.S., and Kovsh, A.R. (2003) Feedback sensitivity of 1.3 μm InAs/GaAs quantum dot lasers. *Electron. Lett.*, **39** (25), 1819–1820.
8. Levine, A.M., van Tartwijk, G.H.M., Lenstra, D., and Erneux, T. (1995) Diode lasers with optical feedback: stability of the maximum gain mode. *Phys. Rev. A*, **52** (5), R3436–R3439. DOI: 10.1103/physreva.52.r3436.
9. Erneux, T. (2000) Asymptotic methods applied to semiconductor laser models. *Proc. SPIE*, **3944** (1), 588–601. DOI: 10.1117/12.391466.
10. Erneux, T. and Glorieux, P. (2010) *Laser Dynamics*, Cambridge University Press, UK.
11. Lüdge, K., Bormann, M.J.P., Malić, E., Hövel, P., Kuntz, M., Bimberg, D., Knorr, A., and Schöll, E. (2008) Turn-on dynamics and modulation response in semiconductor quantum dot lasers. *Phys. Rev. B*, **78** (3), 035316. DOI: 10.1103/physrevb.78.035316.
12. Lüdge, K. and Schöll, E. (2009) Quantum-dot lasers - desynchronized nonlinear dynamics of electrons and holes. *IEEE J. Quantum Electron.*, **45** (11), 1396–1403.
13. Lüdge, K. and Schöll, E. (2010) Nonlinear dynamics of doped semiconductor quantum dot lasers. *Eur. Phys. J. D*, **58**, 167–174. DOI: 10.1140/epjd/e2010-00041-8.
14. Lüdge, K., Aust, R., Fiol, G., Stubenrauch, M., Arsenijević, D., Bimberg, D., and Schöll, E. (2010) Large signal response of semiconductor quantum-dot lasers. *IEEE J. Quantum Electron.*, **46** (12), 1755–1762. DOI: 10.1109/jqe.2010.2066959.
15. Viktorov, E.A., Erneux, T., Mandel, P., Piwonski, T., Madden, G., Pulka, J., Huyet, G., and Houlihan, J. (2009) Recovery time scales in a reversed-biased quantum dot absorber. *Appl. Phys. Lett.*, **94**, 263502.
16. Piwonski, T., Pulka, J., Madden, G., Huyet, G., Houlihan, J., Viktorov, E.A., Erneux, T., and Mandel, P. (2009) Intradot dynamics of inas quantum dot based electroabsorbers. *Appl. Phys. Lett.*, **94** (12), 123504. DOI: 10.1063/1.3106633.
17. Erneux, T., Viktorov, E.A., Mandel, P., Piwonski, T., Huyet, G., and Houlihan, J. (2009) The fast recovery dynamics of a quantum dot semiconductor optical amplifier. *Appl. Phys. Lett.*, **94** (11), 113501.
18. Erneux, T., Viktorov, E.A., Kelleher, B., Goulding, D., Hegarty, S.P., and Huyet, G. (2010) Optically injected quantum-dot lasers. *Opt. Lett.*, **35** (7), 937–939.
19. Otto, C., Lüdge, K., and Schöll, E. (2010) Modeling quantum dot lasers with optical feedback: sensitivity of bifurcation scenarios. *Phys. Stat. Sol. B*, **247** (4), 829. DOI: 10.1002/pssb.200945434.
20. Otto, C., Globisch, B., Lüdge, K., Schöll, E., and Erneux, T. Complex dynamics of semiconductor quantum dot lasers subject to delayed optical feedback. (2012) *Int. J. Bifurcat. Chaos*, Submitted.

21. Sugawara, M., Mukai, K., and Shoji, H. (1997) Effect of phonon bottleneck on quantum-dot laser performance. *Appl. Phys. Lett.*, **71** (19), 2791–2793. DOI: 10.1063/1.120135.
22. Ishida, M., Hatori, N., Akiyama, T., Otsubo, K., Nakata, Y., Ebe, H., Sugawara, M., and Arakawa, Y. (2004) Photon lifetime dependence of modulation efficiency and K factor im 1.3 μm self-assembled *InAs/GaAs* quantum-dot lasers: Impact of capture time and maximum modal gain on modulation bandwidth. *Appl. Phys. Lett.*, **85** (18), 4145.
23. Erneux, T., Viktorov, E.A., and Mandel, P. (2007) Time scales and relaxation dynamics in quantum-dot lasers. *Phys. Rev. A*, **76**, 023819. DOI: 10.1103/physreva.76.023819.
24. Malić, E., Ahn, K.J., Bormann, M.J.P., Hövel, P., Schöll, E., Knorr, A., Kuntz, M., and Bimberg, D. (2006) Theory of relaxation oscillations in semiconductor quantum dot lasers. *Appl. Phys. Lett.*, **89**, 101107. DOI: 10.1063/1.2346224.
25. Malić, E., Bormann, M.J.P., Hövel, P., Kuntz, M., Bimberg, D., Knorr, A., and Schöll, E. (2007) Coulomb damped relaxation oscillations in semiconductor quantum dot lasers. *IEEE J. Sel. Top. Quantum Electron.*, **13** (5), 1242–1248. DOI: 10.119/jqstqe.2007.905148.
26. Lüdge, K., Schöll, E., Viktorov, E.A., and Erneux, T. (2011) Analytical approach to modulation properties of quantum dot lasers. *J. Appl. Phys.*, **109** (9), 103112-1–103112-8.
27. O'Brien, D., Hegarty, S.P., Huyet, G. and Uskov, A.V. (2004) Sensitivity of quantum-dot semiconductor lasers to optical feedback. *Opt. Lett.*, **29** (10), 1072–1074.
28. Uskov, A.V., Boucher, Y., Bihan, J.L., and McInerney, J. (1998) Theory of a self-assembled quantum-dot semiconductor laser with auger carrier capture: Quantum efficiency and nonlinear gain. *Appl. Phys. Lett.*, **73** (11), 1499–1501. DOI: 10.1063/1.122185.
29. Engelborghs, K., Luzyanina, T., and Roose, D. (2002) Numerical bifurcation analysis of delay differential equations using DDE-Biftool. *ACM Trans. Math. Softw.*, **28**, 1–21.

7
Bifurcation Study of a Semiconductor Laser with Saturable Absorber and Delayed Optical Feedback

Bernd Krauskopf and Jamie J. Walker

7.1
Introduction

Semiconductor lasers are a very efficient type of laser that has found numerous applications in recent years – most prominently in optical data storage and in optical telecommunication. Indeed, fiber-optic communication has become the method of choice for transmitting large amounts of information, and semiconductor laser devices are the optical light sources behind today's telecommunication networks. There are many reasons for the popularity of this type of laser as a light source: semiconductor lasers are very efficient in converting electrical energy into coherent light, very small (with cavity lengths of about 1 mm), and inexpensive and easy to manufacture. From a more fundamental perspective, a semiconductor laser is a damped nonlinear oscillator. It is now well known that, apart from stable emission, semiconductor laser systems may show a wealth of other dynamics, including different types of periodic outputs, as well as quasiperiodic and chaotic dynamics. This type of dynamics are brought about by external influences, such as modulation of the electrical pump current, external optical input or optical feedback. See, for example, [1, 2] and other chapters as entry points to the extensive literature on nonlinear laser dynamics.

The focus of this study is a semiconductor laser with saturable absorber (SLSA), which has been shown experimentally [3] and theoretically [4, 5] to be capable of producing self-pulsations. The underlying physical process is called passive Q-switching, and it can be explained as follows; see also [6, 7]. The absorber in (or adjacent to) the laser cavity acts as a store of energy that is supplied to the semiconductor laser by an electrical pump current. Filling this energy store is a relatively slow process (with respect to the internal timescale of the laser dynamics). When the absorber is saturated, the laser is able to overcome its losses, and all the stored energy is released in a very short period of time, leading to a pulse of emitted light. The intensity drops back to zero, and the process repeats. The result is a pulse train with a typical pulse-repetition frequency on the order of several GHz.

It is this property of the SLSA that makes it interesting for use in pulse generation for telecommunication and for optical clocks. However, there is one

Nonlinear Laser Dynamics: From Quantum Dots to Cryptography, First Edition. Edited by Kathy Lüdge.
© 2012 Wiley-VCH Verlag GmbH & Co. KGaA. Published 2012 by Wiley-VCH Verlag GmbH & Co. KGaA.

drawback: the self-sustained oscillations may be quite sensitive to the influence of external or internal noise [8]. More specifically, when the absorber is close to being saturated, the next pulse can be triggered by even quite small amounts of noise. The result of noise is so-called timing jitter of the pulses, meaning that the time in between successive pulses is subject to considerable fluctuations. The system actually displays what is known as coherence resonance: there is a noise level that minimizes the timing jitter of the pulse train of the SLSA; see [9]. Clearly any considerable jitter due to noise is detrimental in the mentioned applications, because they require precise timing of pulses.

It has been shown that the timing jitter of the SLSA is due to the fact that self-sustained pulsations occur close to a region of excitability [5, 9]. Excitability is a well-known concept that comes originally from biology and chemistry [10]. Examples are excitation waves in reaction–diffusion systems, such as cardiac muscle tissue and the Belousov–Zhabotinsky reaction [11]; excitability is also an important concept in neuron and cell modeling, and it is one of the mechanisms that may lead to the spiking of nerve cells [10, 12]. More recently, there has been a surge of interest in excitable laser systems. Indeed, lasers with saturable absorber are not the only class of lasers in which excitability has been found. Other laser systems demonstrating excitability include lasers with optical injection [13, 14] or optical feedback [15, 16], multisection DFB lasers [17], and lasers with integrated dispersive reflectors [18]; see also [19]. Potential applications of excitable lasers include clock recovery, where the laser acts as an optical switch that reacts only to sufficiently large optical input, and pulse reshaping, where a dispersed input pulse can generate a clean large-amplitude output pulse.

In this chapter, we consider the dynamics of the SLSA when part of its output light is fed back after a given delay time τ. This laser system can be realized as is sketched in Figure 7.1. A beam splitter (BS) diverts a part of the laser's output into a feedback loop. The feedback strength κ can be varied by an attenuator, and the delay time τ is determined by a chosen length of fiber optical cable; an optical isolator (ISO) prevents unwanted back-reflections. One motivation for studying this setup is to understand how to operate the SLSA with delayed optical feedback in such a way that it produces a pulse train with desired properties. The naive underlying

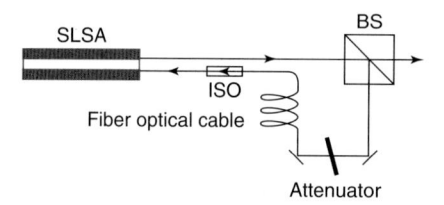

Figure 7.1 Sketch of an SLSA (illustrated by a (white) gain medium surrounded by a (gray) absorber medium) that is fed back a part of its output via an external optical feedback loop with fixed delay time, as determined by a length of fiber optical cable; the feedback loop also includes a beam splitter (BS), an optical isolator (ISO), and an attenuator.

idea is the following. Suppose that the SLSA is in the excitable regime (in the absence of feedback). When a first pulse is triggered, a part of it will travel back to the SLSA via the feedback loop to trigger the next pulse. If this process continues stably, then a train of pulses will be generated. Importantly, the frequency of the pulse train is determined by the delay time τ of the external feedback loop (and not just by the material properties of the SLSA). Hence, the SLSA with delayed optical feedback features two additional control parameters, the delay time τ and the feedback strength κ. The main question is how these two external parameters influence the dynamics of the overall system, given the properties of the SLSA as determined by its internal parameters.

To answer this question, we perform a bifurcation study of the SLSA with delayed optical feedback as modeled by the established Yamada rate equations [20] for the gain G, the absorption Q, and the intensity I, to which a delay term has been added. When written in dimensionless form, one obtains the system

$$\dot{G} = \gamma (A - G - GI), \tag{7.1}$$
$$\dot{Q} = \gamma (B - Q - aQI), \tag{7.2}$$
$$\dot{I} = (G - Q - 1) I + \kappa I (t - \tau) \tag{7.3}$$

where the dot denotes derivation with respect to time. The delay term $\kappa I (t - \tau)$, with feedback strength $\kappa \geq 0$ and delay time $\tau \geq 0$, models the feedback via the external optical loop. Physically, the length l of the feedback loop, which is mainly due to the fiber optical cable, determines the single fixed delay $\tau = l/c > 0$, where c is the speed of light.

For $\kappa = 0$, one recovers the Yamada equations – a system of three first-order ordinary differential equations (ODEs) that describe a single-mode laser with saturable absorber – in the form that was considered for the bifurcation study in [5]. There are four dimensionless parameters: the pump parameter A of the gain, the pump parameter B of the absorption, the cross-saturation coefficient a, and the timescale ratio γ between the relaxation rates (or decay times) of gain and absorber. The derivation of the Yamada equations can be found in [20] and also in the chapter by Vladimirov et al., where they emerge as the limit (for small G and Q) of Eqs. (8.22)–(8.24). Note that, due to scaling choices for variables and parameters, these other versions of the Yamada equations differ somewhat from the one considered here; see also [5].

For physical reasons, the parameter space of Eqs. (7.1)–(7.3) for $\kappa = 0$ is confined to $A \geq 0$, $B \geq 0$, $a \geq 1$, and $\gamma \geq 0$. Furthermore, γ is a small parameter (of the order of 10^{-3}–10^{-4}), which means that gain G and absorption Q evolve on a much slower timescale than the intensity I. Hence, the SLSA is an example of a slow–fast system with an explicit splitting of timescales; see, for example, [21]. A complete bifurcation analysis of Eqs. (7.1)–(7.3) for $\kappa = 0$ can be found in [5]. In particular, just before the onset of naturally occurring self-pulsations, the dynamics of the SLSA is excitable [5, 9]; these results are summarized in Section 7.2, and they form the basis of what is presented here.

To address the question of how the dynamics of the SLSA is influenced by the external feedback loop, we present a bifurcation study of Eqs. (7.1)–(7.3) for

$\kappa \geq 0$. Note that this means that we are dealing with a system of delay differential equations (DDEs) with a single fixed delay. As such, it has as its phase space the space $C([-\tau, 0]; \mathbb{R}^3)$ of continuous functions over the delay interval $[-\tau, 0]$ with values in (G, Q, I)-space; see, for example, [22–24]. It is this element of infinite dimensionality that allows Eqs. (7.1)–(7.3) to show much richer dynamics than the SLSA alone (when $\kappa = 0$). Up until only a few years ago, practical methods for analyzing DDEs were limited to linearization around equilibria of the system and numerical integration of the governing equations. Today, however, numerical tools for the detection and continuation in parameters of equilibria, periodic solutions, and their bifurcations are also available for DDEs in the form of the packages DDE-BIFTOOL [25] and PDDE-CONT [26]; see also the recent surveys [27, 28]. We use here the package DDE-BIFTOOL to carry out a bifurcation study of the full DDE given by Eqs. (7.1)–(7.3). More specifically, we fix $B = 5.8$, $a = 1.8$, $\gamma = 0.04$ throughout and consider bifurcation diagrams in the (τ, κ)-plane, where we consider two main cases for the gain pump rate A: one where the SLSA is off and excitable, and the other where it is is off and not excitable. The transition between these two cases as A is changed is explained in terms of the passage through codimension-three bifurcation points.

In light of the explicit split into slow and fast variables of the system, what is presented here is a case study of a slow–fast system subject to delayed feedback. This more general aspect provides a second motivation, because it may also be of interest for other areas of application. For example, the issue of delayed feedback or coupling also arises in the context of interacting (populations of) neuron cells, which themselves may display dynamics on separate timescales.

The chapter is organized as follows. In Section 7.2, we summarize the results from [5] for $\kappa = 0$. The next four sections are devoted to the study of the full DDE for $\kappa \geq 0$. Section 7.3 presents analytic results on basic bifurcations of equilibria. Sections 7.4 and 7.5 are devoted to bifurcation diagrams in the (τ, κ)-plane for two representative values of the gain pump parameter A, and Section 7.6 discusses the transition between them via different codimension-three bifurcations. Finally, we summarize in Section 7.7.

7.2
Bifurcation Analysis of the SLSA

The Yamada model in the form of Eqs. (7.1)–(7.3) for $\kappa = 0$ describes an SLSA in two different geometric configurations: (i) in a sectional configuration where there are gain and absorber sections inside the laser cavity but with the same carrier lifetime and (ii) in a striped configuration where unpumped side regions act as the absorber (as in Figure 7.1).

Figure 7.2 summarizes the main results of a bifurcation analysis of the Yamada model in [5]. The main object is the two-parameter bifurcation diagram in the (A, γ)-plane of pump parameter A of the gain and timescale separation parameter γ, where the pump parameter B of the absorption and the cross-saturation

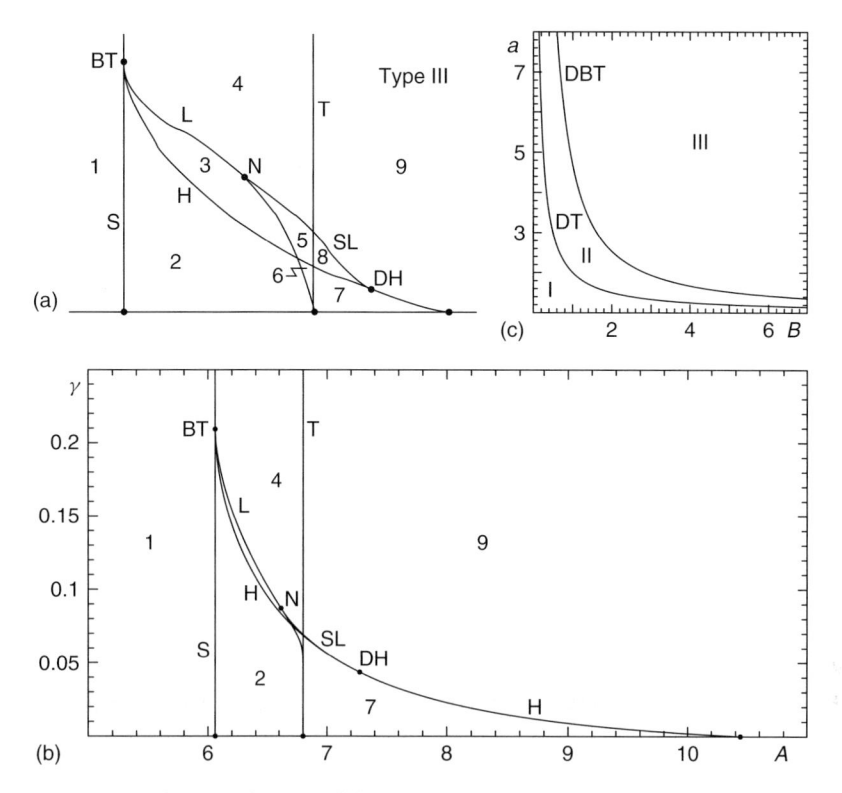

Figure 7.2 Bifurcation diagram of the Yamada model, Eqs. (7.1)–(7.3) for $\kappa = 0$. Panel (a) is a sketch of the bifurcation diagram of type III in the (A, γ)-plane of gain pump parameter A and timescale separation parameter γ; panel (b) shows this bifurcation diagram in the (A, γ)-plane as computed for $(B, a) = (5.8, 1.8)$; see Figure 7.3 for the corresponding phase portraits in regions 1–9. Panel (c) shows the division of the (B, a)-plane of absorption pump parameter B and cross-saturation coefficient a into regions of bifurcation diagrams of types I–III. Reprinted from *Optics Communications* 159(4-6), Selfpulsations of lasers with saturable absorber: dynamics and bifurcations, Dubbeldam and Krauskopf, pp. 325–338, © (1999), with permission from Elsevier.

coefficicent a have fixed values. We are concerned with the bifurcation diagram of type III (in the notation of [5]), because it features all possible dynamics of the SLSA. This bifurcation diagram is sketched diagrammatically in Figure 7.2a, and it is shown in panel (b) as computed with the continuation package AUTO [29] for $(B, a) = (5.8, 1.8)$. The bifurcation diagram of type III is physically relevant because it can be found for any sufficiently large values of B and a. This is illustrated in Figure 7.2c, which shows how the (B, a)-plane is divided (by two curves DT and DBT of two different types of degenerate Bogdanov–Takens bifurcations) into three regions corresponding to bifurcation diagrams of types I–III. The bifurcation diagrams of types I and II (which are not considered here) can be found in [5].

We now discuss the bifurcation diagram of type III in Figure 7.2a,b in more detail; see, for example, [30, 31] as general references to bifurcation theory. Several

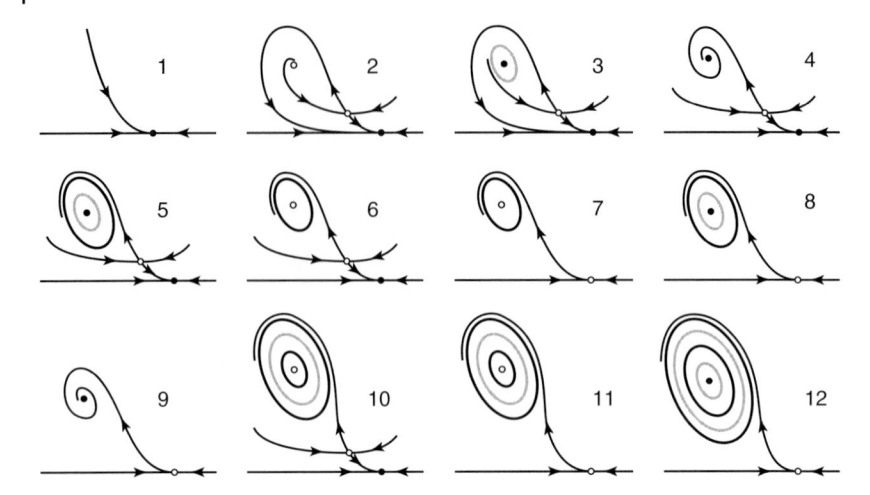

Figure 7.3 Phase portraits corresponding to the numbered regions 1–9 in Figure 7.2 and regions 1–12 in Figures 7.6 and 7.9. Shown are two-dimensional sketches in projection onto the (G, I)-plane, where the invariant set $\{I = 0\}$ is at the bottom; the third direction that is not shown is attracting. Black dots are attracting equilibria; open dots are saddle equilibria, and they have stable and unstable manifolds; black closed curves are attracting periodic orbits, and gray closed curves are saddle periodic orbits.

bifurcation curves divide the (A, γ)-plane into nine regions of topologically different phase portraits, which can be found in panels 1–9 of Figure 7.3. The phase portraits are represented as two-dimensional sketches because, after possible transients, the dynamics takes place in a globally attracting two-dimensional surface that is close to $\{G - Q - 1 = 0\}$; the third direction that is not shown in Figure 7.3 is consequently attracting. Along each bifurcation curve in the (A, γ)-plane, one finds a particular bifurcation (qualitative change of the dynamics), which is said to be of codimension one; specifically, we encounter:

- a saddle-node bifurcation curve S, where two equilibria are created (or disappear); an example is the transition between phase portraits 1 and 2;
- a transcritical bifurcation curve T, where an equilibrium for $I = 0$ changes stability by an equilibrium moving out of (or into) the region for $I > 0$; an example is the transition between phase portraits 4 and 9;
- a Hopf bifurcation curve H, where a periodic orbit (corresponding to self-oscillations) is created (or disappears); an example is the transition between phase portraits 2 and 3;
- a curve L along which one finds a homoclinic loop to a saddle equilibrium; a periodic orbit bifurcates from this homoclinic loop; an example is the transition between phase portraits 3 and 4;
- a curve SL of saddle-node bifurcation of limit cycles, where two periodic orbits (one attracting and one of saddle type) are created (or disappear); an example is the transition between phase portraits 4 and 5.

7.2 Bifurcation Analysis of the SLSA

The bifurcation diagram in Figure 7.2a,b is organized by a number of special points – known as codimension-two bifurcations – where several bifurcation curves come together. The main organizing center is a Bogdanov–Takens point BT on the curve S (characterized by a double-zero eigenvalue of the Jacobian at the equilibrium), from which the curves H and L emerge. The homoclinic loop curve L changes type in a codimension-two point N (where the saddle quantity of the equilibrium in the homoclinic loop becomes zero and changes sign); here the bifurcating periodic orbit changes from being repelling (in region 3) to being attracting (in region 5). From N the curve SL emerges, and it ends on the Hopf curve H at a point DH of degenerate Hopf bifurcation; note from Figure 7.2b that the curve SL is very close to the Hopf curve H. The homoclinic loop curve L ends at the bottom point of the transcritical curve T (where $\gamma = 0$); notice that the curve L follows the curve T extremely closely for $\gamma < 0.05$; see Figure 7.2b.

One conclusion from Figures 7.2 and 7.3 is that, for realistically small values of the timescale separation parameter γ, one finds a unique sequence of bifurcations as the gain pump parameter A is increased. Note in this context that A is the only parameter that can be changed during an experiment. Initially, the laser is off, which is represented in region 1 by an attractor with intensity $I = 0$. When the curve S is crossed, the laser is still off in region 2, but there are now two additional equilibria; both are saddle points in (G, Q, I)-space with $I > 0$. (Recall that the missing direction in Figure 7.3 is attracting.) When A is increased further, the homoclinic loop curve L and the transcritical curve T are practically crossed at the same time; this marks the onset of self-pulsations, which are represented in region 7 by an attracting periodic orbit. We remark that region 6, where one finds coexistence of the stable equilibrium and a stable periodic orbit, is so small that it was not found in studies that changed A for fixed small γ. Yet, region 6 must exist for topological reasons, and it was indeed found only as part of the bifurcation diagram of type III by allowing γ to take larger values. Finally, when A is increased even further, the Hopf bifurcation curve H is crossed. The self-pulsations disappear, and the laser produces light with constant intensity I; in region 9, this is represented by a globally attracting equilibrium.

It is an important realization that the SLSA is excitable in region 2, that is, for a considerable range of the gain pump parameter A. Figure 7.4 shows the underlying mechanism in more detail. The laser is in its off-state, and this is represented by the globally attracting equilibrium with $I = 0$. In response to any sufficiently small perturbation in the intensity I, the SLSA relaxes back to the off-state. On the other hand, any perturbation above a certain threshold results in a large pulse before the SLSA relaxes back to the off-state; one speaks of the excitability threshold, and it is given in this case by the stable manifold of the saddle point that lies close to the stable equilibrium. Physically, the perturbation in I is sufficient in this case to overcome the losses and release the energy stored in the absorber. The SLSA requires what is known as a refractory period (to recharge the absorber) before the next pulse can be triggered. Note that the saddle point moves closer to the attractor as the transcritical curve T is approached, which implies that the

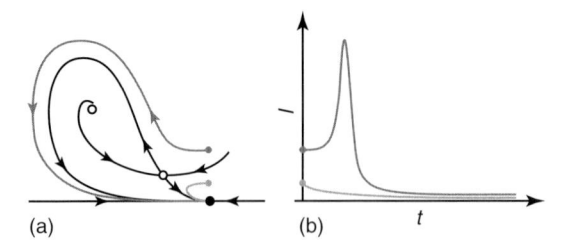

Figure 7.4 The excitable phase portrait 2 (a) gives rise to a pulse for a sufficiently large perturbation from the off-state (b).

excitability threshold decreases. Hence, in region 2, the SLSA becomes more and more likely to produce noise-induced pulses as A is increased.

7.3
Equilibria of the DDE and Their Stability

If we now consider $k \geq 0$, then Eqs. (7.1)–(7.3) are a system of DDEs and, as such, difficult to study by analytical means. Nevertheless, it is possible to find explicit formulas for some of their bifurcations. This information forms the basis of the numerical bifurcation analysis in the next sections.

There are three equilibria, E_1, E_2, and E_3, given by

$$E_1 : (G, Q, I) = (A, B, 0), \tag{7.4}$$

$$E_2, E_3 : (G, Q, I) = \left(\frac{A}{1 + I_\pm}, \frac{B}{1 + aI_\pm}, I_\pm\right) \tag{7.5}$$

where

$$I_\pm = \frac{-aA + b + a + 1 - a\kappa - \kappa}{2a(\kappa - 1)}$$

$$\pm \frac{\sqrt{(aA - B - a - 1 + a\kappa + \kappa)^2 - 4a(\kappa - 1)(A - B - 1 + \kappa)}}{2a(\kappa - 1)}. \tag{7.6}$$

The equilibrium E_1 lies on the invariant plane $\{I = 0\}$; it exists for all values of the parameters and corresponds to the nonlasing solution. The two equilibria E_2 and E_3 have nonzero intensity and, since $2a(\kappa - 1) \leq 0$ for all $\kappa \in [0, 1]$, we find that $I_- > I_+$. That is, the intensity value at E_2 is less than that at E_3. Note that for $\kappa = 0$, the intensity equation for $I_\pm$ reduces to that for the Yamada model [5].

To check the stability and bifurcations of the equilibria, we make use of the fact that Eqs. (7.1)–(7.3) are a DDE with a single fixed delay $\tau > 0$, which has the general form

$$\dot{x} = f(x(t), x(t - \tau), \psi) =: f(u, v, \psi).$$

Here $x(t) \in \mathbb{R}^3$ represents a point in the physical (G, Q, I)-space, and f is a smooth function that depends on the parameter vector ψ. The stability of an equilibrium

7.3 Equilibria of the DDE and Their Stability

x_0 is determined by the roots λ of the characteristic equation

$$\det\left(\lambda(\delta_{ij}) - A_1 - A_2 e^{-\lambda\tau}\right) = 0$$

where

$$A_1 = \frac{\partial f(x_0)}{\partial u}, \quad A_2 = \frac{\partial f(x_0)}{\partial v},$$

and (δ_{ij}) is the identity matrix. The transcendental characteristic equation evaluated at an equilibrium has countably infinitely many roots λ_i, but only finitely many of them have a positive real part. Hence, an equilibrium is either an attractor with infinitely many attracting directions or a saddle point with a finite number of repelling directions and an infinite number of attracting directions; see [22–24].

For Eqs. (7.1)–(7.3), we have

$$A_1 = \begin{bmatrix} -\gamma(1+I) & 0 & -\gamma G \\ 0 & -\gamma(1+aI) & -\gamma aQ \\ I & -I & G - Q - 1 \end{bmatrix}, \quad A_2 = \begin{bmatrix} 0 & 0 & 0 \\ 0 & 0 & 0 \\ 0 & 0 & \kappa \end{bmatrix},$$

and the characteristic equation is

$$\begin{aligned}
0 = \lambda^3 &+ [1 - G + Q + \gamma(2 + aI + I) - \kappa e^{-\lambda\tau}]\lambda^2 + [\gamma^2(1 + I + aI + aI^2) \\
&+ \gamma(2 - 2G + 2Q + I + aI - aGI + QI) - \gamma(2 + I + aI)\kappa e^{-\lambda\tau}]\lambda \\
&+ \gamma^2(1 - G + Q + I + aI + aI^2 - aGI \\
&+ QI - (1 + I + aI + aI^2)\kappa e^{-\lambda\tau}).
\end{aligned} \qquad (7.7)$$

Note that for $\kappa = 0$, this characteristic equation reduces to the eigenvalue equation of the Yamada system [5].

While the analysis of Eq. (7.7) is generally quite difficult, we can make the following observations.

Proposition 1 *In the (τ, κ)-plane, one finds the following two local bifurcations along horizontal lines:*

(i) *the locus T of transcritical bifurcations, given by*

$$\kappa_T(A, B) = 1 - A + B \qquad (7.8)$$

where the equilibria E_1 and E_2 meet on the invariant line $\{I = 0\}$. For $\kappa < \kappa_T$ both E_1 and E_2 exist (in the region where $I \geq 0$), and E_1 is attracting. For $\kappa > \kappa_T$ only E_1 exists (in the region where $I \geq 0$), and it is a saddle point.

(ii) *the locus S of saddle-node bifurcations, given by*

$$\kappa_S(A, B, a) = \frac{-aA + a - B - 1 + 2\sqrt{aAB}}{a - 1} \qquad (7.9)$$

where the equilibria E_2 and E_3 bifurcate. For $\kappa < \kappa_S$ the system possesses just the equilibrium E_1; for $\kappa > \kappa_S$ also the equilibria E_2 and E_3 exist, of which E_2 is always a saddle point (in the region where $I \geq 0$).

Proof. For statement (i) we consider the equilibrium E_1, where $(G, Q, I) = (A, B, 0)$, so that E.q. (7.7) reduces to

$$0 = \lambda^3 + \left[1 - A + B + 2\gamma - \kappa e^{-\lambda \tau}\right]\lambda^2 + \left[\gamma^2 + 2\gamma - 2\gamma A \right.$$
$$\left. + 2\gamma B - 2\gamma \kappa e^{-\lambda \tau}\right]\lambda + \gamma^2 \left(1 - A + B - \kappa e^{-\lambda \tau}\right).$$

The condition that $\lambda = 0$ is a root immediately gives Eq. (7.8). The existence of E_2 with $I \geq 0$ for $\kappa < \kappa_T$ follows from Eq. (7.6), and the stability of E_1 was checked numerically with DDE-BIFTOOL by inspecting the change of roots of the characteristic equation across S.

For statement (ii), one needs to consider a root $\lambda = 0$ of the full characteristic equation, that is, a root of the last term of Eq. (7.7). A simpler alternative is to realize that a saddle-node bifurcation corresponds to the square root in Eq. (7.6) being zero, that is,

$$0 = (aA - B - a - 1 + a\kappa + \kappa)^2 - 4a(\kappa - 1)(A - B - 1 + \kappa),$$

and Eq. (7.9) follows. The stability of the bifurcating equilibria E_2 and E_3 was again checked numerically with DDE-BIFTOOL. ∎

It follows from Eq. (7.8) that the transcritical locus T lies only in the physically relevant part of the (τ, κ)-plane, meaning that $\kappa_T \geq 0$, provided that

$$A \leq B + 1$$

Note that equality above gives exactly the condition that there is a transcritical bifurcation in the Yamada model; see Appendix A in [5]. In other words, the transcritical locus T occurs only for $\kappa \geq 0$ provided that A is chosen to lie to the left of the curve T in Figure 7.2b. Similarly, Eq. (7.9) implies that the saddle-node locus S is such that $\kappa_S \geq 0$ provided that

$$A \leq \frac{(-1 + a + 2\sqrt{(a-1)B} + B)}{a}.$$

Equality above gives exactly the condition that there is a saddle-node bifurcation in the Yamada model; see Appendix A in [5]. Hence, the saddle-node locus S only occurs for $\kappa_S \geq 0$ provided that A is chosen from region 1, to the left of the saddle-node locus S in Figure 7.2b. We remark that $\kappa_S \leq 1$ is always satisfied, which means that Eq. (7.6) does actually not become singular at either E_2 or E_3.

The equilibrium E_3 may lose its stability in a Hopf bifurcation. The ansatz that there is a purely imaginary root $\lambda = i\omega$ of the characteristic Eq. (7.7) leads to a complicated transcendental equation for the locus of Hopf bifurcation. As the bifurcation diagrams in the next sections show, there is not a simple formula for this locus, which may consist of infinitely many disjoint curves of Hopf bifurcation in the (τ, κ)-plane. Rather than computing them numerically from Eq. (7.7), we compute the curves of Hopf bifurcation with the continuation package DDE-BIFTOOL; this is equivalent, because DDE-BIFTOOL also solves the characteristic equation numerically, albeit in implicit form [25, 28].

7.4
Bifurcation Study for Excitable SLSA

We now study the influence of the feedback loop on phase portrait 2 of the Yamada model, where the off-state of the laser is a global attractor and the system is excitable; see Figure 7.4. To this end, we fix the gain pump parameter at $A = 6.5$, to the left of the transcritical curve in Figure 7.2b. (Recall that $B = 5.8$, $a = 1.8$ and $\gamma = 0.04$ are fixed.) For this value of A, the transcritical locus T, but not the saddle-node locus S, can be found in the physically relevant region where $\kappa \geq 0$; see Proposition 1.

Figure 7.5 shows the computed bifurcation diagram of Eqs. (7.1)–(7.3) in the (τ, κ)-plane. Apart from the horizontal curve T at $\kappa_T = 1 - A + B = 0.3$, the figure shows a single, connected Hopf curve H and a homoclinic loop curve L. The curve H has more and more self-intersections for larger values of τ; each such intersection is a Hopf–Hopf bifurcation point where the system possesses two pairs of purely complex conjugate eigenvalues. The homoclinic loop curve L crosses the Hopf curve H twice; see the enlargements in Figure 7.5b,c. The curve L then appears to approach the line $\kappa = 0$ as $\tau \to \infty$.

Figure 7.6 shows two qualitative sketches of this bifurcation diagram in the (τ, κ)-plane. Panel (a) is for quite small values of τ up to about 1; compare with Figure 7.5c. Figure 7.6b is for intermediate values of τ up to about 60; compare with Figure 7.5a. This range of the (τ, κ)-plane is most relevant from the applications

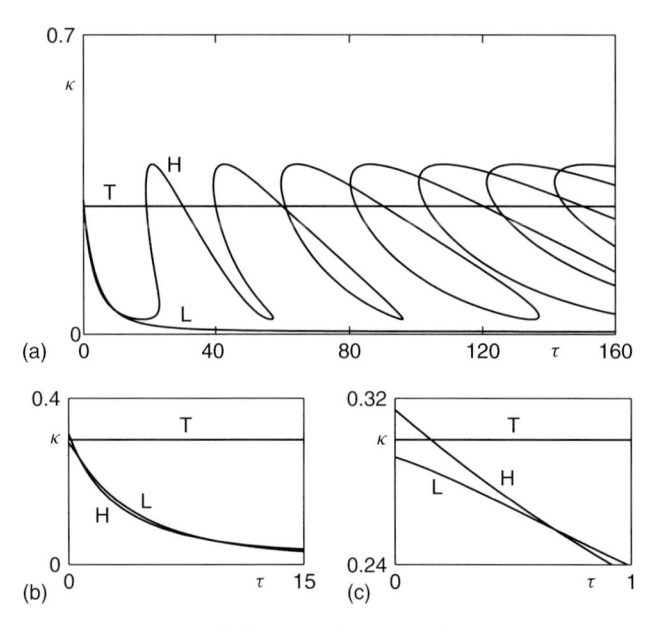

Figure 7.5 Computed bifurcation diagram in the (τ, κ)-plane for $A = 6.5$ (a); panels (b) and (c) are two successive enlargements. Shown are the transcritical curve T, a Hopf curve H, and a homoclinic loop curve L.

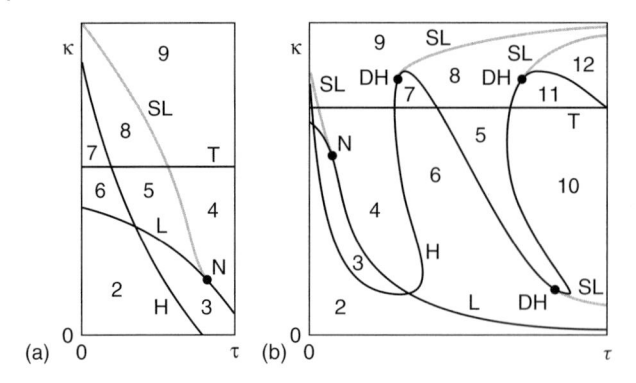

Figure 7.6 Sketches of the bifurcation diagram in the (τ, κ)-plane for $A = 6.5$; panel (a) is up to small and panel (b) up to an intermediate value of τ. The gray curves SL are loci of saddle-node of limit cycle bifurcations; numbered regions correspond to phase portraits in Figure 7.3.

point of view, because we are interested in self-pulsations of high frequency (hence, τ should not be too large). The new features in Figure 7.6 are (gray) curves SL of saddle-node bifurcation of limit cycles: one emanates from a codimension-two point N on the curve L where the saddle is neutral, and the other curves SL emanate from degenerate Hopf points DH on the Hopf curve H. At each such point, the Hopf bifurcation changes criticality, meaning that the bifurcating periodic orbit changes form being attracting to being of saddle type. Because it is quite difficult to follow a saddle-node bifurcation of limit cycles in a DDE, we verified the positions of the curves SL with careful numerical simulations of Eqs. (7.1)–(7.3). We also found that the Hopf bifurcation curve H and the homoclinic loop curve L both change criticality very close to where they cross for lower values of κ; in fact, this happens practically at their intersection point within the accuracy of our investigation, and no additional curve SL could be found.

The sketched bifurcation curves in Figure 7.6 constitute a conjectured partial bifurcation diagram in the shown part of the (τ, κ)-plane; it is complete enough to allow us to identify the numbered regions with different phase portraits that can be found in Figure 7.3. Notice that region 2 of excitable dynamics is immediately adjacent to the line $\{\kappa = 0\}$; this is expected for our choice of A because the DDE for $\kappa > 0$ is a regular perturbation of the ODE for $\kappa = 0$. What is more, for small τ we find a bifurcation structure that is very much like that in the (A, γ)-plane of type III; compare Figure 7.6a with Figure 7.2a. In particular, phase portraits 2–9 (that is, all phase portraits of the Yamada system with the exception of that in region 1) can be found in the corresponding regions. In other words, near the limit $\tau = 0$ the delay time τ and the feedback strength κ unfold the dynamics in a similar way as gain pump parameter A and timescale separation parameter γ when $\tau = \kappa = 0$. Already for intermediate values of τ as in Figure 7.6b, on the other hand, we find a bifurcation structure that is more complicated, with additional dynamics in regions 10–12. Note from Figure 7.3 that these new phase portraits can still be drawn in a two-dimensional plane. All these phase portraits have stable oscillations; their new feature is the existence of additional nested periodic orbits.

For intermediate values of the feedback strength around $\kappa = 0.2$, one finds a characteristic transition as the delay time τ is increased; see Figure 7.3 for the phase portraits that are encountered. For small τ in region 2, the laser is excitable. When the curve H is crossed into region 3 as τ is increased, a periodic orbit of saddle type is created around the bifurcating equilibrium, which is now stable. Hence, the laser is bistable: it is still excitable when it is in the off-state with $I = 0$, but for suitable initial conditions it may also emit light at the constant intensity value of the stable equilibrium with $I > 0$. When the curve L is crossed and region 4 is entered, the unstable periodic orbit disappears; the system is still bistable but no longer excitable: any sufficiently large perturbation of the off-state now brings the laser into the basin of attraction of the stable equilibrium with $I > 0$. When τ is increased further, the curve H is crossed a second time. Now a stable periodic orbit is born along H so that in region 6 we now find bistability between the off-state and self-sustained oscillations. As τ is increased even further, the Hopf bifurcation curve H is crossed several more times; this leads to the creation of more periodic orbits, which are alternatingly attracting and of saddle type. As Figure 7.5a clearly shows, the bifurcation diagram becomes increasingly complicated for even larger values of τ – so much so that it becomes impractical to map out all regions of different dynamics.

An important new aspect of the bifurcation diagram in Figure 7.6b is the fact that region 6, where one finds bistability between the off-state and self-pulsations, is now so large that it becomes experimentally accessible. It can be reached, for example, from region 2 by increasing κ for a fixed intermediate value of the delay time τ, or from region 4 by increasing τ for suitable fixed κ. The relevant region in the (τ, κ)-plane is shown enlarged in Figure 7.7a. Panel (b) shows the period of the attracting periodic orbit Γ as it is continued in the direction of decreasing κ from the point p on the Hopf curve H. As is to be expected, the period of Γ increases and diverges to infinity as the homoclinic loop curve L is approached. Panels (c1)–(c3) of Figure 7.7 show the response of the system to a sufficiently large perturbation (above the stable manifold of the saddle point in region 6). After an initial large pulse, the system settles down to the attracting periodic orbit and, hence, produces regular oscillations. As the curve L is approached, these oscillations take the form of self-pulsations with clearly defined pulses. Indeed, the interspike time (which is the period of the periodic orbit) increases as κ is decreased toward the curve L.

7.5
Bifurcation Study for Nonexcitable SLSA

We now consider the influence of the feedback loop on phase portrait 1 of the Yamada model, where the off-state of the laser is a global attractor, but the system is not excitable; see Figure 7.3. Figure 7.8a shows the relevant computed bifurcation

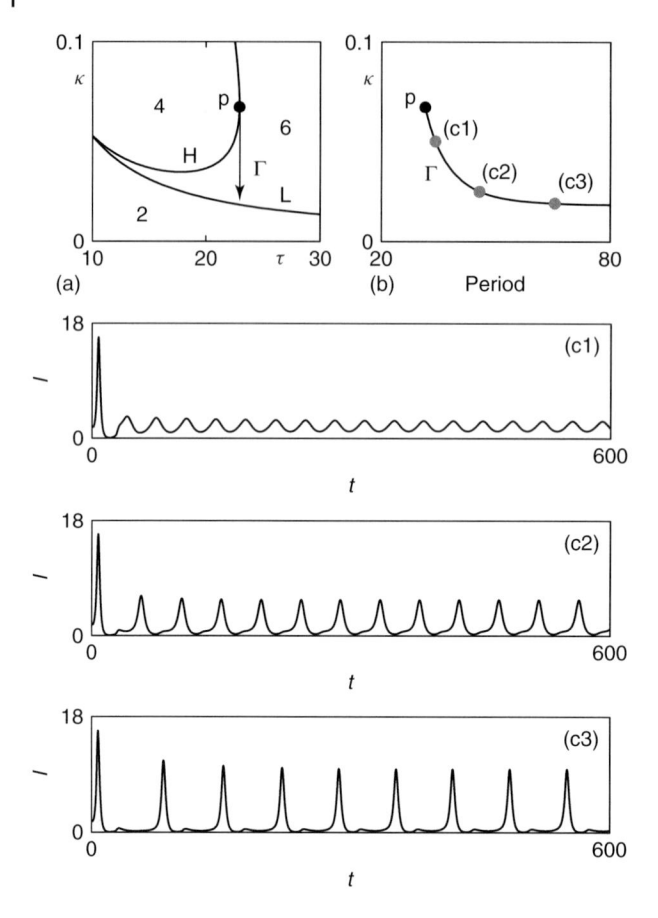

Figure 7.7 Panel (a) shows a region of the (τ, κ)-plane with transition to region 6. Panel (b) shows the period of the periodic orbit Γ in region 6 as continued from the point p of the Hopf curve H toward the homoclinic loop curve L. Panels (c1)–(c3) show associated time series after a sufficiently large perturbation from the stable off-state.

diagram of Eqs. (7.1)–(7.3) in the (τ, κ)-plane for $A = 5.9$, a value somewhat to the left of the saddle-node curve S in Figure 7.2b. (Again, $B = 5.8$, $a = 1.8$, and $\gamma = 0.04$ are fixed.) Hence, according to Proposition 1, we find the saddle-node locus S as the horizontal line at $\kappa_S \approx 0.09578$. The line S bounds a horizontal strip near $\kappa = 0$, where one finds phase portrait 1 for any τ; again, this is to be expected from the fact that the DDE is a regular perturbation of the Yamada system. In contrast to the case for $A = 6.5$ in Figure 7.5a, for $A = 5.9$, the locus of Hopf bifurcation is no longer a single curve. Figure 7.8a actually shows several disjoint Hopf curves that intersect in numerous double Hopf points; along the shown curves H, one finds a Hopf bifurcation at the equilibrium E_3. Note further that

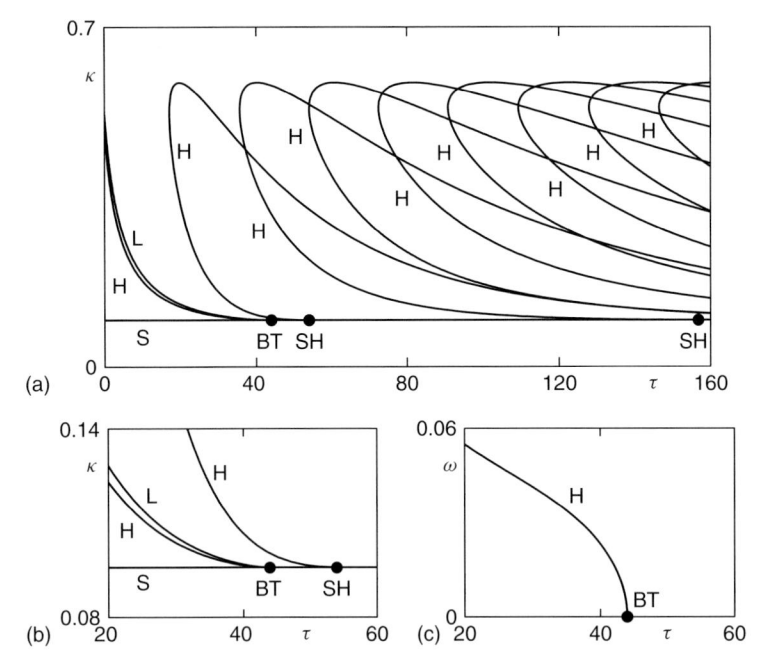

Figure 7.8 Computed bifurcation diagram in the (τ,κ)-plane for $A = 5.9$ (a) and an enlargement near a Bogdanov-Takens point BT and a saddle-node Hopf point SH (b). Shown are the saddle-node curve S, Hopf curves H, and a homoclinic loop curve L. Panel (c) shows the imaginary part ω along the Hopf curve that ends at the point BT.

the transcritical locus T no longer intersects the Hopf curves H; it lies at $\kappa_T = 0.9$, which is outside the range shown in Figure 7.8a.

Another new feature of Figure 7.8a are codimension-two points on the saddle-node line S. There are a Bogdanov–Takens point BT (where the system has a double-zero eigenvalue) and two codimension-two saddle-node Hopf points SH (where there is a zero eigenvalue and a complex conjugate pair of purely imaginary eigenvalues) [30, 31]. These points are end points of shown Hopf curves. Panel (c) shows that the imaginary part ω along the Hopf curve H decreases to zero as the point BT is approached, which is evidence for the fact that one is indeed dealing with a Bogdanov–Takens point. As was also checked, the imaginary part ω along the respective Hopf curves tends to a nonzero limit at the points SH. Notice also that the homoclinic loop curve L ends at the point BT; see Figure 7.8b.

Figure 7.9 shows a qualitative sketch of the partial bifurcation diagram for $A = 5.9$. As before, we added a number of bifurcation curves that are very difficult to continue directly in the DDE, but must exist near the known codimension-two bifurcation points; existence and positions of these curves was again verified by careful numerical simulations of Eqs. (7.1)–(7.3). There are two (gray) curves SL of saddle-node bifurcation of limit cycles emanating from

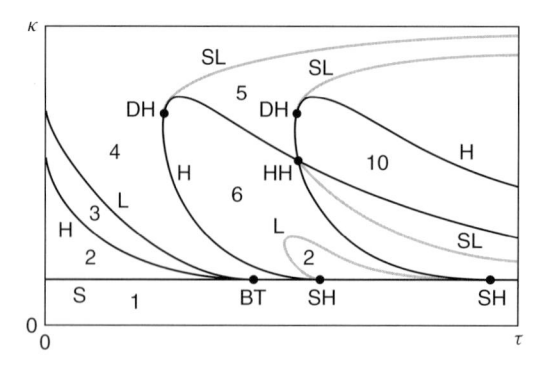

Figure 7.9 Sketch of the bifurcation diagram in the (τ,κ)-plane for $A = 5.9$. The gray curves SL are loci of saddle-node of limit cycle bifurcations, and the gray curve L is another curve of homoclinic loops; for phase portaits in numbered regions, see Figure 7.3.

codimension-two degenerate Hopf points DH and one (gray) curve SL emanating from a codimension-two Hopf–Hopf point HH. Furthermore, we also added an additional homoclinic loop curve L that connects the two points SH and bounds a second region where one finds the excitable phase portrait 2.

Figure 7.9 shows what additional bifurcation curves are involved in the interaction of the three left-most Hopf curves of the bifurcation diagram in Figure 7.8a. Indeed, many more additional Hopf curves and codimension-two points HH exist, meaning that the sketch in Figure 7.9 does not give a complete division of the (τ,κ)-plane into regions of different dynamics. Nevertheless, it is complete enough to allow us to identify the regions where one finds portraits 1–6 and 10 from Figure 7.3.

7.6
Dependence of the Bifurcation Diagram on the Gain Pump Parameter

The two bifurcation diagrams in the (τ,κ)-plane for $A = 5.9$ and $A = 6.5$, in Figures 7.8a and 7.5a, respectively, are clearly qualitatively different. Yet, since they depend only on the gain pump parameter A, changing A from $A = 5.9$ to $A = 6.5$ (or vice versa) transforms the two bifurcation diagrams into one another. We now describe briefly how this happens via the transition through codimension-three bifurcations. Here we take a geometric approach that is supported by numerical computation with DDE-BIFTOOL.

The overall features of this transition when A is increased through the interval $A \in [5.9, 6.5]$ can be described as follows:

- The different Hopf curves for $A = 5.9$ merge into the single Hopf curve for $A = 6.5$.
- The codimension-two points BT and SH disappear toward infinite values of τ.

7.6 Dependence of the Bifurcation Diagram on the Gain Pump Parameter | 177

- The curve S disappears when A reaches the curve S in Figure 7.2b at $A = (-1 + a + 2\sqrt{(a-1)B + B})/a$.
- The transcritical curve T moves down and starts to intersect the curve H.

We concentrate here on the bifurcations of the Hopf curves as the main ingredients in the transition for $A \in [5.9, 6.5]$. To this end, we now consider all Hopf curves for $A = 5.9$ – including those where the bifurcation takes place at E_2, which are not shown in Figures 7.8 and 7.9.

As A is increased, one finds values of A where the connectivity between different branches of Hopf curves in the (τ, κ)-plane changes locally; we speak of a saddle transition of Hopf curves. Figure 7.10 shows two examples of this bifurcation, which is of codimension three. More specifically, Figure 7.10a1, a2 shows the bifurcation diagram before and after a saddle transition of Hopf curves near the points BT and SH; the inset panels show the purely imaginary parts of the bifurcating eigenvalues along the Hopf curves. Notice how two separate Hopf bifurcation curves that end at BT and SH, respectively, connect differently, creating a direct connection from BT to SH. Figure 7.10b1,b2 shows a saddle transition of Hopf curves near the point SH for larger values of τ. The result is again a different connectivity of Hopf curves H near the point SH. A saddle transition of Hopf curves is a bifurcation of codimension three because it changes the topological type of the bifurcation diagram in the (τ, κ)-plane at a single discrete value of A. This bifurcation has been found in transitions between different two-parameter bifurcation diagrams in other laser systems; see, for example, [32, 33]. When seen in (τ, κ, A)-space, a saddle transition of Hopf curves corresponds to the transition of a plane given by

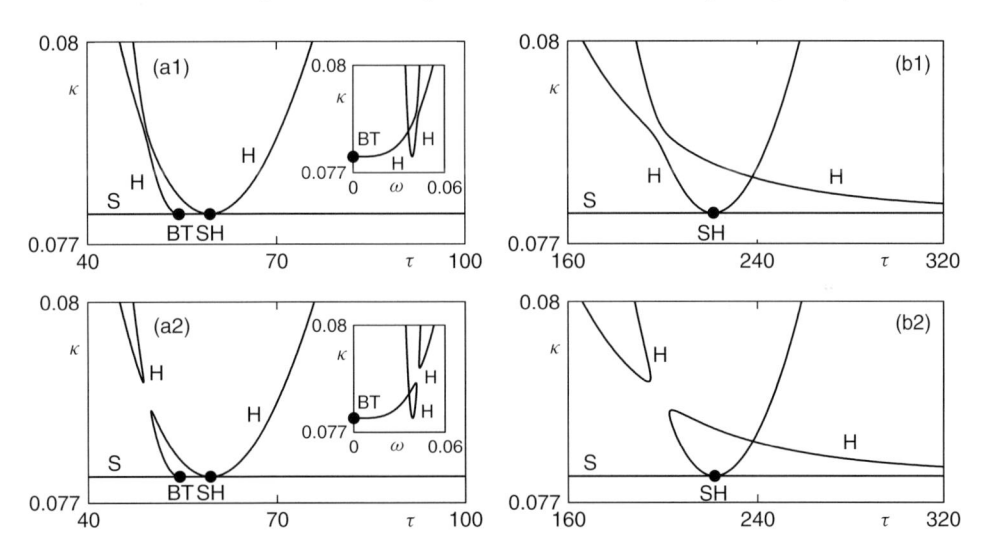

Figure 7.10 Before (top) and after (bottom) a saddle transition of Hopf curves H, near BT and SH (a) and near SH (b); insets in panels (a1) and (a2) show the imaginary part ω at the Hopf curves H. These bifurcation diagrams have been computed with DDE-BIFTOOL, where A has the values 5.9311 in (a1), 5.9312 in (a2), 5.9311 in (b1), and 5.9312 in (b2).

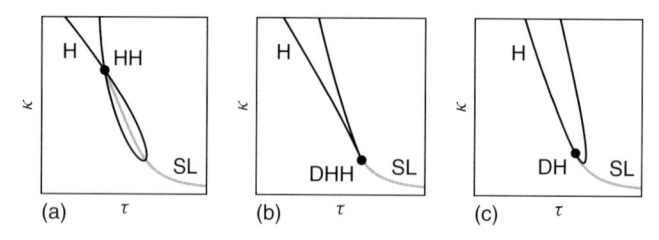

Figure 7.11 Sketch of the transition through a codimension-three degenerate Hopf–Hopf point DHH, where a loop in a Hopf curve H with a point HH transforms into a lobe with a point DH.

$A = const$ through a saddle point of the two-dimensional surface $H(\tau, \kappa)$ of Hopf bifurcation; here, a saddle point is given by the condition that grad $H = 0$ and the determinant of the Hessian is negative.

After the saddle transitions of Hopf curves in Figure 7.10, the codimension-two points BT and SH are "free to move" toward larger values of τ as A is increased further. In the process they "drag" the other bifurcation curves with them toward larger values of τ. In this way, the homoclinic loop curve L ending at BT (not shown in Figure 7.10) becomes the lower boundary of the bifurcation diagram; compare with Figure 7.5a. With increasing A, the saddle-node curve S moves down in κ and finally disappears at $A = (-1 + a + 2\sqrt{(a-1)B} + B)/a$ into the (unphysical) region of negative κ.

Relating the saddle transition in Figure 7.10b1,b2 back to the larger picture in Figure 7.9 shows that this transition results in a Hopf curve H with a loop, where the self-intersection point is the sketched Hopf–Hopf point. Investigation with DDE-BIFTOOL shows that, when A is increased beyond the value 5.9312, this loop changes into the "lobe" with a point DH that one finds in Figure 7.6b. The mechanism is the transition through a codimension-three degenerate Hopf–Hopf bifurcation point DHH, and it is sketched in Figure 7.11. At the moment of bifurcation in panel (b), the Hopf curve has a cusp, which is the point DHH.

7.7
Conclusions

We presented a bifurcation study of the influence of an optical feedback loop on an SLSA as modeled by the Yamada ODE. The resulting DDE model was studied by means of linear stability analysis of its equilibria in combination with a bifurcation analysis with the continuation package DDE-BIFTOOL. More specifically, bifurcation diagrams in the plane of delay time τ and feedback strength κ were presented for two relevant choices of the gain pump parameter A. The transition of the bifurcation diagram in the (τ, κ)-plane with A was discussed in terms of transitions through certain codimension-three bifurcations.

The work presented here can be seen as a case study of how a laser system with delay can be investigated with tools from bifurcation theory and, in particular, with numerical continuation as implemented, for example, in the package DDE-BIFTOOL. The physical motivation for this work is the wish to use the additional control parameters τ and κ to ensure reliable self-pulsations of the overall system with small timing jitter. To this end, the bifurcation diagram in the (τ, κ)-plane was considered for a small range of the gain pump parameter A near the region where the SLSA without delay is excitable. In this way, we identified large and experimentally accessible additional regions where the SLSA produces a stable train of pulses in the presence of the optical feedback loop.

Indeed, this study is far from complete and there are several directions for future research. First, an investigation of the influence of noise on dynamics of the SLSA with delay would be the logical next step to determine the timing properties of the corresponding pulse trains under more realistic conditions. Second, we restricted our attention to a relatively small range of the delay time τ. As a result, the phase portraits we found for the SLSA with delay are all quite special in that they can be drawn by planar phase portraits, where the missing (infinitely many) directions are strongly attracting. A more wide-ranging bifurcation analysis, in dependence of A as well as on other parameters of the Yamada system, would be expected to result in the discovery of more complicated dynamics.

References

1. Krauskopf, B. and Lenstra, D. (eds) (2000) *Fundamental Issues of Nonlinear Laser Dynamics*, AIP Conference Proceedings, vol. 548, American Institute of Physics Publishing, Melville, NY.
2. Kane, D.M. and Shore, K.A. (eds) (2005) *Unlocking Dynamical Diversity: Optical Feedback Effects on Semiconductor Lasers*, John Wiley & Sons, Inc.
3. Antoranz, J.C., Gea, J., and Velarde, M.G. (1981) Oscillatory phenomena and Q-switching in a model for a laser with a saturable absorber. *Phys. Rev. Lett.*, **47**, 1895–1898.
4. Erneux, T. (1988) Q-switching bifurcation in a laser with a saturable absorber. *J. Opt. Soc. Am. B*, **5**, 1063–1069.
5. Dubbeldam, J.L.A. and Krauskopf, B. (1999) Self-pulsations of lasers with saturable absorber: dynamics and bifurcations. *Opt. Commun.*, **159**, 325–338.
6. Mandel, P. (1997) *Theoretical Problems in Cavity Nonlinear Optics*, Cambridge Studies in Modern Optics, Cambridge University Press.
7. Kawaguchi, H. (1994) *Bistabilities and Nonlinearities in Laser Diodes*, Artech House.
8. Georgiou, M. and Erneux, T. (1992) Self-pulsating laser oscillations depend on extremely small amplitude noise. *Phys. Rev. A*, **45**, 6636–6642.
9. Dubbeldam, J.L.A., Krauskopf, B., and Lenstra, D. (1999) Excitability and coherence resonance in lasers with saturable absorber. *Phys. Rev. E*, **60**, 6580–6588.
10. Murray, J.D. (1990) *Mathematical Biology*, Springer-Verlag, New York.
11. Grill, S., Zykov, V.S., and Müller, S.C. (1996) Spiral wave dynamics under pulsatory modulation of excitability. *J. Phys. Chem.*, **100**, 19082–19088.
12. Taylor, D., Holmes, P., and Cohen, A.H. (1997) *Excitable Oscillators as Models for Central Pattern Generators*, Series on stability, Vibration and Control of Systems, Series B, World Scientific Publishing Company, Singapore.
13. Tredicce, J.R. (2000) Excitability in laser systems: the experimental side, *Fundamental Issues of Nonlinear Laser Dynamics*

14. Wieczorek, S.M., Krauskopf, B., and Lenstra, D. (2002) Multipulse excitability in a semiconductor laser with optical injection. *Phys. Rev. Lett.*, **88**, 063901.
15. Mullet, J. and Mirasso, C.R. (1999) Numerical statistics of power dropouts based on the Lang-Kobayashi model. *Phys. Rev. E*, **59**, 5400–5405.
16. Yacomotti, A.M., Eguia, M.C., Aliaga, J., Martinez, O.E., Mindlin, G.B., and Lipsich, A. (1999) Interspike time distribution in noise driven excitable systems. *Phys. Rev. Lett.*, **83**, 292–295.
17. Wünsche, H.-J., Brox, O., Radziunas, M., and Henneberger, F. (2002) Excitability of a semiconductor laser by a two-mode homoclinic bifurcation. *Phys. Rev. Lett.*, **88**, 023901.
18. Tronciu, V.Z., Wünsche, H.-J., Radziunas, M., and Schneider, K.R. (2001) Excitability of lasers with integrated dispersive reflectors. *Proc. SPIE*, **4283**, 347.
19. Krauskopf, B., Schneider, K.R., Sieber, J., Wieczorek, S.M., and Wolfrum, M. (2003) Excitability and self-pulsations near homoclinic bifurcations in semiconductor laser systems. *Opt Commun.*, **215**, 367–379.
20. Yamada, M. (1993) A theoretical analysis of self-sustained pulsation phenomena in narrow stripe semiconductor lasers. *IEEE J. Quantum Electron.*, **QE-29**, 1330–1336.
21. Jones, C.K.R.T. (1995) Geometric Singular Perturbation Theory. *Dynamical Systems (Montecatini Terme, 1994), Springer Lecture Notes in Mathematics* **1609**, Springer-Verlag, New York, pp. 44–118.
22. Hale, J.K. and Verduyn Lunel, S.M. (1993) *Introduction to Functional Differential Equations*, Springer-Verlag, New York.
23. Diekmann, O., Van Gils, S.A., Verduyn Lunel, S.M., and Walther, H.O. (1995) *Delay Equations: Functional-, Complex- and Nonlinear Analysis*, Springer-Verlag, New York.
24. Verduyn Lunel, S.M. and Krauskopf, B. (2000) The mathematics of delay equations with an application to the Lang-Kobayashi equations, *Fundamental Issues of Nonlinear Laser Dynamics* (eds B. Krauskopf and D. Lenstra), AIP Conference Proceedings, vol. 548, American Institute of Physics Publishing, Melville, NY, pp. 66–87.
25. Engelborghs, K., Luzyanina, T., Samaey, G., and Roose, D. (2001) DDE-BIFTOOL: a Matlab package for bifurcation analysis of delay differential equations. Tech. Rep. TW-330, Department of Computer Science, K. U. Leuven, Belguim, available from http://www.cs.kuleuven.ac.be/cwis/research/twr/research/software/delay/ddebiftool.shtml.
26. Szalai, R. (2005) *PDDE-CONT: A Continuation and Bifurcation Software for Delay-differential Equations*, Budapest University of Technology and Economics, Hungary, http://www.mm.bme.hu/ szalai/pdde.
27. Krauskopf, B. (2005) Bifurcation analysis of lasers with delay. *Unlocking Dynamical Diversity: Optical Feedback Effects on Semiconductor Lasers*, John Wiley & Sons, Inc, pp. 147–183.
28. Roose, D. and Szalai, R. (2007) Continuation and bifurcation analysis of delay differential equations, in *Numerical Continuation Methods for Dynamical Systems: Path Following and Boundary Value Problems* (eds B. Krauskopf, H.M. Osinga, and J. Galan-Vioque), Springer-Verlag, New York, pp. 359–399.
29. Doedel, E.J. and Oldeman, B.E. with major contributions from Champneys, A.R., Dercole, F., Fairgrieve, T.F., Kuznetsov, Yu.A., Paffenroth, R.C., Sandstede, B., Wang, X.J., and Zhang, C.H. (2010) AUTO-07p Version 0.7: Continuation and bifurcation software for ordinary differential equations, Department of Computer Science, Concordia University, Montreal, Canada, http://cmvl.cs.concordia.ca/auto/.
30. Guckenheimer, J. and Holmes, P. (1986) *Nonlinear Oscillations, Dynamical Systems and Bifurcations of Vector Fields*, 2nd Printing, Springer-Verlag, New York.

31. Kuznetsov, Yu.A. (1992) *Elements of Applied Bifurcation Theory*, Springer-Verlag, New York.
32. Green, K. and Krauskopf, B. (2004) Bifurcation analysis of a semiconductor laser subject to non-instantaneous phase-conjugate feedback. *Opt. Commun.*, **231**, 383–393.
33. Erzgräber, Krauskopf, B. and Lenstra, D. (2007) Bifurcation analysis of a semiconductor laser with filtered optical feedback. *SIAM J. Appl. Dynam. Syst.*, **6**, 1–28.

8
Modeling of Passively Mode-Locked Semiconductor Lasers

Andrei G. Vladimirov, Dmitrii Rachinskii, and Matthias Wolfrum

8.1
Introduction

Short optical pulses have numerous technological applications including high bit rate communications, optical tomography, spectroscopic measurements, material processing, frequency standards, and so forth. A powerful method for generating such kind of pulses is based on the so-called mode-locking (ML) technique [1]. A mode-locked laser operates simultaneously in a large number of longitudinal modes with equal intermode frequency spacings and a fixed relation between the phases of the modes. This laser usually emits a sequence of short pulses with a repetition period close to the cavity round trip time and minimum possible pulse duration inversely proportional to the number of locked modes. Different approaches have been developed to achieve ML in lasers. Active ML is based on application of an external modulation with the period close to the laser cavity round trip time [2, 3]. An alternative technique known as *passive ML* [4–6] does not require any external source of radiofrequency modulation. In this case, the losses are modulated by the ML pulse itself because of the presence of an additional saturable absorber medium in the cavity. When propagating in the laser cavity, an intensive light pulse bleaches the saturable absorber and, therefore, experiences a reduced saturable loss as compared to regimes with "unlocked" modes.

Semiconductor ML lasers are compact, low-cost, and reliable sources of short optical pulses with high repetition rates suitable for applications in telecommunication technology [7, 8]. In monolithic passively ML semiconductor lasers, saturable absorption is implemented by applying a reverse bias to a short additional laser section built from the same material as the gain section. Since rapid development of the technology requires a constant improvement of the ML characteristics, different techniques are used to improve the quality of ML pulses: addition of passive and spectral filtering (DBR) sections into the laser cavity, application of periodic external modulations (hybrid ML [7, 9]), and so forth. Nowadays, monolithic quantum dot mode-locked lasers [10–14] demonstrate a number of important advantages as compared to standard quantum well devices and, therefore, have become one of the most promising systems for various applications.

Nonlinear Laser Dynamics: From Quantum Dots to Cryptography, First Edition. Edited by Kathy Lüdge.
© 2012 Wiley-VCH Verlag GmbH & Co. KGaA. Published 2012 by Wiley-VCH Verlag GmbH & Co. KGaA.

Analytical theory of ML in lasers was first developed by Haus [1]. Assuming that the gain and losses per round trip are very small such that the pulse circulating in the cavity changes only slightly per cavity round trip, he derived a partial differential equation for the evolution of the electric field amplitude in a mode-locked laser. Within the framework of this approach, Haus considered two practically important limiting situations: the case of fast saturable absorber [6] and the case of slow saturable absorber [5]. Although the Haus theory has proved to be a very efficient tool for modeling of solid state, fiber, and some other types of lasers, the applicability of this theory to ML in semiconductor lasers is rather limited. This is mainly because the small gain and loss approximation does not hold there. Most theoretical studies on ML in these lasers are based on time-consuming numerical simulations of traveling wave equations supplied with proper boundary conditions, for example, [15–17]. In this chapter, we describe an alternative, relatively simple, approach to a theoretical analysis of ML in semiconductor lasers, without invoking the small gain-and-loss approximation. Our model is based on a set of delay differential equations (DDE) proposed in [18–20] and allows for not only fast and easy numerical analysis of ML regimes but also their analytical study. In this chapter, we summarize some results of numerical and analytical analysis of this DDE model.

The chapter is organized as follows. In Section 8.2 a set of three DDEs [18–20] for a passively mode-locked semiconductor laser is derived from the traveling wave equations, which govern spatiotemporal evolution of the electric field amplitude and the carrier densities in the gain and absorber sections. The derivation is performed under the assumption of unidirectional lasing in a ring cavity and Lorentzian line shape of the spectral filtering. Section 8.3 is devoted to numerical analysis of the DDE model. Here, the bifurcations responsible for the appearance and break-up of ML solutions are described in detail. The next two sections are devoted to the analytical analysis of different ML regimes in the limit when spectral filtering in the cavity is very broad [20]. The connection between the DDE model and classical approaches developed by New [4] and Haus [5] are discussed in Section 8.4, where a 3-D map describing the transformation of the ML pulse characteristics after a complete round trip in the cavity is constructed under the slow saturable absorber approximation. Analytical stability analysis of the fundamental ML regime with respect to the Q-switching instability is discussed in Section 8.5 [21], where we use a variational approach for estimation of the characteristics and domains of stability of ML regimes. Finally, in Section 8.6, some modifications and extensions of the DDE ML model are discussed.

8.2
Derivation of the Model Equations

In this section, we derive a system of DDEs for the dynamics of an ML laser with a ring cavity, using the so-called lumped element method [22]. We assume that one of the two counter propagating waves in the laser cavity is suppressed, hence the laser generation is unidirectional. The laser is divided along its longitudinal

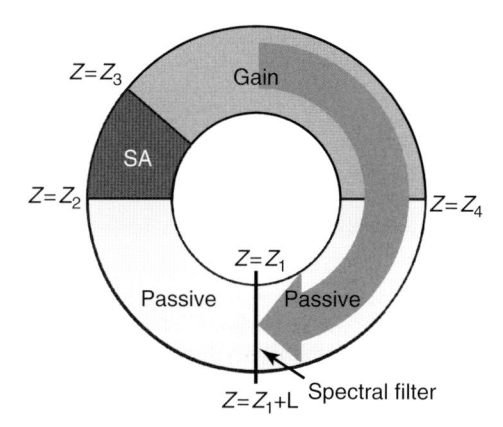

Figure 8.1 Schematic presentation of a ring ML laser (unidirectional lasing) with gain, absorber (SA), and passive sections.

direction z into four sections (Figure 8.1) containing absorber and gain media as well as passive media in the first and the last sections. An additional spectral filter limits the bandwidth of the emitted light. In the first step, we employ in each section the so-called traveling wave model [16, 23, 24]

$$\frac{\partial E_k(t,z)}{\partial z} + \frac{1}{v}\frac{\partial E_k(t,z)}{\partial t} = \frac{g_k \Gamma_k}{2}(1 - i\alpha_k)\left[N_k(t,z) - N_k^{tr}\right] E_k(t,z) \quad (8.1)$$

$$\frac{\partial N_k(t,z)}{\partial t} = J_k - \gamma_k N_k(t,z) - v g_k \Gamma_k \left[N_k(t,z) - N_k^{tr}\right] |E_k(t,z)|^2, \quad (8.2)$$

describing the slow evolution of the complex electric field amplitude envelope $E_k(t,z)$ and the real carrier density $N_k(z,t)$ within the k-th section $z \in [z_k, z_{k+1}]$, $k = 1, \ldots, 4$. The light group velocity v is assumed to be constant and equal in all sections. The line-width enhancement factor α_k, the differential gain (loss) g_k, the transverse modal fill factor Γ_k, the carrier density at transparency N_k^{tr}, and the carrier relaxation rate γ_k depend on z; assume different values in the gain and absorber section; and vanish in the passive sections. The pump parameter J_k is proportional to the injection current and is different from zero only in the gain section. By a lumped element approach, the gain dispersion of the active medium and all other spectral filtering components (e.g. distributed Bragg reflectors) are introduced by the boundary condition

$$\widehat{E}_1(\omega, z_1) = \widehat{f}(\omega)\widehat{E}_4(\omega, z_1 + L),$$

where $\widehat{E}$ is the Fourier transformed field, $\widehat{f}(\omega)$ represents the line shape of the spectral filter, and L is the total length of the laser. Similarly, nonresonant losses and out-coupling are introduced by interface conditions

$$E_{k+1}(t, z_{k+1}) = \sqrt{\kappa_{k+1}} E_k(t, z_{k+1}), \quad k = 1, \ldots, 3.$$

In a passive section, we set $N_k \equiv N_k^{tr} = 0$, and the evolution reduces to

$$\frac{\partial E_k(t,z)}{\partial z} + \frac{1}{v}\frac{\partial E_k(t,z)}{\partial t} = 0. \quad (8.3)$$

Together with the coordinate change $(t, z) \to (\tau, \zeta)$, where $\tau = t - z/v$ is the delayed time and $\zeta = z/v$ is a normalized spatial coordinate, we transform Eqs. (8.1) and (8.2) to the dimensionless form

$$\frac{\partial A_k(\tau, \zeta)}{\partial \zeta} = \frac{1}{2}(1 - i\alpha_k) n_k(\tau, \zeta) A_k(\tau, \zeta), \qquad (8.4)$$

for the rescaled field $A(\tau, \zeta) = E(t, z)\sqrt{v g_g \Gamma_g}$ and

$$\frac{\partial n_g(\tau, \zeta)}{\partial \tau} = j_g - \gamma_g n_g(\tau, \zeta) - n_g(\tau, z)|A(\tau, \zeta)|^2, \qquad (8.5)$$

$$\frac{\partial n_q(\tau, \zeta)}{\partial \tau} = -j_q - \gamma_q n_q(\tau, \zeta) - s' n_q(\tau, \zeta)|A(\tau, \zeta)|^2, \qquad (8.6)$$

for the carrier densities $n_{g,q}(\tau, \zeta) = v g_{g,q} \Gamma_{g,q}\left[N_{g,q}(t,z) - N_{g,q}^{\text{tr}}\right]$ in the gain and absorber sections, respectively. Further parameters are $j_g = v g_g \Gamma_g \left(J_g - \gamma_g N_g^{\text{tr}}\right)$, $j_q = v g_q \Gamma_q \gamma_q N_q^{\text{tr}}$, and $s' = (g_q \Gamma_q)/(g_g \Gamma_g)$ measuring the ratio of saturation intensities in the gain and absorber sections. The subscripts g and q relate to the parameter values in the corresponding sections. For the passive sections, Eq. (8.4) simplifies now to

$$\frac{\partial A(\tau, \zeta)}{\partial \zeta} = 0. \qquad (8.7)$$

Solving Eqs. (8.4)–(8.7) and using Eq. (8.2), one can now obtain explicit expressions for the evolution of the electric field envelope during its propagation through each of the four laser sections. According to Eq. (8.7), the passage of the electric field through the two passive sections leads to the simple relations

$$A_1(\tau, \zeta_2) = A_1(\tau, \zeta_1), \quad A_4(\tau, \zeta_1 + L/v) = \sqrt{\kappa_3} A_3(\tau, \zeta_4) \qquad (8.8)$$

where the losses κ_k concentrated at the interfaces $\zeta_k = z_k/v$ have already been included for the transition through the interface ζ_3. The evolution of the electric field envelope through the gain and absorber sections together with the losses at the corresponding interfaces is given by

$$A_3(\tau, \zeta_4) = \sqrt{\kappa_3} e^{\frac{1-i\alpha_g}{2}G(\tau)} A_2(\tau, \zeta_3), \quad A_2(\tau, \zeta_3) = \sqrt{\kappa_2} e^{-\frac{1-i\alpha_q}{2}Q(\tau)} A_1(\tau, \zeta_2), \qquad (8.9)$$

obtained by integrating Eq. (8.4). Here, the dimensionless quantities

$$G(\tau) = \int_{\zeta_3}^{\zeta_4} n_g(\tau, \zeta) d\zeta, \quad Q(\tau) = -\int_{\zeta_2}^{\zeta_3} n_q(\tau, \zeta) d\zeta,$$

describe the total saturable gain and loss introduced by the corresponding sections [25, 26]. Integrating Eqs. (8.5) and (8.6) with respect to ζ from ζ_3 to ζ_4 and from ζ_2 to ζ_3, respectively, and using the relation

$$\int_{\zeta_{2,3}}^{\zeta_{3,4}} n_{q,g}(\zeta, \tau)|A_{2,3}(\tau, \zeta)|^2 d\zeta = -|A_{2,3}(\zeta_{3,4}, \tau)|^2 + |A_{2,3}(\zeta_{2,3}, \tau)|^2,$$

which follows from Eq. (8.4), we obtain equations for the evolution of gain and loss,

$$\partial_\tau G(\tau) = g_0 - \gamma_g G(\tau) - |A(\tau, \zeta_4)|^2 + |A(\tau, \zeta_3)|^2 \tag{8.10}$$

$$\partial_\tau Q(\tau) = q_0 - \gamma_q Q(\tau) + s' |A(\tau, \zeta_3)|^2 - s' |A(\tau, \zeta_2)|^2. \tag{8.11}$$

Here, $g_0 = \int_{\zeta_3}^{\zeta_4} j_g d\zeta$ and $q_0 = \int_{\zeta_2}^{\zeta_3} j_q d\zeta$ are the total unsaturated gain and loss, respectively. Finally, for the transition through the interface at ζ_1, we have to invoke in addition to the nonresonant losses κ_1, the spectral filtering (Eq. (8.2)) and the periodic boundary conditions that have to be rewritten in the coordinates τ, ζ as $A_4(\tau, \zeta_1 + L/v) = A_1(\tau + T, \zeta_1)$, where $T = L/v$ is the cold cavity round trip time. Expressing the filtering in the time domain we obtain

$$A_1(\tau + T, \zeta_1) = \sqrt{\kappa_1} \int_{-\infty}^{\tau} f(\tau - \theta) A_4(\theta, \zeta_1 + L/v) d\theta, \tag{8.12}$$

where $f(\tau)$ is assumed to decay sufficiently fast as $\tau \to \infty$ to ensure the convergence of the integral in the right-hand side of Eq. (8.12). Wrapping up all transformations (Eqs. (8.8), (8.9), and (8.12)), we obtain the transformation of the electric field amplitude $A(\tau) \equiv A_1(\tau, \zeta_1)$ after a complete cavity round trip,

$$A(\tau + T) = \int_{-\infty}^{\tau} f(\tau - \theta) R(\theta) A(\theta) d\theta, \tag{8.13}$$

where

$$R(\tau) = \sqrt{\kappa} e^{(1-i\alpha_g) G(\tau)/2 - (1-i\alpha_q) Q(\tau)/2}. \tag{8.14}$$

Here, the attenuation factor $\kappa = \kappa_1 \kappa_2 \kappa_3 \kappa_4 < 1$ describes the total linear nonresonant intensity loss per cavity round trip.

Equation (8.13) describes the evolution of the electric field envelope in a ring cavity laser with an arbitrary spectral filter line shape defined by the response function $f(\tau)$. If there is no spectral filtering, this function becomes the delta function, $f(\tau) = \delta(\tau)$, and Eq. (8.13) transforms to the map

$$A(\tau + T) = R(\tau) A(\tau). \tag{8.15}$$

This map is similar to the Ikeda map, which was proposed for modeling multistability and chaos in a ring cavity with a nonlinear medium [27, 28]. Eq. (8.15) describes the time evolution of the electric field envelope A in a laser without spectral filtering, that is, in the approximation studied by New [4]. The solution of this equation corresponding to the ML regime is a T-periodic sequence of delta functions, $|A(\tau)|^2 = \Delta P \sum_{n=-\infty}^{\infty} \delta(\tau - nT)$ where ΔP is the pulse energy. This solution is characterized by infinite bandwidth and infinitely short ML pulses.

Now let us assume that the spectral filtering has a Lorentzian line shape. In this case, the response function has the form $f(\tau) = \gamma \exp[(-\gamma + i\Omega)\tau]$ and Eq. (8.13) can be replaced by a DDE with the delay time given by the cold cavity round trip time T. After the change of the variable $A \to A \exp(i\Omega\tau)$, this DDE takes the form

$$\gamma^{-1} \partial_\tau A(\tau) + A(\tau) = \sqrt{\kappa} e^{(1-i\alpha_g) G(\tau-T)/2 - (1-i\alpha_q) Q(\tau-T)/2 - i\varphi} A(\tau - T), \tag{8.16}$$

with $\varphi = \Omega T$. Equations (8.8) and (8.9) can be used to express the fields $A_2(\tau, \zeta_2)$, $A_3(\tau, \zeta_3)$, and $A_4(\tau, \zeta_4)$ in terms of the field $A(\tau) = A_1(\tau, \zeta_1)$ in Eqs. (8.10) and (8.11). The resulting equations describing the evolution of the saturable gain and loss read

$$\partial_\tau G(\tau) = g_0 - \gamma_g G(\tau) - \kappa_1 \kappa_2 \kappa_3 e^{-Q(\tau)} \left(e^{G(\tau)} - 1\right) |A(\tau)|^2, \tag{8.17}$$

$$\partial_\tau Q(\tau) = q_0 - \gamma_q Q(\tau) - s'\kappa_1\kappa_2 \left(1 - e^{-Q(\tau)}\right)|A(\tau)|^2. \tag{8.18}$$

The system of DDEs (Eqs. (8.16)–(8.18)) establish a closed delay differential model of a ring cavity ML laser with a Lorentzian filter, which that is the central subject of this chapter.

We note that the approach used above to derive the model (Eqs. (8.16)–(8.18)) is similar to that proposed by Gurevich and Khanin to study a passively ML solid state laser [29–31]. However, the delay differential model of these authors has a singularity for zero electric field. Therefore, Eqs. (8.16)–(8.18) are more suitable for studying the limit of a slow saturable absorber as the electric field envelope A is close to zero between the ML pulses in this limit. After a proper rescaling of the electric field envelope, Eqs. (8.16)–(8.18) can be rewritten in the form

$$\gamma^{-1}\partial_t A + A = \sqrt{\kappa} e^{(1-i\alpha_g)G(t-T)/2 - (1-i\alpha_q)Q(t-T)/2 - i\varphi} A(t-T) \tag{8.19}$$

$$\partial_t G = g_0 - \gamma_g G - e^{-Q}\left(e^G - 1\right)|A|^2 \tag{8.20}$$

$$\partial_t Q = q_0 - \gamma_q Q - s\left(1 - e^{-Q}\right)|A|^2. \tag{8.21}$$

Here, $s = s'/\kappa_2$ is the effective saturation parameter, which is inversely proportional to the linear nonresonant losses κ_3 introduced at $z = z_3$ between the gain and the absorber sections. In order to simplify the notations, we have replaced the time variable τ with t. In what follows, we analyze numerically and analytically ML regimes in the delay differential model (Eqs. (8.19)–(8.21)).

The presence of the delayed terms in the right-hand side of Eq. (8.19) reflects the fact that the model Eqs. (8.19)–(8.21) describe a multimode laser. At the end of this paragraph, we consider a reduction of these equations to the model of a single-mode laser with a saturable absorber. When the electric field envelope changes sufficiently slow in time, the first term $\gamma^{-1}\partial_t A$ in Eq. (8.19) can be neglected. Multiplying the resulting equation by its complex conjugate, we obtain $I(t+T) = \kappa e^{G(t)-Q(t)}I(t)$ where $I = |A|^2$ is the optical field power. Now, assuming that the characteristic time scale of the optical field power variation is much smaller than the round trip time T (this is a good approximation for a single-mode generation regime) and using the approximation $I(t+T) \approx I(t) + T\partial_t I(t)$, we arrive at the system

$$T\partial_t I = -I + e^{G-Q+\ln\kappa}I \tag{8.22}$$

$$\partial_t G = g_0 - \gamma_g G - e^{-Q}(e^G - 1)I \tag{8.23}$$

$$\partial_t Q = q_0 - \gamma_q Q - s(1 - e^{-Q})I \tag{8.24}$$

of ordinary differential equations. Note that in the limit of small G, Q, and $\ln\kappa$, this system transforms to a standard model of a single-mode class B laser with a saturable absorber [32–35] in which all the exponential functions are replaced with their linear approximations. Both models assume that atomic polarization can be adiabatically eliminated. In [36], dynamical regimes of laser generation in a laser with saturable absorber were studied without adiabatic elimination of atomic polarization. In particular, the normal form obtained in [36] describes the local dynamics near a bifurcation point with the zero eigenvalue of multiplicity three. In particular, it has been shown that an attractor of Lorenz type can exist near this bifurcation point.

The stationary solution of Eqs. (8.22)–(8.24) with a nonzero optical field intensity is defined by the equations $\kappa e^{G-Q} = 1$, $g_0 = \gamma_g G + e^{-Q}(e^G - 1)I$, and $q_0 = \gamma_q Q + s(1 - e^{-Q})I$. This solution can lose its stability via an Andronov–Hopf bifurcation leading to the so-called Q-switching regime with periodically oscillating optical field intensity. A similar scenario (associated, however, with the destabilization of the ML regime, rather than a stationary generation regime) is observed in ML lasers, where a bifurcation of the ML state leads to the Q-switched ML regime characterized by a sequence of short optical pulses with the peak intensity modulated by a low-frequency envelope. The Q-switching frequency Ω_R of the pulse amplitude modulation is usually much smaller than the pulse repetition frequency, $\Omega_R \ll 2\pi/T$, with a typical value of a few gigahertz for monolithic ML semiconductor lasers. The Q-switching instability of the stationary laser generation regime and that of the ML regime are related. However, further analysis shows that the Andronov–Hopf bifurcation boundary of the stationary solution of Eqs. (8.22)–(8.24) does not provide an accurate estimate of the Q-switching instability boundary of the ML regime. A numerical evidence of this fact is presented below, Figure 8.11. A detailed theoretical analysis of Q-switching behavior in single-mode lasers with saturable absorbers can be found in [33, 34, 37, 38]. Bifurcation analysis of a model of such kind of lasers subjected to a delayed feedback is discussed in Chapter 7 of this book by B. Krauskopf and J. J. Walker.

8.3 Numerical Results

In the remaining part of this chapter, we use the DDE model (Eqs. (8.19)–(8.21)) for a detailed investigation of the dynamical regimes in an ML laser. An important advantage of this model is that it allows for the application of efficient algorithms for simulation and numerical bifurcation analysis that have been developed for systems of DDEs, in particular, the software package DDEBIFTOOL [39]. In this section, we present results based on such numerical investigations. In addition, we present in Sections 8.4, and 8.5 several analytical results concerning the ML pulse stability in the limit $\gamma T \to \infty$.

The simplest stationary solution of Eqs. (8.19)–(8.21) corresponds to zero laser intensity (laser off):

$$A = 0, \quad G = g_0/\gamma_g, \quad Q = q_0/\gamma_q. \tag{8.25}$$

Additionally, there also exist cw solutions of the form $A(\tau) = A_0 e^{i\gamma \omega t}$ with nonzero intensity $|A_0|^2$. Their intensities and frequency shifts ω can be found by solving the equations (cf. Eq. (8.19))

$$\kappa e^{G-Q} - 1 - \omega^2 = 0 \qquad (8.26)$$

$$\omega + \tan\left[\gamma T\omega + (\alpha_g G - \alpha_q Q)/2 - \varphi\right] = 0 \qquad (8.27)$$

where expressions for G and Q in terms of the laser intensity $|A_0|^2$ can be found by equating to zero the right-hand sides of Eqs. (8.20) and (8.21). The transcendental Eqs. (8.26) and (8.27) have multiple solutions, each corresponding to a certain longitudinal mode. In the limit $\gamma T \to \infty$, the frequency interval between two neighboring cw solutions coincides with the cold cavity intermode frequency spacing $\delta\nu = 1/T$. The value of the linear gain parameter g_0 where a cw solution with the frequency detuning ω_k bifurcates from the zero intensity solution (Eq. (8.25)) is given by $g_0 = \gamma_g\left[q_0/\gamma_q - \ln\kappa + \ln\left(1 + \omega_k^2\right)\right]$. Let the cw solution with $\omega = \omega_0$ have minimal detuning $|\omega_0|$ from the central point of the Lorentzian spectral filtering profile. This solution, having minimal effective losses, bifurcates from the solution (Eq. (8.25)) at the linear threshold point:

$$g_0^{Th} = \gamma_g\left[q_0/\gamma_q - \ln\kappa + \ln\left(1 + \omega_0^2\right)\right]. \qquad (8.28)$$

All other cw solutions with $|\omega_k| > |\omega_0|$ appear at larger threshold currents and, therefore, bifurcate from the zero intensity solution when it is already unstable. Hence, in general, only a single cw solution can be stable just above the bifurcation. Some of these unstable cw solutions can, however, become stable with increasing pump parameter g_0 [40].

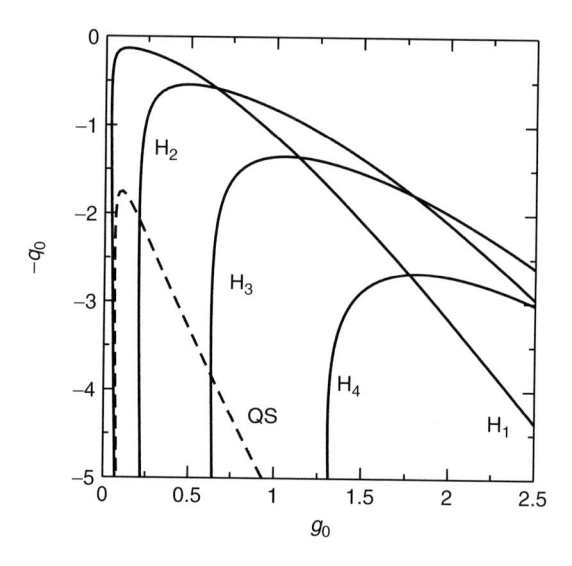

Figure 8.2 Andronov–Hopf bifurcations of the cw solution with $\omega = 0$. Parameter values: $T = 25$ ps, $\gamma = 25$, $\alpha_{g,q} = 0$, $s = 25$, $\gamma_g = 1$ ns, $\gamma_q = 10$ ps, and $\kappa = 0.5$.

The results of a numerical linear stability analysis of the cw solution with $\omega = \omega_0 = 0$ in the two parameters (g_0, q_0) are presented in Figure 8.2. The curves H_n indicate Andronov–Hopf bifurcations leading to the regimes with pulsating laser intensity with the period close to T/n. Thus, at the curve H_1 the fundamental ML solution with the pulse repetition frequency close to $\Omega_1 = 2\pi/T$ appears. The curves H_n with $n = 2, 3, 4$ denote the bifurcations leading to harmonic ML solutions with the repetition rates close to $n\Omega_1$. Finally, the curve QS corresponds to an Andronov–Hopf bifurcation with a much smaller frequency associated with Q-switching.

Starting from the Andronov–Hopf bifurcation curves, we computed branches of solutions with periodic laser intensity and their stability, Figure 8.3a. One can observe that the branch ML_1 corresponding to the fundamental ML regime has two stable parts. The first of them is very narrow and is located near the left Andronov–Hopf bifurcation point at small values of g_0 where the amplitude of the solution ML_1 is small. The second stable part is limited by two bifurcation points. The left point (QP) corresponds to a bifurcation into a regime of Q-switched ML corresponding to quasi-periodic laser intensity. In this regime, the peak power of ML pulses is modulated at the Q-switching frequency. At smaller values of the pump parameter g_0 below the bifurcation point QP, the modulation amplitude strongly increases. At the right-hand side, there is a saddle-node bifurcation (SN) where two ML solutions, stable and unstable, coalesce and disappear. The solutions corresponding to harmonic ML regimes are denoted by ML_2 and ML_3 in Figure 8.3a. These solutions undergo bifurcations similar to that of the fundamental ML solution branch ML_1. According to Figure 8.3, bistability between different ML regimes can exist for certain parameter values.

The results of direct numerical integration of Eqs. (8.19)–(8.21) performed using the FORTRAN routine RADAR5 [41] are presented in Figures 8.3–8.5. Figure 8.3b illustrates the dependence of the local extrema of the laser intensity time trace on the pump parameter g_0. It has been constructed using the following procedure. First, Eqs. (8.19)–(8.21) have been integrated from $t = 0$ till $t \approx 2 \cdot 10^3$ in order to skip the transients. Then, during the time interval $\Delta t \approx 200$ local maxima and minima of the intensity time trace have been plotted for every given value of the parameter g_0. Figure 8.3b shows that for sufficiently small parameter g_0, the laser operates in the Q-switched ML regime. An intensity time trace illustrating this regime is shown in Figure 8.4a. With the increase of the parameter g_0, this regime undergoes an inverse Andronov–Hopf bifurcation leading to a transition to the fundamental ML regime (Figure 8.4c). A stability analysis of the fundamental ML regime with respect to oscillations with Q-switching frequency is performed in Section 8.5. The Q-switching instability of an ML regime was also studied in [42] within the framework of the Haus master equation. When the pump parameter further increases, a transition to harmonic ML regimes with two and three times higher pulse repetition rate takes place (Figure 8.4d,e). These regimes are characterized by smaller pulse peak intensities than the fundamental ML regime. Note that the appearance and break-up of harmonic ML regimes can be viewed as a result of the interaction between neighboring pulses via saturable gain and absorption.

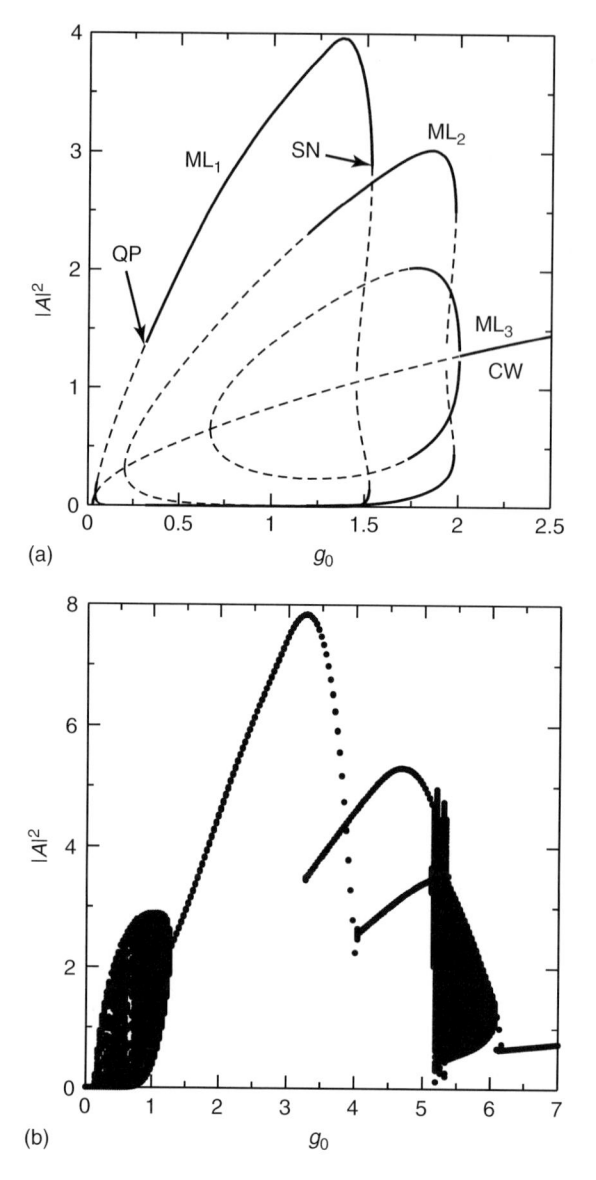

Figure 8.3 (a) Branches of ML solutions (ML$_1$– ML$_3$) appearing from the Andronov–Hopf bifurcations shown in Figure 8.2 from the branch of cw solutions (CW). Solid (dotted) lines denote stable (unstable) solutions. Parameter values are: $q_0 = 2$, $T = 25$ ps, $\gamma^{-1} = 0.4$ ps, $\alpha_{g,q} = 0$, $s = 5$, $\gamma_g^{-1} = 1$ ns, $\gamma_q^{-1} = 10$ ps and $\kappa = 0.5$. (b) Sampled intensity peaks obtained by direct numerical integration of Eqs. (8.19)–(8.21) with $q_0 = 5.0$, $T = 25$ ps, $\gamma^{-1} = 0.286$ ps, $\kappa = 0.1$, $s = 15$, $\gamma_g^{-1} = 1$ ps, $\gamma_q^{-1} = 10$ ns, and $\varphi = 0$.

A detailed analysis of "noncoherent" ML pulse interaction was performed in [43] by a method similar to that described in Section 8.5. The break-up of ML regimes is accompanied by the sudden appearance of a chaotic modulation of the pulse peak power (Figure 8.4b). Finally, at large pumping, the laser operates in the cw regime with time-independent electric field intensity. The parameter scan presented in Figure 8.3b is in qualitative agreement with the experimental results reported in [44, 45], where a gradual transition from Q-switched ML to a stable fundamental ML regime was observed on increasing the injection current. An experimental observation of harmonic ML regime with the pulse repetition period approximately two times smaller than that of the fundamental ML regime was reported in [16, 46].

Figures 8.5a,b have been computed similar to the Figure 8.3a, but with the line-width enhancement factors being used as the bifurcation parameters instead of the parameter g_0. According to Figure 8.5a, corresponding to $\alpha_q = 3.0$, ML pulses with the largest peak intensity are observed when the line-width enhancement factors in the gain and absorber sections are approximately equal, $\alpha_g \approx \alpha_q$. These pulses have the smallest widths also, see Figure 8.5c. With decreasing line-width enhancement factor in the gain section ($\alpha_g < \alpha_q$), the ML pulses become wider, their peak intensity decreases, and, finally, a transition to a regime with quasi-periodic laser intensity takes place. This regime is characterized by additional oscillations at the leading edge of the ML pulse (solid line in Figure 8.5c). An increase of α_g produces an even stronger effect on the ML regime, namely a transition into a chaotic regime. As it follows from the results of our calculations, this transition is accompanied by an intermittency between a regular ML regime and a regime with chaotically pulsating laser intensity. Indeed, just above the transition point, time intervals of regular ML regime alternate with intervals of irregular pulsations. The duration of time intervals of chaotic behavior decreases with α_g, and, finally, a purely chaotic regime establishes. The fact that ML pulses with highest quality were observed for $\alpha_g \approx \alpha_q$ can be understood by recalling that the saturable gain G and absorption Q enter Eq. (8.19) with opposite signs. Therefore, in the case $\alpha_g \alpha_q > 0$, the chirp introduced by the line-width enhancement factor in the gain section is at least partially compensated by the absorber section. According to our numerical results, the largest compensation takes place for $\alpha_g \approx \alpha_q$, that is, when the frequencies ω of the cw solutions are independent of their intensities, Eqs. (8.26) and (8.27). In the case when the line-width enhancement factors in the gain and absorber sections are equal, their increase has almost no effect on the pulse peak power (Figure 8.5b). The break-up of the ML regime at $\alpha_g = \alpha_q > 3.4$ leads to an abrupt transition to a chaotic regime.

Figure 8.6 presents time traces of the electric field intensity and the net gain per cavity round trip $\mathcal{G}(t) = G(t) - Q(t) + \ln \kappa$ for two fundamental ML regimes, corresponding to different pump parameter values. In Figure 8.6a, corresponding to $g_0 = 1.2$, the net gain is negative between pulses and becomes positive only during short time intervals when the pulse intensity is large. Therefore, this solution satisfies New's criterion. This criterion requires that, in order to prevent the growth of perturbations at the tails of an ML pulse, the inequality $G - Q + \ln \kappa < 0$ should

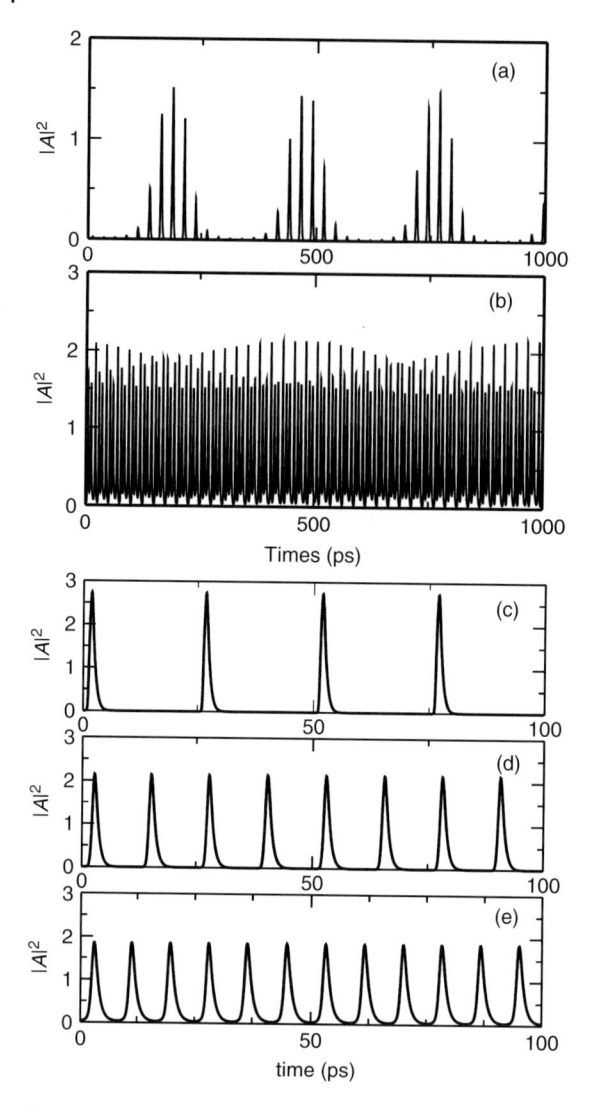

Figure 8.4 Laser intensity time traces. Left: Non-periodic solutions; (a) Q-switched ML regime at $g_0 = 0.5$, (b) irregular ML regime at $g_0 = 5.25$. Right: ML regimes; (c) 40 GHz fundamental ML regime at $g_0 = 3.0$, (d) 80 GHz harmonic ML regime with two pulses in the cavity at $g_0 = 4.0$, (e) 120 GHz harmonic ML regime with three pulses in the cavity at $g_0 = 5.0$. The values of other parameters are the same as in Figure 8.3b.

be fulfilled between the pulses where the intensity $|A|^2$ is close to zero [4]. In other words, small perturbations of the vanishing intensity "background" between the pulses should decay in time. On the contrary, in Figure 8.6b, corresponding to $g_0 = 1.67$, a stable ML solution of Eqs. (8.19)–(8.21) shows a positive net gain at the leading edge. The existence of stable pulses with positive net gain at the trailing

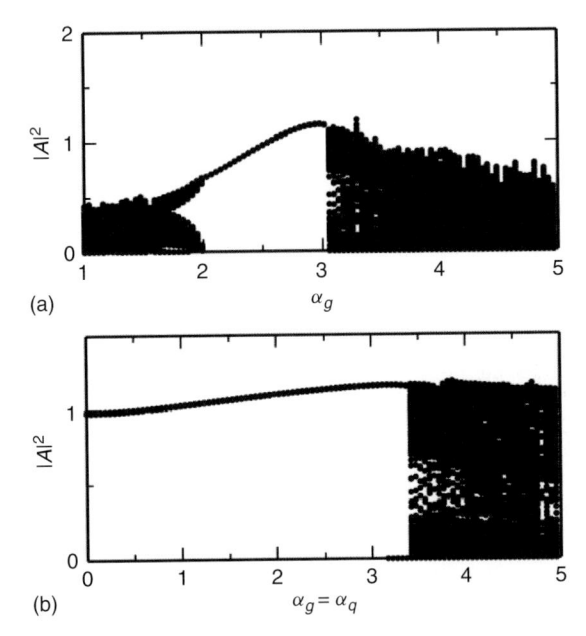

Figure 8.5 (a,b) Sampled intensity peaks for varying line-width enhancement factors in gain and absorber sections, $g_0 = 1.0$, $q_0 = 3.0$, $\gamma^{-1} = 0.5$ ps. (a) intensity peaks time vs α_q, $\alpha_q = 3.0$. (b) variation of both line-width enhancement factors, $\alpha_g = \alpha_q$. Other parameter values are the same as in Figure 8.3b. (c) Fundamental ML pulses calculated for different values of the line-width enhancement factor in the gain section. $\alpha_q = 3$. Solid line: $\alpha_g = 1.8$; dashed line: $\alpha_g = 2.25$; dash-dot line: $\alpha_g = 3.0$.

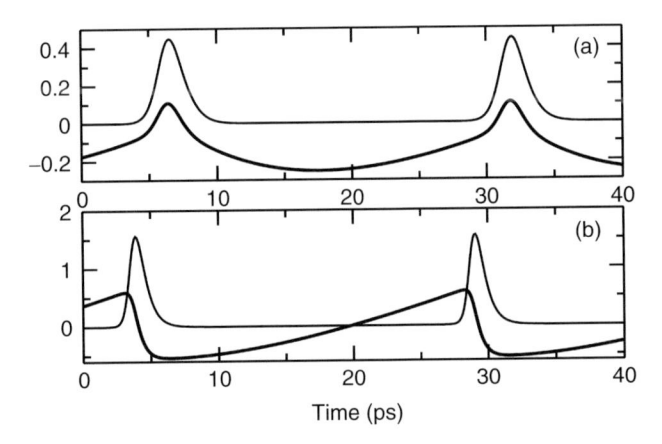

Figure 8.6 Intensity time traces (thin line) and net gain parameter (thick line). $\alpha_g = \alpha_q = 2.0$, $q_0 = 2.67$, $\gamma^{-1} = 0.67$ ps. (a) – ML pulses with negative net gain during the whole slow stage, $g_0 = 1.2$; (b)ML pulses with positive net gain at their leading edge, $g_0 = 1.67$. Other parameter values are the same as in Figure 8.5.

edge in passively ML lasers was reported earlier [47, 48]. Pulses with positive net gain at the leading edge, similar to those shown in Figure 8.6b, were previously reported only for the case of synchronously pumped actively ML lasers [49–51]. Here, the net gain window is opened in the course of the carrier density recovery process. For typical parameter values of semiconductor lasers, the gain recovers much slower than the absorption ($\gamma_g/\gamma_q \ll 1$). Therefore, the gain continues to recover even when the absorption has already almost reached its saturated value. This can lead to the appearance of a net gain window that is not caused by the interplay of the absorption and the returning ML pulse, as it happens in the case of the classical passive ML mechanism [4], or by an external periodic modulation, as in the case of active ML.

As it was already mentioned, the fact that New's criterion is not satisfied does not necessarily mean that the ML regime is unstable. Stable ML pulses with "unstable background" can exist for system (Eqs. (8.19)–(8.21)) because of the difference between the pulse group velocity v_p and group velocity v_0 of perturbations of the background. Let us consider an ML regime with the pulse repetition period $T_p = T + \delta T$, and $\delta T \ll T$. Then the pulse group velocity can be estimated as $v_p = vT/T_p \approx v(1 - \delta T/T)$ where v is the cold cavity group velocity. Now, we estimate the velocity of small perturbations. Taking into account the inequality $\gamma T \gg 1$, the left-hand side of Eq. (8.19) can be rewritten in the form $\partial_t A(t) + \gamma A(t) \approx \gamma A(t + \gamma^{-1})$. By equating this to $\gamma A(t - T)$, we get the cavity round trip time equal approximately to $T + \gamma^{-1}$, which gives the estimate

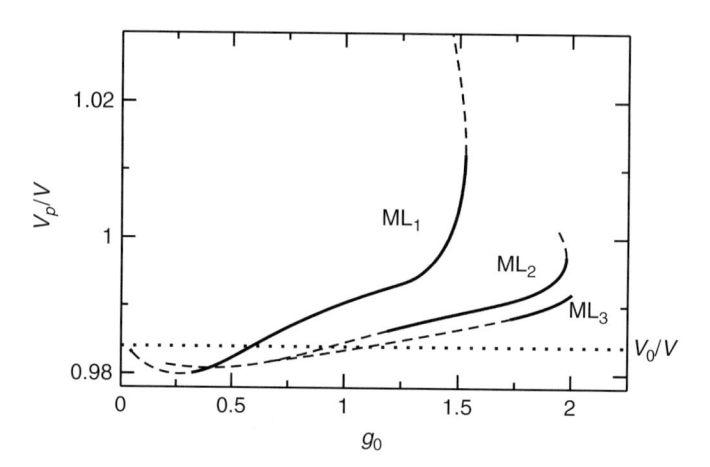

Figure 8.7 Group velocity of ML pulses divided by the cold cavity group velocity v. $ML_{1,2,3}$ correspond to ML solutions shown in Figure 8.3. Solid (dashed) lines correspond to stable (unstable) solutions. Horizontal dotted line: normalized group velocity of small perturbations, v_0/v. Parameter values are the same as in Figure 8.3b.

$v_0 = v\left[1 - (\gamma T)^{-1}\right]$ for the group velocity of small perturbations. ML pulses with positive net gain at their leading edge shown in Figure 8.6b are stable since they turn out to move faster than the perturbations of the background (Figure 8.7). Similarly the pulses with positive net gain at the trailing edge may be stable if they are moving slower than the small perturbations. The latter situation was reported in [48].

8.4
Stability Analysis for the ML Regime in the Limit of Infinite Bandwidth

The number of locked cavity modes can be roughly estimated as the ratio of the bandwidth γ of the spectral filtering element to the intermode frequency spacing T^{-1}. Here we discuss the limit where this number is large, that is $\gamma T \to \infty$. In this limit, the duration of the ML pulse $\tau_p \propto \gamma^{-1}$ tends to zero, its amplitude $A_0 \propto \gamma^{1/2}$ tends to infinity, while the pulse energy $\Delta P \propto |A_0|^2 \tau_p$ remains finite. Physically, this means that the number of locked modes increases while the amplitude of each mode decreases. We assume that the ML pulse duration τ_p is much shorter than the relaxation times of the gain and absorber media, $\tau_p \ll \gamma_{g,q}^{-1}$. This assumption, known as *the slow saturable absorber approximation* [4, 5], is realistic for parameter values, which are typical for monolithic semiconductor lasers. ML in a laser with slow absorber was studied analytically by New [4] and Haus [5]. They distinguished between the slow and fast stages of the evolution of the ML solution within the cavity round trip time. The fast stage corresponds to a short time interval when the electric field pulse amplitude is large. At this stage, the relaxation terms in the right-hand side of Eqs. (8.20) and (8.21) can be neglected. During the slow stage, the electric field is close to zero between the two successive pulses, $|A(t)|^2 \approx 0$ (Figure 8.8). At this stage, we neglect the term proportional to the field intensity $|A|^2$ in Eqs. (8.20)–(8.21).

In this manner, the laser equations can be solved separately for the slow stage and, under additional simplifying assumptions, also for the fast stage. Then, a combined analytic solution can be obtained by matching the slow- and fast-stage solutions. We show that classical results of New [4] and Haus [5] can be reproduced by the DDE model (Eqs. (8.19)–(8.21)) in the limit in which the gain and loss per cavity round trip are small. Furthermore, in Sections 8.4.4 and 8.5.2, we generalize these results to the case of large gain and loss per cavity round trip, that is, for the situation of semiconductor lasers. We see that the generalized approach of Haus, compared to that of New, is more accurate for low pulse intensities but has a more restricted application domain.

8.4.1
New's Stability Criterion

An important criterion for stability of ML pulses was proposed by New [4]. New's criterion requires the net gain $G(t) - Q(t) + \ln \kappa$ to be negative at every time

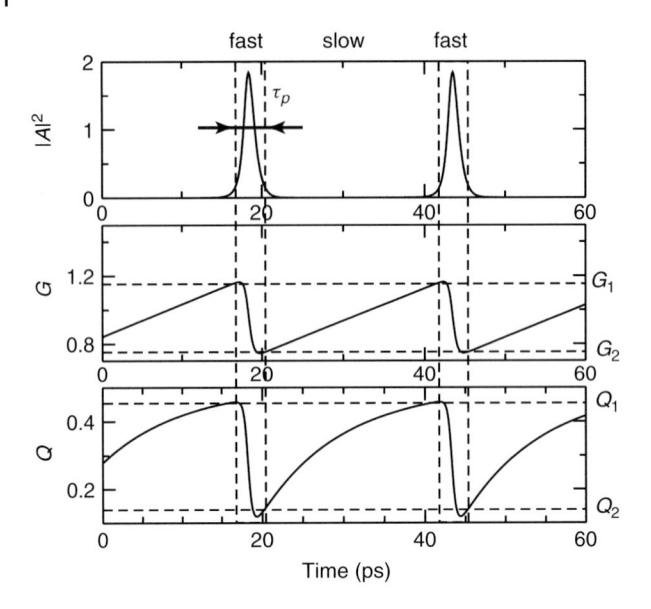

Figure 8.8 Time evolution of the electric field intensity $|A|^2$, the saturable gain G, and the saturable loss Q in an ML laser with a slow absorber. The fast stage is defined by the pulse duration τ_p. G_1 and Q_1 are the saturable gain and loss at the pulse leading edge (transition from slow to fast stage). G_2 and Q_2 are the saturable gain and loss values at the pulse trailing edge (transition from fast to slow stage).

during the slow stage. Physically, it means that small perturbations effecting the low-intensity interval between two subsequent pulses should decay (absolute stability). This criterion can be shown to be satisfied if the net gain is negative at the beginning and end of the slow stage,

$$G_1 - Q_1 + \ln \kappa < 0, \quad G_2 - Q_2 + \ln \kappa < 0. \tag{8.29}$$

Here G_2 and Q_2 (G_1 and Q_1) are the saturable gain $G(t)$ and loss $Q(t)$ evaluated at the beginning (end) of the slow stage (Figure 8.8). As the end of the slow stage coincides with the beginning of the fast stage (and vice versa), the two relations in Eq. (8.29) define stability at the leading and trailing edges of the pulse, respectively. However, this criterion does not take into account that small perturbations can propagate between the pulses and, eventually, be absorbed by either the leading or the trailing edge of the pulse within the time interval of order γ^{-1}. Hence, such perturbations should not necessarily destroy the ML pulses even if New's criterion is violated. Indeed, as we have seen, our numerical computations in Section 8.3 confirm the existence of stable ML regimes with unstable background between the pulses. Naturally, such pulses should be more sensitive to noise and other small perturbations than those satisfying New's criterion.

8.4.2
Slow Stage

For the slow stage of an ML solution where $|A(t)|^2 \approx 0$, Eqs. (8.20) and (8.21) become linear: $\partial_t G = g_0 - \gamma_g G$, $\partial_t Q = q_0 - \gamma_q Q$. Solving these equations, we express the saturable gain G_1 and loss Q_1 at the pulse leading edge through their values G_2 and Q_2 at the pulse trailing edge,

$$G_1 = G_2 e^{-\gamma_g T} + \frac{g_0}{\gamma_g}\left(1 - e^{-\gamma_g T}\right), \tag{8.30}$$

$$Q_1 = Q_2 e^{-\gamma_q T} + \frac{q_0}{\gamma_q}\left(1 - e^{-\gamma_q T}\right). \tag{8.31}$$

Here, the duration of the slow stage is taken equal to the cold cavity round trip time T in the limit $\gamma \to \infty$. Eqs. (8.30) and (8.31) can be further simplified in the following two limiting situations:

1) The absorber relaxes completely between two successive pulses, that is, $\gamma_q T \gg 1$. Then, Eq. (8.31) simplifies to $Q_1 = q_0/\gamma_q$.
2) The relaxation time of the gain medium is much shorter than the cavity round trip time, $\gamma_g T \ll 1$. In this case, Eq. (8.30) can be replaced by the relation $G_1 = G_2 + g_0 T$.

8.4.3
Fast Stage

During the fast stage, we neglect the relaxation terms in the right-hand side of Eqs. (8.20) and (8.21). With this approximation, introducing the differential pulse energy $p(t) = \int_0^t |A(\theta)|^2 \, d\theta$, where $t = 0$ corresponds to the beginning of the fast stage, we rewrite these equations as

$$\partial_p g(p) = -e^{-q(p)}\left(e^{g(p)} - 1\right), \quad \partial_p q(p) = -s\left(1 - e^{-q(p)}\right), \tag{8.32}$$

where $g(p(t)) = G(t)$ and $q(p(t)) = Q(t)$. Integrating Eq. (8.32), we express the saturable gain $G_2 = g(P)$ and loss $Q_2 = q(P)$ at the trailing edge of the pulse in terms of their values $G_1 = g(0)$ and $Q_1 = q(0)$ at the leading edge,

$$G_2 = g(P) = -\ln\left[1 - \frac{1 - e^{-G_1}}{\left(e^{sP - Q_1} - 1 + e^{-Q_1}\right)^{1/s}}\right], \tag{8.33}$$

$$Q_2 = q(P) = \ln\left[1 + e^{-sP}\left(e^{Q_1} - 1\right)\right]. \tag{8.34}$$

Here, $P = p(\tau_p) = \int_0^{\tau_p} |A(t)|^2 \, dt$ is the total energy of the ML pulse. Integrating the squared absolute value of both sides of Eq. (8.19) over the fast stage, we obtain

$$\gamma^{-2} \int_0^{\tau_p} |\partial_t A(t)|^2 \, dt + P = \kappa \int_0^{\tau_p} e^{G(t) - Q(t)} |A(t)|^2 \, dt. \tag{8.35}$$

Using the solution of Eq. (8.32), we calculate the integral at the right-hand side of this equation explicitly and arrive at the relation

$$\gamma^{-2} \int_0^{T_p} |\partial_t A(t)|^2 \, dt + P = \kappa \ln \frac{e^{G_1} - 1}{e^{G_2} - 1}. \tag{8.36}$$

Note that the integral term at the left-hand side of Eq. (8.36), describing the energy loss due to spectral filtering, does not vanish in the limit $\gamma \to \infty$. Hence, strictly speaking, this term can not be neglected in the limit of infinite spectral filtering bandwidth. Indeed, Haus showed that $|\partial_t A(t)|^2 \propto \gamma^2 |A(t)|^2$ for pulses of duration $\tau_p \propto \gamma^{-1}$ [5]. In other words, the spectral width of the ML pulse increases with the spectral filtering bandwidth in such a way that the losses due to filtering remain finite and nonzero in the limit $\gamma \to \infty$. Thus, we have to express the integral term in Eq. (8.36) via the pulse parameters to provide an analytic solution for the fast stage. Two specific cases where this is possible are discussed in the next Sections 8.4.4 and 8.5.2.

8.4.4
Laser Without Spectral Filtering

New's approach is based on the assumption that there is no spectral filtering in the laser cavity, that is, the response function in Eq. (8.13) is given by $f(t) = \delta(t)$. Hence, in New's approximation, the term $\gamma^{-1} \partial_t A(t)$ with the time derivative in Eq. (8.19) is neglected. Consequently, the integral term in Eq. (8.36) vanishes and this equation becomes

$$P = \kappa \ln \frac{e^{G_1} - 1}{e^{G_2} - 1}. \tag{8.37}$$

In this way, Eqs. (8.30), (8.31), (8.33), and (8.34) together with Eq. (8.37) form a closed system, which can be solved with respect to the unknowns $G_{1,2}$, $Q_{1,2}$, and P. As a result, one obtains the dependence of the pulse energy P on the laser parameters. But, as we have seen above, neglecting the integral term for the losses due to spectral filtering in Eq. (8.36) cannot be rigorously justified even for infinitely large γ. In this sense, Eq. (8.37) is a rather rough approximation for Eq. (8.36). However, we will show that this approximation works rather well for a relatively large parameter domain.

Substituting the solution of system (Eqs. (8.30), (8.31), (8.33), (8.34), and (8.37)) in the inequalities (Eq. (8.29)), we obtain the mode-locking stability boundaries according to New's criterion. Figure 8.9 presents the result of this calculation for a solution with period T that corresponds to the fundamental ML regime (lines L_1 and T_1, and for a solution with period $T/2$ corresponding to a harmonic ML regime with two pulses in the cavity (lines L_2 and T_2). The two stability domains intersect, indicating that hysteresis between the ML regimes with different pulse repetition rates is possible in a certain parameter domain (as in the direct numerical simulations shown in Figure 8.3a). According to Figure 8.9, the stability boundaries for the leading and the trailing edges of the pulse meet at a codimension two point

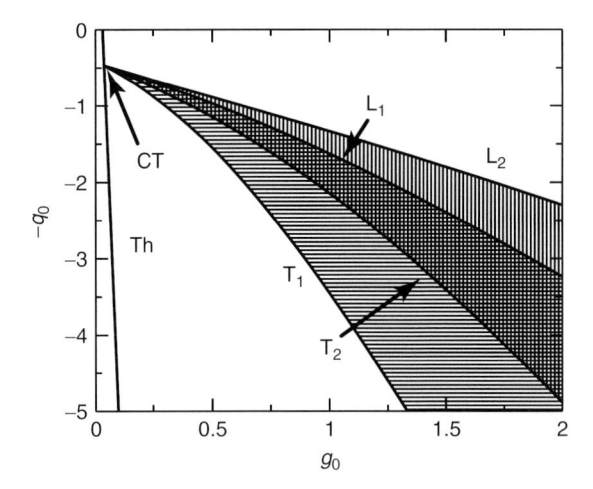

Figure 8.9 ML stability domains according to New's criterion. The area marked by horizontal (vertical) hatching is the stability domain of the 40 GHz fundamental ML regime (ML regime with 80 GHz repetition frequency). The straight line Th is the linear threshold defined by Eq. (8.28) with $\omega_0 = 0$. The lines $L_{1,2}$ and $T_{1,2}$ define the leading and trailing edge stability boundaries, respectively. CT is the codimension two point defined by Eq. (8.38). The parameters are $T = 25$ ps, $s = 25$, $\gamma_g^{-1} = 1$ ns, $\gamma_q^{-1} = 7.5$ ps, $\kappa = 0.1$.

CT for each of the ML solutions. This point, lying on the linear threshold line Th defined by Eq. (8.28) with $\omega_0 = 0$, can be calculated explicitly as

$$q_0 = \ln \frac{\kappa(s-1)}{s\kappa - 1}, \quad g_0 = \frac{\gamma_g}{\gamma_q} \ln \frac{s-1}{s\kappa - 1}. \tag{8.38}$$

It is known that the ML pulses satisfying New's stability criterion can exist only if the absorbing medium saturates faster than the gain medium, that is, the condition $s' > 1$ is satisfied [1, 5]. Eq. (8.38) imply that for large gain and losses per cavity round trip, the existence of such pulses is possible only if a stronger condition is satisfied,

$$s\kappa \equiv \frac{s'\kappa}{\kappa_2} = s'\kappa_1\kappa_3\kappa_4 > 1. \tag{8.39}$$

In the limit of small losses, $\kappa_k \to 1$ ($k = 1, 2, 3, 4$), this condition coincides with the classical condition $s' > 1$. However, for the parameter values typical for semiconductor lasers with large losses, $\kappa_2\kappa_3\kappa_4 \ll 1$, Eq. (8.39) becomes much more restrictive than the condition $s' > 1$. For $s\kappa > 1$, the parameter domain of the existence of ML pulses satisfying New's criterion becomes larger with increasing $s\kappa$. This is in qualitative agreement with the experimental results in [44], which show that the quality of the ML can be improved by decreasing nonresonant losses in the laser cavity.

Equations (8.30), (8.31), (8.33), (8.34), and (8.37) generalize New's model, since they do not assume small gain and losses per cavity round trip as in [4]. To obtain

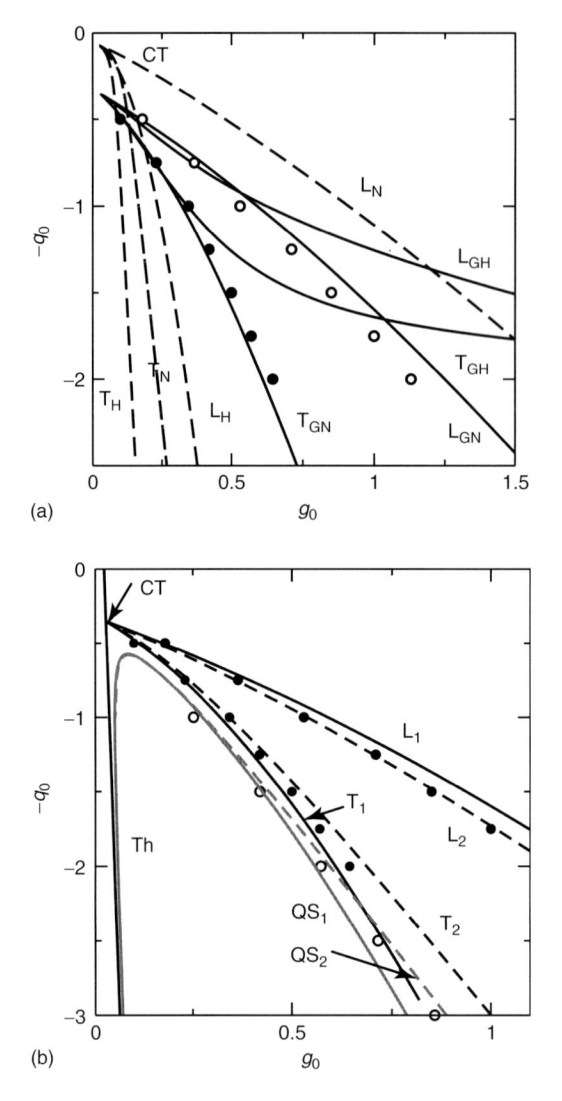

Figure 8.10 (a) Comparison of ML stability boundaries, as defined by New's background stability criterion, computed by four different approaches. Solid lines L_{GN} and T_{GN} (L_{GH} and T_{GH}) show the leading and trailing edge stability boundaries for the generalized New's model (generalized Haus' model). Stability boundaries for the original models of New (Haus) are shown by dashed lines L_N and T_N (L_H and T_H). Dots (circles) show the leading (trailing) edge stability boundaries obtained by direct numerical integration of Eqs. (8.19)–(8.21). Parameter values are the same as in Figure 8.9 (with $\alpha_g = \alpha_q = \varphi = 0$). (b) Stability boundaries according to New's criterion (black) and onset of Q-switching instability (gray) for the fundamental ML regime. $L_{1,2}$ – leading edge, $T_{1,2}$ – trailing edge instability. Solid lines L_1, T_1, and QS_1 are obtained from (8.42), (8.43) and (8.46). Dashed lines L_2, T_2, and QS_2 are obtained from Eqs. (8.42), 8.43), (8.54) and (8.55). Circles show the instability boundaries obtained by direct numerical integration of Eqs. (8.19)–(8.21). The straight line Th is the linear lasing threshold defined by Eq. (8.28) with $\omega_0 = 0$. CT is the co-dimension two point defined by Eq. (8.38). Parameters are $\kappa = 0.1$, $s = 25$, $\gamma_g^{-1} = 1$ ns, $\gamma_q^{-1} = 7.5$ ps, $T = 25$ ps, $\gamma^{-1} = 0.2$ ps.

New's relation for the pulse parameters from these equations, we expand Eqs. (8.33) and (8.34) to the first-order terms with respect to G_1 and Q_1 and neglect the higher order terms,

$$G_2 = G_1 e^{-P}, \quad Q = Q_1 e^{-sP}. \tag{8.40}$$

Then, substituting Eq. (8.33) in Eq. (8.37) and expanding the resulting equation to the first-order terms with respect to G_1, Q_1 and $\ln \kappa$, we obtain the equation

$$G_1 \left(1 - e^P\right) - Q_1 \frac{(1 - e^{sP})}{s} - P \ln \kappa = 0 \tag{8.41}$$

for the pulse energy, which is equivalent to Eqs. (11) and (12) in [4].

In Figure 8.10a we compare the stability regions of ML pulses satisfying New's stability criterion, calculated by four different models. The dashed lines L_N and T_N show the ML pulse leading and trailing edge stability boundaries defined by Eqs. (8.30), (8.31), (8.40), and (8.41), which are equivalent to New's equations [4]. The solid lines L_N and T_N are defined by the generalization of New's model to the case of large gain and losses described above. Dots denote points on New's stability boundaries obtained by direct numerical integration of Eqs. (8.19)–(8.21) with $\gamma^{-1} = 0.3$ ps. Note that the stability domain becomes larger for smaller γ. As one can see from Figure 8.10a, the generalized New's model, despite the fact that it neglects losses due to spectral filtering, is in good agreement with the results of numerical integration of the DDE model. On the other hand, the discrepancy between the stability boundaries defined by New's original model and the numerical results is substantially larger. The reasons are the large gain and losses per cavity round trip that are encountered typically in semiconductor lasers.

When the time derivative of the electric field envelope in Eq. (8.19) is neglected, as in the approach adopted in this section, the stability boundaries defined by New's criterion are independent of the line-width enhancement factors $\alpha_{g,q}$, which play a role only when the spectral filtering is taken into account.

8.5
The Q-Switching Instability of the ML Regime

In this section, we discuss stability of the ML regime with respect to the periodic pulse amplitude modulation by the Q-switching frequency. Using the slow- and fast-stage equations derived from Eqs. (8.19)–(8.21) in Sections 8.4.2 and 8.4.3, one can define a map that describes the transformation of the ML pulse parameters per cavity round trip. A nontrivial fixed point of this map corresponds to a periodic ML solution. We show that this fixed point can lose its stability via a Neimark–Sacker bifurcation in which a pair of complex conjugate Floquet multipliers crosses the unit circle. This bifurcation is responsible for the Q-switching instability of the ML regime, one of the main sources of amplitude noise in semiconductor lasers.

Now, let G_n and Q_n be the saturable gain and loss values at the beginning of the fast stage after n cavity round trips, that is, at the n-th pulse leading edge. The pulse energy is defined by the relation $P_n = \int_0^{\tau_n} |A|^2 dt$ where the integration limits, 0 and τ_n, correspond to the beginning and the end of the fast stage, respectively. Using Eqs. (8.30), (8.31), (8.33), and (8.34), we obtain the map describing the transformation of the saturable gain and loss after a cavity round trip as

$$G_{n+1} = -e^{-\gamma_g T} \ln\left[1 - \frac{1-e^{-G_n}}{(1+e^{sP_n-Q_n}-e^{-Q_n})^{1/s}}\right] + (1-e^{-\gamma_g T})g_0/\gamma_g, \quad (8.42)$$

$$Q_{n+1} = e^{-\gamma_q T} \ln\left[1 + e^{-sP_n}(e^{Q_n}-1)\right] + (1-e^{-\gamma_q T})q_0/\gamma_q. \quad (8.43)$$

Here, G_{n+1} and Q_{n+1} are the saturable gain and loss at the beginning of the fast stage after $n+1$ cavity round trips, that is, at the leading edge of the $(n+1)$-th pulse. In order to complete the definition of the pulse parameters transformation map, Eqs. (8.42) and (8.43) should be complemented by a relationship between the energies P_n and P_{n+1} of two consecutive pulses derived from Eq. (8.19) for the electric field envelope A. We obtain analytic approximations to this relationship using two simplifying approaches. The first approach is based on New's approximation assuming no spectral filtering in the laser cavity [4]. This approximation, already used in Section 8.4.4 to determine the background stability boundaries according to New's criterion, will now be applied for computing the onset of the Q-switching instability. However, some important parameters of the ML solution such as the pulse duration and the deviation of the pulse repetition period from the cavity round trip time T cannot be estimated in the framework of this approximation. We obtain these parameters in Section 8.5.3 using an alternative approach based on variational method, which takes into account the effect of spectral filtering on laser dynamics.

8.5.1
Laser Without Spectral Filtering

Let us rewrite Eq. (8.19) in the form

$$\gamma^{-1}\partial_t A_{n+1}\left(t - \gamma^{-1}\delta_n\right) + A_{n+1}\left(t - \gamma^{-1}\delta_n\right) = \sqrt{\kappa}e^{\frac{1-i\alpha_g}{2}G_n(t) - \frac{1-i\alpha_q}{2}Q_n(t)} A_n(t) \quad (8.44)$$

with $A_{n+1}(t) \equiv A_n(t+T_n)$ and $\delta_n = \gamma(T_n - T)$ where T_n is the time interval between the two successive pulses. Multiplying Eq. (8.47) with its complex conjugate and integrating over the cold cavity round trip time T, we obtain

$$\gamma^{-2}\int_0^{t_{n+1}} |\partial_t A_{n+1}|^2 dt + P_{n+1} = \kappa \int_0^{P_n} e^{G_n(p)-Q_n(p)} dp, \quad (8.45)$$

where the integrals at both sides are over the fast stage only as the electric field intensity during the slow stage is negligibly small. Eq. (8.45) describes the energy balance over a round trip time in the cavity. It extends Eq. (8.35) to the case when successive pulses have different amplitudes. The integral term at the left-hand side of Eq. (8.45) describes the energy losses due to spectral filtering. As in this section, the effect of spectral filtering is completely neglected, we omit this term and, integrating explicitly the right-hand side, obtain

$$P_{n+1} = \kappa \ln\left[1 - e^{G_n} + e^{G_n}(1 + e^{sP_n-Q_n} - e^{-Q_n})^{1/s}\right]. \quad (8.46)$$

The 3-D map defined by Eqs. (8.42), (8.43), and (8.46) describes the transformation of the ML pulse parameters G_n, Q_n, and P_n after a complete round trip in the laser cavity. The fixed point $(g_0/\gamma_g, q_0/\gamma_q, 0)$ of this map with the zero pulse energy is stable for $\eta = g_0/\gamma_g - q_0/\gamma_q + \ln \kappa < 0$ (linear lasing threshold). It loses stability in a transcritical bifurcation at the linear lasing threshold $\eta = 0$. The fixed point (G_*, Q_*, P_*) with $P_* > 0$, which appears as a result of the transcritical bifurcation, corresponds to the fundamental ML regime of the DDE model (Eqs. (8.19)–(8.21)). Depending on the parameter values, the bifurcating branch with $P_* > 0$ can be either stable or unstable. In the latter case, a bistability can exist between the solution with zero electric field intensity and the solution corresponding to the ML regime. However, we assume the laser parameters to satisfy the relation

$$(\kappa^{-1} - e^{-q_0/\gamma_q}) \tanh \frac{\gamma_q T}{2} > s(1 - e^{-q_0/\gamma_q}) \tanh \frac{\gamma_g T}{2},$$

which implies the bifurcation of a stable branch with $P_* > 0$. The fixed point (G_*, Q_*, P_*) can be computed numerically. Linear stability analysis shows that this point can lose stability via a Neimark–Sacker bifurcation (an Andronov–Hopf bifurcation for maps) when two complex conjugate Floquet multipliers cross the unit circle. The bifurcating periodic solution corresponds to the so-called Q-switched ML regime, characterized by a periodically modulated pulse energy. Hence, the Neimark–Sacker bifurcation curve, shown in Figure 8.10b by the solid line QS, defines the boundary separating the domains of stable periodic ML and Q-switched ML. The fixed point (G_*, Q_*, P_*) exists to the right of the linear threshold line Th defined by Eq. (8.28) with $\omega_0 = 0$ and is stable above the line QS. The stability boundaries defined by New's criterion and computed according to the method of Section 8.4.4 are shown in Figure 8.10b by the solid lines L and T. One can see that the lower stability boundary T is separated from the bifurcation line QS by a narrow strip where the periodic ML pulses are stable with respect to the Q-switching instability but do not satisfy New's stability condition because they have a positive net gain at their trailing edge. The circles in Figure 8.10b show the ML regime stability boundaries obtained by direct numerical integration of Eqs. (8.19)–(8.21). The empty circles show the Q-switching instability boundary, while the black circles show the instability boundary defined by New's criterion. One can see that these numerical results are in good agreement with the analytical results obtained in the limit of no spectral filtering in the laser cavity.

Figure 8.11 presents the dependence of the Q-switching stability boundary of the fundamental ML regime and the ML instability boundary defined by New's criterion on the attenuation factor κ and the ratio s of saturation intensities in the gain and absorbing sections. The instability boundaries shown by solid and dashed lines were obtained from Eqs. (8.42), (8.43), (8.46), and (8.29). Our numerical computations imply that these boundaries depend mainly on the product $s\kappa$ rather than on each of these parameters separately. According to Figure 8.11, this observation is particularly true for large losses that are present in semiconductor lasers. The domain of Q-switched ML moves in the direction of larger values of the linear gain and loss parameters g_0 and q_0, and becomes wider with decreasing $s\kappa$.

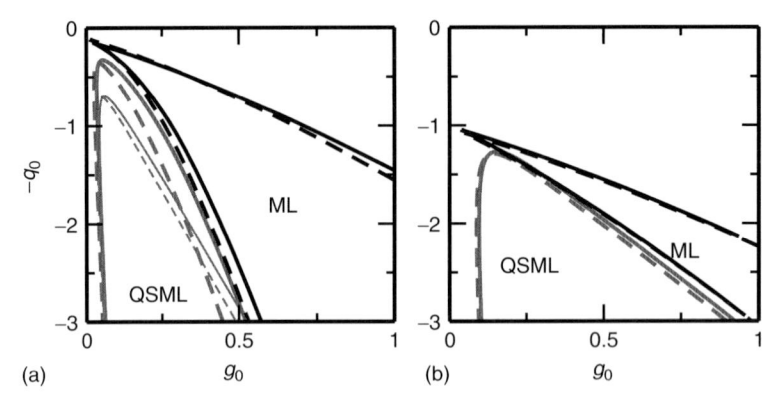

Figure 8.11 Domains of stable periodic mode locking (ML) and Q-switched mode locking (QSML). (a) – $s\kappa = 5$; (b) – $s\kappa = 1.3$. Solid and dashed black lines show the stability boundaries of the fundamental ML solution calculated for $s = 35$ and $s = 15$, respectively, using the criterion of New. Gray lines indicate Q-switching instability boundaries of the the ML solution. Thin gray lines in (a) show the Q-switching instability boundary of the cw solution. This instability does not exist for $s\kappa = 1.3$. Other parameters are as in Figure 8.10b.

This behavior is in agreement with experimental results in [44], where it was shown that the domain of the stable periodic ML regime of a semiconductor laser increases with the reflectivity of the facets. The thin gray lines in Figure 8.11a show the stability boundary of the cw state of Eqs. (8.22)–(8.24) for $s\kappa = 5$ (the solid thin line corresponds to $s = 35$, the dashed line corresponds to $s = 15$). For these parameter values, the Q-switching instability boundaries of the domains of the stable ML and cw states are close to each other. However, for the parameter values of Figure 8.11b ($s\kappa = 1.3$) the cw state does not undergo an Andronov–Hopf bifurcation, which, therefore, can not be used for approximating the Q-switching instability boundary of the ML regime.

Qualitatively, the effect of the parameters s and κ on the ML dynamics admits a simple interpretation. The ratio s' of the saturation energies of the two laser sections controls the main nonlinear mechanism responsible for the compression of the ML pulse. Hence, one expects that the quality of ML improves and the pulse becomes shorter with the increase of this parameter. The attenuation factor κ in Eqs. (8.19)–(8.21) measures the strength of the feedback (per round cavity trip), thus controlling another important mechanism of the ML pulse formation. If κ is too small, then the amplitude of the pulse, which comes back after a round trip in the laser cavity, is not sufficient for creating the net gain window necessary for supporting ML. Therefore, one expects that the increase of κ should favor the ML regime. Experimental results confirm the validity of this qualitative argument [52]. We note, however, that the results of our quantitative analysis, such as the important role of the product $s\kappa$ discussed above, are beyond the scope of this simple qualitative argument.

8.5.2
Weak Saturation Limit

In this section, we assume that the gain and absorbing media are weakly saturated by the ML pulses. This assumption allows us to obtain an explicit expression for the pulse shape by solving analytically the fast-stage equations. Using the fast time variable $\zeta = \gamma t$, we rewrite Eq. (8.47) as

$$\partial_\zeta a_{n+1}(\zeta - c_n) + a_{n+1}(\zeta - c_n) = F_n(p_n) a_n(\zeta), \tag{8.47}$$

where $a_n(\zeta) = \gamma^{-1/2} A_n(t)$, $p_n(\zeta) = \int_{-\infty}^{\zeta} |a_n(s)|^2 \, ds$ and $c_n = \lim_{\gamma \to \infty} \delta_n$. The function $F_n(p_n)$ is obtained by substituting the solution of Eq. (8.32) into Eq. (8.47),

$$F(P) = \sqrt{\kappa} \left[1 + e^{-sp(\zeta)}(e^{Q_1} - 1)\right]^{-1/2} \left[1 - \frac{1 - e^{-G_1}}{\left(e^{sp(\zeta)-Q_1} - e^{-Q_1} + 1\right)^{1/s}}\right]^{-1/2} \tag{8.48}$$

Eqs. (8.47) and (8.48) describe the ML pulse shape in the limit of infinite Lorentzian bandwidth. For a laser operating close to the threshold, the pulse energy is small, $p(\zeta) \leq P \ll 1/s$, that is, both the gain and absorbing media are weakly saturated. In this approximation, which underpins the theory of Haus [5], Eqs. (8.33) and (8.34) and the function $F(P)$ in Eq. (8.47) can be Taylor expanded with respect to $P(\zeta)$ up to second order. Substituting this expansion in Eq. (8.47) and omitting the higher order terms, we obtain

$$-a_{n+1} - (1 - c_n) \partial_\zeta a_{n+1} + c_n \left(\frac{c_n}{2} - 1\right) \partial_{\zeta\zeta} a_{n+1} + \left[F_0 + F_0' p_n + \frac{F_0''}{2} p_n^2\right] a_n = 0, \tag{8.49}$$

with F_0, F_0', and F_0'' denoting the function $F(p)$ and its derivatives evaluated at the point $p = 0$. Here we assumed $a(\zeta - c) \approx a(\zeta) - c a_\zeta(\zeta) + c^2 \partial_{\zeta\zeta} a(\zeta)/2$, that is, a parabolic gain dispersion, which is a good approximation for a laser operating near threshold. In a periodic ML regime, when $a_{n+1}(\zeta) = a_n(\zeta) = a(\zeta)$, $c_n = c$, and $p_n(\zeta) = p(\zeta)$, Eq. (8.49) has the solution [1, 53]:

$$a(\zeta) = \sqrt{\frac{P}{2\zeta_p}} \operatorname{sech}\left(\frac{\zeta}{\zeta_p}\right), \tag{8.50}$$

where P is the pulse energy, $\zeta_p = \gamma \tau_p$ is the normalized pulse width. Substituting Eq. (8.50) into Eq. (8.49) and equating the coefficients by different powers of the hyperbolic secant, we obtain the following quadratic equation for the pulse energy:

$$2(F_0 - 1) + F_0' P + \frac{3 F_0''}{8} P^2 = 0. \tag{8.51}$$

Solving Eqs. (8.51), (8.30), (8.31), (8.33), and (8.34) with respect to the pulse parameters $G_{1,2}$, $Q_{1,2}$, and P and substituting the solution in Eq. (8.29), we obtain the stability boundaries of solution (8.50) defined by New's criterion. This result extends analytic results of Haus [5] to the case of large gain and losses per cavity round trip.

From Eq. (8.48), it follows that the net gain $G - Q + \ln\kappa$ is zero at the beginning of the fast stage if $F_0 = 1$. Hence, the relation $F_0 = 1$ defines the pulse leading-edge stability boundary. Furthermore, Eq. (8.51) implies that $F_0 = 1$ and $F'_0 = 0$ at the codimension two point where the leading edge stability boundary meets the linear threshold line defined by Eq. (8.28) with $\omega_0 = 0$. Solving the system of equations $F_0 = 1$ and $F'_0 = 0$ with respect to G_1 and Q_1 and taking into account that $G_1 = G_2 = g_0/\gamma_g$ and $Q_1 = Q_2 = q_0/\gamma_q$ if $P = 0$, we obtain the same codimension two point (Eq. (8.38)) as in the previous section. Hence, the ML stability boundaries, which we compute using the generalized approaches of New and Haus, start from the same point of the linear threshold in the parameter space.

The generalized model of this section coincides with the original model of Haus in the limit of small gain and losses per cavity round trip. Indeed, expanding Eq. (8.51) to second order with respect to G_1, Q_1, and $\ln\kappa$, we obtain the equation

$$G_1 - Q_1 + \ln\kappa - \frac{1}{2}(G_1 - sQ_1)P + \frac{3}{16}(G_1 - s^2Q_1)P^2 = 0 \tag{8.52}$$

for the pulse energy. For $G_1 \ll s^2 Q_1$, this equation is equivalent to (Eq. (36)) in [5].

The leading and trailing edge stability boundaries defined by the original equations of Haus are shown in Figure 8.10a by the dashed lines L_H and T_H, respectively. The same boundaries defined by the generalized Haus model presented in this section are shown by the solid lines L_H and T_H. The prediction of the original Haus model, as that of original model of New, substantially deviates from the stability domain obtained by the direct numerical integration of the delay differential Eqs. (8.19)–(8.21) for the parameter values of semiconductor lasers. Figure 8.10a shows that the stability domain defined by the generalized Haus model is in good agreement with the numerical results if the pulse energy is sufficiently small. The agreement becomes worse for higher pulse energies. On the other hand, the stability domain defined by the generalized New's model is in good agreement with the numerical results even for strong saturation.

In the case when gain and loss are very small and a pulse changes only very slightly after one cavity round trip, Eq. (8.49) can be reduced to a partial differential equation known in the Haus theory [1]. In this case, we have $F_0 \approx 1$, $a_n(\zeta) \equiv a(\zeta, n)$, $p_n(\zeta) \equiv p(\zeta, n)$, $\partial_\zeta a_{n+1}, \partial_{\zeta\zeta} a_{n+1} \approx \partial_\zeta a(\zeta, n), \partial_{\zeta\zeta} a(\zeta, n)$, and $a_{n+1}(\zeta) - a_n(\zeta) \approx \partial_n a(\zeta, n)$. Using these relations Eq. (8.49) can be rewritten in the form

$$\partial_n a = (1-c)\partial_\zeta a - c\left(\frac{c}{2} - 1\right)\partial_{\zeta\zeta} a + \left[F'_0 + \frac{F''_0}{2}p\right]pa.$$

Note that the analysis performed in this paragraph can be easily generalized to the case when $\alpha_g, \alpha_q, \varphi \neq 0$. In this case, the solution of the Haus master equation has the form of a chirped hyperbolic secant sech$^{1+i\beta}$. A detailed derivation of one of the modifications of the Haus model from the DDE one was performed in [42]. A general method of a reduction of DDEs with large delay to equations of Ginzburg–Landau type is discussed in [54–57].

8.5.3
Variational Approach

The simplified map model defined by Eqs. (8.42), (8.43), and (8.46) is based on interpreting the ML solution as a T-periodic sequence of δ-pulses of equal energy P_*. The definition of this map does not include the line-width enhancement factors $\alpha_{g,q}$ and, hence, does not take into account their effect. Furthermore, the model does not provide information about such important parameters of the ML regime as the pulse width and the deviation of the pulse repetition period T_* from the cavity round trip time T. In order to estimate these parameters, we modify the definition of the map using the following variational approach. We look for solutions of the fast stage Eq. (8.19) after n cavity round trips in the form

$$A_n(\tau) = \sqrt{\frac{P_n \gamma}{2\tau_n}} \operatorname{sech}\left(\frac{\gamma t}{\tau_n}\right), \tag{8.53}$$

where P_n is the pulse energy and τ_n/γ is the pulse duration. It was shown in [5] that a hyperbolic secant formula is an exact solution of the Haus master equation derived in the limit of weak saturation, that is when the nonlinearities can be replaced with the Taylor expansion up to the second-order terms with respect to the pulse energy P (see also Section 8.5.2). However, we use here the ansatz (Eq. (8.53)) for large saturation too.

For simplicity, we assume zero line-width enhancement factors $\alpha_g = \alpha_q = 0$. Substituting Eq. (8.53) in Eq. (8.45) and taking into account that the right-hand side of this equation equals the right-hand side of Eq. (8.46), we obtain

$$\frac{P_{n+1}}{3\tau_{n+1}^2} + P_{n+1} = \kappa \ln\left[1 - e^{G_n} + e^{G_n}(1 + e^{sP_n - Q_n} - e^{-Q_n})^{1/s}\right]. \tag{8.54}$$

Importantly, since in the limit of infinite bandwidth $\gamma T \to \infty$ the normalized pulse width τ_n is finite and nonzero, the two terms in the left-hand side of Eq. (8.54) are of the same order of magnitude, while the approach of New is based on the assumption that the first term is much smaller than the second one. Hence, Eq. (8.54) assuming the Lorentzian spectral filtering line shape in the limit of infinite bandwidth, and Eq. (8.46) assuming no spectral filtering lead to different estimates of the ML pulse energy.

Thus, we take into account the spectral filtering by replacing Eq. (8.46) with Eq. (8.54), while Eqs. (8.42) and (8.43), which do not depend on the pulse shape, remain the same. Since Eq. (8.54) contains as an additional pulse parameter the normalized pulse width τ_n, an additional equation for the transformation of this parameter after a cavity round trip is needed. Such an equation can be obtained by integrating Eq. (8.44) from zero to the cavity round trip time T. We assume that the optical field intensity is negligibly small during the slow stage, hence the interval of integration is reduced to the fast stage only, and we use the Eqs. (8.33), (8.34), (8.53), and $A_{n+1}(t) = \sqrt{\gamma P_{n+1}/(2\tau_{n+1})}\operatorname{sech}(\gamma t/\tau_{n+1})$ for the variables G, Q, A_n, and A_{n+1} when integrating Eq. (8.44). Squaring the resulting equation and passing to the limit $\gamma T \to \infty$, we obtain

$$\tau_{n+1} P_{n+1} = \kappa \tau_n P_n \left(\frac{1}{\pi} \int_0^{P_n} \frac{\Phi(p, Q_n, G_n)}{\sqrt{p(P_n - p)}} dp \right)^2 \tag{8.55}$$

with

$$\Phi(p, Q_n, G_n) = \left[1 + e^{-sp}(e^{Q_n} - 1) \right]^{-1/2} \left[1 - \frac{1 - e^{-G_n}}{(1 + e^{sp - Q_n} - e^{-Q_n})^{1/s}} \right]^{-1/2}.$$

Note that since the ansatz (Eq. (8.53)) is not an exact solution of the model, there is a freedom in choosing the fourth equation, which completes the definition of the pulse parameters transformation map. We show below that our choice of defining Eq. (8.55) is justified by the reasonable results obtained in this way.

The 4-D map defined by Eqs. (8.42), (8.43), (8.54), and (8.55) can be analyzed in the same way as the 3-D map in the previous section. Again, the stable fixed point (G_*, Q_*, P_*, τ_*) with a positive pulse energy component P_* is interpreted as a solution corresponding to the fundamental ML regime and the Neimark–Sacker bifurcation line is used to approximate the Q-switching instability boundary of the domain of stable ML. This boundary is shown in Figure 8.11 together with the background stability boundaries obtained by New's criterion from the 4-D map (dashed lines) and 3-D map (solid lines). The results of numerical direct integration of Eqs. (8.19)–(8.21) are shown by dots. One can see that, as expected, the results obtained with the 4-D map model are in better agreement with the results of direct numerical integration of the DDE model than those obtained with the 3-D map model. However, the difference between the stability boundaries obtained with and without taking into account spectral filtering is relatively small for the parameter values of Figure 8.10b. A more important advantage of the 4-D map model over the 3-D map one is that the former allows one to estimate the normalized pulse width τ_* and difference $\delta_* = \gamma(T_* - T)$ between the pulse repetition period and the cavity round trip time. The first of these quantities, τ_*, can be obtained by computing the fixed point (Q_*, G_*, P_*, τ_*) of the map (Eqs. (8.42), (8.43), (8.54), and (8.55)). The second quantity, δ_*, can be evaluated by means of an algorithm, which is similar to the one we used to derive Eq. (8.55). Namely, for the periodic ML solution of period T_* Eq. (8.44) can be rewritten as

$$\gamma^{-1} \partial_t A(t - \gamma^{-1} \delta_*) + A(t - \gamma^{-1} \delta_*) = \sqrt{\kappa} e^{G(t)/2 - Q(t)/2} A(t).$$

Substituting the fast stage Eqs. (8.33), (8.34), and (8.53) into this equation, multiplying it by t, and integrating over the round trip time, we obtain

8.5 The Q-Switching Instability of the ML Regime

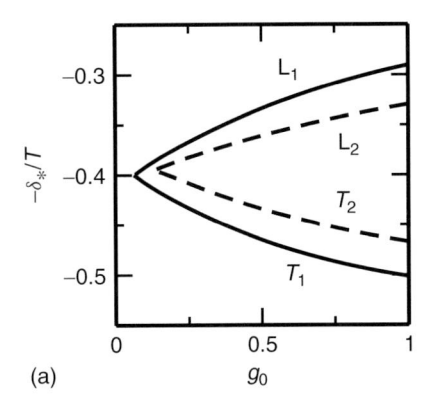

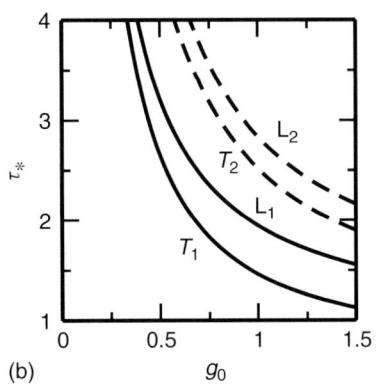

Figure 8.12 (a) Normalized difference between the ML pulse repetition period T_* and the round trip time T. (b) Normalized ML pulse width τ_*. The lines $L_{1,2}$ ($T_{1,2}$) corresponds to the leading edge (trailing edge) background stability boundary shown in Figure 8.10b. Solid and dashed lines are calculated for $s\kappa = 5$ and $s\kappa = 1.3$, respectively. Parameter values are the same as in Figure 8.10b.

$$\delta_* = 1 + \frac{\tau_* \sqrt{\kappa}}{\pi} \int_0^{P_*} \frac{\Phi(p, Q_*, G_*)}{\sqrt{p(P_* - p)}} \operatorname{arctanh}\left(\frac{2p}{P_*} - 1\right) dp.$$

Figure 8.12a,b shows how the quantities τ_* and δ_* change along the ML stability boundaries defined by New's criterion with increasing pump parameter g_0. We found that these two quantities are not sensitive to simultaneous variations of the parameters s and κ, provided that the product $s\kappa$ is fixed (this finding is similar to the observation we made above about the Q-switching, leading edge, and trailing edge instability boundaries of the ML domain, which mainly depend on the product $s\kappa$, rather than on the parameters s and κ separately). The plots of the ML pulse parameters τ_* and δ_* evaluated at the background stability boundaries in Figure 8.12 are obtained for two different values of the product $s\kappa$. The lines L and T correspond to the pulse leading and trailing edge stability boundary, respectively. From Figure 8.12b, it follows that the pulse width is smaller at the trailing edge stability boundary, which is close to the instability threshold indicated by gray lines in Figure 8.11. The quantity $-\delta_*$ increases (decreases) with increasing pump parameter g_0 along the line L (T), that is, the pulse repetition rate increases with g_0 along the pulse leading edge instability boundary and decreases at the pulse trailing edge instability boundary. The reason is that near the line L the net gain window is shifted toward the leading edge of the pulse, hence the ML pulse is accelerated by the nonlinear media. Similarly, near the trailing edge instability boundary, the nonlinear effects retard the ML pulse as the net gain window is shifted toward its trailing edge in this case. The meeting point of the lines L and T in Figure 8.12a lies on the linear threshold line and is characterized by an infinitely small pulse energy. The quantity $-\delta_*$ at this point is negative due to the dispersive losses introduced by the spectral filtering element.

8.6
Conclusion

In this chapter we have derived and analyzed both numerically and analytically a model of a passively mode-locked semiconductor laser based on a system of three differential equations with time delay (Eqs. (8.19–8.21)) [18–20]. In the limit of small gain and loss per cavity round trip using the asymptotic approach described in [42, 54–57] the equation for the electric field envelope (Eq. (8.19)) can be reduced to a partial differential equation of Ginzburg–Landau type (see also Section 8.5.2). This reduction reveals a relation between the DDE model and the Haus master equation. An important characteristic feature of the DDE model is, however, that unlike different modifications of the Haus master equation, this model does not assume small gain and loss per cavity round trip. Therefore, this model is more suitable for studying ML in semiconductor lasers. In particular, the DDE model is capable of describing the ML pulse asymmetry (Figures 8.4 and 8.5c) observed experimentally in [17, 58, 59].

A comparison of the results obtained using the DDE model with those from the linear cavity traveling wave model was performed in [60, 61]. It was shown that qualitative results obtained with these two models are very similar. According to the relation $s = s'/\kappa_2$, where $\kappa_2 < 1$ is the attenuation factor introduced in Eq. (8.9), in order to get a better agreement with the traveling wave model, the saturation parameter s in Eq. (8.21) should be taken larger than the ratio of the saturation intensities of the gain and absorber sections, $s > s'$. In the "multisection" version of the DDE model proposed in [61] the laser sections are treated as sets of smaller subsections each described by its own set of material variables. In this version, distributed losses in the gain and absorber sections are modeled by introducing attenuation coefficients between two neighboring subsections. Numerical simulations performed in [61] indicate that a reasonable quantitative agreement can be achieved between the traveling wave and the DDE model.

The DDE model can be easily modified for a description of active and hybrid ML as well as of the effect of delayed feedback and various types of optical injection on the characteristics of ML regime. In particular, in [9], this model was used for a theoretical study of the ML pulse repetition frequency locking by an external periodic modulation of the reverse voltage V applied to the absorber section. In this article, it was assumed that the escape rate from the ground state to the exited state in the absorber section exponentially depends on V. An analytical approach to the calculation of the locking cone width in a hybrid mode-locked laser [62] and in a mode-locked laser subjected to a single-mode injection [63] was developed on the basis of the DDE model. The effect of single-mode [14] and dual-mode [64] optical injection on the characteristics of ML regime was studied numerically. In [65], the DDE model was generalized to include a photonic crystal element inside the laser cavity.

Generalizations of the DDE model developed for a description of ML in quantum dot lasers [22] deserve particular attention. A modification of the DDE model that takes into account carrier exchange between the 2D carrier reservoir (wetting layer) and a single (ground) discrete level in quantum dots was proposed in [66]. A bifurcation analysis of this model was performed in [67]. It was shown that the dynamics of a passively mode-locked quantum dot laser strongly depends on the relative lengths of the gain and absorber sections. A laser with a relatively short absorber section can exhibit Q-switched ML and, at sufficiently small absorber voltages, even a pure Q-switching regime (the latter was demonstrated experimentally in [67]). On the contrary, in a laser with relatively long absorber section, the Q-switching instability can be completely suppressed and a bistability appears between the laser off state and different ML regimes. Some more sophisticated modifications of the DDE model, which take into account both the ground and the first exited state in quantum dots as well as the possibility of lasing at the exited state, were studied in [61, 68].

Acknowledgments

The authors from the Weierstrass Institute acknowledge the support of the SFB project 787 of the DFG. Andrei G. Vladimirov also acknowledges the support of the EU FP7 Marie Curie Initial Training Network, Grant No. 264687 and the program "Research and Pedagogical Cadre for Innovative Russia," Grant No. 2011-1.5-503-002-038. Dmitrii Rachinskii acknowledges the support of the Russian Foundation for Basic Research, grant 10-01-93112.

References

1. Haus, H. (2000) Modelocking of lasers. *IEEE J. Sel. Top. Quantum Electron.*, **6**, 1173–1185.
2. Kuizenga, D.J. and Siegman, A.E. (1970) FM and AM mode locking of the homogeneous laser - Part I: theory. *IEEE J. Quantum Electron.*, **6**, 694.
3. Haus, H.A. (1975) A theory of forced mode locking. *IEEE J. Quantum Electron.*, **QE-11**, 323.
4. New, G.H.C. (1974) Pulse evolution in mode-locked quasi-continuous lasers. *IEEE J. Quantum Electron.*, **10**, 115–124.
5. Haus, H. (1975) Theory of mode locking with a slow saturable absorber. *IEEE J. Quantum Electron.*, **11**, 736–746.
6. Haus, H. (1975) Theory of mode locking with a fast saturable absorber. *J. Appl. Phys.*, **46**, 3049–3058.
7. Kaiser, R. and Hüttl, B. (2007) Monolithic 40-GHz mode-locked MQW DBR lasers for high-speed optical communication systems. *IEEE J. Sel. Top. Quantum Electron.*, **13**, 125–135.
8. Jiang, L.A., Ippen, E.P., and Yokoyama, H. (2007) Semiconductor mode-locked lasers as pulse sources for high bit rate data transmission, *Ultrahigh-speed Optical Transmission Technology*, Springer, pp. 21–51.
9. Fiol, G., Arsenijević, D., Bimberg, D., Vladimirov, A.G., Wolfrum, M., Viktorov, E.A., and Mandel, P. (2010) Hybrid mode-locking in a 40 ghz monolithic quantum dot laser. *Appl. Phys. Lett.*, **96**, 011104.
10. Kuntz, M., Fiol, G., Lämmlin, M., Bimberg, D., Thompson, M.G., Tan, K.T., Marinelli, C., Penty, R.V., White, I.H., Ustinov, V.M., Zhukov, A.E., Shernyakov, Y.M., and Kovsh, A.R. (2004) 35 GHz mode-locking of 1.3 μm quantum dot lasers. *Appl. Phys. Lett.*, **85**, 843.

11. Rafailov, E.U., Cataluna, M.A., and Sibbett, W. (2007) Mode-locked quantum-dot lasers. *Nat. Photon.*, **1**, 395–401.
12. Kuntz, M., Fiol, G., Laemmlin, M., Meuer, C., and Bimberg, D. (2007) High-speed mode-locked quantum-dot lasers and optical amplifiers. *Proc. IEEE*, **95**, 1767–1778.
13. Fiol, G., Meuer, C., Schmeckebier, H., Arsenijević, D., Liebich, S., Laemmlin, M., Kuntz, M., and Bimberg, D. (2009) Quantum-dot semiconductor mode-locked lasers and amplifiers at 40 GHz. *IEEE J. Quantum Electron.*, **45**, 1429–1435.
14. Rebrova, N., Habruseva, T., Huyet, G., and Hegarty, S.P. (2010) Stabilization of a passively mode-locked laser by continuous wave optical injection. *Appl. Phys. Lett.*, **97**, 101105.
15. Avrutin, E., Marsh, J., and Portnoi, E. (2000) Monolithic and multi-GigaHerz mode-locked semiconductor lasers: Constructions, experiments, models, and applications. *IEE Proc. Optoelectron.*, **147**, 251.
16. Bandelow, U., Radziunas, M., Vladimirov, A.G., Huettl, B., and Kaiser, R. (2006) 40 GHz modelocked semiconductor lasers: Theory, simulations and experiment. *Opt. Quantum Electron.*, **38** (4), 495–512.
17. Radziunas, M., Vladimirov, A.G., Viktorov, E.A., Fiol, G., Schmeckebier, H., and Bimberg, D. (2011) Strong pulse asymmetry in quantum-dot mode-locked semiconductor lasers. *Appl. Phys Lett.*, **98**, 031104.
18. Vladimirov, A., Turaev, D., and Kozyreff, G. (2004) Delay differential equations for mode-locked semiconductor lasers. *Opt. Lett.*, **29**, 1221–1223.
19. Vladimirov, A. and Turaev, D. (2004) A new model for a mode-locked semiconductor laser. *Radiophys. Quantum Electron.*, **47**, 769–776.
20. Vladimirov, A. and Turaev, D. (2005) Model for passive mode-locking in semiconductor lasers. *Phys. Rev. A*, **72**, 033808 (13 pages).
21. Rachinskii, D., Vladimirov, A., Bandelow, U., Hüttl, B., and Kaiser, R. (2006) Q-switching instability in a mode-locked semiconductor laser. *J. Opt. Soc. Am. B*, **23** (4), 663–670.
22. Rafailov, E.U., Cataluna, M.A., and Avrutin, E. (2011) *Ultrafast Lasers Based on Quantum Dot Structures*, Wiley-VCH Verlag GmbH.
23. Tromborg, B., Lassen, H., and Olesen, H. (1994) Travelling wave analysis of semiconductor lasers. *IEEE J. Quantum Electron.*, **30**, 939–956.
24. Bandelow, U., Radziunas, M., Sieber, J., and Wolfrum, M. (2001) Impact of gain dispersion on the spatio-temporal dynamics of multisection lasers. *IEEE J. Quantum Electron.*, **37**, 183–188.
25. Agrawal, G.P. and Olsson, N.A. (1989) Self-phase modulation and spectral broadening of optical pulses in semiconductor laser amplifiers. *IEEE J. Quantum Electron.*, **25**, 2997–2306.
26. Khalfin, V.B., Arnold, J.M., and Marsh, J.H. (1995) A theoretical model of synchronization of a mode-locked semiconductor laser with an external pulse stream. *IEEE J. Sel. Top. Quantum Electron.*, **1**, 523–527.
27. Ikeda, K. (1979) Multiple-valued stationary state and its instability of the transmitted light by a ring cavity. *Opt. Commun.*, **30**, 257–261.
28. Ikeda, K., Daido, H., and Akomoto, O. (1980) Optical turbulence: Chaotic behavior of transmitted light from a ring cavity. *Phys. Rev. Lett.*, **45**, 709–712.
29. Gurevich, G.L. (1970) *Izvestiya VUZ. Radiofizika*, **13**, 1019.
30. Gurevich, G.L. and Khanin, Y.I. (1970) *Z. Tech. Fiz.*, **40**, 1566.
31. Khanin, Y.I. (1999) *Fundamentals of Laser Dynamics*, Fizmatlit, Moscow.
32. Haus, H.A. (1976) Parameter ranges for CW passive mode locking. *IEEE J. Quantum Electron.*, **12**, 169–176.
33. Mandel, P. and Erneux, T. (1984) Stationary, harmonic, and pulsed operations of an optically bistable laser with saturable absorber. *Phys. Rev. A*, **30**, 1893–1901.
34. Erneux, T. (1988) Q-switching bifurcation in a laser with a saturable absorber. *J. Opt. Soc. Am. B*, **5**, 1063–1069.
35. Yamada, M. (1993) A theoretical analysis of self-sustained pulsation phenomena

in narrow stripesemiconductor lasers. *IEEE J. Quantum Electron.*, **QE-29**, 1330.
36. Vladimirov, A. and Volkov, D. (1993) Low-intensity chaotic operations of a laser with a saturable absorber. *Opt. Commun.*, **100** (1–4), 351–360.
37. Mandel, P. (1995) *Theoretical Problems in Cavity Nonlinear Optics*, Cambridge Studies in Modern Optics, Cambridge University Press.
38. Dubbeldam, J.L.A. and Krauskopf, B. (1999) Selfpulsations of lasers with saturable absorber: dynamics and bifircations. *Opt. Commun.*, **159**, 325–338.
39. Engelborghs, K., Luzyanina, T., and Samaey, G. (2001) DDE-BIFTOOL V. 2.00: A matlab package for bifurcation analysis of delay differential equations. Tech. Rep. TW-330, Department of Computer Science, K.U.Leuven, Leuven, Belgium.
40. Pérez-Serrano, A., Javaloyes, J., and Balle, S. (2011) Longitudinal mode multistability in Ring and Fabry-Pérot lasers: the effect of spatial hole burning. *Opt. Express*, **19**, 3284–3289.
41. Guglielmi, N. and Hairer, E. (2000) Users' Guide for the Code RADAR5.
42. Kolokolnikov, T., Nizette, M., Erneux, T., Joly, N., and Bielawski, S. (2006) The q-switching instability in passively mode-locked lasers. *Physica D*, **219**, 13–21.
43. Nizette, M., Rachinskii, D., Vladimirov, A.G., and Wolfrum, M. (2006) Pulse interaction via gain and loss dynamics in passive mode-locking. *Physica D*, **218** (1), 95–104.
44. Palaski, J. and Lau, K. (1991) Parameter ranges for ultrahigh frequency mode locking of semiconductor lasers. *Appl. Phys. Lett.*, **59**, 7–9.
45. Yu, S., Krauss, T.F., and Laybourn, P.J.R. (1991) Mode locking in large monolithic semiconductor ring lasers. *Opt. Eng.*, **37**, 1164–1168.
46. Hohimer, J.P. and Vawter, G.A. (1993) Passive mode-locking of monolithic semiconductor ring lasers at 86 GHz. *Appl. Phys. Lett.*, **63**, 1598–1600.
47. Dubbeldam, J.L.A., Leegwater, J.A., and Lenstra, D. (1997) Theory of mode-locked semiconductor lasers with finite relaxation times. *Appl. Phys. Lett.*, **70**, 1938–1940.
48. Paschotta, R. and Keller, U. (2001) Passive mode locking with slow saturable absorbers. *Appl. Phys. B*, **73**, 653–662.
49. Catherall, J.M. and New, G.H.C. (1986) Role of spontaneous emission in dynamics of mode locking by synchronous pumping. *IEEE J. Quantum Electron.*, **QE-22**, 1593–1599.
50. Catherall, J.M., New, G.H.C., and Radmore, P. (1982) Approach to the theory of mode locking by sinchronous pumping. *Opt. Lett.*, **7**, 319–321.
51. New, G.H.C. (1990) Self-stabilization of synchronously mode-locked lasers. *Opt. Lett.*, **15**, 1306–1308.
52. Hüettl, B., Kaiser, R., Kindel, C., Fidorra, S., Rehbein, W., Stolpe, H., Sahin, G., Bandelow, U., Radziunas, M., A.G. Vladimirov, and Heidrich, H. (2006) Monolithic 40 GHz mqw mode-locked lasers on GaInAsP/InP with low pulse widths and controlled Q-switching. *Appl. Phys. Lett.*, **88** (22), 221104 (3 pages).
53. Chen, J., Haus, H., and Ippen, E. (1993) Stability of lasers mode locked by two saturable abssorbers. *IEEE J. Quantum Electron.*, **QE-29**, 1228–1232.
54. Giacomelli, G. and Politi, A. (1996) Relationship between delayed and spatially extended dynamical systems. *Phys. Rev. Lett.*, **76**, 2686–2689.
55. Kaschenko, S. (1998) The Ginzburg-Landau equation as a normal form for a second-order difference-differential equation with a large delay. *Comput. Math. Math. Phys.*, **38**, 443–451.
56. Grigorieva, E., Haken, H., Kaschenko, S., and Pelster, A. (1999) Travelling wave dynamics in a nonlinear interferometer with spatial field transformer in feedback. *Physica D*, **125**, 123–141.
57. Nizette, M. (2004) Stability of square oscillations in a delayed-feedback system. *Phys. Rev. E*, **70**, 056204.
58. Salvatore, R.A., Schrans, T., and Yariv, A. (1995) Pulse characteristics of passively mode-locked diode lasers. *Opt. Lett.*, **20**, 737–739.
59. Schmeckebier, H., Fiol, G., Meuer, C., Arsenijević, D., and Bimberg, D.

(2010) Complete pulse characterization of quantum dot mode-locked lasers suitable for optical communication up to 160 gbit/s. *Opt. Express*, **18**, 3415–3425.
60. Vladimirov, A.G., Pimenov, A.S., and Rachinskii, D. (2009) Numerical study of dynamical regimes in a monolithic passively mode-locked semiconductor laser. *IEEE J. Quantum Electron.*, **45**, 462–468.
61. Rossetti, M., Bardella, P., and Montrosset, I. (2011) Modeling passive mode-locking in quantum dot lasers: A comparison between a finite-difference travelling-wave model and a delayed differential equation approach. *IEEE J. Quantum Electron.*, **47**, 569–576.
62. Vladimirov, A.G., Wolfrum, M., Fiol, G., Arsenijevic, D., Bimberg, D. Viktorov, E., Mandel, P., and Rachinskii, D. (2010) Locking characteristics of a 40-GHz hybrid mode-locked monolithic quantum dot laser, *Semiconductor Lasers and Laser Dynamics IV* (eds K. Panajotov, M. Sciamanna, A.A. Valle, and R. Michalzik) Proceedings of SPIE, 7720.
63. Rebrova, N., Huyet, G., Rachinskii, D., and Vladimirov, A. (2010) An Optically Injected Mode Locked Laser, WIAS Preprint 1561.
64. Habruseva, T., O'Donoghue, S., Rebrova, N., Reid, D., Barry, L., Rachinskii, D., Huyet, G., and Hegarty, S. (2010) Quantum-dot mode-locked lasers with dual-mode optical injection. *IEEE Photon. Technol. Lett.*, **22**, 359–361.
65. Heuck, M., Blaaberg, S., and Moerk, J. (2010) Theory of passively mode-locked photonic crystal semiconductor lasers. *Opt. Express*, **18**, 18003–18014.
66. Viktorov, E., Mandel, P., Vladimirov, A.G., and Bandelow, U. (2006) A model for mode-locking in quantum dot lasers. *Appl. Phys. Lett.*, **88**, 201102.
67. Vladimirov, A.G., Bandelow, U., Fiol, G., Arsenijević, D., Kleinert, M., Bimberg, D., Pimenov, A., and Rachinskii, D. (2010) Dynamical regimes in a monolithic passively mode-locked quantum dot laser. *J. Opt. Soc. Am. B*, **27**, 2102–2109.
68. Cataluna, M.A., Nikitichev, D.I., Mikroulis, S., Simos, H., Simos, C., Mesaritakis, C., Syvridis, D., Krestnikov, I., Livshits, D., and Rafailov, E.U. (2010) Dual-wavelength mode-locked quantum-dot laser, via ground and excited state transitions: experimental and theoretical investigation. *Opt. Express.*, **18**, 121113.

9
Dynamical and Synchronization Properties of Delay-Coupled Lasers

Cristina M. Gonzalez, Miguel C. Soriano, M. Carme Torrent, Jordi Garcia-Ojalvo, and Ingo Fischer

9.1
Motivation: Why Coupling Lasers?

For some decades already, people have optically coupled semiconductor lasers with each other. Initially, the main motivation has been to superpose the emission of several lasers coherently, thereby boosting the output power. Different strategies have been followed, such as injection-locking high-power lasers with low-power coherent seed lasers, or to build laser arrays of edge-emitting lasers with laterally coupled lasers (see e.g., [1] and references therein). Later also, two-dimensional arrays of vertical-cavity surface-emitting lasers (VCSELs) have been realized [1]. In the injection-locking approach, the coupling was mostly unidirectional and via the coherent optical field. In the case of laser arrays, the coupling can originate from different mechanisms, including coupling via a shared carrier reservoir and/or the spatial overlap of the optical fields. In either case, the coupling times have been negligible or irrelevant for the observed behavior. Nevertheless, besides the intended injection-locked stable emission, both configurations also exhibited dynamical instabilities in the laser emission. For an overview of the history and the physics of injection-locking instabilities, see [2]. The laterally coupled laser arrays can also exhibit dynamical instabilities (see e.g., [3, 4]). Comparing the emission of the individual stripes in the laser arrays, Winful *et al.* demonstrated one of the first examples for the possibility of synchronizing deterministic chaos [5]. In 1997 Hohl *et al.* found that weakly coupling two nonidentical edge-emitting lasers face-to-face at a significant distance could lead not only to locking of their optical frequencies but also to the synchronization of their relaxation oscillations, thereby affecting their dynamical behavior [6]. They found that the coupled lasers can exhibit localized synchronization characterized by low-amplitude oscillations in one laser and large oscillations in the other. The laser intensities exhibited periodic or quasiperiodic oscillations. A few years later, Heil *et al.* and Fujino *et al.* found, that the nonnegligible delay in the coupling of face-to-face, mutually coupled lasers induces characteristic instabilities in their emission dynamics and particular synchronization properties [7, 8]. This has inspired many studies on the influence

Nonlinear Laser Dynamics: From Quantum Dots to Cryptography, First Edition. Edited by Kathy Lüdge.
© 2012 Wiley-VCH Verlag GmbH & Co. KGaA. Published 2012 by Wiley-VCH Verlag GmbH & Co. KGaA.

of delayed coupling on laser dynamics, as well as delay-coupled systems in general. One key criterion for the classification of the dynamical behavior is the length of the coupling delay. Qualitatively different behavior has been found for the cases of short [9] and long delays [7]. If the coupling delay τ is of the same order as the relaxation oscillation period τ_{RO} of the laser ($\tau \sim \tau_{RO}$), it is referred to as the *short delay regime*. If $\tau \gg \tau_{RO}$, it is denoted as the *long delay regime*. In the following, we concentrate on the long delay regime. This chapter covers aspects of delay-induced instabilities, synchronization properties, modulation characteristics, influence of noise, and the potential application of delay-coupled lasers.

9.2
Dynamics of Two Mutually Delay-Coupled Lasers

9.2.1
Dynamical Instability

The starting point for the study of delay-coupled lasers has been the configuration of two longitudinally delay-coupled semiconductor lasers in the long delay regime. A sketch of this configuration is depicted in Figure 9.1.

The long delay regime, defined by $\tau \gg \tau_{RO}$, is typically represented by geometric coupling distances of $l > 30$ cm, corresponding to coupling delays of $\tau > 1$ ns. We first consider the symmetric situation, meaning very similar lasers with adjusted wavelengths, identical operating conditions, and symmetric bidirectional coupling. In [7], it has been found that the delayed coupling induces chaotic intensity dynamics on timescales ranging from subnanoseconds to microseconds. Figure 9.2 depicts a typical intensity time series of one of the mutually coupled lasers.

The dynamics resembles the dynamics found for delayed optical feedback. In some sense, these studies can be seen as an extension of the investigations of lasers with delayed feedback. The dynamics comprises similar low-frequency fluctuation (LFF) behavior. However, here it does not originate from passive feedback, but from delayed coupling because of the respective other laser. The signal of one of the lasers is injected into the other with a delay of τ, while the injection of the latter laser is injected back into the former with a delay of 2τ after its initial emission. Consequently, the dynamics of the delay-coupled lasers is comparable to the dynamics of a single laser subject to optical feedback with a delay of 2τ, as opposed to τ, which one would expect from simple symmetry upon reflection considerations.

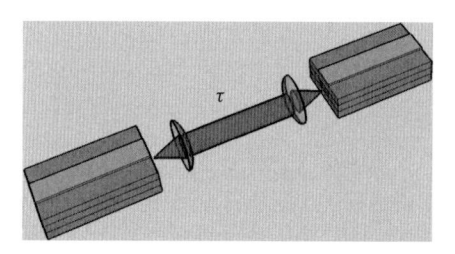

Figure 9.1 Two face-to-face delay-coupled semiconductor lasers. From [10].

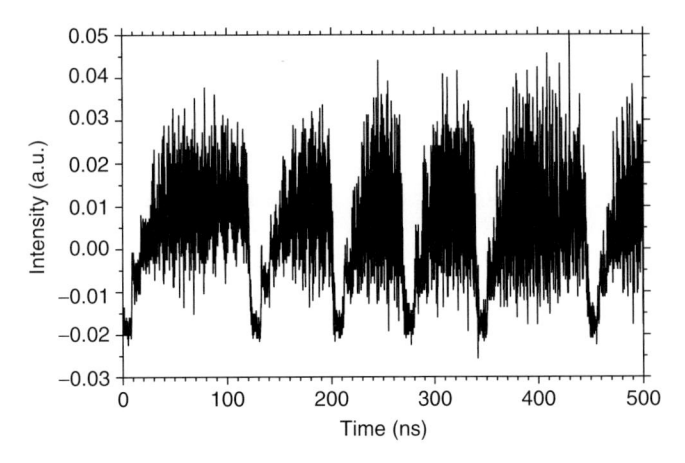

Figure 9.2 Intensity dynamics induced by mutual delayed coupling under symmetric conditions. From [10].

The large coupling time and the chaos synchronization phenomenon are the key elements leading to the coupling-induced instabilities.

In this LFF-like regime, the lasers are biased close to their solitary (stand-alone) threshold and subjected to a moderate amount of coupling. The output power of the lasers then shows irregular oscillatory cycles, which consist of an abrupt dropout, leading to a turnoff of the lasers followed by a gradual buildup of the powers, which is again followed by a dropout. The duration of the whole cycle is usually about 10τ up to 100τ. High temporal resolution measurements of the LFF cycle show that during this cycle, the laser produces picosecond pulses, which gradually grow in amplitude after a dropout [11, 12].

The signature of these delay-induced instabilities can also be seen in the optical frequency domain. The optical emission line can typically broaden from several MHz in solitary lasers far into the GHz range in semiconductor lasers coupled with delay. For this reason, delayed feedback and delayed coupling are referred to as the cause of a collapse in the optical coherence of the lasers.

The resemblance between a single laser with feedback and two mutually coupled lasers can be further extended to a ring of unidirectionally coupled lasers. In the special case of $N=1$ and $N=2$, the ring configuration is reduced to a laser subject to delayed self-feedback and two mutually delay-coupled identical lasers, respectively, as shown in Figure 9.3. The dynamics, correlation scaling, and synchronization behavior of N elements coupled in a unidirectional ring with an evenly spaced delay τ/N can be understood and predicted by those of a single element subject to self-feedback [13].

As an example of the dynamics in the ring, we present in Figure 9.4 numerical results for one laser with delayed optical feedback, and two bidirectionally coupled lasers, $N=4$ and $N=100$. While the time traces (left panels) show no apparent change in the dynamical behavior when the number of lasers is increased, the power spectra (middle panels), and normalized intensity autocorrelation functions

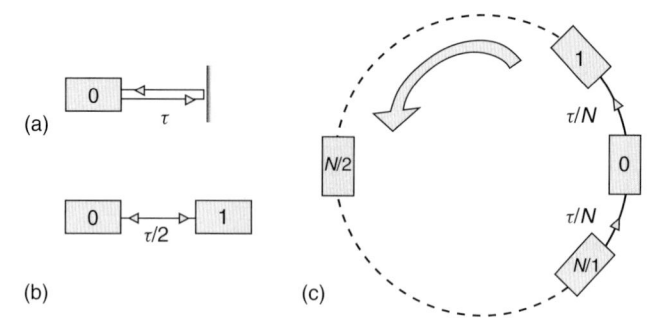

Figure 9.3 (a) Laser with delayed feedback. (b) Two bidirectionally delay-coupled lasers. (c) A ring configuration of N unidirectionally delay-coupled lasers.

(right panels) show evidence of clear changes. Besides a broadband component, the spectrum of one laser subjected to feedback exhibits discrete frequency peaks related to external cavity modes. It is striking that the peaks in the spectrum become less defined when the number of lasers in the ring increases. The corresponding normalized intensity autocorrelation functions shed more light on this apparent damping of the compound cavity modes. We have analyzed the heights of the correlation peaks and found that the peak around $t = 2\tau$ (panel c) in the case of one laser with delayed feedback is exactly reproduced around $t = \tau$ in the case of two bidirectionally coupled lasers (panel f). In both cases, the signals have passed twice through a nonlinear element. Also, the peak around $t = 2\tau$ (panel f) is reproduced at $t = \tau$ when $N = 4$ (panel i) and can also be found at $t = 2\tau$ (panel c). We have verified that any correlation peak of one laser with delayed feedback at $t = M\tau$ is reproduced in the autocorrelation function of one laser in a ring with $N = M$ lasers at $t = \tau$. In general, we find that the shape and position of a correlation peak is defined by the number of passes through a laser nonlinearity. We can exactly reconstruct the autocorrelation function of a laser in a ring of N elements by selecting the corresponding peaks in the autocorrelation function of one laser with delayed feedback.

9.2.2
Instability of Isochronous Solution

As a next step, the intensity fluctuations of the two lasers have been compared with respect to each other. Figure 9.5 depicts the intensity dynamics of two mutually delay-coupled lasers.

For ease of comparison, the second time series is vertically flipped. The two lasers exhibit correlated power dropouts and also correlated subnanosecond oscillations. However, these oscillations do not occur at the same time. The maximum correlation peak (reaching values of $C > 0.9$) is obtained for a relative time shift, roughly given by τ or $-\tau$. Although the configuration is completely symmetric, the behavior is not. The lasers are not identically synchronized and show different

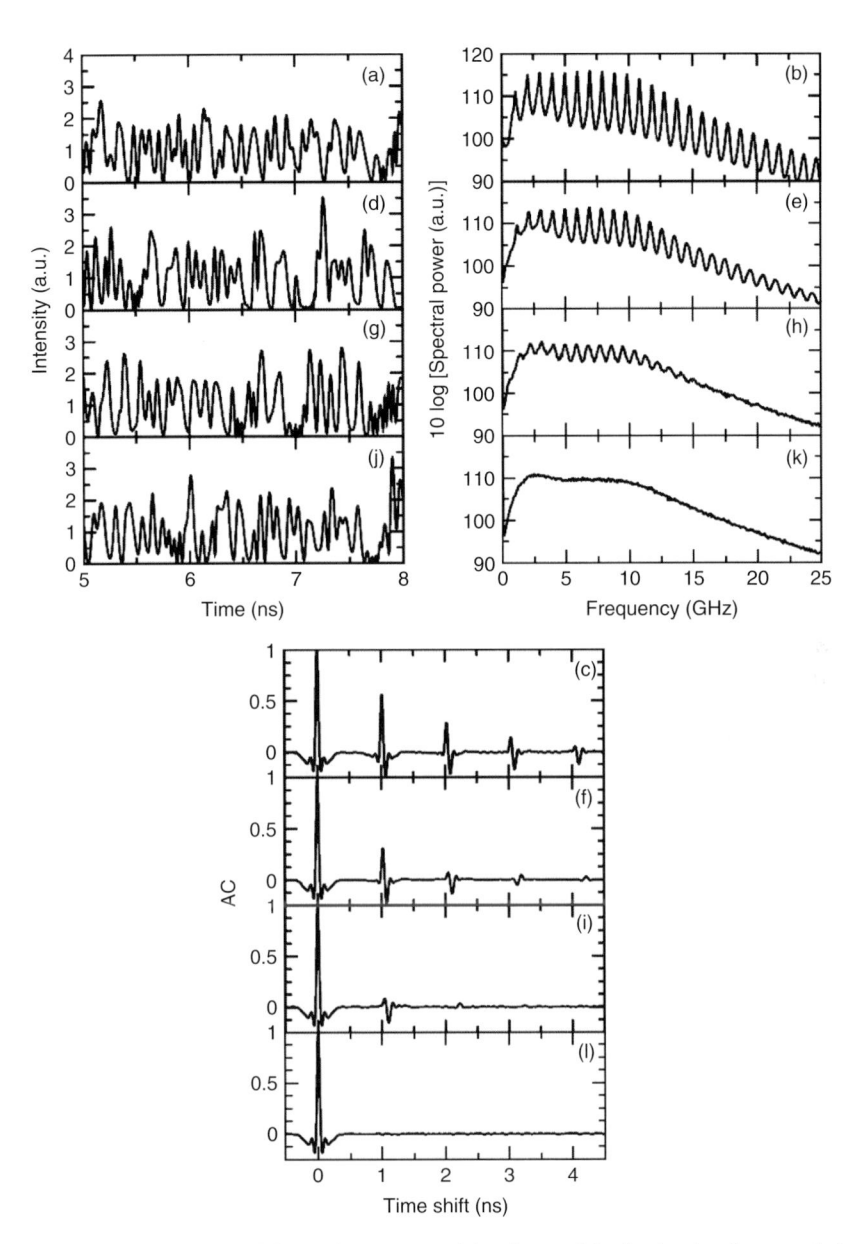

Figure 9.4 Time traces (left panel), power spectra (middle panel), and normalized intensity autocorrelation functions (right panel) of the emission of one laser for different values of N. (a)–(c) correspond to a laser with delayed optical feedback, (d)–(f) to two bidirectionally coupled lasers, (g)–(i) to 4 lasers in a ring configuration, and (j)–(l) to 100 lasers in a ring configuration.

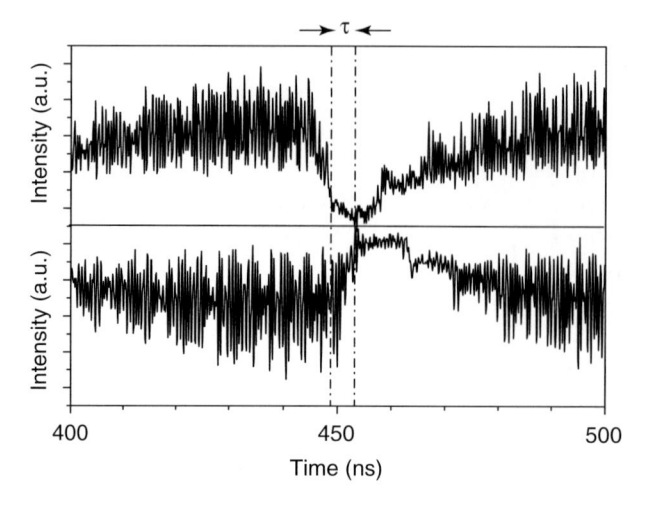

Figure 9.5 Comparison of the intensity dynamics of two delay-coupled lasers under symmetric conditions. From [10].

temporal dynamics. They exhibit a form of generalized synchronization in which their behaviors are determined by the dynamics of the respective other laser, but how the dynamics between the two lasers relates to each other has not been identified yet. It is not given by a simple functional relationship. The maxima in the cross-correlation function occurring at $\pm\tau$ indicate that one laser follows the respective other laser with a delay, however, only showing similar, not identical, dynamics. Therefore, this type of generalized synchronization has also been referred to as leader–laggard-type synchronization. While for completely symmetric conditions leader and laggard roles emerge spontaneously and can even change in time, the role can also be externally controlled by the introduction of slight asymmetries. One way to achieve this is by introducing a relative spectral detuning of the emission of the two lasers. Already, for nominal detunings of only about 1 GHz, being small to typical optical locking ranges of larger than 10 GHz, the leader and laggards roles can be fixed. For edge-emitting lasers, the laser with higher frequency becomes the leader in the dynamics [7, 14].

The emission dynamics of delay-coupled laser configurations – here, in particular, for two mutually delay-coupled lasers – can be modeled (assuming single solitary laser mode emission and low-to-moderate coupling) via a set of rate equations, resembling the Lang–Kobayashi equations [15] for the laser with feedback:

$$\dot{E}_{1,2}(t) = \mp i\Delta E_{1,2} + \frac{1}{2}(1 - i\alpha)\left[g_{n,i}n_i\right]E_{1,2} + \kappa_c E_{2,1}(t - \tau), \tag{9.1}$$

with κ_c being the coupling strength (see also chapter 6). E represents the slowly varying electric fields around the symmetric reference frame $(\Omega_2 + \Omega_1)/2$, and Δ the detuning of the lasers in this reference frame. $\Omega_{1,2}$ is the free running optical frequency of laser 1 and 2, respectively. α refers to the line-width enhancement factor, and $g_{n,i}$ the differential gain of laser i. The complementary equations for the excess carrier densities n_i read:

$$\dot{n}_i = (p-1)\frac{I_{\text{th},i}}{e} - \gamma_e n_i - \left(\Gamma_0 + g_{n,i} n_i(t)\right) \|E_i\|^2, \tag{9.2}$$

where, as before, $E_i(t)$ refers to the optical field generated by laser i, γ_e the carrier decay rate and Γ_0 the photon decay rate. $I_{\text{th},i}$ is the bias current at the solitary threshold of laser i, e is the electron charge, and p the pump parameter. $\|\cdots\|$ denotes the amplitude of the complex field.

From the experimental studies, one might assume that asymmetries in the setup or laser equations are the origin of the symmetry breaking, resulting in the observed leader–laggard behavior. This can be tested in the modeling, where one can choose perfectly symmetric conditions. As a result of the modeling, the same leader–laggard behavior is found, corresponding to the generalized synchronization solution. Still, because of the symmetry of the system, a symmetric solution has to exist, and in the modeling one can even prepare the system to start in this solution. Without noise, the system might even prevail in this state for some time. However, as soon as one laser experiences a tiny perturbation, the system escapes to the generalized synchronized solution. The symmetric solution is unstable. This is shown in Figure 9.6.

Remarkably, the unstable character of the symmetric, isochronously synchronized solution exhibited by delay-coupled oscillators holds not only for the chosen parameter conditions but also for all considered parameter situations, and even for a large class of delay-coupled oscillators in general. It is only recently that this general property has been understood [16, 17].

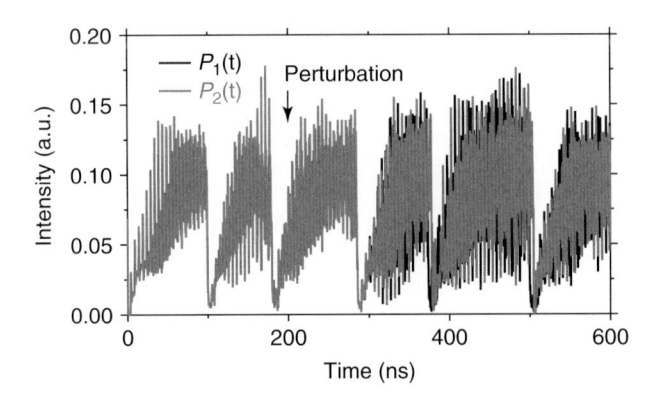

Figure 9.6 Intensity dynamics of two mutually delay-coupled lasers, obtained by modeling. At $t = 0$, the system is prepared in the isochronously synchronized state. At $t = 200$ ns, a small perturbation is applied, resulting in the emergence of the leader–laggard state. Courtesy of Claudio Mirasso.

9.3
Properties of Leader–Laggard Synchronization

9.3.1
Emergence of Leader–Laggard Synchronization

As we have seen above, when two mutually coupled lasers having the same optical frequency operate, they synchronize with a lag, with a random change in the leader and laggard roles. To understand the emergence of this symmetry breaking in the system, one can analyze the transition from unidirectional to bidirectional injection. This can be accomplished, for instance, with the experimental setup shown in the left panel of Figure 9.7, in which the directionality of the coupling is varied in a controlled way, by separating the coupling path into two unidirectional paths and adding a neutral-density filter of varying transmittance to one of the paths. This allows one to see the transition from stable unidirectional injection to chaotic synchronization with a leader in the dynamics, and how this chaotic lag synchronization arises in the system.

In the unidirectional case, the receiver laser (LD2) is stable at very low injection levels, with an optical power close to that of the emitter laser. When the (unidirectional) coupling is increased, the receiver laser goes from stable to oscillatory output. This oscillation becomes more and more unstable if the coupling is increased further, or if a reverse injection is added. When we depart from the unidirectional coupling state by gradually increasing the injected light coming from the reverse path, chaos arises in the system, with a clear symmetry breaking introduced by the time delay of the coupling paths.

The transition to chaos can be observed in the output intensities of the two lasers. The right panel of Figure 9.7 shows the time traces of the two lasers (LD1 at the top and LD2 at the bottom) and the corresponding cross-correlation functions for increasing back injection. In the case of purely unidirectional injection (Figure 9.7a,b), the emitter laser is naturally stable, and the receiver laser exhibits small oscillations as a result of the injection. The respective cross-correlation function has its maximum at $-\tau_{1,2}$ (the flight time from LD1 to LD2). When the back injection is nonzero, but small, the cross-correlation function reveals a quasiperiodic state due to the very high asymmetry in the couplings. The highest peak appears at $-\tau_{1,2}$, but several higher harmonics occur at lags $\tau_{1,2} + \tau_{2,1}$. Quasiperiodicity is revealed by growing peaks at those lags in the cross-correlation function (Figure 9.7c,d). The chaotic dynamics typical of symmetric coupling is observed in the weakly assymmetric coupling case (Figure 9.7e,f), and is characterized by a quick decrease to zero of the cross-correlation function away from its maximum. The difference in the rates at which the envelope of the cross-correlation peak decays characterizes the transition from a quasiperiodic to a chaotic behavior. These results show that the symmetry-breaking behavior underlying the leader–laggard dynamics [7] emerges from a quasiperiodic state that later transforms into a chaotic state with a well-defined leader in

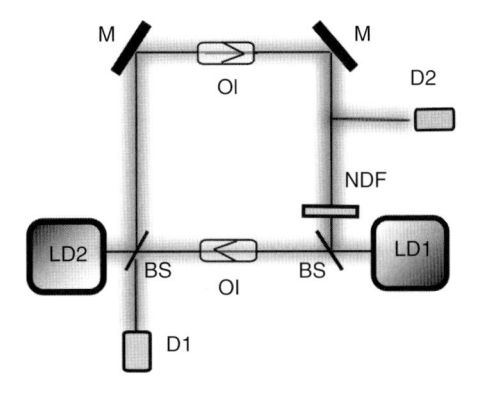

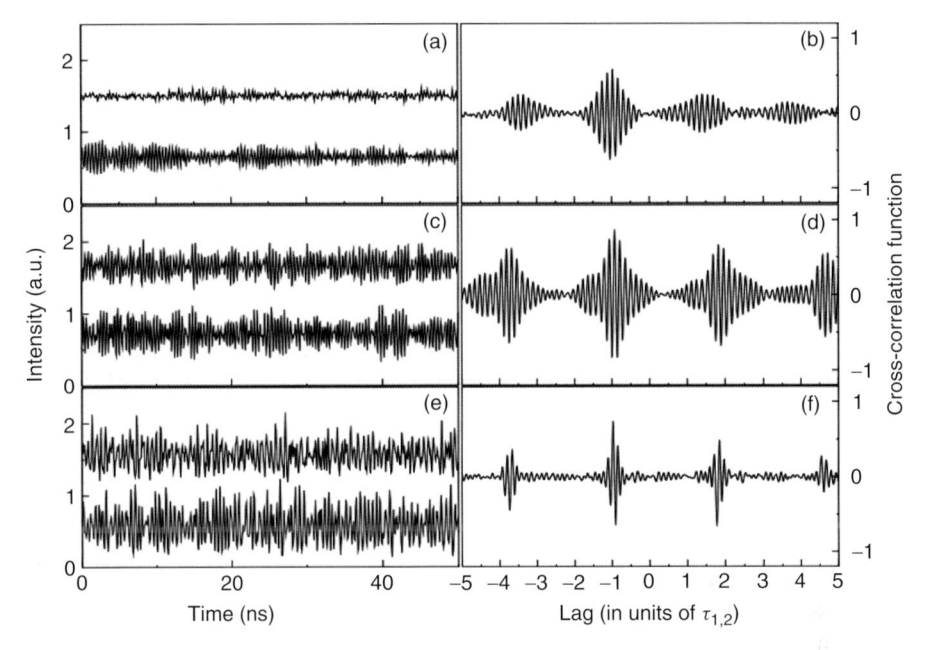

Figure 9.7 Left panel: experimental setup to examine the emergence of lag synchronization. Two lasers LD1 and LD2 are optically coupled through two unidirectional pathways running in opposite directions. A neutral-density filter in the path from LD2 to LD1 allows to tune the coupling from purely unidirectional from LD1 to LD2, to purely bidirectional. Right panel: (a,c,e) output intensities of LD1 (top) and LD2 (bottom) and the corresponding cross-correlation functions (b,d,f) for increasing back injection from LD2 to LD1 (from top to bottom). Abbreviations: NDF, neutral density filter; OI, optical isolator; M, mirror; D1, D2, detectors.

the dynamics. The quasiperiodic state is characterized by out-of-phase synchronized outputs, with a cross-correlation function exhibiting a clear maximum at $-\tau_{1,2}$, and secondary peaks at a distance equal to the sum of the external cavities, while in the chaotic case the secondary peaks suffer a loss of correlation.

9.3.2
Control of Lag Synchronization

When the output of a semiconductor laser with feedback, operating in the LFF regime, is introduced into a second laser, power dropouts are also induced in the latter, provided the two lasers are similar enough in their physical properties. The dropouts are synchronized between the two lasers and, in general, the emitter laser leads the dynamics (i.e., the dropouts in the emitter precede those in the transmitter) [18, 19] with a time lag equal to the coupling time. If the coupling and feedback strengths are tuned such that the total injection (feedback + coupling) is equal for the two lasers, and if the feedback time is larger than the coupling delay, the emitter laser can anticipate the receiver [20]. This synchronization state is, nevertheless, much less common and more difficult to reach than the usual lag synchronization state discussed here. Interestingly, a similar dynamics is observed in the case of two *bidirectionally* coupled lasers, even in the absence of an external mirror, as we have seen above. We remind the reader that when the two lasers have the same frequencies, the leader and laggard roles alternate randomly between the two lasers, whereas in the presence of frequency detuning the laser with higher frequency is the one leading the dynamics, again with a time lag equal to the coupling time. In the bidirectional case, a well-defined leader also exists when one of the lasers is subject to feedback [21]; this behavior can again be attributed to the existence of a frequency detuning between the lasers, which is, in this case, induced by the feedback itself [22].

The question then is how the transition between the unidirectional and bidirectional coupling schemes comes about. To that end, one can use the experimental setup shown in the left panel of Figure 9.8. In this scheme, two semiconductor lasers, one of them subject to optical feedback from an external mirror, are coupled optically via two distinct paths through which light is made to travel in opposite directions with suitable optical isolators. The directionality of the coupling can be varied in a controlled way by tuning a neutral-density filter in one of the two paths.

The right panel of Figure 9.8 shows the dynamical behavior of this system when coupling varies from purely unidirectional to purely bidirectional. In the absence of coupling from any of the two paths, laser LD2 is stable, while laser LD1 operates in the LFF regime due to the optical feedback from the external mirror M. When a sufficient amount of light from LD1 is injected into LD2, the latter exhibits power dropouts as well, following those of LD1 with a certain time lag (see plot (a) in the right panel of Figure 9.8). The time lag can be determined by comparing the times at which synchronized power dropouts occur in the two lasers. A histogram of the time differences between synchronized power dropouts corresponding to this regime is shown in plot (b). The lag is calculated as the difference between the dropout times in LD1 and LD2. Therefore, a negative value corresponds to an *advance* of LD1 over LD2, as expected and evident from the vertical dashed lines in plot (a). Intuitively, this lag is produced by the time needed by the light of one laser to affect the dynamics of the other. We note that another synchronized state is possible in this setup, in which the lasers are synchronized at zero-lag (provided

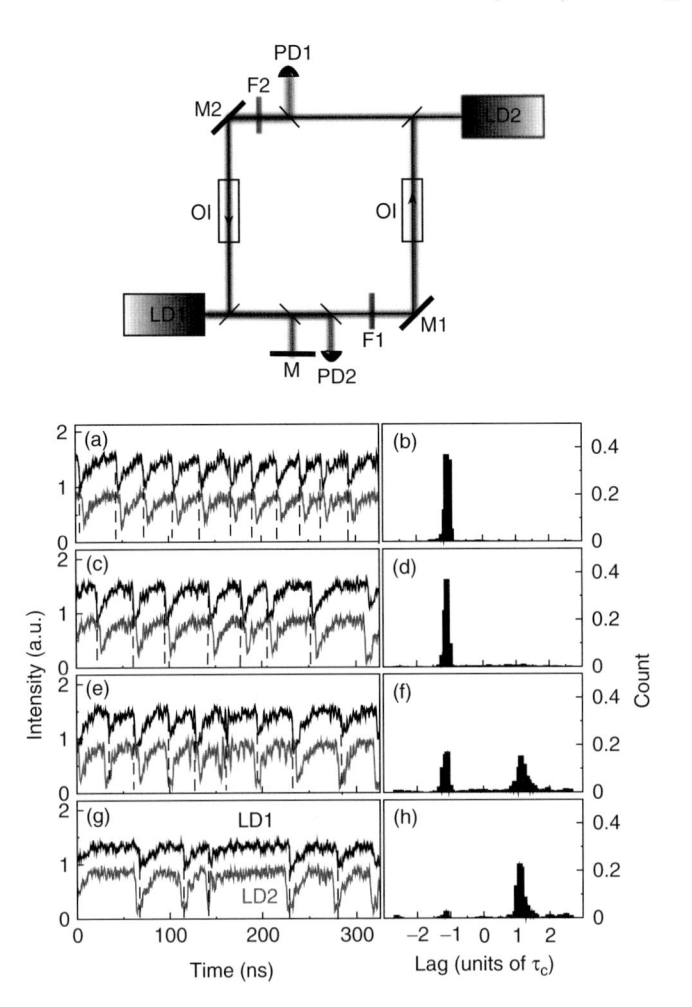

Figure 9.8 Left panel: experimental arrangement of two semiconductor lasers coupled via two independent unidirectional paths. Laser LD1 receives optical feedback from mirror M. Right panel: experimental output intensities (left column) and the corresponding histogram of time differences between 1000 synchronized dropouts in the two lasers (right column) for increasing transmittance of the filter F2 (from top to bottom). The time traces in the left plots have been shifted vertically for clarity, with LD1 corresponding to the top trace and LD2 to the bottom trace in each plot. Vertical dashed lines in those plots signal the occurrence of a dropout in laser LD1. Abbreviations: M, mirror; PD1, PD2, photo diode; OI, optical isolator. Adapted from [23].

the feedback and coupling times are equal) [24], but this requires careful tuning to make the coupling and feedback strengths equal, and extremely similar lasers [18]; this regime is not shown here.

When the light emitted from LD2 is allowed to reach LD1, it becomes possible to control the strength of that coupling, varying the transmittivity of filter F2, while keeping the amount of light injected from LD1 into LD2 constant. Plot (c) in the right panel of Figure 9.8 shows that for moderate transmittivities, the situation does not change much with respect to the purely unidirectional case (LD1 leads the dynamics a time $\sim\tau_c$), even though a substantial amount of light from LD2 is already entering LD1. For larger back coupling, however, (plot e) laser LD2 begins to have a certain influence and takes over the leader role randomly, leading to a bimodal and symmetric histogram of time differences between dropouts (plot f). The situation resembles that of two mutually coupled lasers without mirrors described above [7], even though that case is perfectly symmetrical and the present one is not, since one of the lasers (LD1) is subject to feedback but not the other. Finally, when the the amount of light being coupled back from LD2 into LD1 is large enough, until the coupling is purely bidirectional (plots g,h), laser LD2 takes over the leader role permanently: its dropouts consistently precede those of LD1, again a time $\sim\tau_c$.

So, naturally, the question arose whether the zero-lag solution always has to be unstable, or whether this can be overcome by modifying the coupling configuration. This question resulted in the extension to a chain of three mutually delay-coupled semiconductor lasers, as discussed in the following section.

9.4
Dynamical Relaying as Stabilization Mechanism for Zero-Lag Synchronization

9.4.1
Laser Relay

We have shown in the previous sections that the natural synchronized state of two delay-coupled lasers is one in which one of the lasers leads the other one a time equal to the flight time between the lasers. However, in many natural situations, oscillations between distant dynamical elements can be isochronous even in the presence of nonnegligible coupling delays. A specially important example of this phenomenon arises in the nervous system, where zero-lag synchronization has been observed between distant cortical regions [25, 26] and pairwise recordings of neuronal signals [27]. The mechanism of this phenomenon, through which two distant dynamical elements can synchronize at zero-lag even in the presence of nonnegligible delays in the transfer of information between them, has been debated for many years in the field of neuroscience. Complex mechanisms and neural architectures have been proposed to answer this question, [28–30] which, however, exhibit limitations in the maximum synchronization range (see e.g., [29]), and rely on complex network architectures [30].

Coupling in mutually injected semiconductor lasers is naturally subject to delay. Thus, coupled lasers can be used to explore potential mechanisms for zero-lag synchronization. The left panel of Figure 9.9 shows a simple mechanism, consisting of three similar dynamical elements coupled bidirectionally in a series, in such a way that the central element acts as a *relay* of the dynamics between the outer elements [31]. The central laser, which does not need to be carefully matched to the other two, mediates their dynamics. Without coupling, the three lasers

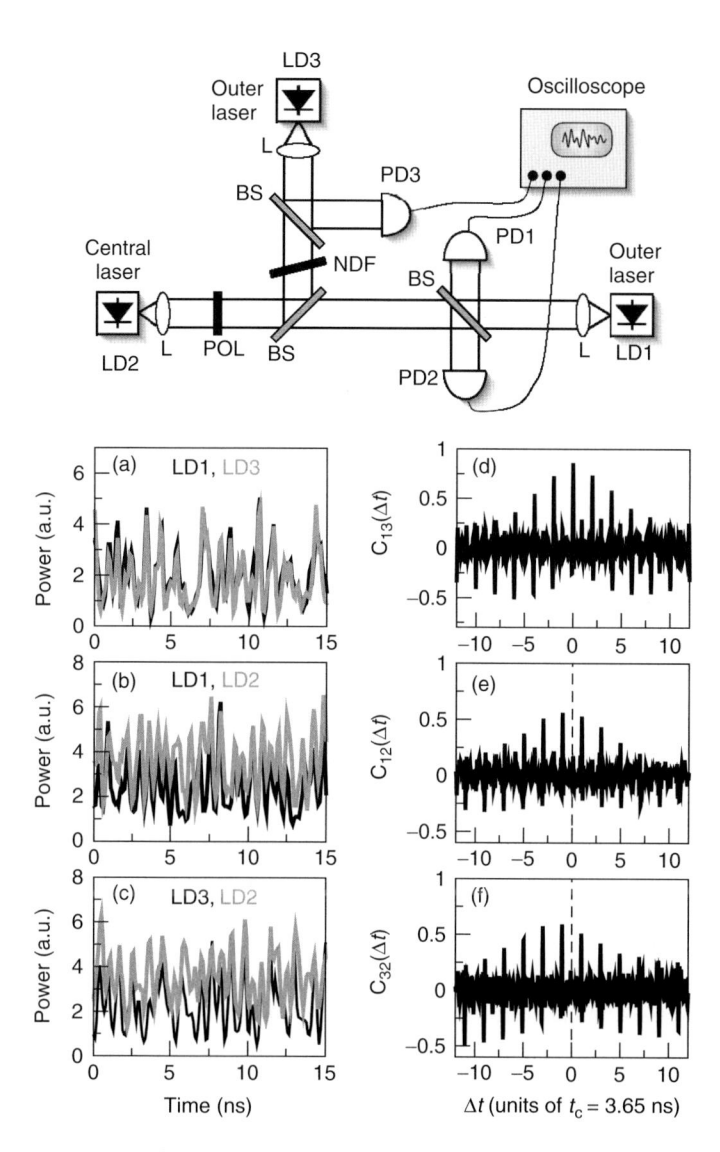

Figure 9.9 Left panel: a central laser (LD2) exchanges information between the other two (LD1 and LD3). The coupling time between the central and outer lasers are matched to each other between both branches. Right panel: time series (a–c) and cross-correlation functions (d–f) of the output intensity of the three laser pairs, for a negatively detuned central laser. The time series of the central laser are shifted τ_c for an easier comparison. Abbreviation: NDF, neutral-density filter. Abbreviations: NDF, neutral density filter; PD1, PD2, PD3, photo diode; BS, beam splitter; POL, polarizer; L, lens. Adapted from [31].

emit constant power, representing damped relaxation oscillators. In the presence of coupling, the lasers exhibit chaotic outputs that, remarkably, are synchronized with zero-lag between the outer lasers, while the central laser either leads or lags the outer lasers. The right panel of Figure 9.9 shows the time series of the output intensities (left column), in pairs, and the corresponding cross-correlation functions $C_{ij}(\Delta t)$, defined in such a way that a maximal cross-correlation at a positive time difference Δt_{max} indicates that element j is *leading* element i with a time advance Δt_{max}, and vice versa. In the situation shown in the figure, the optical frequency of the central laser was slightly decreased with respect to the outer lasers (negatively detuned) by adjusting its temperature, for optimal synchronization quality. Zero-lag synchronization between the intensities of the outer lasers can be clearly seen in Figure 9.9a, and also manifests itself in the cross-correlation function shown in Figure 9.9d, which presents an absolute maximum at $\Delta t_{max} = 0$ (i.e., at zero-lag). The correlation between the central laser and the outer ones (Figure 9.9b,c) is not as high, and presents a nonzero time lag, as can be seen from the cross-correlation functions shown in Figure 9.9e,f. This lag coincides with the coupling time between the lasers. The fact that Δt_{max} is negative means that the central laser dynamically lags the two outer lasers. Therefore, the outer lasers are not simply driven by the central one. This zero-lag synchronization is quite robust against spectral detuning of the lasers, even for positive detuning (in which case the central laser leads the dynamics).

9.4.2
Mirror Relay

The main argument, why zero-lag synchronization could be observed in the laser chain, is that the relay laser in the center is redistributing the signals of the respective outer lasers symmetrically. Consequently, the question arises whether the central laser could be replaced by a semi transparent mirror as relay element. The corresponding scheme is depicted in Figure 9.10.

The two semiconductor lasers are mutually coupled through a partially transparent mirror placed in the coupling path between both lasers. Therefore, the light injected into each laser is the sum of its delayed feedback from the mirror and the light coming from the respective other laser. For the numerical studies, coupling coefficients and feedback strengths were chosen such that the lasers operate in a chaotic regime. In numerical investigations of this configuration, identical synchronization between the dynamics of both lasers was obtained for arbitrary coupling distances between the lasers. For the mirror being precisely in the center, zero-lag synchronization was found. Changing the position of the mirror turned

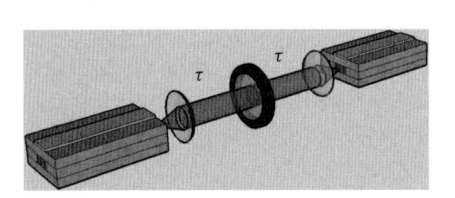

Figure 9.10 Scheme of two mutually delay-coupled lasers with a semitransparent mirror as relay.

out not to be relevant for the synchronization quality. Even for strongly asymmetric positioning of the mirror, identical synchronization was still observed, then with a temporal offset given by the difference of the corresponding delay times. Thus, identical and even zero-lag synchronization can be achieved with different realizations of the relay element. A parameter that, however, turned out to be critical for the semi transparent mirror configuration for obtaining good synchronization quality was phase difference between the optical coupling and feedback phases. While the experiments in the all-optical scheme are not easy to sufficiently control the synchronization quality [32], using electro-optic systems successful identical synchronization could be demonstrated experimentally [33].

The results discussed above show that zero-lag synchronization can be achieved in delay-coupled lasers with a relay element in the center. However, it is not clear under which conditions this solution is stable, or whether it is even unconditionally stable because of the common driving of the outer lasers through the relay element. A detailed stability analysis showed the existence of unstable regimes, in particular for not sufficiently strong coupling [34]. In addition, it was found that even in large regimes where the synchronization manifold is transversely stable, characterized by negative transverse Lyapunov exponents, bubbling can still occur. Bubbling is the phenomenon of eventual escapes from a synchronization manifold due to an invariant set being transversely unstable. Responsible for the occurrence of bubbling are saddle points, corresponding to destructive interference conditions of the optical field in the outer lasers and the incoupled fields. These saddle points are not only crucial for the onset of the coupling-induced dynamical instabilities but also for eventual escapes from the synchronization manifold, resulting in bubbling behavior.

9.5
Modulation Characteristics of Delay-Coupled Lasers

9.5.1
Periodic Modulation

The power dropouts exhibited by a single semiconductor laser with optical feedback, when operating in the regime of low-frequency fluctuations, have been shown to become periodic when an external modulation is applied to the injection current [35, 36] or to the feedback strength [37]. The laser response to an external harmonic modulation, however, is greatly enhanced by coupling [38] (similar to what is found in general models of nonlinear media [39]). Coupling leads to a very efficient entrainment, which means that less pump current modulation is needed, and thus the modulation is practically absent in the output of the coupled system (in contrast to what happens in single lasers [35]). Thus, in the presence of coupling the low-frequency dropouts are not distorted, but only entrained.

Figure 9.11 shows the response of a system of two optically coupled lasers to a variation of the coupling strength, when one of the lasers is subject to a periodic modulation of its pump current. At the maximum coupling (i.e., when the

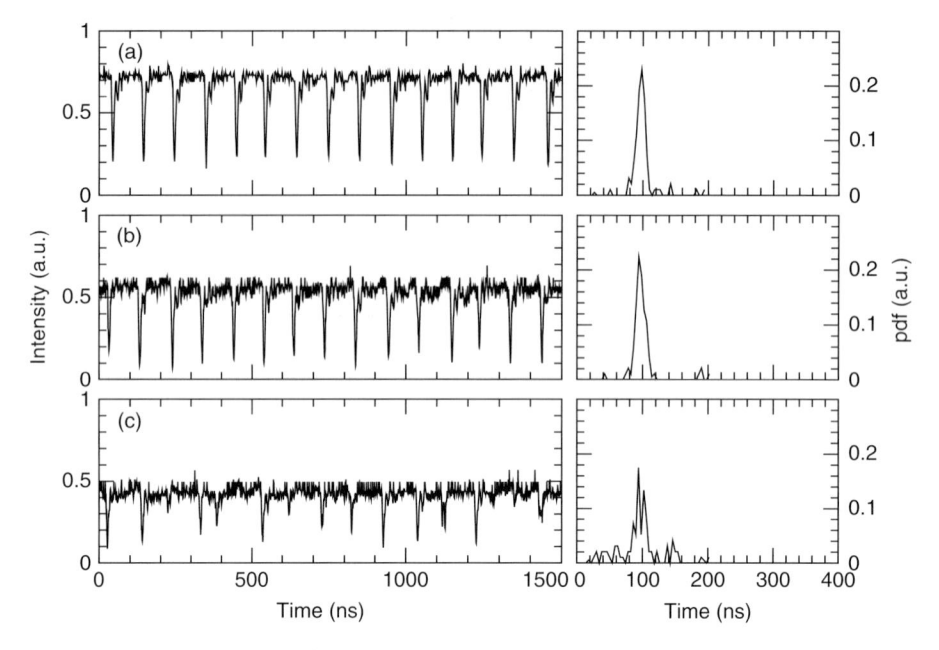

Figure 9.11 Experimental time series of one of two optically coupled lasers, when the coupling between the two lasers is decreased, which is accomplished by placing a neutral-density filter between the two lasers. Relative to the maximum coupling: (a) 100%, (b) 83.9%, and (c) 45.8%. The right panels show the corresponding probability distribution functions of the intervals between dropouts. From [40].

amount of light that is injected in one laser from the other is maximum given the experimental conditions, top row in the figure), the two lasers are synchronized (only the nonmodulated laser output is shown) and perfectly entrained to the periodic signal. For intermediate values of coupling (middle row), the entrainment persists, and only when the coupling strength is reduced more than 50% of its maximum value is the quality of the entrainment noticeably degraded (bottom row). The results are also given in terms of the probability distribution function of the time intervals between consecutive dropouts. The irregular shape of this function in Figure 9.11c indicates loss of entrainment. Thus, for large enough coupling, the response of the two lasers to a pump modulation of one of them is a perfect entrainment, with no direct evidence of the current modulation in the output intensity of either laser, in contrast with the case of a single modulated laser with feedback, in which case the laser is also fully entrained but the current modulation is strongly present in the laser's output [35, 40].

It is also interesting to examine the situation in which both lasers are subject to external pump modulation with different frequencies. Let us consider, for instance, the case of two harmonics of a common fundamental f_0, defined by $f_1 = kf_0$ and $f_2 = (k+1)f_0$ with $k > 1$. This is the simplest example of a complex signal, and previous experimental and theoretical studies have shown that certain nonlinear systems subject to this type of complex signal respond at the fundamental frequency, which is not present in the input. This phenomenon is known as the *missing fundamental illusion*, and has recently been interpreted in terms of an optimal response of excitable systems to a suitable amount of noise, under the name of *ghost stochastic resonance* [41].

Ghost resonant behavior occurring in isolated dynamical elements has been reported experimentally in lasers [42, 43] and electronic circuits [44]. Experiments have shown that the phenomenon also arises in two coupled lasers, both when the lasers are stable in the absence of coupling [45] (so that the power dropouts are induced by coupling), and when the lasers exhibit power dropouts even without coupling [46] (so that the isolated lasers behave as bona fide excitable systems [47, 48]). Recent studies in neuronal systems, both theoretical [49, 50] and experimental [51], show that coupling is able to mediate the processing of distributed inputs in networks of neurons (which possess independent dynamics even in the absence of coupling). The experiments that we describe in what follows confirm the existence of this emerging property of excitable networks, using semiconductor lasers with optical feedback as highly controllable excitable systems.

The behavior of a system of two bidirectionally coupled lasers for $k = 2$ and $f_0 = 5$ MHz is shown in Figure 9.12 for increasing amplitudes of the modulation, assumed equal for both signals. The figure shows the time trace of the intensity of one of the two lasers on the left, and the probability distribution of the interval between dropouts on the right (the results are basically identical for the other laser, since both lasers are synchronized). The interdropout probability distribution is computed from a collection of 1000 dropouts in each case. For a small modulation amplitude (top row in Figure 9.12), the dropouts occur infrequently at different periods. As the amplitude grows (middle row), most interpulse intervals occur at a definite period T_0 corresponding to the fundamental frequency f_0, which is not present in either of the input signals. For larger amplitudes (bottom row), the input signals take over and dropouts begin to occur at the (larger) input frequencies, reducing the response of the system at the missing fundamental frequency. Therefore, a resonant behavior is observed with respect to the modulation strength: for an intermediate modulation amplitude, the system optimally processes the distributed inputs. We note that this resonance is nontrivially arising from the interplay between the direct electrical modulation of the pump current and the indirect optical driving coming from the other laser.

In the experimental conditions used, the lasers are detuned such that one of them consistently leads the dynamics, with a time lag equal to the coupling time [7]. The behavior of the system does not change if the input modulations are switched between the leader and laggard lasers. It is remarkable that the distributed signals are processed irrespective of this underlying asymmetry in the coupled dynamics.

The subharmonic resonance presented above can also be observed at the level of the RF spectrum of the lasers' outputs, as shown in the right panels of Figure 9.12. Peaks of the three frequencies involved, the two (higher) input frequencies $f_1 = 10$ MHz and $f_2 = 15$ MHz and the fundamental frequency $f_0 = 5$

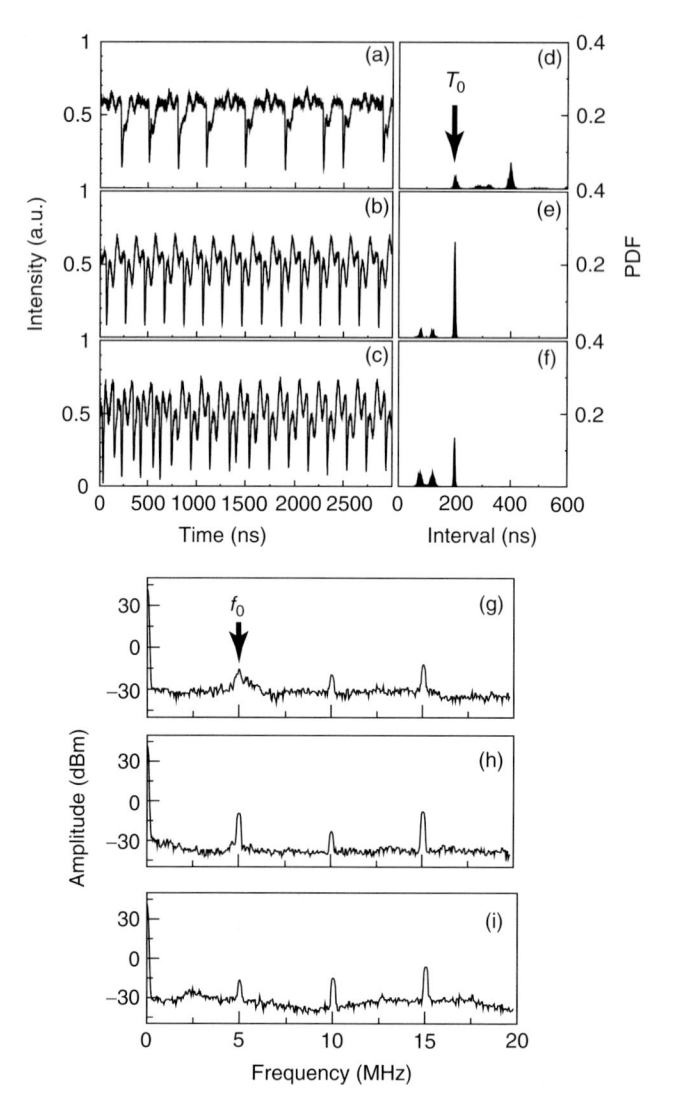

Figure 9.12 Experimental output intensity of one of two coupled lasers (a–c), the corresponding probability distribution functions (PDF) (d–f) of the time intervals between consecutive dropouts (right column), and the corresponding RF spectrum of the output intensity (g–i) for increasing values of the modulation amplitude: (a,d,g) $A_1 = A_2 =$ 0.285 mA; (b,e,h) $A_1 = A_2 = 0.643$ mA; (c,f,i) $A_1 = A_2 = 0.750$ mA. The input frequencies are $f_1 = 10$ MHz and $f_2 = 15$ MHz, corresponding to interpulse periods $T_1 = 100$ ns and $T_2 = 66.7$ ns. The ghost frequency is $f_0 = 5$ MHz, corresponding to a period $T_0 = 200$ ns. Adapted from [46].

MHz, are clearly observed in the spectrum. The height of the peaks at f_1 and f_2 increases monotonically with the modulation amplitude (from top to bottom), while the peak at f_0 is highest at an intermediate amplitude, which is a clear indicator of a resonance occurring at the missing fundamental frequency [41].

9.5.2
Noise Modulation

Over the past decades, much attention in the field of stochastic processes has been paid to the question of how noise can lead to order [52–54] (see also chapter 11). In a seminal work, Bryant and Segundo [55] showed that the introduction of white noise in a neuron model produced an invariance in the firing times. In that study, repeating stimulations of the neuron with the same segment of Gaussian-white noise current resulted in a reproducible interspike time response. A similar study was made in neocortical neurons of rats by Mainen and Sejnowski [56]. They showed that for constant stimuli the spike trains were imprecise, whereas the introduction of fluctuations in the stimuli, resembling synaptic activity, produced spike trains with reproducible timing. In lasers, a good example of the regularity introduced by noise was given by Uchida et al. [57], who showed the reproducibility of a laser's response to a noisy drive signal. Specifically, a noisy signal was sent repeatedly to a Nd:YAG microchip laser and the system was capable, after a transient, of producing identical response outputs. For small amplitude of the added noise, the outputs are not identical, because the relaxation oscillations driven by internal noise dominate the laser output. There is an optimal noise level for which the outputs are identical, because the common-noise-driven signal overcomes the internal noise.

Another constructive effect of noise is inducing the synchronization of coupled systems. This topic has been studied theoretically in coupled chaotic systems [58–60], and experimentally in chaotic circuits [61]. The common feature in all these works is that when a certain amount of common noise is introduced, the coupled systems are driven to collapse onto the same trajectory. This property can be used to achieve the isochronal solution in a symmetrical bidirectionally coupled semiconductor laser system, by applying a common source of external noise to the pump current of both lasers (Figure 9.13). For large enough noise intensity, the system reaches a common output without lag between them, stabilizing the isochronal solution.

The left panel of Figure 9.14 shows the correlated dynamics of the two lasers when the amount of common noise increases. The temperatures of the lasers are adjusted such that their frequencies are as similar as possible, in order to optimize the mutual injection. The pump currents and the bidirectional alignment are then optimized by looking for the maximum enhancement of the output power due to the mutual injection. The pump currents are fixed slightly above their solitary threshold, for which the lasers operate in the LFF regime. This regime allows for an easy observation and measurement of the isochrony during the experiments.

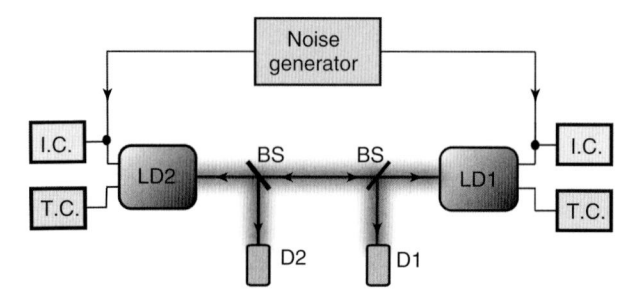

Figure 9.13 Experimental setup leading to noise-induced synchronization. Two semiconductor lasers LD1 and LD2 are coupled by mutual injection, and subject to a common noise source being applied to their pump currents. Abbreviations: I.C., injection current; T.C., temperature control; BS, beam splitter.

With the system symmetrically injected, the same noise source is introduced simultaneously into the pump current of the two lasers through an internal bias-T of the laser mounts. In that way, the noise is superimposed to the DC operating level set by the current controller. The left panel of Figure 9.14 displays the output intensities and the corresponding cross-correlation functions for increasing values of the noise level. The output intensities are displaced vertically for clarity, with the top trace representing the output of LD1 and the bottom trace representing LD2. Without noise (plots a and b), LFF dynamics can be observed in the output intensities, and the correspondent cross-correlation function shows a maximum at a lag equal to the flight time τ_c between the lasers. The LFF dynamics starts to disappear as the noise level increases (plot g), even though the cross-correlation still has its maximum at τ_c (plot h). Finally, for a large enough noise level (plot i), a correlation peak arises at zero-lag (plot j). These results show that common noise is able to induce zero-lag synchronization in mutually coupled lasers.

The zero-lag synchronized state represented in plots (i,j) of the left panel of Figure 9.14 are nevertheless markedly different from the intrinsic dynamics of the lasers. In particular, the cross-correlation of the signals shows a broadening of its maximum peak. In order to determine the origin of this broadening, one can turn to numerical simulations of the system, which allow for an arbitrarily large temporal resolution of the dynamics and an infinite bandwidth of the noise being added to the lasers' pump currents. Indeed, experimental monitoring of the dynamics has a resolution that is strongly limited by the bandwidth of the photodetectors and oscilloscope, and the bandwidth of the common noise is also limited by the frequency filtering characteristics of the bias-T and laser mount. If we ignore bandwidth limitations in our experimental system, we can simulate the output intensities for different noise correlation times and compare the cross-correlation functions for filtered and unfiltered signals, to find a value that shows isochrony for both kinds of signals. The correlation time of the noise is known to play an important role in the dynamics of chaotic lasers [62]. Numerical simulations can be performed on a Lang–Kobayashi-type model of the two mutually coupled lasers [7].

9.5 Modulation Characteristics of Delay-Coupled Lasers | 237

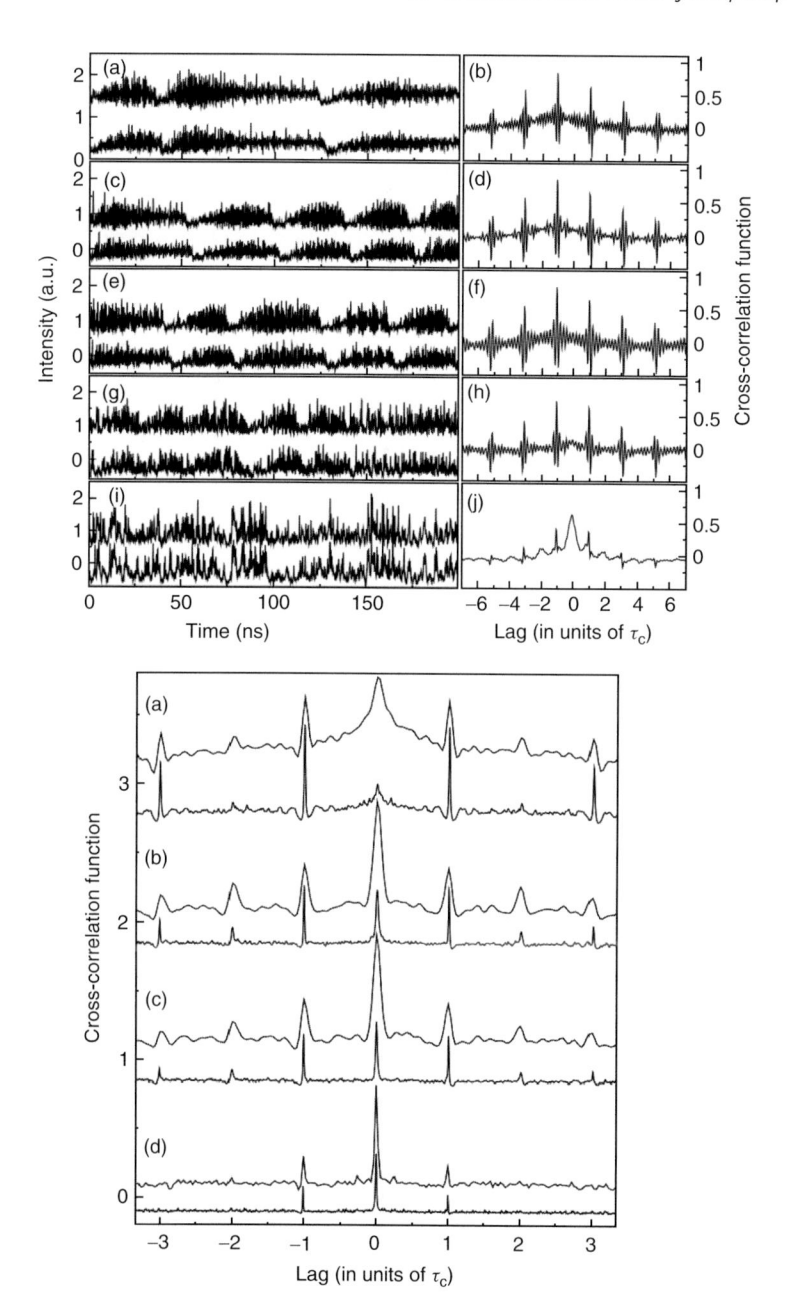

Figure 9.14 Left panel: experimental output intensities (a,c,e,g,i) and corresponding cross-correlation functions (b,d,f,h,j) for different values of injected noise level (increasing from top to bottom). Right panel: numerical cross-correlation functions for filtered (top) and unfiltered (bottom) signals for a fixed noise intensity and varying values of the noise correlation time (decreasing from top to bottom).

The right panel of Figure 9.14 shows the cross-correlation functions of the filtered (top traces) and unfiltered (bottom traces) time series of the laser intensities, for decreasing correlation times of the noise. The top trace in Figure 9.14a corresponds to parameters that approximately match the conditions of the experimental results shown in plots (i,j) of the left panel of Figure 9.14. The first thing that can be noted is that the zero-lag peak in the cross-correlation is very small in the case of the unfiltered signals (bottom trace in plot (a) of the right panel of Figure 9.14). This shows that the noise acts only in the slow dynamics of the system. As the bandwidth of the noise increases (i.e., its correlation time decreases, from top to bottom in the right panel of Figure 9.14), the zero-lag peak in the cross-correlation of the *unfiltered* time series starts to increase in amplitude with respect to the side peaks, until finally for a small enough time correlation (plot d) zero-lag synchronization arises not only at slow timescales but also in the fast dynamics. This confirms that the nonzero correlation time of the noise is the cause of the differences between both types of cross-correlations. For high correlation time of the noise, the system only reacts to the fluctuations in its slow dynamics, whereas in the limit of very low noise correlation time both dynamics can respond, for the same noise strength.

9.5.3
Application: Key Exchange Protocol

Coupled semiconductor lasers play an important role in many applications. Among them longitudinal coupled-cavity lasers (e.g., C^3 lasers) have been used for spectral selection; coherent coupling allows for high output power with good spectral and beam properties; laterally coupled laser arrays have been realized to also achieve coherent coupling. As presented in the previous sections, a delay in the coupling path introduces dynamical instabilities and particular synchronization properties that can be harnessed for applications. Delayed-coupling configurations are being considered for applications in encrypted communication (see also chapter 14). In [63], a novel key exchange protocol has been suggested, utilizing the synchronization properties of two mutually delay-coupled semiconductor lasers with the semitransparent mirror as relay element, as depicted in Figure 9.15.

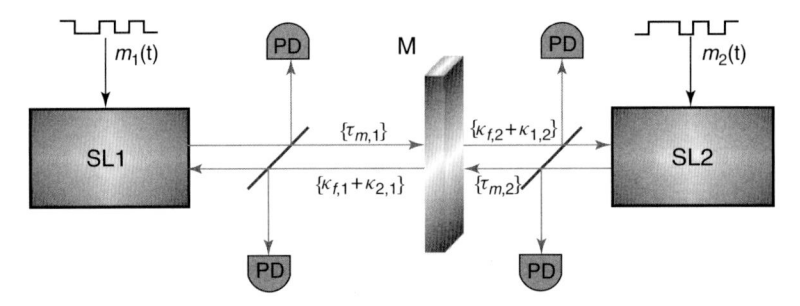

Figure 9.15 Scheme of two mutually coupled semiconductor lasers with a partially transparent mirror as relay element. $m_{1,2}$ encoding messages. Adapted from [63].

9.5 Modulation Characteristics of Delay-Coupled Lasers

In the system of Figure 9.15, an identical synchronization between the two coupled lasers is possible, that is, the cross-correlation function exhibits the maximum peak at a time lag that amounts to the difference between the coupling times of both lasers with the mirror $\Delta t = \tau_{m,2} - \tau_{m,1}$. The maximum of the cross-correlation can be zero when the mirror is displaced to the center ($\tau_{m,2} = \tau_{m,1}$). This scheme allows for a bidirectional transmission of information since a small, short perturbation of the bias current of any of the lasers only deteriorates momentarily the synchronization between the optical output powers P_1 and P_2 [63]. The communication process is illustrated in Figure 9.16 where the bias currents of the lasers are simultaneously modulated with different pseudorandom digital messages of small amplitude. When both parties of the communication send the same bit of information by modulating the bias current, the synchronization difference between the powers emitted by the two lasers with a time lag $[P_1(t) - P_2(t + \Delta t)]$ is ideally zero. In this way, both sides of the link can negotiate and exchange an encrypted key through a public channel. The security of the communication is guaranteed by the fact that an eavesdropper can only monitor the difference $P_1(t) - P_2(t + \Delta t)$ and has no clue as to which are the bits being sent when this difference is zero.

Also, further schemes for public channel cryptography based on chaos synchronization of mutually delay-coupled systems have been proposed. In principle, they might suffer from two different kinds of attacks. Hardware attacks utilize a chaotic setup similar to those of the synchronized chaotic partners, whereas software attacks might be based on the mathematical manipulation of the recorded signal. In

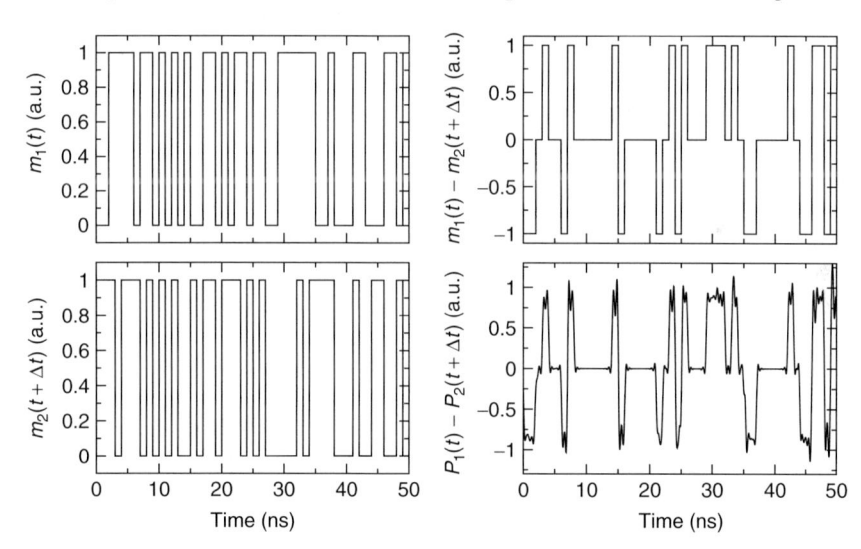

Figure 9.16 Illustration of the message decryption process. $m_{1,2}$ are the messages encoded by $SL_{1,2}$, $m_1(t) - m_2(t + \Delta t)$ is the subtraction of the messages with a time lag Δt, and this message subtraction is reconstructed by the synchronization error $P_1(t) - P_2(t + \Delta t)$. The synchronization error has been filtered with a fifth-order Butterworth filter with a cutoff frequency of 0.8 GHz.

order to exclude advanced software attacks, the task of the attacker can be mapped onto a NP-complete problem. An elegant method utilizing private commutative filters, transmission of integer signals, additional nonlinear terms to the transmitted signal, and periods of cutoffs in communication, has been introduced in [64].

Finally, networks of delay-coupled lasers are currently being studied for the realization of novel information processing concepts, being inspired by neuronal systems. They are expected to represent a suitable reservoir for the realization of a liquid state machine. This illustrates that delay effects might become very useful, improving existing applications or allowing for novel applications.

9.6
Conclusion

Delay-coupled systems are ubiquitous in both natural and technological settings. The signals that connect dynamical systems do not propagate instantaneously, and when those dynamical systems are fast enough, the propagation time of the coupling signals cannot be neglected. This occurs in diverse systems ranging from the brain (whose neurons operate in timescales of the order of milliseconds, whereas neuronal signals take tens of milliseconds to propagate over distances of centimeters) to semiconductor lasers in communication networks (which operate on timescales of tens of picoseconds, whereas light takes nanoseconds to travel over distances of centimeters). Besides their undeniable technological interest, delay-coupled semiconductor lasers constitute well-controllable test beds to study the effect of delay coupling. In particular, delay-coupled semiconductor lasers show a rich phenomenology of dynamical properties and synchronization scenarios, some of which have been reviewed in this chapter. In addition, we have illustrated here the application potential of different delay-coupling configurations. For instance, we have presented how bidirectionally coupled lasers might be used for implementing a key exchange protocol based on synchronized chaotic behavior.

The configurations discussed in this chapter represent only a few possibilities among the many more that are possible. Motivated by the coupling configurations described here, more complicated network arrangements of many delay-coupled lasers, or other delay-coupled oscillators, could be realized (see also chapter 10). This perspective, in particular, represents a very promising and challenging field of study, which might help us to understand, for instance, certain aspects of brain dynamics, and even more to explore and realize bioinspired concepts of information processing, such as reservoir computing.

Acknowledgments

The authors thank many collaborators and institutions, who contributed to the reported work, through their intellectual, practical, and financial help. In particular, they express many thanks to Claudio Mirasso, Michael Peil, Tilmann Heil,

Wolfgang Elsäßer, Otti D'Huys, Jan Danckaert, Raul Vicente, Guy Van der Sande, Thomas Erneux, Javier Martin-Buldu, Rajarshi Roy, and Atsushi Uchida. Part of the work was funded via the different research programs from the European Community in FP5, FP6, and FP7 (Projects OCCULT, GABA, PHOCUS), the Spanish MICINN under project Nos. TEC2009-14101 DeCoDicA and FIS2009 13360, the ICREA Academia programme, the German Research Foundation, the Volkswagen Foundation, and the German BMBF.

References

1. Botez, D. and Scifres, D. (eds) (1994) *Laser Diode Arrays*, Cambridge Studies in Modern Optics, Cambridge University Press.
2. Wieczorek, S., Krauskopf, B., Simpson, T., and Lenstra, D. (2005) The dynamical complexity of optically injected semiconductor lasers. *Phys. Rep.*, **416**, 1–128.
3. DeFreez, R.K., Bossert, D.J., YU, N., Hartnett, K, Elliott, R.A., and Winful, H. (1988) Spectral and picosecond temporal properties of flared guide y-coupled phase-locked laser arrays. *Appl. Phys. Lett.*, **53**, 2380–2382.
4. Merbach, D., Hess, O., Herzel, H., and Schöll, E. (1995) Injection-induced bifurcations of transverse spatiotemporal patterns in semiconductor laser arrays. *Phys. Rev. E*, **52**, 1571–1578.
5. Winful, H.G. and Rahman, L. (1990) Synchronized chaos and spatiotemporal chaos in arrays of coupled lasers. *Phys. Rev. Lett.*, **65**, 1575–1578.
6. Hohl, A., Gavrielides, A., Erneux, T., and Kovanis, V. (1997) Localized synchronization in two coupled nonidentical semiconductor lasers. *Phys. Rev. Lett.*, **78** (25), 4745–4748.
7. Heil, T., Fischer, I., Elsäßer, W., Mulet, J., and Mirasso, C.R. (2001) Chaos synchronization and spontaneous symmetry-breaking in symmetrically delay-coupled semiconductor lasers. *Phys. Rev. Lett.*, **86** (5), 795–798.
8. Fujino, H. and Ohtsubo, J. (2001) Synchronization of chaotic oscillations in mutually coupled semiconductor lasers. *Opt. Rev.*, **8**, 351–357.
9. Wünsche, H.J., Bauer, S., Kreissl, J., Ushakov, O., Korneyev, N., Henneberger, F., Wille, E., Erzgräber, H., Peil, M., Elsäßer, W., and Fischer, I. (2005) Synchronization of delay-coupled oscillators: a study of semiconductor lasers. *Phys. Rev. Lett.*, **94**, 163 901 1–4.
10. Larger, L. and Fischer, I. (2011) Optical Delay Dynamics and its Applications, *The Complexity of Dynamical Systems: A Multi-disciplinary Perspective*, Wiley-VCH Verlag GmbH, pp. 63–98.
11. Fischer, I., van Tartwijk, G., Levine, A., Elsäßer, W., Göbel, E., and Lenstra, D. (1996) Fast pulsing and chaotic itinerancy with a drift in the coherence collapse of semiconductor lasers. *Phys. Rev. Lett.*, **76** (2), 220–223.
12. Fischer, I., Heil, T., and Elsäßer, W. (2000) Emission dynamics of semiconductor lasers subject to delayed optical feedback: an experimentalist's perspective. AIP Conference Proceedings, vol. 548, pp. 218, Texel (The Netherlands).
13. van der Sande, G., Soriano, M.C., Fischer, I., and Mirasso, C.R. (2008) Dynamics, correlation scaling, and synchronization behavior in rings of delay-coupled oscillators. *Phys. Rev. E*, **77**, 055202.
14. Mulet, J., Mirasso, C., Heil, T., and Fischer, I. (2004) Synchronization scenario of two distant mutually coupled semiconductor lasers. *J. Opt. B: Quantum Semiclassical Opt.*, **6**, 97–105.
15. Lang, R. and Kobayashi, K. (1980) External optical feedback effects on semiconductor laser properties. *IEEE J. Quantum Electron.*, **QE-16**, 347–355.
16. D'Huys, O., Vicente, R., Danckaert, J., and Fischer, I. (2010) Amplitude and phase effects on the synchronization of delay-coupled oscillators. *Chaos*, **20** (4), 043127.

17. Flunkert, V., Yanchuk, S., Dahms, T., and Schöll, E. (2010) Synchronizing distant nodes: a universal classification of networks. *Phys. Rev. Lett.*, **105**, 254101.
18. Locquet, A., Masoller, C., and Mirasso, C.R. (2002) Synchronization regimes of optical-feedback-induced chaos in unidirectionally coupled semiconductor lasers. *Phys. Rev. E*, **65**, 056205.
19. Locquet, A., Masoller, C., MÈgret, P., and Blondel, M. (2002) Comparison of two types of synchronization of external-cavity semiconductor lasers. *Opt. Lett.*, **27**, 31–33.
20. Masoller, C. (2001) Anticipation in the synchronization of chaotic semiconductor lasers with optical feedback. *Phys. Rev. Lett.*, **86**, 2782–2785.
21. Sivaprakasam, S., Shahverdiev, E.M., Spencer, P.S., and Shore, K.A. (2001) Experimental demonstration of anticipating synchronization in chaotic semiconductor lasers with optical feedback. *Phys. Rev. Lett.*, **87**, 154101.
22. Avila, J.F.M., Vicente, R., Rios Leite, J.R., and Mirasso, C.R. (2007) Synchronization properties of bidirectionally coupled semiconductor lasers under asymmetric operating conditions. *Phys. Rev. E*, **75**, 066202.
23. Gonzalez, C.M., Torrent, M.C., and Garcia-Ojalvo, J. (2007) Controlling the leader-laggard dynamics in delay-synchronized lasers. *Chaos*, **17** (3), 033122.
24. Uchida, A., Rogister, F., GarcÌa-Ojalvo, J., Roy, R., and Wolf, E. (2005) *Synchronization and Communication with Chaotic Laser Systems*, vol. 48, Elsevier, pp. 203–341.
25. Engel, A., Konig, P., Kreiter, A., and Singer, W. (1991) Interhemispheric synchronization of oscillatory neuronal responses in cat visual cortex. *Science*, **252** (5009), 1177–1179.
26. Roelfsema, P.R., Engel, A.K., Konig, P., and Singer, W. (1997) Visuomotor integration is associated with zero time-lag synchronization among cortical areas. *Nature*, **385** (6612), 157–161.
27. Schneider, G. and Nikolic, D. (2006) Detection and assessment of near-zero delays in neuronal spiking activity. *J. Neurosci. Methods*, **152** (1-2), 97–106.
28. Traub, R.D., Whittington, M.A., Stanford, I.M., and Jefferys, J.G.R. (1996) A mechanism for generation of long-range synchronous fast oscillations in the cortex. *Nature*, **383** (6601), 621–624.
29. König, P., Engel, A.K., and Singer, W. (1995) Relation between oscillatory activity and long-range synchronization in cat visual cortex. *Proc. Natl. Acad. Sci. U.S.A.*, **92** (1), 290–294.
30. Bibbig, A., Traub, R.D., and Whittington, M.A. (2002) Long-range synchronization of γ and β oscillations and the plasticity of excitatory and inhibitory synapses: a network model. *J. Neurophysiol.*, **88** (4), 1634–1654.
31. Fischer, I., Vicente, R., Buldú, J.M., Peil, M., Mirasso, C.R., Torrent, M.C., and García-Ojalvo, J. (2006) Zero-lag long-range synchronization via dynamical relaying. *Phys. Rev. Lett.*, **97** (12), 123902.
32. Rosenbluh, M., Aviad, Y., Cohen, E., Khaykovich, L., Kinzel, W., Kopelowitz, E., Yoskovits, P., and Kanter, I. (2007) Spiking optical patterns and synchronization. *Phys. Rev. E*, **76**, 046207.
33. Peil, M., Larger, L., and Fischer, I. (2007) Versatile and robust chaos synchronization phenomena imposed by delayed shared feedback coupling. *Phys. Rev. E*, **76**, 045201–4(R).
34. Flunkert, V., D'Huys, O., Danckaert, J., Fischer, I., and Schöll, E. (2009) Bubbling in delay-coupled lasers. *Phys. Rev. E*, **79** (6), 1–4.
35. Sukow, D.W. and Gauthier, D.J. (2000) Entraining power-dropout events in an external-cavity semiconductor laser using weak modulation of the injection current. *IEEE J. Quantum Electron.*, **36** (2), 175–183.
36. Buldú, J.M., García-Ojalvo, J., Mirasso, C.R., and Torrent, M.C. (2002) Stochastic entrainment of optical power dropouts. *Phys. Rev. E*, **66** (2), 021106.
37. Lam, W., Guzdar, P.N., and Roy, R. (2003) Effect of spontaneous emission noise and modulation on semiconductor lasers near threshold with optical feedback. *Int. J. Mod. Phys. B*, **17**, 4123–4138.

38. Buldu, J.M., Vicente, R., Perez, T., Mirasso, C.R., Torrent, M.C., and Garcia-Ojalvo, J. (2002) Periodic entrainment of power dropouts in mutually coupled semiconductor lasers. *Appl. Phys. Lett.*, **81** (27), 5105–5107.
39. Lindner, J.F., Meadows, B.K., Ditto, W.L., Inchiosa, M.E., and Bulsara, A.R. (1995) Array enhanced stochastic resonance and spatiotemporal synchronization. *Phys. Rev. Lett.*, **75** (1), 3–6.
40. Buldú, J.M., García-Ojalvo, J., Torrent, M.C., Vicente, R., Pérez, T., and Mirasso, C.R. (2003) Entrainment of optical low-frequency fluctuations is enhanced by coupling. *Fluctuat. Noise Lett.*, **3** (2), L127–L136.
41. Chialvo, D.R. (2003) How we hear what is not there: a neural mechanism for the missing fundamental illusion. *Chaos*, **13** (4), 1226–1230.
42. Buldú, J.M., Chialvo, D.R., Mirasso, C.R., Torrent, M.C., and García-Ojalvo, J. (2003) Ghost resonance in a semiconductor laser with optical feedback. *Europhys. Lett.*, **64** (2), 178.
43. Van der Sande, G., Verschaffelt, G., Danckaert, J., and Mirasso, C.R. (2005) Ghost stochastic resonance in vertical-cavity surface-emitting lasers: experiment and theory. *Phys. Rev. E*, **72** (1), 016113.
44. Calvo, O. and Chialvo, D.R. (2006) Ghost stochastic resonance in an electronic circuit. *Int. J. Bifurcat. Chaos*, **16** (3), 731–735.
45. Buldu, J.M., Gonzalez, C.M., Trull, J., Torrent, M.C., and Garcia-Ojalvo, J. (2005) Coupling-mediated ghost resonance in mutually injected lasers. *Chaos*, **15** (1), 013103.
46. González, C.M., Buldú, J.M., Torrent, M.C., and García-Ojalvo, J. (2007) Processing distributed inputs in coupled excitable lasers. *Phys. Rev. A*, **76** (5), 053 824.
47. Giudici, M., Green, C., Giacomelli, G., Nespolo, U., and Tredicce, J.R. (1997) Andronov bifurcation and excitability in semiconductor lasers with optical feedback. *Phys. Rev. E*, **55** (6), 6414–6418.
48. Mulet, J. and Mirasso, C.R. (1999) Numerical statistics of power dropouts based on the lang-kobayashi model. *Phys. Rev. E*, **59** (5), 5400–5405.
49. Balenzuela, P. and Garcia-Ojalvo, J. (2005) Neural mechanism for binaural pitch perception via ghost stochastic resonance. *Chaos*, **15** (2), 023903.
50. Balenzuela, P., Garcia-Ojalvo, J., Manjarrez, E., Martínez, L., and Mirasso, C.R. (2007) Ghost resonance in a pool of heterogeneous neurons. *Biosystems*, **89** (1-3), 166–172.
51. Manjarrez, E., Balenzuela, P., García-Ojalvo, J., Vásquez, E.E., Martínez, L., Flores, A., and Mirasso, C.R. (2007) Phantom reflexes: Muscle contractions at a frequency not physically present in the input stimuli. *Biosystems*, **90** (2), 379–388.
52. Horsthemke, W. and Lefever, R. (1984) *Noise Induced Transitions*, Springer.
53. Garcia-Ojalvo, J. and Sancho, J. (1999) *Noise in Spatially Extended Systems*, Springer.
54. Lindner, B., García-Ojalvo, J., Neiman, A., and Schimansky-Geier, L. (2004) Effects of noise in excitable systems. *Phys. Rep.*, **392** (6), 321–424.
55. Bryant, H.L. and Segundo, J.P. (1976) Spike initiation by transmembrane current: a white-noise analysis. *J. Phys.*, **260** (2), 279–314.
56. Mainen, Z.F. and Sejnowski, T.J. (1995) Reliability of spike timing in neocortical neurons. *Science*, **268** (5216), 1503–1506.
57. Uchida, A., McAllister, R., and Roy, R. (2004) Consistency of nonlinear system response to complex drive signals. *Phys. Rev. Lett.*, **93** (24), 244102.
58. Maritan, A. and Banavar, J.R. (1994) Chaos, noise, and synchronization. *Phys. Rev. Lett.*, **72** (10), 1451–1454.
59. Toral, R., Mirasso, C.R., Hernández-García, E., and Piro, O. (2001) Analytical and numerical studies of noise-induced synchronization of chaotic systems. *Chaos*, **11**, 665–673.
60. Zhou, C. and Kurths, J. (2002) Noise-induced phase synchronization and synchronization transitions in

chaotic oscillators. *Phys. Rev. Lett.*, **88**, 230602.

61. Sánchez, E., Matías, M.A., and Pérez-Mu nuzuri, V. (1997) Analysis of synchronization of chaotic systems by noise: an experimental study. *Phys. Rev. E*, **56**, 4068–4071.

62. Buldú, J.M., García-Ojalvo, J., Mirasso, C.R., Torrent, M.C., and Sancho, J.M. (2001) Effect of external noise correlation in optical coherence resonance. *Phys. Rev. E*, **64** (5), 051109.

63. Vicente, R., Mirasso, C.R., and Fischer, I. (2007) Simultaneous bidirectional message transmission in a chaos-based communication scheme. *Opt. Lett.*, **32** (4), 403–405.

64. Kanter, I., Kopelowitz, E., and Kinzel, W. (2008) Public channel cryptography: Chaos synchronization and hilbert's tenth problem. *Phys. Rev. Lett.*, **101** (8), 084102.

10
Complex Networks Based on Coupled Two-Mode Lasers
Andreas Amann

10.1
Introduction

The topic of semiconductor lasers attracts the attention of experts across a number of different fields. On one hand the commercial success of semiconductor lasers, for example, in medicine or telecommunication drives the development of highly optimized devices with very specific output characteristics. On the other hand, lasers are paradigmatic examples of nonlinear systems and play a decisive role in the development of nonlinear dynamics into a cross disciplinary subject over the last 40 years [1].

Already a free running laser represents a nontrivial nonlinear system [2]. Even more interesting are the phenomena that arise when the light of one laser (the *master*) influences the dynamics of a second laser (the *slave*) [3]. This *injection dynamics* can, for example, result in the locking of the frequency of the slave to the master [4, 5], in excitability [6, 7], coexistence of complex attractors [8, 9], switching between bistable states [10–12], or chaotic synchronization [13, 14]. In a closed loop system, where either the light is reflected back into a laser or two or more lasers are mutually coupled, the time delay due to the round-trip time induces external-cavity modes [15], compound laser modes [16] or may be able to stabilize unstable steady states noninvasively [17].

In the case of *single-mode* lasers, the dynamical properties due to optical injection have been widely studied in the past and are today, in general, well understood. For a comprehensive review, we refer to [18]. The objective of the current chapter is to explore some of the properties of *multimode* and, in particular, *two-mode* lasers. This chapter is organized as follows. In order to motivate our interest in multimode lasers, we discuss the new possibilities, which exist for the construction of complex networks using multimode lasers in Section 10.2. In Section 10.3, we give an overview of the design of multimode lasers, which involves the solution of an inverse problem. We then study the basic properties of the dynamics of two-mode lasers under optical injection in Section 10.4, and finally give a summary and outlook in Section 10.5.

Nonlinear Laser Dynamics: From Quantum Dots to Cryptography, First Edition. Edited by Kathy Lüdge.
© 2012 Wiley-VCH Verlag GmbH & Co. KGaA. Published 2012 by Wiley-VCH Verlag GmbH & Co. KGaA.

10.2
Complex Networks on the Basis of Two-Mode Lasers

While over the last decade the technology for transmitting information by photonic means has been considerably improved, the problem of convenient all-optical signal processing remains a major challenge in photonics. In order to process information within the optical domain, it would, in general, be desirable to implement a relatively sophisticated connection topology, that is, a complex network of coupled lasers. Ideally such complex networks would be integrated on a single chip, which seems to be difficult to achieve on the basis of single-mode lasers. The reason is that each coupling needs to be implemented via a physical optical-fiber link between the individual lasers. Since today no scalable technology is available that allows for arbitrary connections between lasers on a chip, the optical-fiber links need to be connected "by hand," and only networks with a relatively small number of single-mode lasers have been realized so far.

Surprisingly, the use of *multimode* lasers is potentially able to overcome some of the limitations faced by networks of single-mode lasers. In order to understand the basic principle, let us consider the setup of four individual two-mode lasers as in Figure 10.1. On the level of optical-fiber coupling the four lasers are coupled

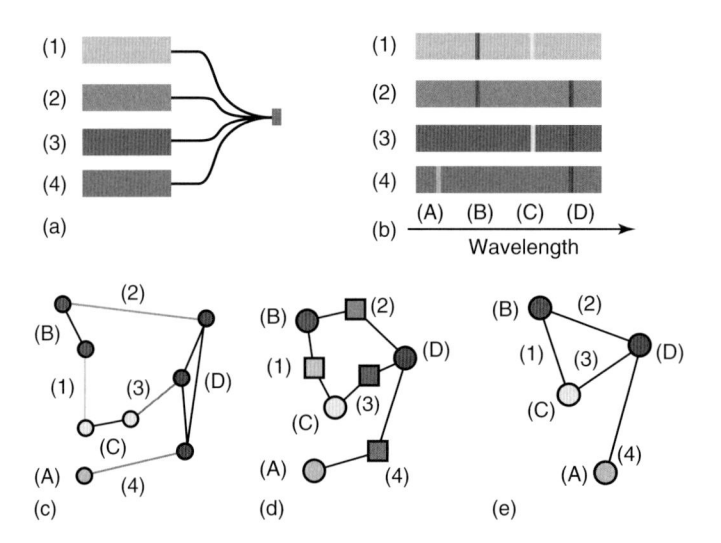

Figure 10.1 (a) Physical-coupling scheme via optical-fiber coupling for four two-mode lasers. The all-to-all coupling is realized via a wavelength multiplexer, which collects the light from the individual lasers into a single wave guide, which is terminated by a mirror. (b) Schematic view of the selected modes for each laser diode. (c) The resulting abstract network topology due to the physical all-to-all coupling as in panel (a) and the mode spectrum of panel (b). (d) By representing the individual two-mode lasers with square-shaped nodes and individual modes with the circular nodes the network from (c) is converted into a bipartite network. In the current example of two-mode lasers, the equivalent conventional network with links between nodes is shown in (e).

in the way as depicted in Figure 10.1a. The output of each laser is collected by an optical multiplexer into a single optical fiber, which is then terminated by an optical mirror. As a consequence, light from each laser is injected back into every other laser and also into itself. In contrast to an arbitrary optical fiber network topology, a simple all-to-all connection of lasers as shown in Figure 10.1a is feasible using today's optical integration technology.

In spite of the trivial coupling on the optical-fiber level, the setup shown in Figure 10.1a, nevertheless, realizes a nontrivial optical network topology. The reason is that the four lasers in this example are in fact *two-mode* lasers. Each individual laser is able to emit light at two different wavelengths simultaneously. The crucial point is now that the emission wavelengths (modes) can be chosen differently for each laser. This is schematically shown in Figure 10.1b. In this case, for example, laser (1) is chosen to emit light at modes (B) and (C), while laser (2) emits at modes (B) and (D). Owing to the mutual coupling between the lasers (Figure 10.1a), laser (1) therefore, in particular, injects light at modes (B) and (C) into laser (2). Since laser (2) is tuned to emit at modes (B) and (D), only light with frequency close to these modes will be able to affect the local dynamics on laser (2). Let us now assume that the frequency difference between the mode (C) and either of the modes (B) or (D) is much greater than the relaxation oscillation frequency, which sets the scale for the local dynamics on laser (2). Then the light of mode (C) from laser (1) is effectively ignored by laser (2), and only the light emitted at mode (B) will be able to affect the dynamics of laser (2). In the network topology graph shown in Figure 10.1c, this is reflected by the link labeled with (B). The nodes in this graph represent the modes that are selected as the lasing modes for an individual laser. The modes which are selected for a given laser are both affected by the local carrier dynamics on that laser. This establishes a dynamical connection between the selected modes through the laser itself. In Figure 10.1c, the lasers labeled from (1) to (4) are therefore represented as links between the nodes.

While Figure 10.1c gives in principle the complete information about the network topology, this type of graph does not follow the usual conventions of network theory, since it contains links of two different types. We can transform the topology of the graph into the network shown in Figure 10.1d. Here the individual modes are represented by circular nodes, while the lasers are represented by square-shaped nodes. A link, for example, from the square node (1) to the circular node (B) corresponds to the fact that mode (B) is selected on laser (1). A network with two types of nodes and links between different types of nodes is called a *bipartite* network. In the current example, we have used two-mode lasers. In the language of the bipartite network (Figure 10.1d), this means that every square node has exactly two links. In this case of two-mode lasers, the network topology can be represented by a conventional network graph as shown in Figure 10.1e. We can interpret this graph as a network in *mode space*, since the nodes in this network represent the optical modes in frequency space. The links represent the individual lasers that select the corresponding modes.

Note that although we have used only four two-mode lasers with trivial optical-fiber coupling as shown in Figure 10.1a, the resulting network topology (Figure 10.1c) shows a surprising complexity. In particular, we have one node (node (D)) with degree 3, since there are three lasers which select that particular node. In principle nodes with arbitrarily large degree are possible by simply adding more lasers that lase on that particular wavelength. The topology of the network is encoded in the way the modes are selected on the individual lasers. From the technological point of view, this means that in order to implement the ideas of Figure 10.1, a scalable method is required that is able to reliably yield a large number of lasers with individually predefined active optical modes. We will briefly introduce a method that is able to achieve this goal in the next section.

From the modeling point of view, a system as shown in Figure 10.1b is a system of two-mode lasers with mutual optical injection. The first step to understand this system will therefore consist of understanding the injection dynamics of two-mode lasers, in general. Some results in this direction will be presented in Section 10.4.

While one motivates to implement a network along the ideas presented in this section is the desire to process optical signals without converting them into electronic signals, a successful implementation would also be relevant for the theory of complex networks itself, which is currently vigorously developing [19–21]. Currently the predictions of complex network theory are almost exclusively verified using data, which is either generated numerically on a computer or collected from preexisting networks, such as the brain or the Internet, whose properties or topologies cannot be easily altered. Using the ideas of the current section, a nonlinear complex network with a predefined topology could in principle be implemented, which would provide an important and flexible test bed for the general theory of complex networks.

10.3
The Design Principles of Two-Mode Lasers

Multimode lasers are able to emit light at more than just one single wave length simultaneously. A simple example of a multimode laser is the Fabry–Perot laser, which typically lases at several tens of optical wave lengths (modes). A schematic view of a Fabry–Perot laser is shown in Figure 10.2, with an active region of

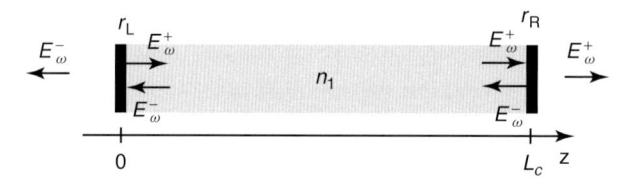

Figure 10.2 Schematic view of a Fabry–Perot laser.

refractive index n_1 and effective gain g enclosed by two mirrors of reflectivity r_L and r_R at both ends. Denoting by $E_\omega^+(z)$ and $E_\omega^-(z)$ the right and left moving electric fields at position z for fixed frequency ω, we can introduce a transfer matrix T that connects the electric fields at both ends of the cavity by

$$\begin{pmatrix} E_\omega^+(\epsilon_z) \\ E_\omega^-(\epsilon_z) \end{pmatrix} = T \begin{pmatrix} E_\omega^+(L_c - \epsilon_z) \\ E_\omega^-(L_c - \epsilon_z) \end{pmatrix}. \tag{10.1}$$

Here ϵ_z and $L_c - \epsilon_z$ denote positions just inside the optical cavity at the right hand end and left hand end, respectively. We are interested in the situation of a free running device without any external optical injection and therefore we set the incoming fields to zero, that is, $E_\omega^+(-\epsilon_z) = E_\omega^-(L_c + \epsilon_z) = 0$. This translates into the conditions

$$E_\omega^+(\epsilon_z) = r_R E_\omega^+(L_c - \epsilon_z), \qquad r_L E_\omega^-(\epsilon_z) = E_\omega^-(L_c - \epsilon_z).$$

Then Eq. (10.1) can be written in the form

$$\mu \begin{pmatrix} r_L \\ 1 \end{pmatrix} = T \begin{pmatrix} 1 \\ r_R \end{pmatrix}, \tag{10.2}$$

with some arbitrary complex number μ. In the case of the Fabry–Perot laser, we have a homogeneous medium of refractive index n_1 and effective gain g, and the transfer matrix is explicitly given by

$$T = \begin{pmatrix} \exp\left(-\left(i\frac{n_1\omega}{c} - g\right)L_c\right) & 0 \\ 0 & \exp\left(\left(i\frac{n_1\omega}{c} - g\right)L_c\right) \end{pmatrix} = \exp(-ik_1 L_c \sigma_z) \tag{10.3}$$

where the complex wave vector is given by $k_1 = \frac{\omega}{c} n_j - ig$, and the Pauli matrices are defined as

$$\sigma_x = \begin{pmatrix} 0 & 1 \\ 1 & 0 \end{pmatrix} \qquad \sigma_y = \begin{pmatrix} 0 & -i \\ i & 0 \end{pmatrix} \qquad \sigma_z = \begin{pmatrix} 1 & 0 \\ 0 & -1 \end{pmatrix}.$$

Inserting T from Eq. (10.3) into Eq. (10.2) and eliminating μ then yields

$$r_L r_R = \exp(-2ik_1 L_c). \tag{10.4}$$

Requiring purely real ω, we obtain the familiar text book lasing condition [22]

$$r_L r_R \exp(2gL_c) = 1.$$

It is now useful to consider for a given g the solutions of Eq. (10.4) in the complex ω plane which can be expressed as

$$\tau_R \omega_m^0 = 2m\pi + i\left(\ln(r_L r_R) + 2gL_c\right) \qquad \text{for } m = 1, 2, \ldots \tag{10.5}$$

where we define the round-trip time $\tau_R = 2nL_c/c$. The imaginary parts of the Fabry–Perot frequencies ω_m^0 determine the decay times of the individual modes, and are negative below lasing threshold. In principle g and to some extend also τ_R also depend on ω, and therefore Eq. (10.5) is an implicit equation for ω. However, it turns out that this dependence is usually only weak and the imaginary part of ω_m^0 is almost constant for a large number of modes as schematically shown in Figure 10.3. This is the reason why for a typical Fabry–Perot laser many modes reach threshold

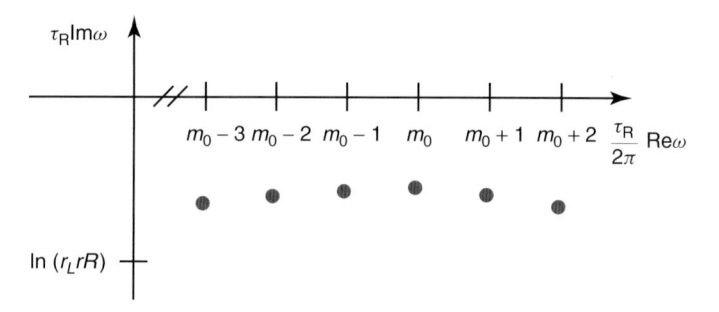

Figure 10.3 Schematic plot of the Fabry–Perot frequencies ω_m^0 according to Eq. (10.5) close to some m_0 with maximal gain g.

almost simultaneously. In practice, this behavior is often not desired, and in the light of the discussion in Section 10.2, one would ideally like to control which of the many candidate modes reach the threshold first.

An innovative method to achieve this goal was proposed by O'Brien and O'Reilly in [23]. This method introduces a number of perturbative features along the cavity in such a way that the refractive index is changed at well-defined positions. Technologically such features are realized using slots on the cavity wave guides, which can be positioned with subwavelength accuracy. These slot features alter the threshold of the *already existing* Fabry–Perot modes. This is in contrast to the commonly used distributed feedback (DFB) or distributed Bragg reflector (DBR) methods of mode selection, which introduce modes that are, in general, incommensurate with the existing Fabry–Perot modes. By using an inverse method, the positions of the features can be calculated to fit a desired gain profile. This was used in [23] to demonstrate the operation of a single-mode laser based on this principle, and was subsequently generalized for the two-mode [24] and multimode case [25]. For more details on the theoretical approach, see [26].

To understand the basic idea of this method, let us first consider the effect of a single feature of length L and refractive index n_2 centered at some position z_1 along the cavity as depicted in Figure 10.4. Our aim is now to calculate the change induced by this feature in linear order of $\delta n = \frac{n_2 - n_1}{n_1}$. The transfer matrix T can be

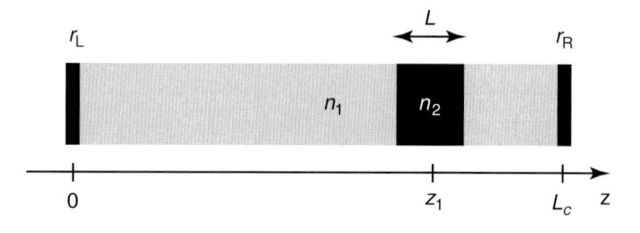

Figure 10.4 Fabry–Perot cavity with one additional feature of length L at position z_1.

calculated as [26]

$$T = P_1\left(z_1 - \frac{L}{2}\right) F(L) P_1\left(L_c - \left(z_1 + \frac{L}{2}\right)\right)$$

where

$$F(L) = T_{12} P_2(L) T_{21}$$

is the transfer matrix for the feature of length L. Here T_{jk} denotes the transition matrix between media with different refractive indices n_j and n_k, and $P_j(z)$ denotes the propagation within a medium with homogeneous refractive index n_j. Explicitly they are given by

$$T_{21} = \frac{1}{2}\begin{pmatrix} 1 + \frac{n_1}{n_2} & 1 - \frac{n_1}{n_2} \\ 1 - \frac{n_1}{n_2} & 1 + \frac{n_1}{n_2} \end{pmatrix}, \quad T_{12} = T_{21}^{-1}, \quad P_j(z) = \exp(-ik_j z \sigma_z).$$

Using straightforward algebra, $F(L)$ can be calculated as

$$F(L) = T_{12} P_2(L) T_{21} = \cos(k_2 L) - \sin(k_2 L)\left(q^- \sigma_y + q^+ \sigma_z\right) \quad (10.6)$$

where $q^+ = (q + q^{-1})/2$ and $q^- = (q - q^{-1})/2$ with $q = n_1/n_2$. Let us now assume that the length of the feature fulfills the half wavelength condition $\sin(k_2 L) = 1$, then in first-order δn the transfer matrix for the feature can be written as

$$F(L) \approx \delta n \sigma_y + \sigma_z = P_2\left(\frac{L}{2}\right)(1 + \delta n \sigma_y) P_2\left(\frac{L}{2}\right)$$

$$\approx P_1\left(\frac{L}{2}\right)(1 + \delta n \sigma_y) P_1\left(\frac{L}{2}\right)$$

where in the last approximation we have assumed that the influence of the feature on the optical path length can be neglected. This allows us to write the transfer matrix for the Fabry–Perot cavity with one additional feature as

$$T \approx P_1(L_c) + \delta n \sigma_y P_1(L_c - 2z_0)$$

where we have used the useful relation

$$P_1(z) \sigma_y = \sigma_y P_1(-z),$$

which follows from the anticommutation properties of the Pauli matrices. The condition of zero incoming field Eq.(10.2) can again be used to find a condition for the available modes. In this case, we obtain

$$Q(\omega, \delta n) = T_{11} + T_{12} r_R - T_{21} r_L - T_{22} r_L r_R = 0$$

and of course for $\delta n = 0$, we know that the Fabry–Perot modes ω_m^0 defined in Eq. (10.5) to fulfill this equation, that is,

$$Q(\omega_m^0, 0) = 0.$$

The question is now, how do the modes ω_m change as we increase δn. Let us therefore assume that $\omega_m(\delta n)$ are functions of the strength of the index perturbation δn. Then we can write in linear order

$$\omega_m(\delta n) = \omega_m^0 + \left.\frac{\partial \omega_m}{\partial \delta n}\right|_{\delta n=0} \delta n + O(\delta n^2).$$

Using the implicit function theorem, we obtain

$$\left.\frac{\partial \omega_m (\delta n)}{\partial \delta n}\right|_{\delta n=0} = -\frac{\left.\frac{\partial Q}{\partial \delta n}\right|_{\delta n=0;\, \omega=\omega_m^0}}{\left.\frac{\partial Q}{\partial \omega}\right|_{\omega=\omega_m^0}}. \qquad (10.7)$$

Since we are mostly interested in changes of the threshold of individual modes and less in their frequency modulation, we concentrate on the imaginary part of Eq. (10.7) and using the normalized feature position $Z_1 = z_1/L_c$, we obtain

$$\mathrm{Im}\left.\frac{\partial \omega_m (\delta n)}{\partial \delta n}\right|_{\delta n=0} = -\frac{2}{\tau_R} \sinh\left[Z_1 (-\ln r_L) + (1 - Z_1) \ln r_R\right] \sin(2\pi m Z_1). \qquad (10.8)$$

This means that the influence of a single feature on the threshold of individual modes m is essentially given by a product of two factors. The "sinh" factor shows that features close to the border of the cavity tend to be more efficient than features toward the center of the cavity. For equal mirror reflectivities $r_L = r_R$, we even find that a feature at the center of the cavity does not change the threshold of any mode in the linear order of δn. While the "sinh" factor is identical for all modes m, the "sin" factor in Eq. (10.8) is crucial for differentiating between modes. Let us write this factor using the notation $m = m_0 + \Delta m$, where m_0 is a central mode as shown in Figure. 10.3. Then we obtain

$$\sin(2\pi m Z_1) = \sin 2\pi m_0 Z_1 \cos 2\pi \Delta m Z_1 + \cos 2\pi m_0 Z_1 \sin 2\pi \Delta m Z_1.$$

Each term is a product of a fast varying and a slowly varying factor. The feature will be placed most efficiently for the mode m_0 and close by modes if we choose $\sin 2\pi m_0 Z_1 = 0$, which can always be achieved by a very small variation of Z_1, since $m_0 \gg 1$, and therefore $\cos 2\pi m_0 Z_1 = \pm 1$. With this stipulated "fine tuning" of the position Z_1, we obtain

$$\mathrm{Im}\left.\frac{\partial \omega_{m_0+\Delta m}}{\partial \delta n}\right|_{\delta n=0} = \pm\frac{2}{\tau_R} \sinh\left[Z_1 (-\ln r_L) + (1 - Z_1) \ln r_R\right] \sin 2\pi \Delta m Z_1.$$

So far we have only considered the effect of a single feature. However, since we only consider effects which are linear in δn, the transition to many features is simply a sum over the effect of N independent features at normalized positions Z_k, and we obtain

$$q_{\Delta m} = \mathrm{Im}\left.\frac{\partial \omega_{m_0+\Delta m} (\delta n)}{\partial \delta n}\right|_{\delta n=0} \qquad (10.9)$$

$$= -\frac{2}{\tau_R} \sum_{k=1}^{N} \sinh\left[Z_k (-\ln r_L) + (1 - Z_k) \ln r_R\right] s_k \sin 2\pi \Delta m Z_k \qquad (10.10)$$

with $s_k = \pm 1$. This transition to many features in linear order of δn is equivalent to a first-order Born approximation in scattering theory. For given s_k and $Z_k \in [0, 1]$, we can use Eq. (10.9) to calculate how much a mode $m_0 + \Delta m$ moves closer to the threshold in linear order of δn.

The practical problem that we have to solve is, however, not how the gain profile is modulated for given slot positions, but how to obtain the slot positions if the gain profile is given. This means that we need to solve the inverse problem of Eq. (10.9). If we are given a series of numbers $q_{\Delta m}$ representing a desired gain profile for a range of values of Δm, then we have to find the positions Z_k which yield those numbers as closely as possible. The key to the solution of this *inverse problem* is the observation that Eq. (10.9) is in a form which resembles a Fourier series. Writing formally the sum on the right hand side as an integral over a feature density function $f(Z) = \sum_k \delta(Z - Z_k) s_k$, we obtain

$$q_{\Delta m} = -\frac{2}{\tau_R} \int f(Z) \sinh\left[Z(-\ln r_L) + (1-Z) \ln r_R\right] \sin(2\pi \Delta m Z) \, dZ.$$

For given $q_{\Delta m}$, we can then invert the Fourier transform to obtain the feature density function $f(Z)$. From $f(Z)$ the positions Z_k and signs s_k can then be obtained via sampling, that is, solving the equation of $\int_0^{Z_k} |f(Z)| \, dZ = k/M$, where M is an appropriate normalization constant and $s_k = \mathrm{sign} f(Z_k)$. For further details on how to choose the gain profile $q_{\Delta m}$ and a more careful consideration of higher order terms, we refer to [26].

Using this method it is, for example, possible to design a single-mode laser [23] or a two-mode laser [24] with side mode suppression ratios of more than 40 dB and a mode spacing in the terahertz regime. For a larger number of selected modes, phenomena, such as mode locking, have been observed [25].

10.4
The Dynamics of Two-Mode Lasers Under Optical Injection

In the previous Section 10.3, we have introduced one practical way to select a number of predefined modes. Let us now return to our original motivation of constructing a complex network using two-mode lasers, which was elaborated in Section 10.2. The general dynamics of mutually coupled multimode lasers is complicated and the subject of current research. However, for the simpler case of optical injection from a single-mode master laser with constant output into a two-mode slave laser, many of the fundamental dynamical features are theoretically well understood and experimentally verified. It is the purpose of the current section to review the theoretical description in this case.

10.4.1
The Model Equations

A suitable model that can be used to describe the optical injection into the two-mode laser is of the form [27]

$$\dot{E}_1 = \frac{1}{2}(1+i\alpha)(g_1(2n+1)-1)E_1, \tag{10.11}$$

$$\dot{E}_2 = \left[\frac{1}{2}(1+i\alpha)(g_2(2n+1)-1) - i\Delta\omega\right] E_2 + K, \tag{10.12}$$

$$T\dot{n} = J - n - (1+2n)\left(g_1|E_1|^2 + g_2|E_2|^2\right), \tag{10.13}$$

where the nonlinear gain terms are given by

$$g_1 = \left[1 + \epsilon\left(|E_1|^2 + \beta_{12}|E_2|^2\right)\right]^{-1}, \quad g_2 = \left[1 + \epsilon\left(|E_2|^2 + \beta_{21}|E_1|^2\right)\right]^{-1}. \tag{10.14}$$

The dynamical quantities in the Eqs. (10.11)–(10.13) are the complex electric field variables $E_1(t)$ and $E_2(t)$ for each mode and the real variable $n(t)$, which characterizes the number of electron-hole pairs available for both modes. The optical injection is given by the injection strength K and the detuning parameter $\Delta\omega$. Further device-specific parameters are the line width enhancement factor α, the ratio of carrier and photon lifetime T and the pump current J. We also include self and cross saturation terms, which are characterized by parameters ϵ, β_{12}, and β_{21}. All dynamical quantities and parameters are rescaled to convenient dimensionless units. Typical parameter values are $\alpha = 2.6$, $T = 800$, $J = 0.5$, $\epsilon = 0.01$, and $\beta_{12} = \beta_{21} = 2/3$ [27]. The time variable in Eqs. (10.11)–(10.13) is measured in units of the photon cavity lifetime τ_c, where we use the value of $\tau_c^{-1} = 9.8 \times 10^{11}\text{s}^{-1}$ to compare with experimental data.

Note that the complex phase of E_1 does not couple to any other quantity, and therefore the phase space of the system (Eqs. (10.11)–(10.13)) is a four-dimensional system of ordinary differential equations and can be parametrized by the four real coordinates $\{|E_1|, \text{Re}\,E_2, \text{Im}\,E_2, n\}$. Since Eq. (10.11) implies that $\frac{d}{dt}|E_1| = 0$, whenever $|E_1|$ vanishes, the single-mode manifold with $|E_1| = 0$ is an invariant submanifold of the system (Eqs. (10.11)–(10.13)).

At this point it is useful to compare our system with the well-studied system for an optically injected single-mode laser [18], which can be written in the form

$$\dot{E} = (1 + i\alpha)nE - i\Delta\omega E + K, \tag{10.15}$$
$$T\dot{n} = J - n - (1+2n)|E|^2, \tag{10.16}$$

where now E refers to the complex field of the single mode, and the other parameters are as before. By comparing both systems, we find that if we neglect the nonlinearity in the gain term (i.e., put $\epsilon = 0$) the system (Eqs. (10.11)–(10.13)) is a minimal extension of the well-tested single-mode system (Eqs. (10.15) and (10.16)) and contains all of its dynamics within the invariant manifold $|E_1| = 0$. We therefore want to first discuss some of the analytic properties of the model (Eqs. (10.11)–(10.13)) in the case of $\epsilon = 0$

10.4.2
The $\epsilon = 0$ Case

Using the notation $E_j = A_j e^{i\phi_j}$, we can then rewrite the Eqs. (10.15) and (10.16) in explicitly four-dimensional form

$$\dot{A}_1 = nA_1, \tag{10.17}$$
$$\dot{A}_2 = nA_2 + K\cos\phi_2, \tag{10.18}$$

10.4 The Dynamics of Two-Mode Lasers Under Optical Injection

$$\dot{\phi}_2 = \alpha n - \Delta\omega - \frac{K}{A_2}\sin\phi_2, \tag{10.19}$$

$$T\dot{n} = J - n - (1 + 2n)\left(A_1^2 + A_2^2\right). \tag{10.20}$$

Let us first consider the equilibria of this system of equations. From Eq. (10.17), it follows that either (i) $A_1 = 0$ or (ii) $n = 0$ holds at the equilibrium.

In case (i), we are in the single-mode submanifold, and due to Eq. (10.17) equilibria with $A_1 = 0$ can only be stable if $n < 0$. With this restriction the analysis proceeds along the lines of the analysis of the single-mode case [28, 29]. Using the Routh–Hurwitz stability criteria for the characteristic polynomial of the Jacobi matrix of Eqs. (10.17)–(10.20), we then obtain the region in the $K - \Delta\omega$ plane for which stable single-mode equilibria exist. While parametric representations $(K(n), \Delta\omega(n))$ of the boundaries of this region can be obtained, it is more convenient to consider explicit expressions of the boundaries in the form of $K(\Delta\omega)$ which are valid in the limit of $T \gg 1$ [28]:

$$K = \frac{\sqrt{J}}{\alpha^2 - 1}\left[\frac{(1+\alpha^2)}{T^2}(1+2J)^2 - \frac{4\alpha\Delta\omega}{T}(1+2J) + (1+\alpha^2)\Delta\omega^2\right]^{1/2}$$

$$\text{for } \Delta\omega \leq \frac{1+2J}{\alpha T}, \tag{10.21}$$

$$K = -\sqrt{\frac{J}{1+\alpha^2}}\Delta\omega \quad \text{for } \Delta\omega \leq 0, \tag{10.22}$$

$$K = \sqrt{J}\Delta\omega \quad \text{for } 0 \leq \Delta\omega \leq \frac{1+2J}{\alpha T}. \tag{10.23}$$

Here Eq. (10.21) parameterizes the single-mode Hopf bifurcation, and Eq. (10.22) yields the saddle-node bifurcation. The line of the transcritical bifurcation Eq. (10.23) is a consequence of solving the system (Eqs. (10.17)–(10.20)) for equilibria with $n = 0$ and $A_1 = 0$.

While case (i) still deals essentially with the bifurcations of the single-mode system, the bifurcations of case (ii) where $n = 0$ and $A_1 \neq 0$ are of genuine two-mode character. The two-mode equilibria are of the form

$$A_2 = \frac{K}{|\Delta\omega|}, \quad \phi_2 = \pm\frac{\pi}{2}, \quad A_1 = \sqrt{J - \left(\frac{K}{|\Delta\omega|}\right)^2}. \tag{10.24}$$

The Jacobian DF of Eqs. (10.17)–(10.20) is then given by

$$DF = \begin{pmatrix} 0 & 0 & 0 & A_1 \\ 0 & 0 & A_2(\Delta\omega) & A_2 \\ 0 & -A_2^{-1}(\Delta\omega) & 0 & \alpha \\ -T^{-1}2A_1 & -T^{-1}2A_2 & 0 & -T^{-1}(1+2J) \end{pmatrix},$$

and the characteristic polynomial is of the form

$$P(\lambda) = \lambda^4 + T^{-1}(1+2J)\lambda^3 + \left(\Delta\omega^2 + T^{-1}2J\right)\lambda^2$$

$$+ T^{-1}\left((1+2J)\Delta\omega^2 + 2\frac{K^2}{\Delta\omega}\alpha\right)\lambda + T^{-1}\left(J^2\Delta\omega^2 - K^2\right). \tag{10.25}$$

We can then use the ansatz $P(\lambda = i\omega_h) = 0$ for the Hopf bifurcation and obtain from the imaginary part of Eq. (10.25)

$$\omega_h^2 = \left(\Delta\omega^2 + \frac{2\alpha K^2}{\Delta\omega(1+2J)}\right).$$

Using the real part of Eq. (10.25) then yields the desired explicit expression for the Hopf boundary of the two-mode state

$$\frac{2\alpha K^2}{1+2J} = -\Delta\omega^3 + \frac{1+2J}{\alpha T}\Delta\omega^2 + \frac{2}{T}J\Delta\omega. \tag{10.26}$$

Note that in contrast to the expressions in the single-mode case (Eqs. (10.21) and (10.22)), this expression is exact.

In Figure 10.5, the analytically obtained bifurcation lines (Eqs. (10.21)–(10.23)) and Eq. (10.26) are shown. We note that the Hopf bifurcation line for the two-mode equilibria has two branches. Interestingly, the two-mode Hopf branch for negative $\Delta\omega$ crosses the single-mode saddle-node bifurcation line at strongly negative $\Delta\omega$. This means that in the region denoted by "TM + SM" in Figure 10.5, a two-mode and a single-mode equilibrium state coexist. This theoretical prediction bistability was indeed found experimentally in [31] and it turned out that switching between the bistable states can be achieved by purely optical means. This allowed

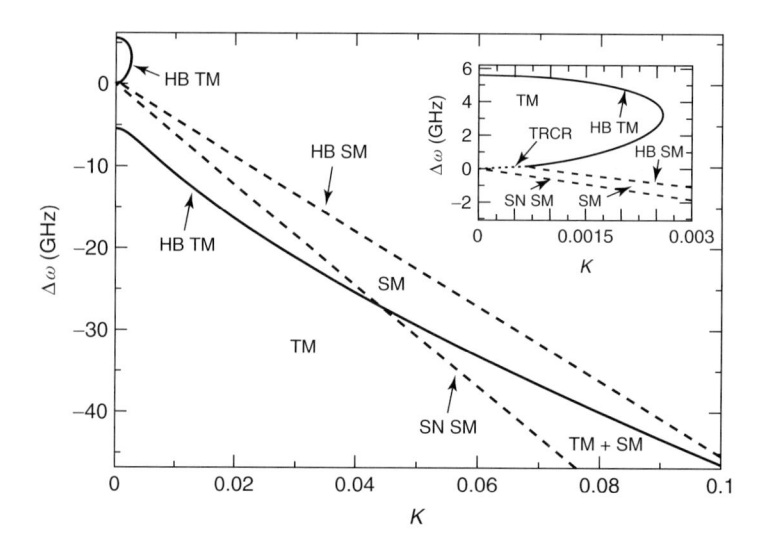

Figure 10.5 Bifurcation diagram for $\epsilon = 0$ with analytic bifurcation lines. SM and TM indicate the single-mode and two-mode region, respectively. In the region denoted by TM + SM single-mode and two-mode equilibria coexist. The Hopf and saddle-node bifurcations of the single-mode region are labeled by HB SM and SN SM, respectively, and are given by the analytic expressions (Eq. (10.21)) and (Eq. (10.22)). The transcritical bifurcation line (Eq. (10.23)) is labeled TRCR in the inset. The two solid lines HB TM show the two branches of the two-mode Hopf bifurcation (Eq. (10.26)).

10.4.3
The Finite ϵ Case

A further feature of the analytic bifurcation diagram for $\epsilon = 0$, which is evident from Figure 10.5 is that the two-mode equilibrium state extends up to $K = 0$. Since according to Eq. (10.24), however, A_2 is proportional to K, this means that for sufficiently negative $\Delta\omega$, and vanishingly small K, the second mode of the laser locks and is at the same time strongly suppressed. This is, however, not what is observed experimentally, where in all cases a finite injection strength is needed to obtain locking, and the injected mode cannot be suppressed arbitrarily. One way to solve this problem is to introduce cross- and self-saturation term proportional via a nonvanishing nonlinear gain parameter ϵ in Eq. (10.14).

While in the case of the single-mode model, it is common wisdom [18, 32] that the nonlinear gain term is not essential to explain the observed dynamical features, this is different in the two-mode case. This is demonstrated in Figure (10.6), which shows the bifurcation diagram in the $(K, \Delta\omega)$ plane for $\epsilon = 0.01$, which was calculated using AUTO [30]. We observe that although ϵ is small, the two-mode Hopf bifurcation line is severely affected, and does not extend to arbitrary small values of K anymore. This is in good agreement with experimental data, and we

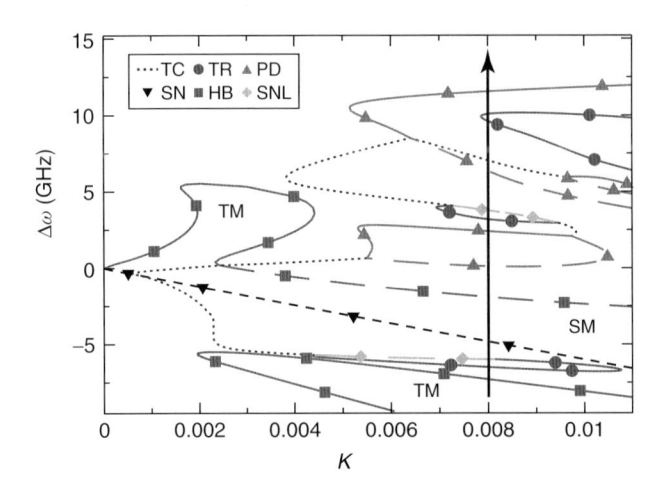

Figure 10.6 Bifurcation diagram for $\epsilon = 0.01$ obtained with the numerical continuation software AUTO [30]. SM and TM correspond to regions with single-mode and two-mode equilibria, respectively. The dotted lines (TC) denote transcritical transitions from single-mode to two-mode solutions. Solid lines and dashed lines mark two-mode and single-mode bifurcations, respectively. The bifurcation types shown are torus (TR), period doubling (PD), saddle node (SN), Hopf (HB), and saddle node of limit cycle (SNL) bifurcation (after [27]).

therefore conclude that the nonlinear gain terms are an important feature of the model equations (Eqs. (10.11)–(10.14)) [27, 31] at least for small injection K.

Using AUTO, it is possible to also study the bifurcations of limit cycles. In Figure 10.6, we show a number of bifurcation lines for limit cycles. We thereby restrict ourselves to the bifurcations that involve at least one stable object, which becomes unstable. We observe that bubbles of period-doubling bifurcations occur which are in a similar way known in the single-mode context [18]. The interesting feature is the transition from the single-mode to two-mode states, which occurs along the transcritical dotted lines.

Let us now study the dynamics at a fixed injection strength $K = 0.008$ and varying values of $\Delta\omega$ along the arrow in Figure 10.6. In Figure 10.7, we show the calculated power spectrum of the total light output, which has been experimentally confirmed with astonishing accuracy in [27]. The many interesting features in the power spectrum of Figure 10.7 can be explained by considering the time traces at different points along the arrow in Figure 10.6. In Figure 10.8, the time trace in the region of the two- mode locked state at $\Delta\omega = -8$ GHz is shown. Evidently both modes are constant and nonzero and the power spectrum in Figure 10.7 shows no features at this position below the first Hopf bifurcation denoted by HB. As we increase the detuning to $\Delta\omega = -7$ GHz, we pass the two-mode Hopf bifurcation and observe the oscillating dynamics of a stable limit cycle in the time trace shown

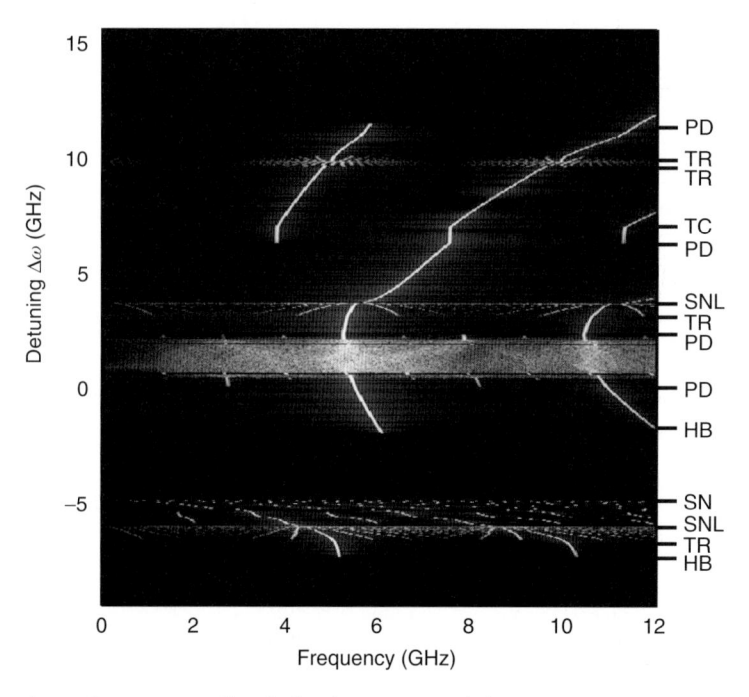

Figure 10.7 Numerically calculated power spectral density for the total field at $K = 0.008$ and varying $\Delta\omega$ (after [27]).

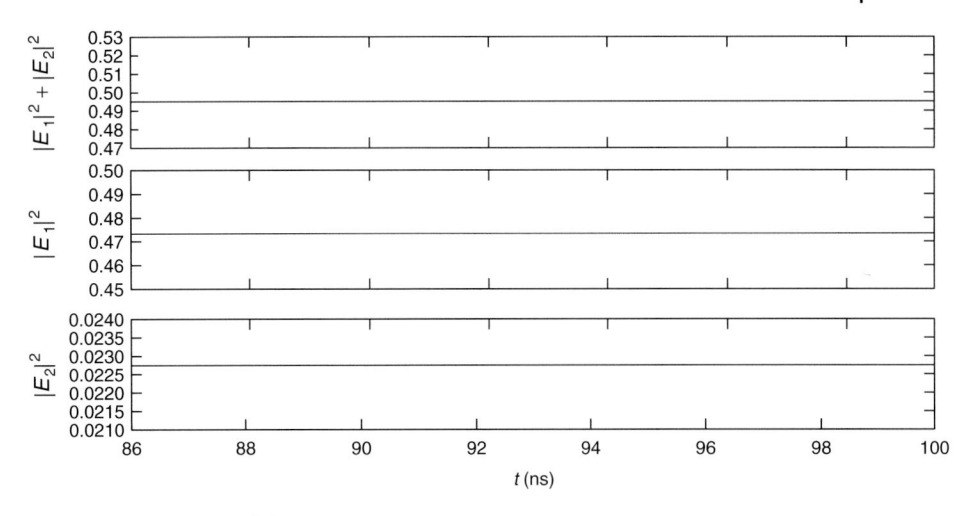

Figure 10.8 Time trace of the system (Eqs. (10.11)–(10.13)) for $K = 0.008$ and $\Delta\omega = -8$ GHz.

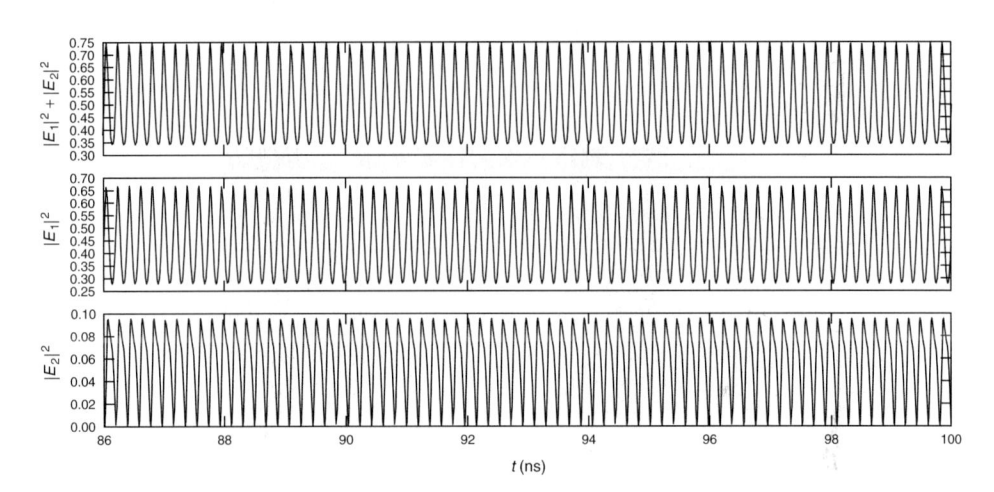

Figure 10.9 As shown in Figure 10.8 except for $\Delta\omega = -7$ GHz.

in Figure 10.9. Consequently, in the power spectrum shown in Figure 10.7, a new peak close to the relaxation frequency $\omega_{RO} \approx 5.5$ GHz appears, which is also reflected at twice this value due to the appearance of second-order harmonics.

As we increase the detuning to $\Delta\omega = -6.2$ beyond the two mode torus bifurcation, a low frequency (<1 GHz) appears in the frequency spectrum and gives rise to a seemingly quasi periodic dynamics as shown in Figure 10.10. This low frequency also shows up as satellite peaks around the still present relaxation frequency component. This frequency then tends toward zero as we further increase $\Delta\omega$ and gives rise to the typical starlike feature in the spectrum as we approach the

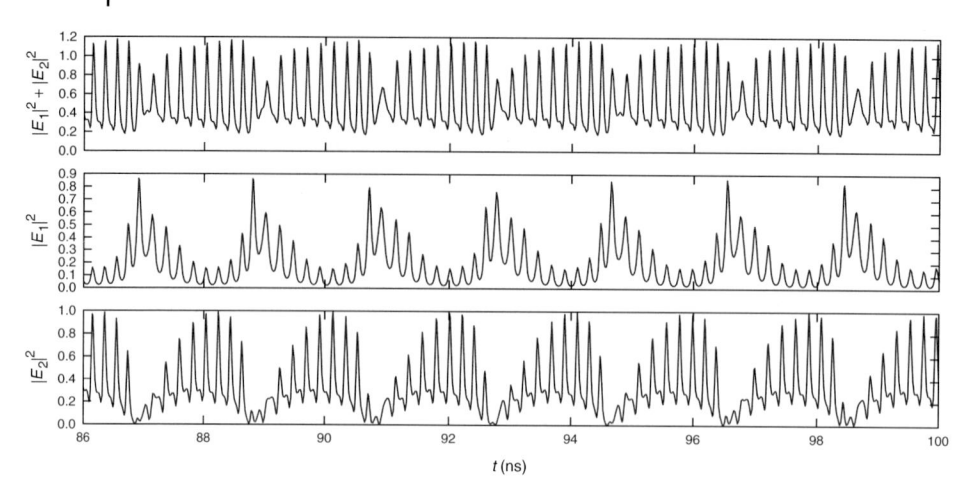

Figure 10.10 As shown in Figure 10.8 except for $\Delta\omega = -6.2$ GHz.

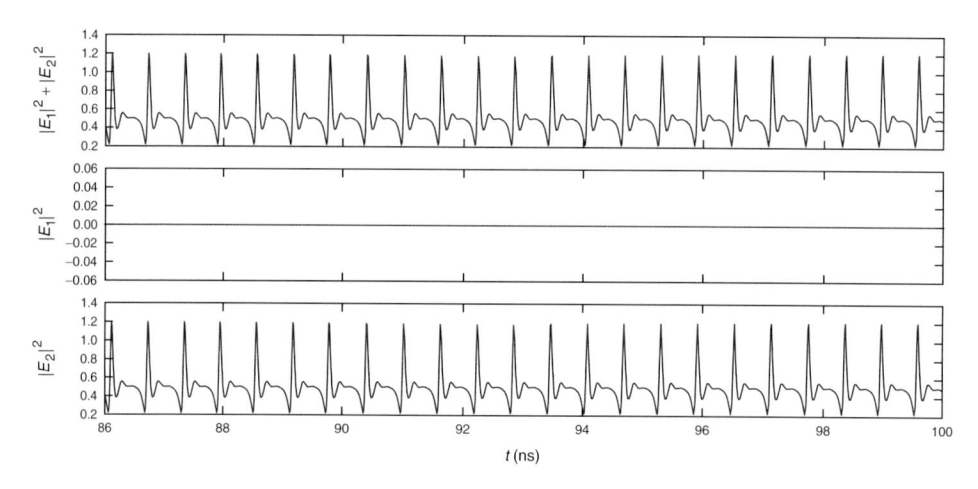

Figure 10.11 As shown in Figure 10.8 except for $\Delta\omega = -5$ GHz.

saddle-node of the limit cycle (SNL) bifurcation line from below. Beyond the SNL line, the uninjected mode is switched of, but the dynamics in the injected mode remains complicated, as shown, for example, in Figure 10.11.

Further increase in $\Delta\omega$ finally leads to the single-mode locked state (time trace not shown) via the well-known single-mode saddle-node (SN) bifurcation. The locked state persists until the single-mode Hopf bifurcation (HB) occurs. At this point a stable single-mode limit cycle is born with a dominating frequency close to the relaxation frequency. A time trace of a limit cycle in the region is shown in Figure 10.12. Further increase in the detuning frequency leads to the well-known period-doubling route to chaos in the single mode. An example for the complicated

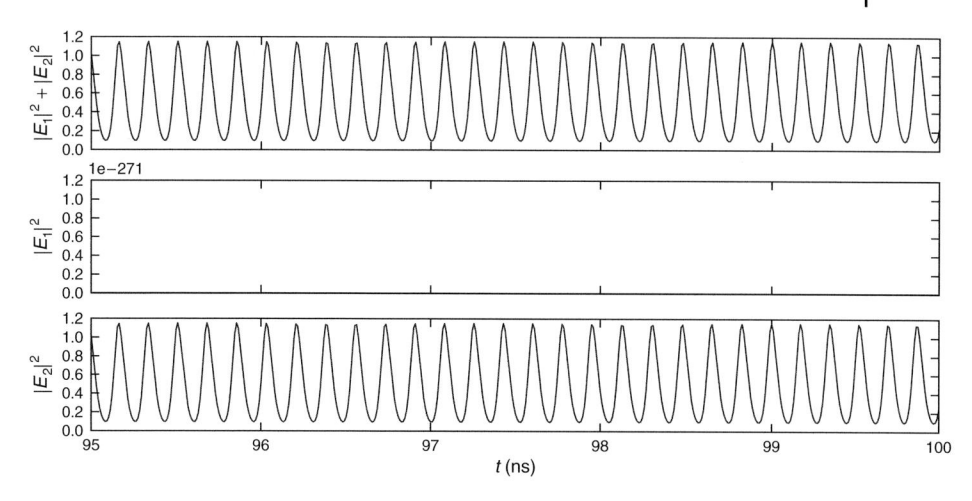

Figure 10.12 As shown in Figure 10.8 except for $\Delta\omega = -0.5$ GHz.

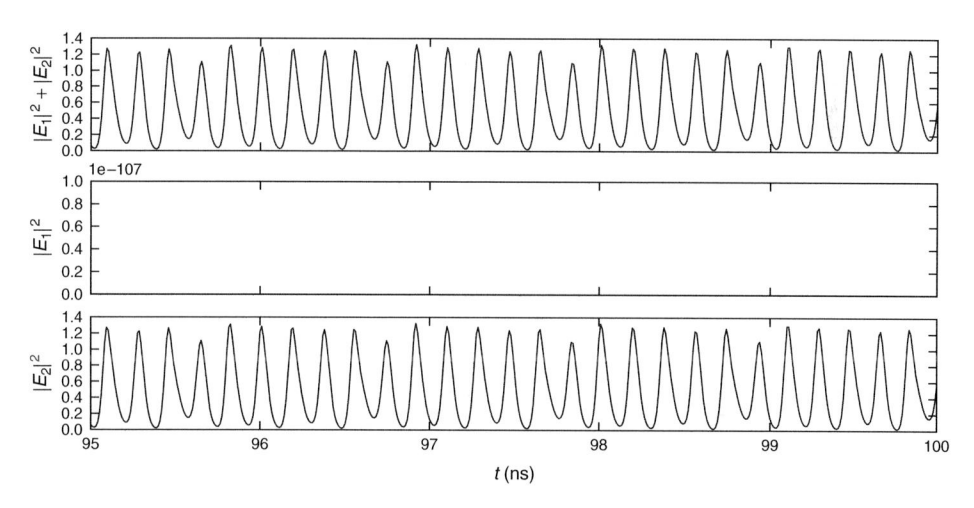

Figure 10.13 As shown in Figure 10.8 except for $\Delta\omega = 0.5$ GHz.

single-mode dynamics in this regime is shown in Figure 10.13. The higher order bifurcations inside the period-doubling bubble around slightly positive detuning in Figure 10.6 are not resolved. However, we observe that within the chaotic regime, a transition from single-mode to two-mode dynamics occurs. An example for chaotic two-mode behavior is shown in Figure 10.14.

With further increase of $\Delta\omega$, we then leave the period-doubling bubble via an inverse period-doubling bifurcation (PD) and are left with a two-mode limit cycle as shown in Figure 10.15. As before, this limit cycle undergoes a torus bifurcation that introduces a low frequency in the power spectrum. The time trace after the torus

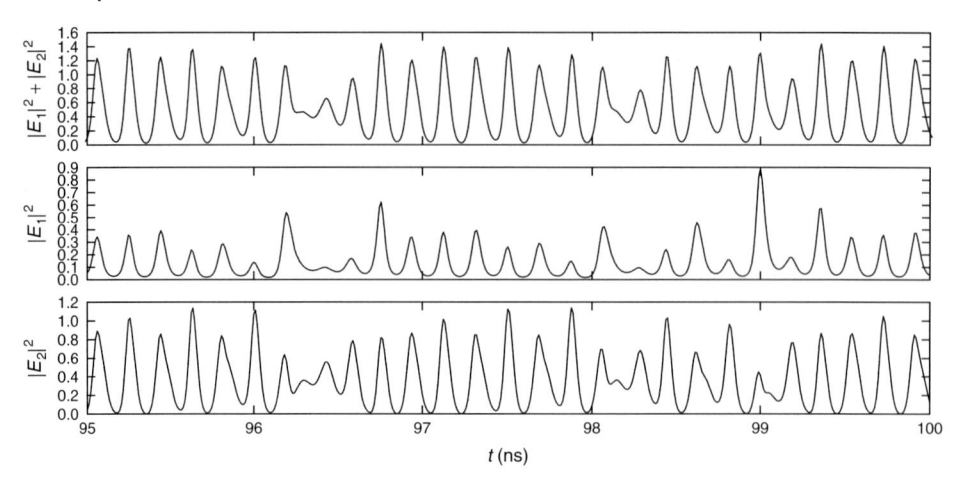

Figure 10.14 As shown in Figure 10.8 except for $\Delta\omega = 2.0$ GHz.

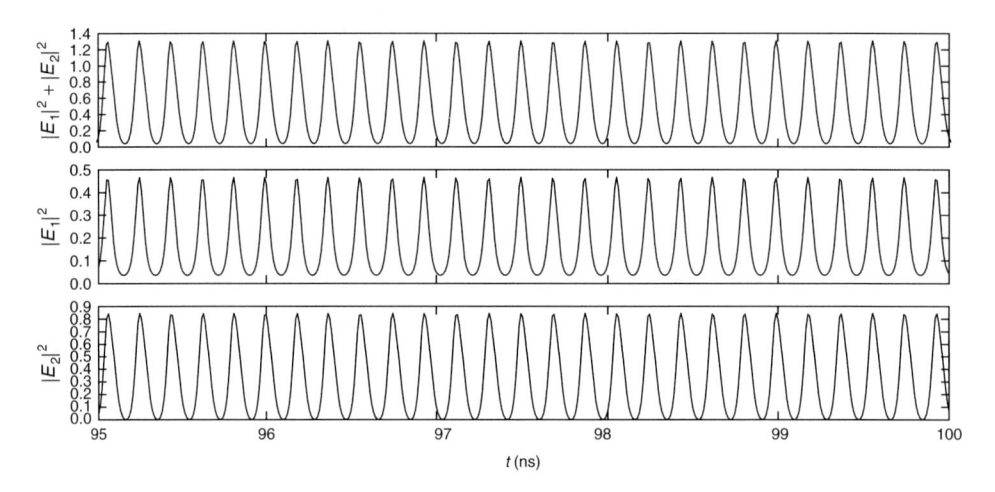

Figure 10.15 As shown in Figure 10.8 except for $\Delta\omega = 2.7$ GHz.

bifurcation is shown in Figure 10.16. Further increase in the detuning frequency leads us back to a single-mode limit cycle as shown in Figure 10.17 via a single-mode saddle-node of limit cycle (SNL) bifurcation. As before, this scenario gives rise to the characteristic starlike features in the power spectrum of Figure 10.7.

Increasing $\Delta\omega$ further leads into a new bubble bounded by period-doubling bifurcation lines around $\Delta\omega \approx 6.2$ GHz. In contrast to the previous period-doubling bubble, there is now, however, no chaotic dynamics evident along the path at which we transverse the bubble. Instead the transition from a single-mode (Figure 10.18) to a two-mode (Figure 10.19) period two limit cycle is mediated via a transcritical

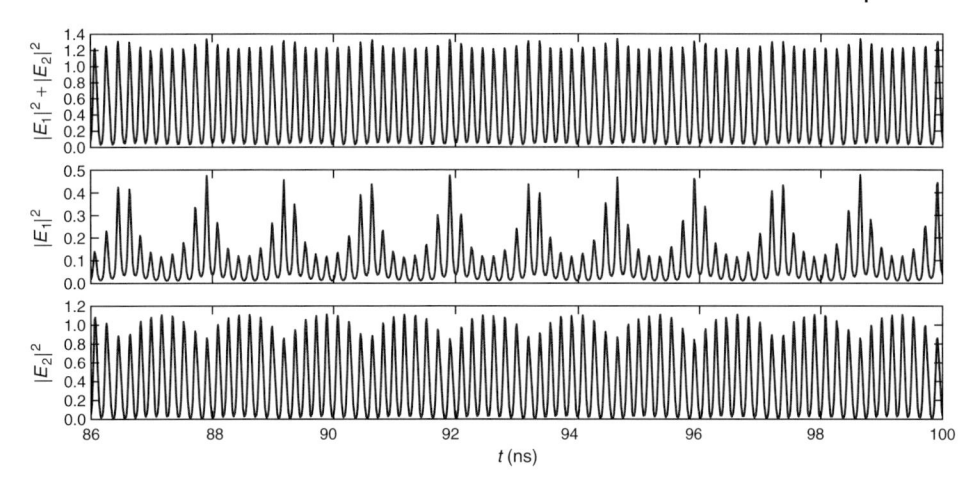

Figure 10.16 As shown in Figure 10.8 except for $\Delta\omega = 3.5$ GHz.

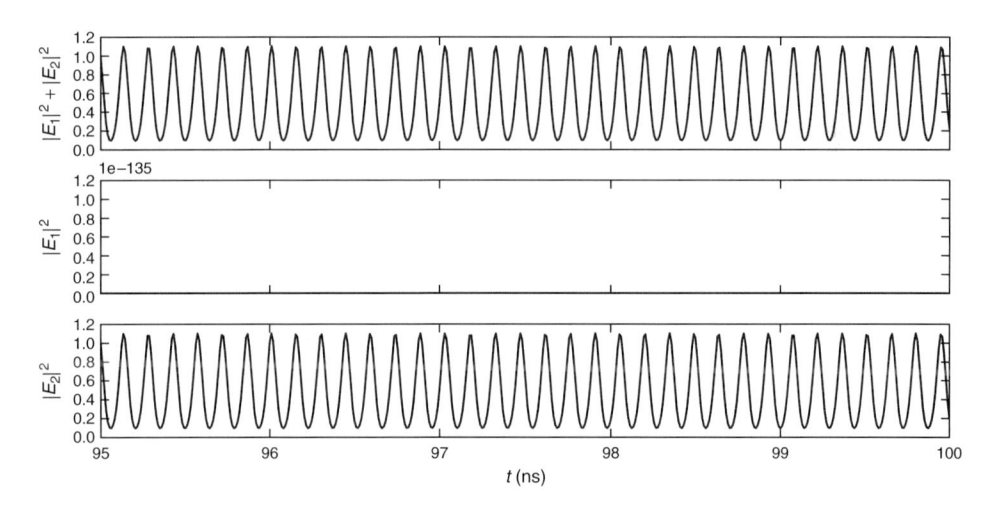

Figure 10.17 As shown in Figure 10.8 except for $\Delta\omega = 5$ GHz.

bifurcation of limit cycles (TC). With further increase in $\Delta\omega$, we briefly cut through a torus region (Figure 10.20) with a particular low additional frequency. A further inverse PD bifurcation finally leads to a simple two-mode limit cycle corresponding to a two-mode wave mixing state shown in Figure 10.21.

From the power spectrum shown in Figure 10.7 and the time traces shown in Figures 10.8–10.21, we conclude that although we have varied only one parameter ($\Delta\omega$), we have observed a large number of different bifurcation scenarios. It has been shown in [27] that these scenarios can be confirmed experimentally with high accuracy.

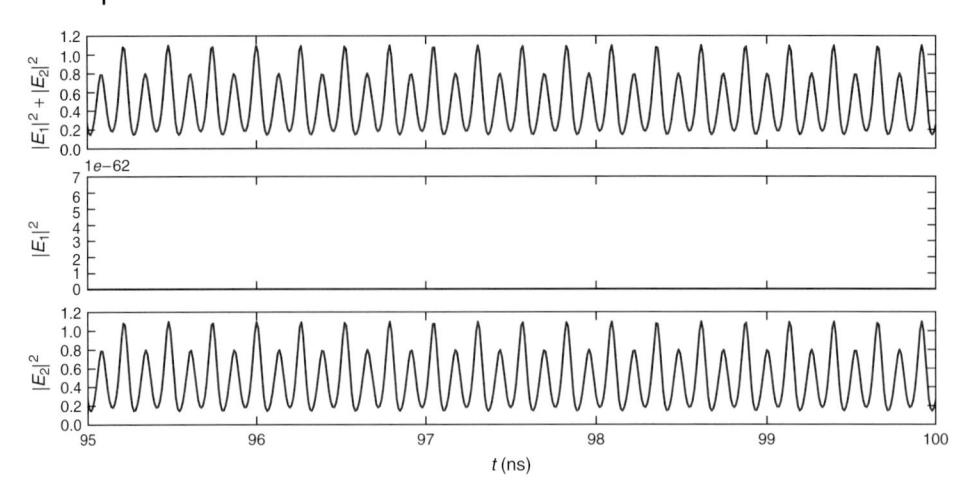

Figure 10.18 As shown in Figure 10.8 except for $\Delta\omega = 6.6$ GHz.

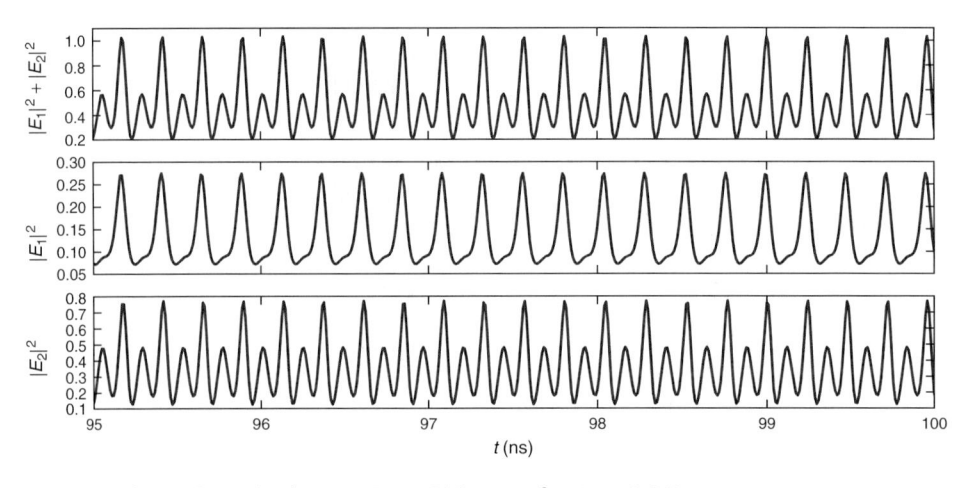

Figure 10.19 As shown in Figure 10.8 except for $\Delta\omega = 8$ GHz.

10.5
Conclusions

In summary, we have reviewed the design principles for Fabry–Perot laser with arbitrarily modified mode spectrum and we have examined basic dynamical features of the injected two-mode laser, which is theoretically and experimentally fairly well understood.

We have argued that it may be possible to design complex optical networks from an array of multimode lasers by selecting the lasing modes differently at each laser. These sophisticated optical connection topologies are, in particular, interesting in connection with all-optical signal processing but would also provide a test bed for complex network theory itself.

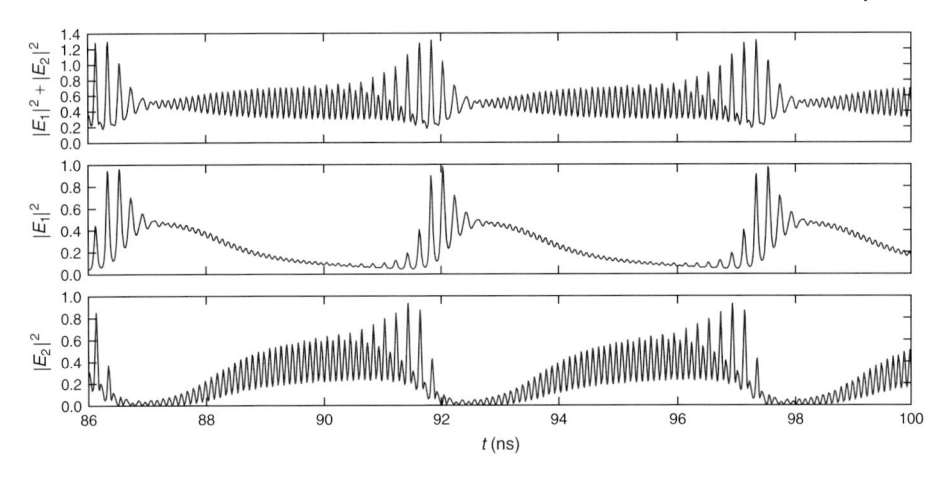

Figure 10.20 As shown in Figure 10.8 except for $\Delta\omega = 9.8$ GHz.

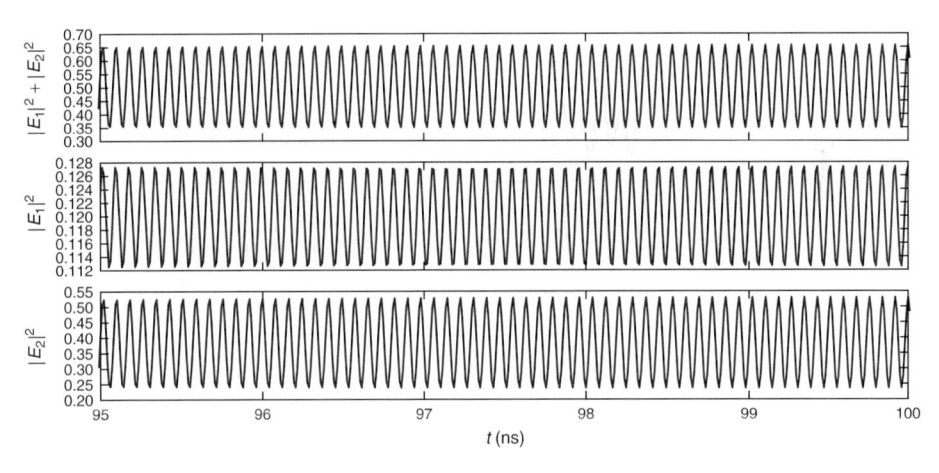

Figure 10.21 As shown in Figure 10.8 except for $\Delta\omega = 12$ GHz.

Acknowledgments

The author thanks S. O'Brien and S. Osborne for collaboration and many fruitful discussions on the topics reviewed in this chapter. This work was funded by Science Foundation Ireland under Grant Number 09/SIRG/I1615.

References

1. Strogatz, S.H. (1994) *Nonlinear Dynamics and Chaos*, Perseus Books.
2. Haken, H. (1975) Analogy between higher instabilities in fluids and lasers. *Phys. Lett. A*, **53**, 77.
3. Nishizawa, J. and Ishida, K. (1975) Injection-induced modulation of laser light by the interaction of laser diodes. *IEEE J. Quantum Electron.*, **11**, 515.

4. Lang, R. (1982) Injection locking properties of a semiconductor laser. *IEEE J. Quantum Electron.*, **18** (6), 976.
5. Möhrle, M., Sartorius, B., Bornholdt, C., Bauer, S., Brox, O., Sigmund, A., Steingrüber, R., Radziunas, H., and Wünsche, H. (2001) Detuned grating multisection-RW-DFB lasers for high-speed optical signal processing. *IEEE J. Sel. Top. Quan. Electron.*, **7**, 217.
6. Goulding, D., Hegarty, S.P., Rasskazov, O., Melnik, S., Hartnett, M., Greene, G., McInerney, J.G., Rachinskii, D., and Huyet, G. (2007) Excitability in a quantum dot semiconductor laser with optical injection. *Phys. Rev. Lett.*, **98**, 153903.
7. Plaza, F., Velarde, M.G., Arecchi, F.T., Boccaletti, S., Ciofini, M., and Meucci, R. (1997) Excitability following an avalanche-collapse process. *Europhys. Lett.*, **38**, 85.
8. Gavrielides, A., Kovanis, V., Varangis, P.M., Erneux, T., and Lythe, G. (1997) Coexisting periodic attractors in injection-locked diode lasers. *Quantum Semiclassical Opt. J. Eur. Opt. Soc. Part B*, **9**, 785.
9. Hwang, S.K. and Liu, J.M. (1999) Attractors and basins of the locking-unlocking bistability in a semiconductor laser subject to strong optical injection. *Opt. Commun.*, **169**, 167.
10. Pan, Z.G., Jiang, S., Dagenais, M., Morgan, R.A., Kojima, K., Asom, M.T., Leibenguth, R.E., Guth, G.D., and Focht, M.W. (1993) Optical injection induced polarization bistability in vertical-cavity surface-emitting lasers. *Appl. Phys. Lett.*, **63**, 2999.
11. Gatare, I., Sciamanna, M., Buesa, J., Thienpont, H., and Panajotov, K. (2006) Nonlinear dynamics accompanying polarization switching in vertical-cavity surface-emitting lasers with orthogonal optical injection. *Appl. Phys. Lett.*, **88**, 101106.
12. Pérez, T., Scirè, A., Van der Sande, G., Colet, P., and Mirasso, C.R. (2007) Bistability and all-optical switching in semiconductor ring lasers. *Opt. Express*, **15**, 12941.
13. Fischer, I., Liu, Y., and Davis, P. (2000) Synchronization of chaotic semiconductor laser dynamics on subnanosecond time scales and its potential for chaos communication. *Phys. Rev. A*, **62**, 11801.
14. Roy, R. and Thornburg, K.S. Jr. (1994) Experimental synchronization of chaotic lasers. *Phys. Rev. Lett.*, **72**, 2009.
15. Lang, R. and Kobayashi, K. (1980) External optical feedback effects on semiconductor injection laser properties. *IEEE J. Quantum Electron.*, **16** (3), 347.
16. Erzgräber, H., Krauskopf, B., and Lenstra, D. (2006) Compound laser modes of mutually delay-coupled lasers. *SIAM J. Appl. Dyn. Syst.*, **5**, 30.
17. Schikora, S., Hövel, P., Wünsche, H.J., Schöll, E., and Henneberger, F. (2006) All-Optical Noninvasive Control of Unstable Steady States in a Semiconductor Laser. *Phys. Rev. Lett.*, **97**, 213902.
18. Wieczorek, S., Krauskopf, B., Simpson, T.B., and Lenstra, D. (2005) The dynamical complexity of optically injected semiconductor lasers. *Phys. Rep.*, **416**, 1.
19. Bornholdt, S., Schuster, H.G., and Wiley, J. (2003) *Handbook of Graphs and Networks*, Wiley-VCH Verlag GmbH. Online Library.
20. Boccaletti, S., Latora, V., Moreno, Y., Chavez, M., and Hwang, D.U. (2006) Complex networks : structure and dynamics. *Phys. Rep.*, **424**, 175.
21. Caldarelli, G. (2007) *Scale-free Networks: Complex Webs in Nature and Technology*, Oxford University Press, USA.
22. Yariv, A. and Yeh, P. (2006) *Photonics: Optical Electronics in Modern Communications (The Oxford Series in Electrical and Computer Engineering)*, Oxford University Press, Inc., New York, NY.
23. O'Brien, S. and O'Reilly, E.P. (2005) Theory of improved spectral purity in index patterned Fabry-Pérot lasers. *Appl. Phys. Lett.*, **86**, 201101.
24. O'Brien, S., Osborne, S., Buckley, K., Fehse, R., Amann, A., O'Reilly, E.P., Barry, L.P., Anandarajah, P., Patchell, J., and O'Gorman, J. (2006) Inverse scattering approach to multiwavelength Fabry-Pérot laser design. *Phys. Rev. A*, **74**, 063814.

25. Bitauld, D., Osborne, S., and O'Brien, S. (2010) Passive harmonic mode locking by mode selection in Fabry–Perot diode lasers with patterned effective index. *Opt. Lett.*, **35**, 2200.
26. O'Brien, S., Amann, A., Rondinelli, J.M., Fehse, R., Osborne, S., and O'Reilly, E.P. (2006) Spectral manipulation in Fabry-Pérot lasers: Perturbative inverse scattering approach. *J. Opt. Soc. Am. B*, **23**, 1046.
27. Osborne, S., Amann, A., Buckley, K., Ryan, G., Hegarty, S.P., Huyet, G., and O'Brien, S. (2009) Antiphase dynamics in an optically injected two-color laser diode. *Phys. Rev. A*, **79**, 023834.
28. Gavrielides, A., Kovanis, V., and Erneux, T. (1997) Analytical stability boundaries for a semiconductor laser subject to optical injection. *Opt. Commun.*, **136**, 253.
29. Erneux, T. (2009) *Optically Injected Two-mode Laser*, Private communication.
30. Doedel, E.J., Champneys, A.R., Fairgrieve, T., Kuznetsov, Y., Oldeman, B., Pfaffenroth, R., Sandstede, B., Wang, X., and Zhang, C. (2007) AUTO-07P: continuation and bifurcation software for ordinary differential equations. Tech. Rep. Concordia University Montreal, Canada.
31. Osborne, S., Buckley, K., Amann, A., and O'Brien, S. (2009) All-optical memory based on the injection locking of a two-color laser diode. *Opt. Express*, **17**, 6293.
32. Yamada, M. (1983) Transverse and longitudinal mode control in semiconductor injection lasers. *IEEE J. Quantum Electron.*, **19**, 1365.

Part III
Synchronization and Cryptography

Nonlinear Laser Dynamics: From Quantum Dots to Cryptography, First Edition. Edited by Kathy Lüdge.
© 2012 Wiley-VCH Verlag GmbH & Co. KGaA. Published 2012 by Wiley-VCH Verlag GmbH & Co. KGaA.

11
Noise Synchronization and Stochastic Bifurcations in Lasers
Sebastian M. Wieczorek

11.1
Introduction

Synchronization of nonlinear oscillators to irregular external signals is an interesting problem of importance in physics, biology, applied science, and engineering [1–10]. The key difference to synchronization to a periodic external signal is the lack of a simple functional relationship between the input signal and the synchronized output signal, making the phenomenon much less evident [1, 11]. Rather, synchronization is detected when two or more identical uncoupled oscillators driven by the same external signal, but starting at different initial states have identical long-term responses. This is equivalent to obtaining reproducible long-term response from a single oscillator driven repeatedly by the same external signal, each time starting at a different initial state. Hence, synchronization to irregular external signals is also known as *reliability* [3] or *consistency* [7], and represents the ability to encode irregular signals in a reproducible manner.

Recent studies have shown that nonlinear oscillators can exhibit interesting responses to stochastic external signals. Typically, a small amount of external noise causes synchronization [[1], Chapter 7], [2, 4]. However, as the strength of external noise increases, there can be a loss of synchrony in oscillators with amplitude–phase coupling (also known as *shear, nonisochronicity,* or *amplitude-dependent frequency*) [[1], Chapter 7], [8–10, 12–14]. Mathematically, loss of noise synchrony, consistency, or reliability is a manifestation of a *stochastic bifurcation of a random attractor*.

This chapter gives a definition of noise synchronization in terms of random pullback attractors and studies synchronization–desynchronization transitions as purely noise-induced stochastic bifurcations. This is in contrast to the effects described in Chapter 2, where noise is used to control or regulate the dynamics that is already present in the noise-free system. We focus on a single-mode class-B laser model and the Landau–Stuart model (Hopf normal form with shear [1]). In Section 11.6, numerical analysis of the locus of the stochastic bifurcation in a three-dimensional parameter space of the "distance" from Hopf bifurcation, amount of amplitude–phase coupling, and external signal strength reveals a simple power law for the Landau–Stuart model, but quite different behavior for the laser model. In Section 11.7, the analysis of the shear-induced stretch-and-fold action that creates horseshoes gives an intuitive explanation for the observed loss of synchrony, and for the deviation from the simple power law in the laser model. Experimentally, stochastic external forcing can be realized by optically injecting noisy light into a (semiconductor) laser as described in Section 11.4. While bifurcations of random pullback attractors and the associated synchronization–desynchronization transitions have been studied theoretically, single-mode semiconductor lasers emerge as interesting candidates for the experimental testing of these phenomena.

11.2
Class-B Laser Model and Landau–Stuart Model

A class-B single-mode laser [15] without noise can be modeled by the rate equations [16]:

$$\frac{dE}{dt} = i\Delta E + g\gamma(1 - i\alpha)NE, \tag{11.1}$$

$$\frac{dN}{dt} = J - N - (1 + gN)|E|^2, \tag{11.2}$$

which define a three-dimensional dynamical system with a normalized electric-field amplitude, $E \in \mathbb{C}$, and normalized deviation from the threshold population inversion, $N \in \mathbb{R}$, such that $N = -1$ corresponds to zero population inversion. Parameter J is the normalized deviation from the threshold pump rate such that $J = -1$ corresponds to zero pump rate. The linewidth enhancement factor, α, quantifies the amount of amplitude–phase coupling, Δ is the normalized detuning (difference) between some conveniently chosen reference frequency and the natural laser frequency, $\gamma = 500$ is the normalized decay rate, and $g = 2.765$ is the normalized gain coefficient [16].

System (Eqs. (11.1) and (11.2)) is $\mathbb{S}^1$-equivariant, meaning that it has rotational symmetry corresponding to a phase shift $E \to Ee^{i\phi}$, where $0 < \phi \leq 2\pi$. For $J \in \mathbb{R}$, there is an equilibrium at $(E, N) = (0, J)$, which represents the "off" state of the laser. This equilibrium is globally stable if $J < 0$ and unstable if $J > 0$. At $J = 0$, there is a Hopf ($\Delta \neq 0$) or pitchfork ($\Delta = 0$) bifurcation defining the laser threshold. Moreover, if $J > 0$, the system has a stable group orbit in the form of periodic orbit for $\Delta \neq 0$ or a circle of infinitely many nonhyperbolic (neutrally stable) equilibria for $\Delta = 0$. In this chapter, we refer to this circular attractor as the *limit cycle*. The limit cycle is given by $(|E|^2, N) = (J, 0)$ and represents the "on" state of the laser. Owing to the $\mathbb{S}^1$-symmetry, the Floquet exponents of the limit cycle can be calculated analytically as eigenvalues of one of the nonhyperbolic equilibria for $\Delta = 0$. Specifically, if

$$0 < J < \frac{4\gamma\left(1 - \sqrt{1 - 1/(2\gamma)}\right) - 1}{g} \approx 9 \times 10^{-5},$$

the overdamped limit cycle has three real Floquet exponents

$$\mu_1 = 0, \quad \mu_{2,3} = -a \pm b, \tag{11.3}$$

and if

$$\frac{4\gamma\left(1-\sqrt{1-1/(2\gamma)}\right)-1}{g} < J < \frac{4\gamma\left(1+\sqrt{1-1/(2\gamma)}\right)-1}{g} \approx 1446,$$

the underdamped limit cycle has one real and two complex-conjugate Floquet exponents

$$\mu_1 = 0, \quad \mu_{2,3} = -a \pm ib, \tag{11.4}$$

where

$$a = \frac{1}{2}(1+gJ) > 0 \quad \text{and} \quad b = \sqrt{|a^2 - 2g\gamma J|} > 0.$$

In the laser literature, the decaying oscillations found for pump rate in the realistic range $J \in (9 \times 10^{-5}, 20)$ are called *relaxation oscillations*. (This should not be confused with a different phenomenon of self-sustained, slow–fast oscillations.) Finally, even though the laser model (Eqs. (11.1) and (11.2)) is three dimensional, it cannot admit chaotic solutions because of the restrictions imposed by the rotational symmetry.

Using center manifold theory [17], the dynamics of Eqs. (11.1) and (11.2) near the Hopf bifurcation can be approximated by the two-dimensional invariant center manifold

$$W^c = \{(E, N) \in \mathbb{R}^3 : N = J - |E|^2\},$$

on which Eqs. (11.1) and (11.2) reduce to

$$\frac{1}{g\gamma}\frac{dE}{dt} = \left[J + i\left(\frac{\Delta}{g\gamma} - \alpha(J - |E|^2)\right)\right] E - E|E|^2.$$

After rescaling time and detuning,

$$\tilde{t} = tg\gamma \quad \text{and} \quad \tilde{\Delta} = \Delta/(g\gamma),$$

we obtain the Landau–Stuart model

$$\frac{dE}{d\tilde{t}} = \left[J + i\left(\tilde{\Delta} - \alpha(J - |E|^2)\right)\right] E - E|E|^2, \tag{11.5}$$

which is identical to the Hopf normal form [17] except for the higher-order term, $i\alpha(J - |E|^2)E$, representing amplitude–phase coupling. Since this term does not affect stability properties of Eq. (11.5), it does not appear in the Hopf normal form. However, in the presence of an external forcing, $f_{\text{ext}}(t)$, this term has to be included because it gives rise to qualitatively different dynamics for different values of α. If $J > 0$, the Landau–Stuart model has a stable limit cycle with two Floquet exponents

$$\mu_1 = 0 \quad \text{and} \quad \mu_2 = -2J. \tag{11.6}$$

11.3
The Linewidth Enhancement Factor and Shear

The linewidth enhancement factor, α, quantifying the amount of amplitude–phase coupling for the complex-valued electric field, E, is absolutely crucial to our analysis. Its physical origin is the dependence of the semiconductor refractive index, and hence the laser-cavity resonant frequency, on the population inversion [15, 18]. A change in the electric-field intensity, $\delta|E|^2$, induces a change, δN, in population inversion (Eq. (11.2)). The resulting change in the refractive index shifts the cavity resonant frequency. The ultimate result is a change of $-\alpha g \gamma \delta N$ in the *instantaneous frequency* of the electric field defined as $d(\arg(E))/dt$.

Mathematically, amplitude–phase coupling is best illustrated by an invariant set associated with each point, q, on the limit cycle. For a point $q(0)$ on a stable limit cycle in a n-dimensional system, this set is defined as

$$\{x(0) \in \mathbb{R}^n : x(t) \to q(t) \text{ as } t \to \infty\}, \tag{11.7}$$

and is called an *isochrone* [19]. In the laser model (Eqs. (11.1) and (11.2)) and Landau–Stuart model (Eq. (11.5)), isochrones are logarithmic spirals that satisfy

$$\arg(E) + \alpha \ln|E| = C, \quad \text{where} \quad C \in (0, 2\pi]. \tag{11.8}$$

To see this, define a phase

$$\Psi = \arg(E) + \alpha \ln|E|, \tag{11.9}$$

and check that $d\Psi/dt$ is constant and equal to Δ for Eqs. (11.1) and (11.2) and $\tilde{\Delta}$ for Eq. (11.5). This means that trajectories for different initial conditions with identical initial phase, $\Psi(0)$, will retain identical phase, $\Psi(t)$, for all time t. Since the limit cycle is stable, all such trajectories will converge to the limit cycle, where they have the same $|E(t)|$. Then, Eq. (11.9) implicates that all such trajectories have the same $\arg(E(t))$ and hence converge to just one special trajectory along the limit cycle as required by Eq. (11.7).

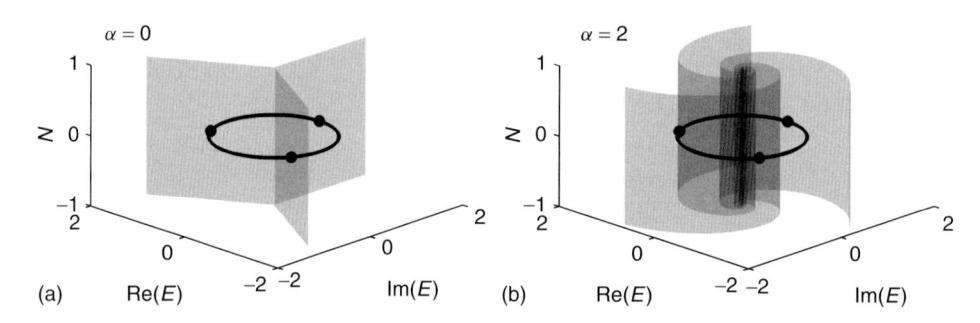

Figure 11.1 (Black) The limit cycle representing the "on" state of the laser for $J = 1$ and (gray) isochrones for three different points on the limit cycle as defined by Eq. (11.7) for (a) $\alpha = 0$ and (b) $\alpha = 2$.

Isochrones of three different points on the laser limit cycle are shown in Figure 11.1. Isochrone inclination to the direction normal to the limit cycle at $q(0)$ indicates the strength of phase space stretching along the limit cycle. If $\alpha = 0$, trajectories with different $|E| > 0$ rotate around the origin of the E-plane with the same angular frequency giving no isochrone inclination and hence no phase-space stretching (Figure 11.1a). However, if $|\alpha| > 0$, trajectories with larger $|E|$ rotate with higher angular frequency giving rise to isochrone inclination and phase-space stretching (Figure 11.1b). Henceforth, we refer to amplitude–phase coupling as *shear*.

11.4
Detection of Noise Synchronization

There are at least two approaches to detecting synchronization of a semiconductor laser to an irregular external signal. One approach involves a comparison of the responses of two or more identical and uncoupled lasers that are driven by the same external signal. The other approach involves a comparison of the responses of a single laser driven repeatedly by the same external signal [7]. Here, we consider responses of M uncoupled lasers with intrinsic spontaneous emission noise that are subjected to *common optical external forcing*, $f_{\text{ext}}(t)$, [16, 20]:

$$\frac{dE_j}{dt} = i\Delta E_j + g\gamma(1 - i\alpha)N_j E_j + f_{Ej}(t) + f_{\text{ext}}(t), \quad (11.10)$$

$$\frac{dN_j}{dt} = J - N_j - (1 + gN_j)|E_j|^2 + f_{Nj}(t), \quad (11.11)$$

$$j = 1, 2, \ldots, M.$$

The lasers are identical except for the intrinsic spontaneous emission noise that is represented by random Gaussian processes

$$f_{Ej}(t) = f_{Ej}^R(t) + if_{Ej}^I(t) \quad \text{and} \quad f_{Nj}(t),$$

that have zero mean and are delta correlated

$$\langle f_{Ej}(t)\rangle = \langle f_{Nj}(t)\rangle = 0,$$
$$\langle f_{Ej}^R(t)f_{Ej}^I(t)\rangle = 0,$$
$$\langle f_{Ei}^R(t)f_{Ej}^R(t')\rangle = \langle f_{Ei}^I(t)f_{Ej}^I(t')\rangle = D_E \delta_{ij}\delta(t - t'), \quad (11.12)$$
$$\langle f_{Ni}(t)f_{Nj}(t')\rangle = 2D_N \delta_{ij}\delta(t - t').$$

Here, δ_{ij} is the Kronecker delta and $\delta(t - t')$ is the Dirac delta function. In the calculations, we use $D_E = 0.05$ and $D_N = 3.5 \times 10^{-8}$ [16].

To measure the quality of synchronization, we introduce the *order parameter*, $I_M(t)$, and the *average order parameter*, $\langle I_M \rangle$, as

$$\langle I_M \rangle = \lim_{T \to \infty} \frac{1}{T} \int_0^T I_M(t)\, dt = \lim_{T \to \infty} \frac{1}{T} \int_0^T \left|\sum_{j=1}^M E_j(t)\right|^2 dt. \quad (11.13)$$

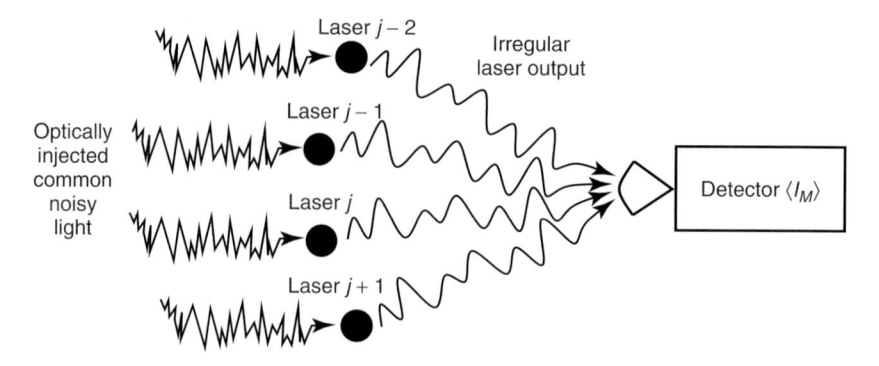

Figure 11.2 An experimental setup for detecting noise synchronization in lasers.

The physical meaning of $I_M(t)$ and $\langle I_M \rangle$ is illustrated in Figure 11.2. If M identical lasers are placed at an equal distance from a small (the order of a wavelength) spot and their light is focused onto this spot, then $I_M(t)$ and $\langle I_M \rangle$ are the instantaneous and average light intensity at the spot, respectively. A single laser oscillates with a random phase owing to spontaneous emission noise so that, for independent lasers, $\langle I_M \rangle$ is proportional to M times the average intensity of a single laser. This follows directly from Eq. (11.13) assuming lasers with identical amplitudes, $|E_j(t)|$, and uncorrelated random phases, $\arg(E_j(t))$. However, when the lasers oscillate in phase, one expects $\langle I_M \rangle$ to be equal M^2 times the average intensity of a single laser. This follows directly from Eq. (11.13) assuming lasers with identical amplitudes and phases. We speak about synchronization when $\langle I_M \rangle \approx M^2$, different degrees of partial synchronization when $M < \langle I_M \rangle < M^2$, and lack of synchronization when $\langle I_M \rangle \approx M$. Note that $\langle I_M \rangle > M^2$ indicates trivial synchronization, where the external forcing term, $f_{ext}(t)$ becomes "larger" than the oscillator terms on the right-hand side of Eq. (11.10). For comparability reasons, we now briefly review the case of a monochromatic forcing and then move on to the case of stochastic forcing.

Let us consider a *monochromatic external forcing*

$$f_{ext}(t) = Ke^{i\nu_{ext} t}$$

where $K \in \mathbb{R}$ is the forcing strength and ν_{ext} is the detuning (difference) between the reference frequency chosen for Δ in Eq. (11.1) and the forcing frequency. Such an external forcing breaks the $\mathbb{S}^1$-symmetry and can force each laser to fluctuate in the vicinity of the well-defined external forcing phase, $\nu_{ext} t$, as opposed to a random walk. This phenomenon was studied in [21] as a thermodynamic phase transition. Figure 11.3a shows $\langle I_M \rangle$ versus K, for an external forcing resonant with the laser,

$$\nu_{ext} = \Delta.$$

Because of the intrinsic spontaneous emission noise, the forcing amplitude has to reach a certain threshold before synchronization occurs. For $\alpha = 0$, a sharp onset of synchronization at $K \approx 10^{-3}$ is followed by a wide range of K with

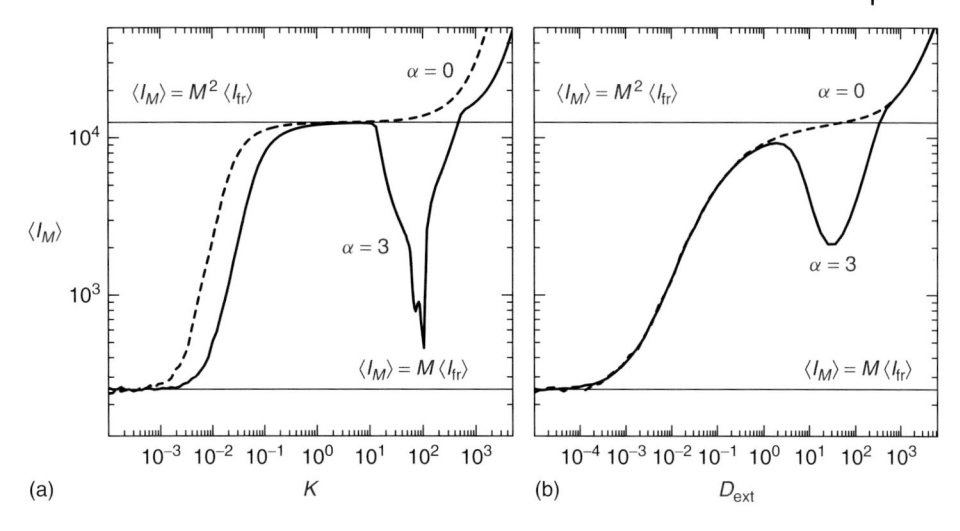

Figure 11.3 The average order parameter as defined by Eq. (11.13) for $M = 50$ uncoupled lasers with common (a) monochromatic and (b) white noise external forcing versus the forcing strength for (dashed) $\alpha = 0$ and (solid) $\alpha = 3$; $J = 5$ and $\nu_{ext} = \Delta$ in panel (a). (Adapted from [12] with permission.)

synchronous behavior, where $\langle I_M \rangle = M^2 \langle I_{fr} \rangle$. Here, $\langle I_{fr} \rangle$ is the average intensity of a single laser without forcing. At around $K = 10^2$, $\langle I_M \rangle$ starts increasing above $M^2 \langle I_{fr} \rangle$. While lasers still remain synchronized, this increase indicates that the external forcing is no longer "weak." Rather, it becomes strong enough to cause an increase in the average intensity of each individual laser. A very different scenario is observed for $\alpha = 3$. There, the onset of synchronization is followed by an almost complete loss of synchrony just before $\langle I_M \rangle$ increases above $M^2 \langle I_{fr} \rangle$. The loss of synchrony is caused by externally induced bifurcations and ensuing chaotic dynamics. These bifurcations have been studied in detail, both theoretically [15, 22–24] and experimentally [25], and are well understood.

The focus of this work is synchronization to *white noise external forcing* represented by the complex random process, that is, Gaussian, has zero mean, and is delta correlated

$$f_{ext}(t) = f_{ext}^R(t) + i f_{ext}^I(t),$$
$$\langle f_{ext}(t) \rangle = \langle f_{ext}^R(t) f_{ext}^I(t) \rangle = 0, \quad (11.14)$$
$$\langle f_{ext}^R(t) f_{ext}^R(t') \rangle = \langle f_{ext}^I(t) f_{ext}^I(t') \rangle = D_{ext} \delta(t - t').$$

White noise synchronization is demonstrated in Figure 11.3b, where we plot $\langle I_M \rangle$ versus D_{ext}. For $\alpha = 0$, a clear onset of synchronization at around $D_{ext} = 10^{-3}$ is followed by synchronous behavior at larger D_{ext}. In particular, there exists a range of D_{ext} where white noise external forcing is strong enough to synchronize phases of intrinsically noisy lasers, but weak enough so that each individual laser has small intensity fluctuations and its average intensity remains

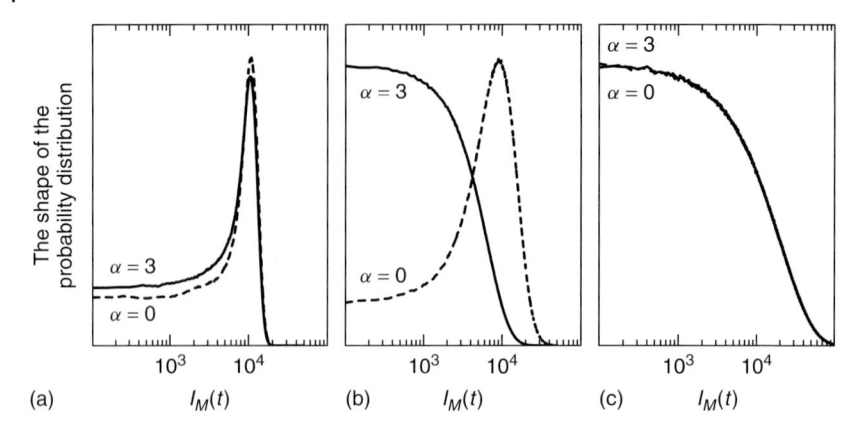

Figure 11.4 The shape of the probability distribution of $I_M(t)$ for white noise external forcing; (dashed) $\alpha = 0$ and (solid) $\alpha = 3$. From (a–c) $D_{ext} = 10^0, 10^1,$ and 10^3. (Adapted from [12] with permission.)

unchanged. In the probability distributions for $I_M(t)$ in Figure 11.4a and b, the distinct peak at $I_M(t) \approx M^2 \langle I_{fr} \rangle$ and a noticeable tail at smaller $I_M(t)$ indicate synchronization that is not perfect. Rather, synchronous behavior is occasionally interrupted with short intervals of asynchronous behavior owing to different intrinsic noise within each laser. For $D_{ext} > 10^2$ the external forcing is no longer "weak" and causes an increase in the intensity fluctuations and the average intensity of each individual laser. Although the lasers remain in synchrony, $\langle I_M \rangle$ increases above $M^2 \langle I_{fr} \rangle$ (Figure 11.3b) and exhibits large fluctuations (Figure 11.4c) as in the asynchronous case. A very different scenario is observed again for $\alpha = 3$. There, the onset of synchronization is followed by a significant loss of synchrony for $D_{ext} \in (2, 200)$. In this range of the forcing strength, one finds qualitatively different dynamics for $\alpha = 0$ and $\alpha = 3$ as revealed by different probability distributions in Figure 11.4b.

Interestingly, comparison between (a) and (b) in Figure 11.3 shows that some general aspects of synchronization to a monochromatic and white noise external forcing are strikingly similar. In both cases, there is a clear onset of synchronization followed by a significant loss of synchrony for sufficiently large α, and subsequent revival of synchronization for stronger external forcing. However, the dynamical mechanism responsible for the loss of synchrony in the case of white noise external forcing has not been fully explored.

11.5
Definition of Noise Synchronization

The previous section motivates further research to reveal the dynamical mechanism responsible for the loss of synchrony observed in Figure 11.3b. To facilitate the analysis, we define synchronization to irregular external forcing within the framework of random dynamical systems. Let us consider a n-dimensional, nonlinear, dissipative, autonomous dynamical system, referred to as *unforced system*

$$\frac{dx}{dt} = f(x, p) \tag{11.15}$$

where $x \in \mathbb{R}^n$ is the state vector and $p \in \mathbb{R}^k$ is the parameter vector that does not change in time. An *external forcing* is denoted with $f_{ext}(t)$, and the corresponding nonautonomous *forced system* reads

$$\frac{dx}{dt} = f(x, p) + f_{ext}(t). \tag{11.16}$$

Let $x(t, t_0, x_0)$ denote a *trajectory* or *solution* of Eq. (11.16) that passes through x_0 at some initial time t_0. In situations where explicitly displaying the initial condition is not important, we denote the trajectory simply as $x(t)$. For an infinitesimal displacement $\delta x(0)$ from $x(0, t_0, x_0)$, the *largest Lyapunov exponent* along $x(t, t_0, x_0)$ is given by

$$\lambda_{max} = \lim_{t \to \infty} \frac{1}{t} \ln \frac{|\delta x(t)|}{|\delta x(0)|}. \tag{11.17}$$

If the external forcing, $f_{ext}(t)$, is stochastic, Eq. (11.16) defines a *random dynamical system* where λ_{max} does not depend on the noise realization, f_{ext} [26]. Furthermore, we define

Definition 1. *An (self-sustained) oscillator is an unforced system (Eq. (11.15)) with a stable hyperbolic limit cycle.*

Definition 2. *An attractor for the forced system (Eq. (11.16)) with stochastic forcing $f_{ext}(t)$ is called a random sink (rs) if $\lambda_{max} < 0$, and a random strange attractor (rsa) if $\lambda_{max} > 0$.*

Definition 3. *A stochastic d-bifurcation is a qualitative change in the random attractor when λ_{max} crosses through zero [[26], Chapter 9].*

Definition 4. *An oscillator is synchronized to stochastic forcing $f_{ext}(t)$ on a bounded subset $D \subset \mathbb{R}^n$ if the corresponding forced system (Eq. (11.16)) has a random sink in the form of a unique attracting trajectory, $a(t, f_{ext})$, such that*

$$\lim_{t_0 \to -\infty} |x(t, t_0, x_0) - a(t, f_{ext})| \to 0,$$

for fixed $t > t_0$ and all $x_0 \in D$.

By Definition 1, an unforced oscillator has zero λ_{max} on an open set of parameters. In the presence of stochastic external forcing, λ_{max} becomes either positive or negative for typical parameter values [10, 26] and remains zero only at some special parameter values defining *stochastic d-bifurcations*. Synchronization in Definition 4 is closely related to and implies *generalized synchronization* [11, 27] or *weak synchronization* [28] – a phenomenon that requires a time-independent functional relationship between the measured properties of the forcing and the oscillator [29]. Following [26, Chapter 9] and [30], we used in Definition 4 the notion of *pullback convergence* where the asymptotic behavior is studied for $t_0 \to -\infty$ and fixed t. (For a study of different notions of convergence in random dynamical systems, we refer the reader to [31].) While $\lambda_{\max}$ does not depend on the noise realization, f_{ext}, random sinks and random strange attractors do depend on f_{ext}. Hence the f_{ext} dependence in $a(t, f_{\text{ext}})$ in Definition 4. Since $\lambda_{\max} < 0$ does not imply a unique attracting trajectory, it is not sufficient to show synchronization as defined in Definition 4. In general, there can be a number of coexisting (locally) attracting trajectories that belong to a *global pullback attractor* [30]. In such cases, one can choose D to lie in the basin of attraction of one of the locally attracting trajectory and speak of synchronization on D.

11.6
Synchronization Transitions via Stochastic d-Bifurcation

To facilitate the analysis we make use of Definitions 1–3 and, henceforth, consider noise synchronization in the laser model with white noise external forcing

$$\frac{dE}{dt} = i\Delta\, E + g\gamma(1 - i\alpha)NE + f_{\text{ext}}(t), \tag{11.18}$$

$$\frac{dN}{dt} = J - N - (1 + gN)|E|^2, \tag{11.19}$$

but without the intrinsic spontaneous emission noise. Now, owing to the absence of intrinsic noise, Definition 4 is equivalent to the synchronization detection scheme chosen in Section 11.4. More specifically, the evolution of M trajectories starting at different initial conditions for a single laser with external forcing is the same as the evolution of an ensemble of M identical uncoupled lasers with the same forcing, where each laser starts at a different initial condition. A random sink for Eqs. (11.18) and (11.19) in the form of a unique attracting trajectory, $a(t, f_{\text{ext}})$, makes trajectories for different initial conditions converge to each other. In an ensemble of M identical uncoupled lasers with common forcing, this means that $\langle I_M \rangle \to M^2 \langle I_{\text{fr}} \rangle$ in time so that synchronization is detected. A random strange attractor for Eqs. (11.18) and (11.19), where nearby trajectories separate exponentially fast because $\lambda_{\max} > 0$, implies $\langle I_M \rangle < M^2 \langle I_{\text{fr}} \rangle$ so that incomplete synchronization or lack of synchronization is detected.

Figure 11.5 shows effects of white noise external forcing on the sign of the otherwise zero $\lambda_{\max}$ in an unforced laser. For $\alpha = 0$, external forcing always shifts $\lambda_{\max}$ to negative values meaning that the system has a random sink for $D_{\text{ext}} > 0$ and $J > 0$ (Figure 11.5a). Additionally, this random sink is a unique trajectory, $a(t, f_{\text{ext}})$, meaning that the laser is synchronized to white noise external forcing. However, for $\alpha = 3$, there are two curves of stochastic d-bifurcation where $\lambda_{\max}$ crosses through

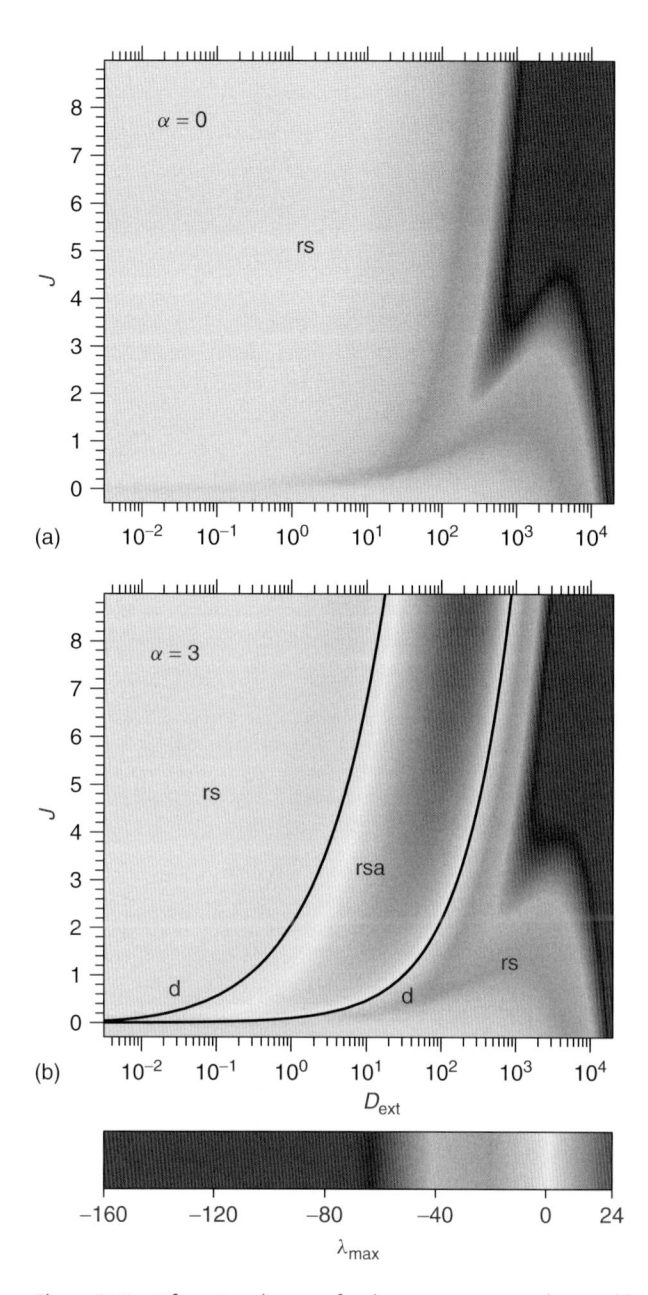

Figure 11.5 Bifurcation diagram for the white noise forced laser model (Eqs. (11.18) and (11.19)) in the (D_{ext}, J)-plane for (a) $\alpha = 0$ and (b) $\alpha = 3$. "rs" stands for random sink, "rsa" stands for random strange attractor, the two black curves "d" denote stochastic d-bifurcation, and color coding is for λ_{max}. An unforced laser has a Hopf bifurcation at $(D_{ext} = 0, J = 0)$. (Adapted from [12] with permission.)

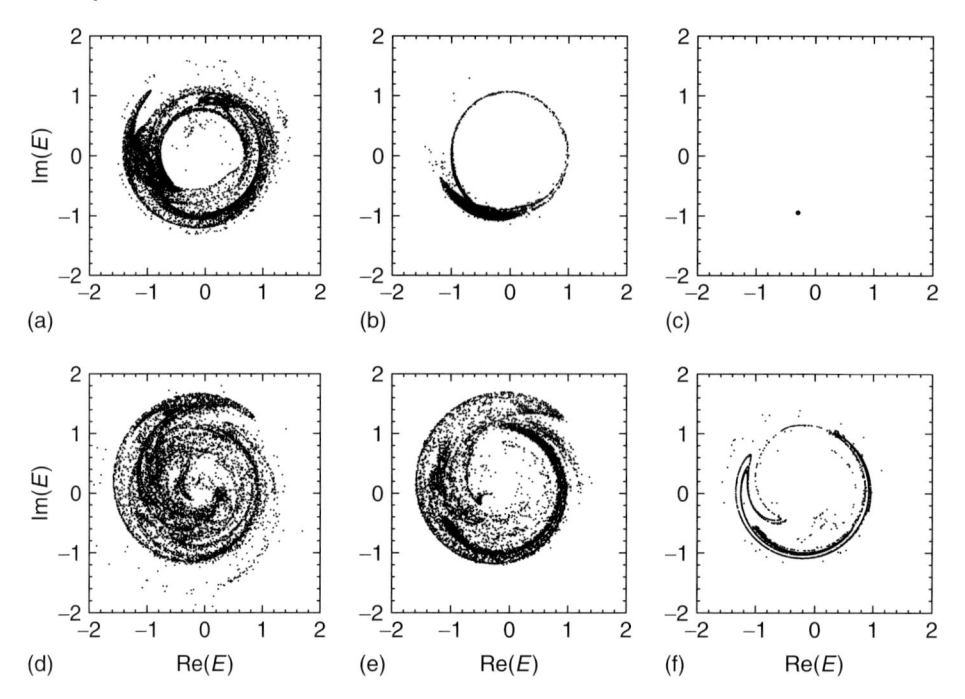

Figure 11.6 Pullback convergence to (a–c) a random sink for $D_{ext} = 0.1$ and (d–f) a random strange attractor for $D_{ext} = 0.5$ in the white noise forced laser model (Eqs. (11.18) and (11.19)) with $\alpha = 3$ and $J = 1$. Shown are snapshots of 10 000 trajectories in projection onto the complex E-plane at time $t = 30$. The initial conditions are uniformly distributed on the E-plane with $N = 0$ at different initial times $t_0 =$ (a,d) 29; (b,e) 28; and (c,f) 0.

zero (Figure 11.5b). Noise synchronization is lost for parameter settings between these two curves, where $\lambda_{max} > 0$ indicates a random strange attractor. Pullback convergence to two qualitatively different random attractors found for $\alpha = 3$ is shown in Figure 11.6. At fixed time $t = 30$, we take snapshots of 10 000 trajectories for a grid of initial conditions with different initial times t_0. In Figure 11.6a–c, trajectories converge in the pullback sense to a random sink. The random sink appears in the snapshots as a single dot whose position is different for different t or different noise realizations f_{ext}. In Figure 11.6d–f, trajectories converge in the pullback sense to a random strange attractor that appears in the snapshots as a fractallike structure. This structure remains fractallike, but is different for different t or different noise realizations f_{ext}.

11.6.1
Class-B Laser Model Versus Landau–Stuart Equations

The stochastic d-bifurcation uncovered in the previous section has been reported in biological systems [8–10, 13, 14], and should appear in a general class of oscillators with stochastic forcing. Here, we use the laser model in conjunction with the Landau–Stuart model to address its dependence on the three parameters: D_{ext}, J, and α, and to uncover its universal properties. With an exception of certain approximations [10], this problem is beyond the reach of analytical techniques and so numerical analysis is the tool of choice.

To help identifying effects characteristic to the more complicated laser model, we first consider the Landau–Stuart model with white noise external forcing

$$\frac{dE}{d\tilde{t}} = \left[J + i\left(\tilde{\Delta} - \alpha(J - |E|^2)\right)\right] E - E|E|^2 + f_{ext}(\tilde{t}), \tag{11.20}$$

where, for the rescaled time $\tilde{t}$, the external forcing correlations become

$$\langle f^R_{ext}(\tilde{t}) f^R_{ext}(\tilde{t}')\rangle = \langle f^I_{ext}(\tilde{t}) f^I_{ext}(\tilde{t}')\rangle = \frac{D_{ext}}{g\gamma} \delta(\tilde{t} - \tilde{t}').$$

Figure 11.7a and b shows the dependence of the d-bifurcation on J, D_{ext}, and α in Eq. (11.20). In the three-dimensional (J, D_{ext}, α) parameter space, the two-dimensional surface of d-bifurcation appears to originate from the half line ($D_{ext} = 0, J = 0, \alpha > 5.3$) of the deterministic Hopf bifurcation, has a ridge at $\alpha_{min} \approx 5.3$, and is asymptotic to $\alpha \approx 9$ with increasing D_{ext} (Figure 11.7b). Furthermore, numerical results in Figure 11.7b suggest that the shape of the d-bifurcation curve in the two-dimensional section (D_{ext}, α) is independent of J. As a consequence, for fixed α within the range $\alpha \in (5.3, 9)$, one finds two d-bifurcation curves in the (D_{ext}, J)-plane (Figure 11.7a) that are parametrized by

$$J_j = C_j(\alpha)\sqrt{2D_{ext}}, \quad \text{where} \quad j = 1, 2, \tag{11.21}$$

and bound the region with a random strange attractor. Since $C_1(\alpha_{min}) = C_2(\alpha_{min}) = 1$, these two curves merge into a single curve

$$J = \sqrt{2D_{ext}}, \tag{11.22}$$

when $\alpha = \alpha_{min}$. On the one hand, for $\alpha \leq \alpha_{min}$, the region with a random strange attractor disappears from the (D_{ext}, J)-plane. On the other hand, for $\alpha > 9$, there is just one d-bifurcation curve in the (D_{ext}, J)-plane, meaning that the region with a random strange attractor becomes unbounded toward increasing D_{ext} (Figure 11.7b).

Similar results are expected for any white noise forced oscillator with shear that is near a Hopf bifurcation, and for "weak" forcing. This claim is supported with numerical analysis of the laser model (Eqs. (11.18) and (11.19)) in Figure 11.7c and d. For a fixed α, Eq. (11.20) and Eqs. (11.18)–(11.19) give identical results if the forcing is weak enough, but significant discrepancies arise with increasing forcing strength. First of all, it is possible to have one-dimensional sections of the (D_{ext}, J)-plane for fixed J with multiple uplifts of λ_{max} to positive values (black dots for $J < 10^{-6}$ in Figure 11.7c). Secondly, the parameter region with a random strange attractor for Eqs. (11.18) and (11.19) expands toward much lower values of $\alpha > \alpha_{min} \approx 1$ (Figure 11.7d). Thirdly, the shape of the two-dimensional surface of d-bifurcation in the laser model becomes dependent on J and has a minimum rather than a ridge. As a consequence, although the stochastic d-bifurcation seems

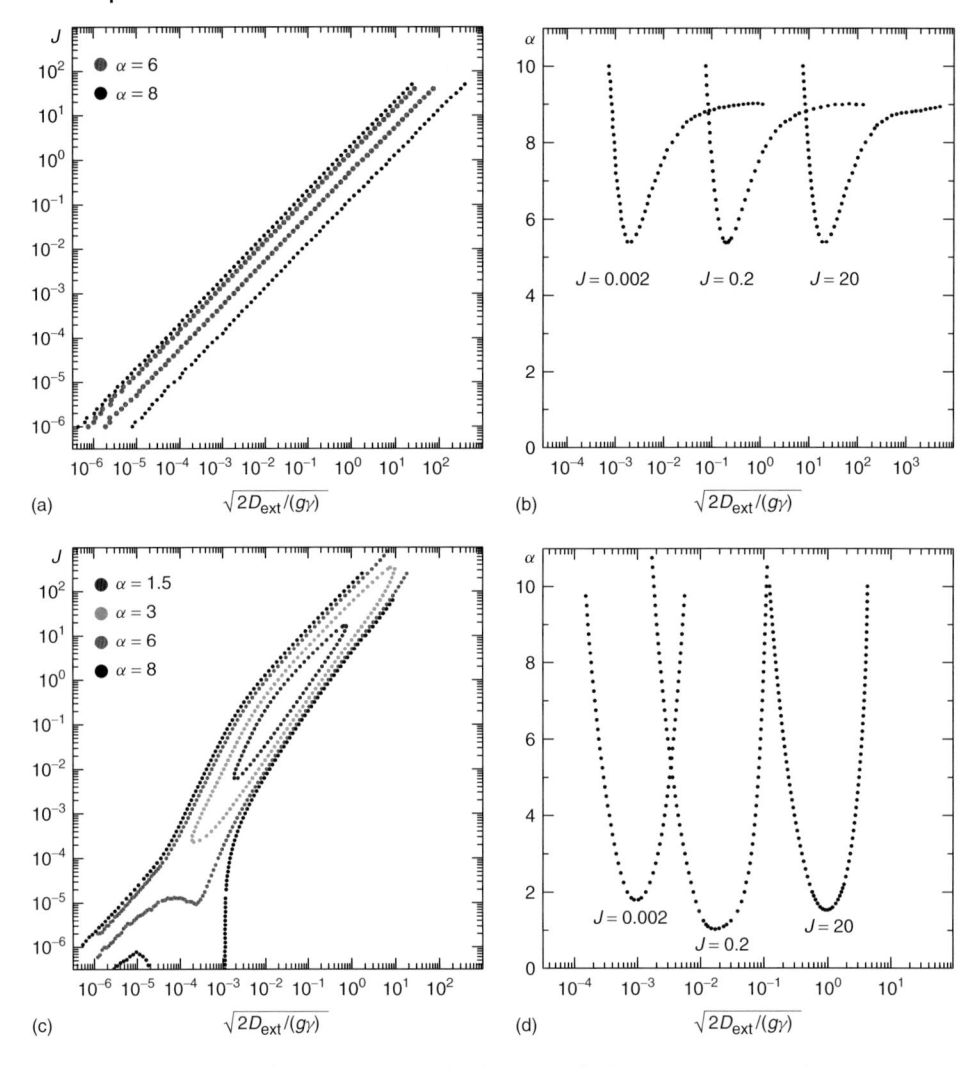

Figure 11.7 A three-parameter study of stochastic d-bifurcation (a,b) in the white noise forced Landau–Stuart model (Eq. (11.20)) and (c,d) the white noise forced laser model (Eqs. (11.18) and (11.19)). To facilitate the comparison between Eqs. (11.20) and (11.18)–(11.19), we plotted results for Eqs. (11.18) and (11.19) with the rescaled forcing strength, $\sqrt{2D_{ext}/g\gamma}$. (Adapted from [12] with permission.)

to originate from the half line ($D_{ext} = 0, J = 0, \alpha > 5.3$), it will appear in the (D_{ext}, J)-plane even for $\alpha \in (1, 5.3)$ as a closed and isolated curve away from the origin of this plane (Figure 11.7c). Finally, the region of random strange attractor remains bounded in the (D_{ext}, J)-plane even for large α.

To unveil the link between the transient dynamics of unforced systems and the forcing-induced stochastic d-bifurcation, we plot J versus Lyapunov exponents in Figure 11.8; note that Lyapunov exponents, λ_i, and Floquet exponents, μ_i, are

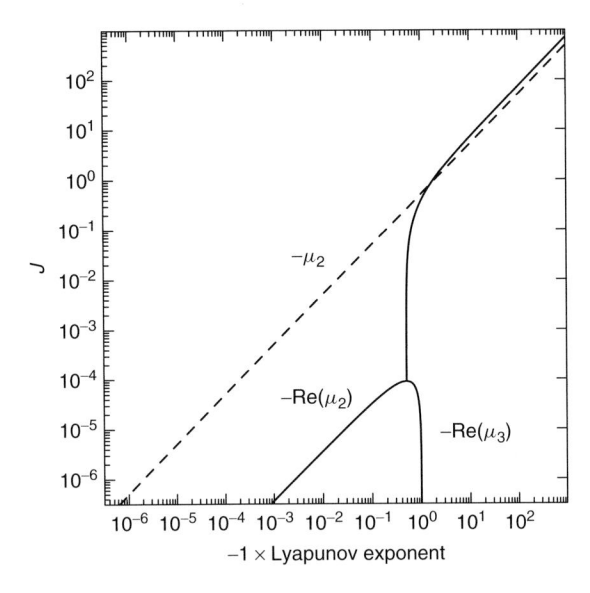

Figure 11.8 J versus (solid) the two nonzero Lyapunov exponents, $\text{Re}(\mu_2)$ and $\text{Re}(\mu_3)$, for the limit cycle in the laser model (Eqs. (11.1) and (11.2)), and (dashed) the nonzero Lyapunov exponent, μ_2, for the limit cycle in the Landau–Stuart model (Eq. (11.5)). (Adapted from [12] with permission.)

related by $\lambda_i = \text{Re}[\mu_i]$. A comparison between Figures 11.7 and 11.8 shows strong correlation between the relaxation toward the limit cycle and the d-bifurcation. In the Landau–Stuart model (Eq. (11.5)), the linear relation (Eq. (11.6)) between J and the nonzero Lyapunov exponent, μ_2 (dashed line in Figure 11.8), results in a linear parametrization (Eq. (11.21)) of d-bifurcation curves in the $(J, \sqrt{2D_{\text{ext}}})$-plane (Figure 11.7a). In the laser model (Eqs. (11.1) and (11.2)), the nonlinear relation (Eqs. (11.3) and (11.4)) between J and the nonzero Lyapunov exponents, $\text{Re}(\mu_2)$ and $\text{Re}(\mu_3)$ (solid curves in Figure 11.8), results in a very similar nonlinear parametrization of d-bifurcation curves in the $(J, \sqrt{2D_{\text{ext}}})$-plane (Figure 11.7c). The splitting up of the chaotic region bounded by the black dots for $J < 9 \times 10^5$ in Figure 11.7c is related to two different eigendirections normal to the limit cycle with significantly different timescales of transient dynamics toward the limit cycle (the two corresponding Lyapunov exponents are shown in solid in Figure 11.8). Finally, the appearance of relaxation oscillations in the laser system is associated with a noticeable expansion of the chaotic region, in particular, toward smaller α.

11.7
Noise-Induced Strange Attractors

Complicated invariant sets, such as strange attractors, require a balanced interplay between phase-space expansion and contraction [17]. If phase-space expansion in certain directions is properly compensated by phase-space contraction in some other directions, nearby trajectories can separate exponentially fast ($\lambda_{max} > 0$) and yet remain within a bounded subset of the phase space.

It has been recently proven that, when *suitably* perturbed, any stable hyperbolic limit cycle can be turned into "observable" chaos (a strange attractor) [32]. This result is derived for periodic discrete-time perturbations (kicks) that deform the stable limit cycle of the unkicked system. The key concept is the creation of horseshoes via a stretch-and-fold action due to an interplay between the kicks and the local geometry of the phase space. Depending on the degree of shear, quite different kicks are required to create a stretch-and-fold action and horseshoes. Intuitively, it can be described as follows. In systems without shear, where points in phase-space rotate with the same angular frequency about the origin of the complex E-plane independent of their distance from the origin, the kick alone has to create the stretch-and-fold action. This is demonstrated in Figure 11.9. Horseshoes are formed as the system is suitably kicked in both radial and angular directions and then relaxes back to the attractor (the circle in Figure 11.9a) of the unkicked system. Repeating this process reveals chaotic invariant sets. However, showing rigorously that a specific kick results in "observable" chaos is a nontrivial

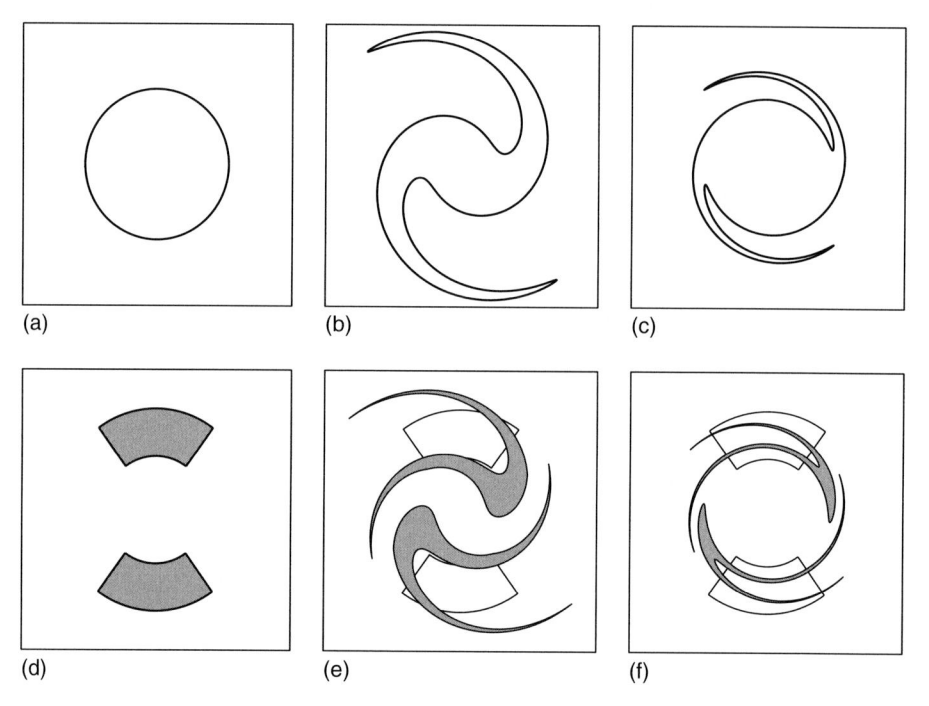

Figure 11.9 Time evolution of sets of initial conditions showing the creation of horseshoes in the phase space of a suitably kicked laser model with no shear ($\alpha = 0$). The sets of initial conditions are (a) the stable circle and (d) boxes containing parts of the circle. Shown are phase portraits (a,d) before, (b,e) immediately after the first kick, and (c,f) some time after the first kick.

task [32]. In the presence of shear, where points in phase space rotate with different angular frequencies depending on their distance from the origin, the kick does not have to be so specific or carefully chosen. In fact, it may be sufficient to kick nonuniformly in the radial direction alone, and rely on natural forces of shear to provide the necessary stretch-and-fold action.

These effects are illustrated in Figure 11.10 for the single laser model (Eqs. (11.1)–(11.2)) with nonuniform kicks in the radial direction alone for $\alpha = 0$ (no shear) and $\alpha = 2$ (shear). There, we set $\Delta = 0$ and refer to the stable limit cycle (dashed circle in Figure 11.10a) as Γ. Kicks modify the electric-field amplitude, $|E|$, by a factor of $0.8\sin[4\arg(E)]$ at times $t = 0$, 0.25, 0.5, and 0.75, but leave the phase, $\arg(E)$, unchanged. For $\alpha = 0$ each point on the black curve spirals onto Γ in time, but remains within the same isochrone defined by a constant electric-field phase, $\arg(E) = \arg(E(0))$. Hence, the black curve does not have any folds at any time. However, for $\alpha = 2$, a kick moves most points on the black curve to different isochrones so that points with larger amplitudes $|E|$ rotate with larger angular frequencies. This gives rise to an intricate stretch-and-fold action. Folds and horseshoes can be formed under the evolution of the flow even though the kicks are in the radial direction alone.

In the laser model, stretch-and-fold action is significantly enhanced by the spiraling transient motion about Γ. For $J > 10^{-1}$, the laser model (Eqs. (11.1) and (11.2)) and the Landau–Stuart model (Eq. (11.5)) have nearly identical relaxation timescales toward Γ (Figure 11.8). However, owing to one additional degree of freedom and oscillatory relaxation (Eq. (11.4)), the instantaneous stretching along Γ in the three-dimensional laser vector field (Eqs. (11.1) and (11.2)) can be much stronger compared to the planar vector field (Eq. (11.5)), especially at short times after the perturbation. This effect is illustrated in Figure 11.11 by the time evolution of the phase difference, $\arg(E_1(t)) - \arg(E_2(t))$, between two trajectories, 1 and 2, starting at different isochrones for $\alpha = 3$. Since both vector fields have identically shaped isochrones (Eq. (11.8)), the phase difference converges to the same value as time tends to infinity. However, at small t, the oscillatory phase difference for Eqs. (11.1) and (11.2) can exceed significantly the monotonically increasing phase difference for Eq. (11.5) (compare solid and dotted curves in Figure 11.11).

It is important to note that the rigorous results for turning stable limit cycles into chaotic attractors are derived for periodic discrete-time perturbations. Stochastic forcing is a continuous-time perturbation, meaning that the analysis in [32] cannot be directly applied to our problem. Nonetheless, such analysis gives a valuable insight as to why random chaotic attractors appear for α sufficiently large, and it helps to distinguish the effects of stochastic forcing.

Here, we demonstrated that purely additive white noise forcing is sufficient to induce random strange attractors in limit cycle oscillators. Furthermore, numerical analysis in Section 11.6.1 shows that, in the case of stochastic forcing, creation of strange attractors requires a different balance between the amount of shear, relaxation rate toward the limit cycle, and forcing strength, as compared to periodic forcing. Unlike in the case of discrete-time periodic forcing, the shear has to be strong enough, $|\alpha| > C > 0$, to allow sufficient stretch-and-fold action.

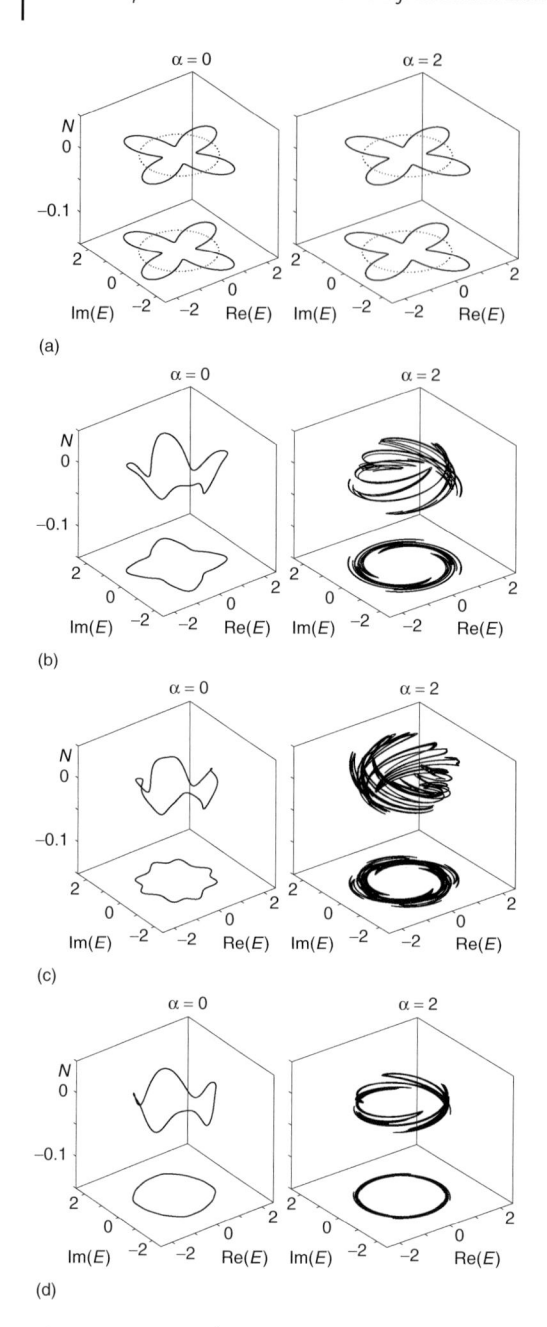

Figure 11.10 Snapshots at times (a) $t = 0$, (b) $t = 0.35$, (c) $t = 0.8$, and (d) $t = 1$ showing the time evolution of 15 000 trajectories with initial conditions distributed equally over the stable circle for Eqs. (11.1) and (11.2). Kicks in the radial direction alone are applied at times $t = 0, 0.25, 0.5, 0.75$. A comparison between $\alpha = 0$ and $\alpha = 2$ illustrates the α-induced stretch-and-fold action in the laser phase space.

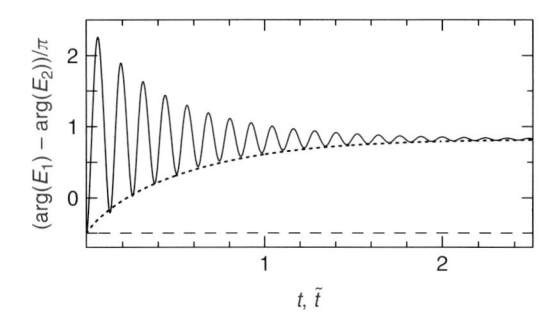

Figure 11.11 Phase-space stretching along the limit cycle shown as time evolution of the phase difference between two trajectories starting at different isochrones. The three curves are obtained using (dashed) Eqs. (11.1) and (11.2) with $\alpha = 0$, (solid) Eqs. (11.1) and (11.2) with $\alpha = 3$, and (dotted) Eq. (11.5) with $\alpha = 3$. (Adapted from [12] with permission.)

Provided that the shear is strong enough, the stochastic forcing strength needs to be at least comparable to the relaxation rate toward the limit cycle to allow formation of random strange attractors. Furthermore, we demonstrated that in higher dimensional systems, different eigendirections with distinctly different relaxation rates toward the limit cycle could give rise to more than one region in the $(J, \sqrt{2D_{ext}})$-plane with a random strange attractor. Last but not least, we revealed that the enhancement in the instantaneous stretch-and-fold action arising from laser relaxation oscillations results in a larger parameter region with a random strange attractor.

11.8 Conclusions

We used the class-B laser model in conjunction with the Landau–Stuart model (Hopf normal form with shear) to study noise synchronization and loss of synchrony via shear-induced stochastic d-bifurcations.

We defined noise synchronization in terms of pullback convergence of random attractors and showed that a nonlinear oscillator can synchronize to stochastic external forcing. However, the parameter region with synchronous dynamics becomes interrupted with a single or multiple intervals of asynchronous dynamics if amplitude–phase coupling or shear is sufficiently large. Stability analysis shows that the synchronous solution represented by a random sink loses stability via stochastic d-bifurcation to a random strange attractor. We performed a systematic study of this bifurcation with dependence on the three parameters: the Hopf bifurcation parameter (laser pump), the amount of shear (laser linewidth enhancement factor), and the stochastic forcing strength.

In this way, we uncovered a vast parameter region with random strange attractors that are induced purely by stochastic forcing. More specifically, in the three-dimensional parameter space, the two-dimensional surface of the stochastic bifurcation originates from the half line of the deterministic Hopf bifurcation. In the plane of stochastic forcing strength and Hopf bifurcation parameter, one finds stochastic d-bifurcation curve(s) bounding region(s) of random strange attractors if the amount of shear is sufficiently large. The shape of d-bifurcation curves is determined by the type and rate of the relaxation toward the limit cycle in the unforced oscillator. Near the Hopf bifurcation and provided that stochastic forcing is weak enough, the d-bifurcation curves satisfy the numerically uncovered power law (Eq. (11.21)). However, as the stochastic forcing strength increases, there might be deviations from this law. The deviations arise because different oscillators experience different effects of higher-order terms and additional degrees of freedom on the relaxation toward the limit cycle. In the laser example, the d-bifurcation curves deviate from the simple power law (Eq. (11.21)), so that the region of a random strange attractor splits up and expands toward smaller values of shear as the forcing strength increases. We intuitively explained these results by demonstrating that the shear-induced stretch-and-fold action in the oscillator's phase space facilitates creation of horseshoes and strange attractors in response to external forcing. Furthermore, we showed that the stretch-and-fold action can be greatly enhanced by damped relaxation oscillations in the laser model, causing the deviation from the simple power law.

References

1. Pikovsky, A.S., Rosenblum, M., and Kurths, J. (2001) *Synchronisation-A Unified Approach to Nonlinear Science*, Cambridge University Press, Cambridge.
2. Pikovsky, A.S. (1984) in *Nonlinear and Turbulent Processes in Physics* (ed. R.Z. Sagdeev), Harwood Academic, Singapore, pp. 1601–1604.
3. Mainen, Z.F. and Sejnowski, T.J. (1995) Reliability of spike timing in neocortical neurons. *Science*, **268**, 1503–1506.
4. Jensen, R.V. (1998) Synchronization of randomly driven nonlinear oscillators. *Phys. Rev. Lett.*, **58**, R6907–R6910.
5. Zhou, C.S., Kurths, J., Allaria, E., Boccaletti, S., Meucci, R., and Arecchi, F.T. (2003) Constructive effects of noise in homoclinic chaotic systems. *Phys. Rev. Lett.*, **67**, 066220.
6. Teramae, J. and Tanaka, D. (2004) Robustness of the noise-induced phase synchronization in a general class of limit cycle oscillators. *Phys. Rev. Lett.*, **93**, 204103.
7. Uchida, A., McAllister, R., and Roy, R. (2004) Consistency of nonlinear system response to complex drive signals. *Phys. Rev. Lett.*, **93**, 244102.
8. Kosmidis, E. and Pakdaman, K. (2003) An analysis of the reliability phenomenon in the fitzhugh-nagumo model. *J. Comput. Neurosci.*, **14**, 5–22.
9. Goldobin, D.S. and Pikovsky, A. (2005) Synchronisation and desynchronisation of self-sustained oscillators by common noise. *Phys. Rev. E*, **71**, 045201(R).
10. Goldobin, D.S. and Pikovsky, A. (2006) Antireliability of noise-driven neurons. *Phys. Rev. E*, **73**, 061906.
11. Rulkov, N.F., Sushchik, M.M., Tsimring, L.S., and Abarbanel, H.D.I. (1995) Generalised synchronisation of chaos in directionally coupled chaotic systems. *Phys. Rev. E*, **51**, 980–994.
12. Wieczorek, S. (2009) Stochastic bifurcation in noise-driven lasers and Hopf oscillators. *Phys. Rev. E*, **79**, 036209.
13. Lin, K.K., Shea-Brown, E., and Young, L.-S. (2009) Reliability of coupled oscillators. *J. Nonlin. Sci.*, **19**, 497–545.

14. Lin, K.K. and Young, L.-S. (2008) Shear-induced chaos. *Nonlinearity*, **21**, 899–922.
15. Wieczorek, S., Krauskopf, B., Simpson, T.B., and Lenstra, D. (2005) The dynamical complexity of optically injected semiconductor lasers. *Phys. Rep.*, **416**, 1–128.
16. Wieczorek, S. and Chow, W.W. (2009) Bifurcations and chaos in a semiconductor laser with coherent or noisy optical injection. *Opt. Commun.*, **282** (12), 2367–2379.
17. Guckenheimer, J. and Holmes, P. (1983) *Nonlinear Oscillations, Dynamical Systems, and Bifurcations of Vector Fields*, Springer-Verlag, New York.
18. Henry, C.H. (1982) Theory of the linewidth of semiconductor laser. *IEEE J. Quantum Electron.*, **QE-18**, 259.
19. Guckenheimer, J. (1974/75) Isochrons and phaseless sets. *J. Math. Biol.*, **3**, 259–273.
20. Lang, R. (1982) Injection locking properties of a semiconductor laser. *IEEE J. Quantum Electron.*, **18**, 976.
21. Chow, W.W., Scully, M.O., and van Stryland, E.W. (1975) Line narrowing in a symmetry broken laser. *Opt. Commun.*, **15**, 6–9.
22. Erneux, T., Kovanis, V., Gavrielides, A., and Alsing, P.M. (1996) Mechanism for period-doubling bifurcation in a semiconductor laser subject to optical injection. *Phys. Rev. A*, **53**, 4372–4380.
23. Chlouverakis, K.E. and Adams, M.J. (2003) Stability map of injection-locked laser diodes using the largest Lyapunov exponent. *Opt. Commun.*, **216**, 405–412.
24. Bonatto, C. and Gallas, J.A.C. (2007) Accumulation horizons and period adding in optically injected semiconductor lasers. *Phys. Rev. E*, **75**, 055204 R.
25. Wieczorek, S., Simpson, T.B., Krauskopf, B., and Lenstra, D. (2002) Global quantitative predictions of complex laser dynamics. *Phys. Rev. E*, **65**, 045207 R.
26. Arnold, L. (1998) *Random Dynamical Systems*, Springer-Verlag, Berlin Heidelberg.
27. Abarbanel, H.D.I., Rulkov, N.F., and Sushchik, M.M. (1996) Generalised synchronisation of chaos: The auxiliary system approach. *Phys. Rev. E*, **53** (5), 4528–4535.
28. Pyragas, K. (1996) Weak and strong synchronisation of chaos. *Phys. Rev. E*, **54** (5), R4508–R4511.
29. Brown, R. and Kocarev, L. (2000) A unifying definition of synchronisation for dynamical systems. *Chaos*, **10** (2), 344–349.
30. Langa, J.A., Robinson, J.C., and Súarez, A. (2002) Stability, instability, and bifurcation phenomena in non-autonomous differential equations. *Nonlinearity*, **15** (2), 1–17.
31. Ashwin, P. and Ochs, G. (2003) Convergence to local random attractors. *Dyn. Syst.*, **18** (2), 139–158.
32. Wang, Q. and Young, L.-S. (2003) Strange attractors in periodically-kicked limit cycles and Hopf bifurcations. *Commun. Math. Phys.*, **240**, 509–529.

12
Emergence of One- and Two-Cluster States in Populations of Globally Pulse-Coupled Oscillators

Leonhard Lücken and Serhiy Yanchuk

12.1
Introduction

Networks of coupled dynamical systems play an important role for all branches of science [1–4]. In the neuroscience, for instance, there is a need for modeling large populations of coupled neurons in order to approach problems connected with the synchronization of neural cells or other types of collective behavior [4–6]. The investigation of the dynamics of coupled lasers [7–10] is important for many purposes including secure communication [11, 12] or high-power generation. The interacting biological, mechanical, or electrical oscillators [13, 14] belong already to classical models for studying various aspects of collective dynamics. In neural networks, the synchronous activity might be pathological [15], and hence, there was recently an increasing effort to control the desynchronization of populations of coupled oscillators. In particular, the coordinated reset stimulation technique [4, 16] proposes to establish a cluster state in the network, in which the oscillator's phases split into several subgroups. This example illustrates the importance of the analysis of cluster formation in coupled systems. This chapter investigates the connection between the properties of a single oscillator, that is, its sensitivity to stimulations, and the formation of clusters in a globally coupled system of such oscillators. We show that by altering the shape of the sensitivity function, called the *phase-response function*, different clusters in a network can be stabilized. More precisely, we study a family of the phase-response curves (PRCs), which are unimodal and turn to zero at the spiking moment. This choice is motivated by several well-known neuron models. It appears that the position of the maximum of the unimodal sensitivity function with respect to the spiking point plays an important role for determining whether the system will synchronize or approach a two-cluster state (Figure 12.1). In particular, when the maximum of the sensitivity function is located in the second half of the period, the one-cluster (or completely synchronized) state acts as a global attractor. In the case, when the sensitivity function reaches its maximum in the first half of the period, various two-cluster states become stable.

Nonlinear Laser Dynamics: From Quantum Dots to Cryptography, First Edition. Edited by Kathy Lüdge.
© 2012 Wiley-VCH Verlag GmbH & Co. KGaA. Published 2012 by Wiley-VCH Verlag GmbH & Co. KGaA.

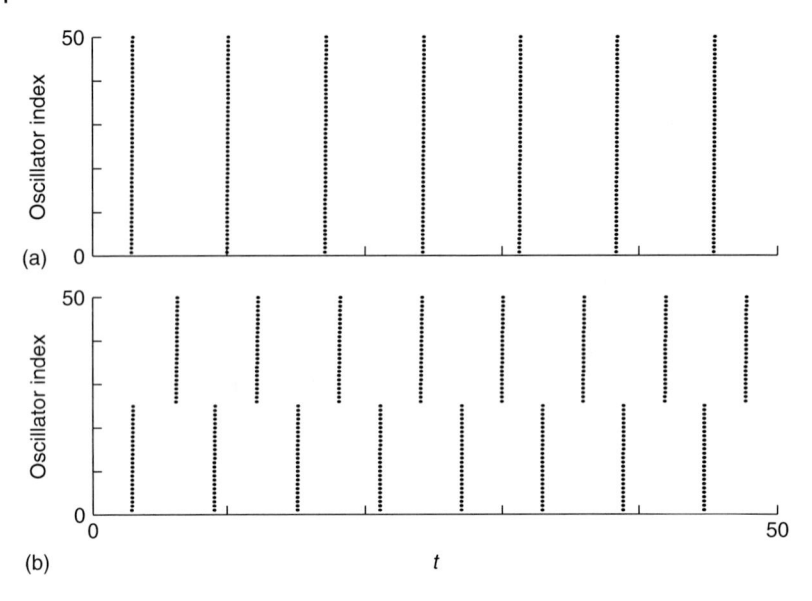

Figure 12.1 Clusters in a population of 50 phase oscillators. Dots indicate the times when an oscillator reaches the threshold; Panel (a) shows the firing pattern of a complete in-phase synchronized population (one cluster), while (b) shows the firings in a symmetric two cluster.

12.1.1
Pulse-Coupled Oscillators

In some coupled systems, for example, neuron populations, the time during which the interaction effectively takes place is much smaller than the characteristic period of oscillations. In such cases, it is reasonable to approximate the interaction by an impact, that is, by assuming that the interaction is immediate. This approximation leads to models of pulse-coupled oscillators, which have been widely used in the literature. For example, Mirollo and Strogatz [17] have shown that the complete synchronization (in this case, it is equivalent to the phase locking) is stable and attracts almost all initial conditions in the network of globally coupled integrate-and-fire (IF) oscillators of the form

$$\frac{dx_j}{dt} = S_0 - \gamma x_j, \quad x_j \in [0, 1), \quad j = 1, \ldots, N \qquad (12.1)$$

with constants $S_0 > \gamma > 0$. One might refer to S_0 as input current and to γ as the dissipation constant. The following additional condition describes the interaction: when kth oscillator reaches the threshold $x_k(t^-) = 1$, then positions of all remaining oscillators are shifted accordingly to the rule

$$x_j(t^+) = \min\{x_j(t) + \varkappa, 1\}, \quad j \neq k \qquad (12.2)$$

with some small $\varkappa > 0$ and the kth oscillator resets to $x_k(t^+) = 0$. It is shown in [17] that complete synchronization is achieved after a finite transient time. The synchronization in a more general model of IF neurons has been shown in [18]. Tsodyks et al. have demonstrated in [19] that the phase-locked state is unstable with respect to inhomogeneity in the local frequencies, that is, when the oscillators become nonidentical.

A larger class of pulse-coupled models was studied in [20–22]. In particular, Goel and Ermentrout [20] obtained sufficient conditions for the stability of a completely synchronous solution. We introduce this class of models in Section 12.1.2.

The dynamics of pulse-coupled oscillators has been studied also for the systems with different topologies, that is, ring topology [23], as well as for delayed interactions [24]. Transient phenomena of randomly diluted networks have been analyzed in [25]. Globally pulse-coupled IF oscillators with a finite pulse-width have been considered in [26–28], where the interaction pulse is assumed to have a shape $\frac{\alpha^2 t}{N} e^{-\alpha t}$ with the width α.

12.1.2
Phase-Response Curve as a Parameter

In this section, we introduce a general class of pulse-coupled phase oscillators [20, 29]. The oscillator's motion between the spikes is described by the rule

$$\frac{d\varphi_j}{dt} = \omega \qquad (12.3)$$

where $\varphi_j \in [0, 2\pi]$. When kth oscillator reaches the threshold at time t, that is, $\varphi_k(t^-) = 2\pi$, it emits a spike to all other oscillators of the network, which are immediately resetted according to

$$\varphi_k(t^+) = 0, \quad \varphi_j(t^+) = \varphi_j(t^-) + \varkappa Z(\varphi_j(t^-)), \quad j \neq k, \qquad (12.4)$$

where $Z(\varphi)$ is called PRC. Effectively, this means that there is no coupling between two consecutive spiking events. The coupling occurs only during the spike and acts through the resetting, since the time of the resetting of the oscillator j depends on the phase position of the oscillator k. The size of the phase jump, that an oscillator performs, when stimulated by an incoming spike depends on its sensitivity to stimulation in its present state. See Figure 12.2 for an illustration.

Let us firstly show that IF oscillators (Eq. (12.1)) can be written in a form similar to Eqs. (12.3) and (12.4) [20]. For this, we rewrite Eq. (12.1) with respect to the phase coordinate instead of the voltage coordinate. Indeed, the coordinate x_j in system (Eq. (12.1)) is supposed to describe the voltage difference across the membrane of a neuron [30]. The phase coordinate φ_j should behave according to Eq. (12.3) with the frequency $\omega = 2\pi/T$, where T is the period of oscillations without interaction and can be found from Eq. (12.1)

$$T = -\frac{1}{\gamma} \ln\left(1 - \frac{\gamma}{S_0}\right).$$

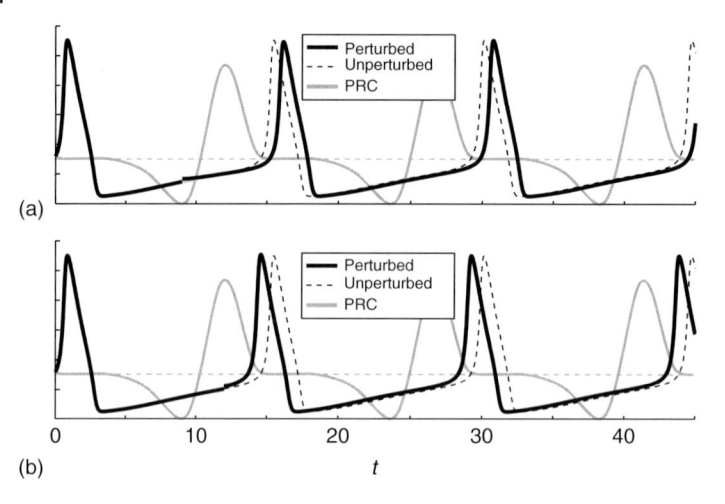

Figure 12.2 Periodic spiking in a Hodgkin–Huxley neuron model. Solid black lines show the evolution of the voltage component of the model when perturbed by a weak pulse at $t = 9$ in (a) and $t = 12$ in (b). The dashed black lines show the unperturbed oscillations. The PRC $Z(\varphi(t))$ of the unperturbed model is plotted solid gray and the dashed gray line corresponds to $Z = 0$. The outcome of the perturbing pulse depends on the time t, or equivalently on the phase $\varphi(t)$, of its application. Either the phase is delayed as in (a), that is, $Z(\varphi(t)) < 0$, or it is forwarded as in (b), that is, $Z(\varphi(t)) > 0$.

The corresponding transformation of variables $x = f(\varphi)$ can be found from the condition

$$\frac{dx}{dt} = \frac{df}{d\varphi}\frac{d\varphi}{dt} = \frac{df}{d\varphi}\omega = S_0 - \gamma f(\varphi),$$

that is, from the initial value problem

$$\frac{df(\varphi)}{d\varphi} = \frac{T}{2\pi}(S_0 - \gamma f(\varphi)), \quad f(0) = 0. \tag{12.5}$$

This gives the function

$$f(\varphi) = \frac{S_0}{\gamma}\left(1 - \exp\left(-\frac{\gamma T}{2\pi}\varphi\right)\right),$$

which maps the interval $0 \leq \varphi \leq 2\pi$ into $0 \leq x \leq 1$. In the transformed coordinates φ_j, the dynamics between the spikes is described by Eq. (12.3). It remains to specify the dynamics at the threshold. Taking into account Eq. (12.2), when kth oscillator reaches the threshold $\varphi_k(t^-) = 2\pi$, its phase φ_k resets to $\varphi_k(t^+) = 0$ and all other oscillators have the impact

$$\varphi_j(t^+) = f^{-1}(x_j(t^+)) = f^{-1}(x_j(t) + \varkappa\Delta(\varkappa, x_j(t)))$$
$$= f^{-1}(f(\varphi_j(t)) + \varkappa\Delta(\varkappa, f(\varphi_j(t))))$$

where $\Delta(\varkappa, x) = \min\{1, (1-x)/\varkappa\} \leq 1$. In the case of small $\varkappa$, that is, the assumption of weak coupling holds, the resetting rule can be approximated as

$$\varphi_j(t^+) = \varphi_j(t) + \varkappa \min\{Z_{IF}(\varphi_j(t)), (2\pi - \varphi_j(t))/\varkappa\} \qquad (12.6)$$

where

$$Z_{IF}(\varphi) := \frac{d(f^{-1})}{dx}(f(\varphi)) = \frac{2\pi T}{S_0} \exp\left(\frac{T\gamma}{2\pi}\varphi\right). \qquad (12.7)$$

Thus, with respect to the phase coordinates, the IF model (Eq. (12.1)) has the form (Eq. (12.3)) and (Eq. (12.6)). In particular, the resetting rule is given by the function

$$Z_{IF,\varkappa}(\varphi) = \min\{Z_{IF}(\varphi_j(t)), (2\pi - \varphi_j(t))/\varkappa\}, \qquad (12.8)$$

which depends on the amplitude of the perturbation $\varkappa$. Figure 12.3 illustrates this function for $\varkappa = 0.05$. Practically, the PRC measures the sensitivity of the phase to external perturbations.

We have shown above the specific example of pulse-coupled IF models and their reduction to pulse-coupled phase oscillators (Eqs. (12.3) and (12.4)). In fact, this procedure is also possible for higher-dimensional smooth systems, whenever the oscillations correspond to a hyperbolic limit cycle, that is, in a generic case. More details can be found in [20, 29, 31]. When the coupling is acting along one component, for example, the voltage variable, as often assumed in the case of neural populations, the PRC appears as a scalar function of the phase. In the case of a higher-dimensional interaction, it should be considered more generally as a vector.

Examples of PRCs for different neuron models are shown in Figure 12.4. Some more numerically and experimentally obtained PRCs can be found in [20, 29, 32]. The remarkable feature of many of such PRCs is that, contrary to the IF model, their PRCs are independent on $\varkappa$ and admit zero values at $\varphi = 0$ and $\varphi = 2\pi$. The conditions $Z(0) = Z(2\pi) = 0$ are also reasonable from the neuroscientific point of view, since they reflect the fact that the neurons are not sensitive to perturbations during the spike (Figure 12.4). Generally speaking, system (Eqs. (12.3) and (12.4))

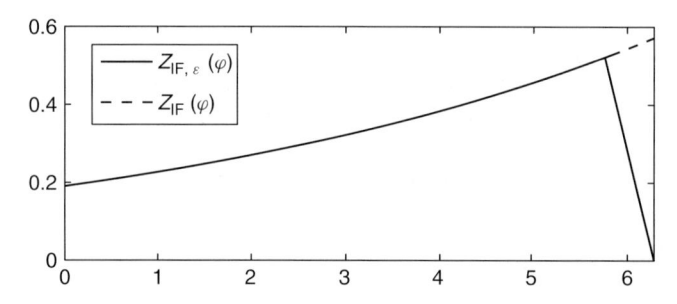

Figure 12.3 Phase-response curve for IF model (Eq. (12.1)). The function $Z_{IF}(\varphi)$ measures the sensitivity of the system to a small external perturbation at different positions φ. The corrected function $Z_{IF,\varkappa}(\varphi)$ does not allow the oscillators to be moved over the threshold by a spike. The values of Z_{IF} are plotted along the vertical axis and φ along the horizontal.

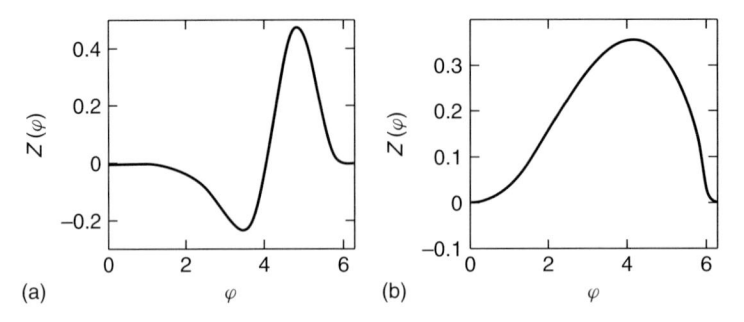

Figure 12.4 Examples of different PRCs: (a) Hodgkin–Huxley model and (b) Connor model. Note that the functions and their derivatives are zero at the ends of the interval $\varphi = 0$ and $\varphi = 2\pi$. (Adapted from [32]).

is a useful model, which possesses quite a big generality by including the PRC as some "infinite-dimensional" parameter.

12.1.3
System Description

Our main object of study is the following system of globally pulse-coupled phase oscillators of the form

$$\frac{d\varphi_j}{dt} = 1 \tag{12.9}$$

with the resetting rule

$$\varphi_k(t^+) = 0, \quad \varphi_j(t^+) = \varphi_j(t^-) + \frac{\varkappa}{N} Z(\varphi_j(t^-)), \quad j \neq k, \tag{12.10}$$

where the velocity of the phase is assumed to be 1 without loss of generality. We assume a fixed, positive overall coupling strength $\varkappa > 0$. The impact is rescaled taking into account the number of oscillators [28]. In this study, we consider a one-parametric family of the PRCs, which are positive and unimodal as shown in Figure 12.5. The parameter $\beta \in [0, 1]$ controls the position of the maximum, namely, for larger β, the maximum is located in the domain of small φ, which corresponds to a more sensitive excitatory response of the system just after spike. For smaller β, the system is more sensitive to perturbations shortly before the spike. The value $\beta = 0.5$ corresponds to an intermediate situation. We assume also that $Z'(0) = Z'(2\pi) = 0$, which is appropriate for a broad class of experimentally and analytically obtained PRCs (Figure 12.4).

We note that the qualitative results reported in the chapter are independent on the exact expression for the PRC, but rather on the shape of the PRC and its behavior at $\varphi = 0$ and $\varphi = 2\pi$. Our particular choice is

$$Z_\beta(\varphi) = 1 - \cos \vartheta_\beta(\varphi), \quad \beta \in [0, 1], \tag{12.11}$$

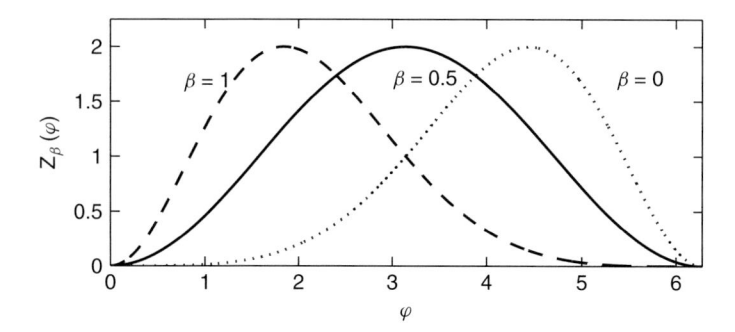

Figure 12.5 Family of the unimodal PRCs $Z_\beta(\varphi)$, see Eq. (12.11).

where

$$\vartheta_\beta(\varphi) = (1-\beta)\frac{\varphi^2}{2\pi} + \beta\left(2\pi - \frac{(\varphi - 2\pi)^2}{2\pi}\right).$$

In particular, $Z_{0.5}(\varphi) = 1 - \cos\varphi$.

System (Eqs. (12.9) and (12.10)) is equivalent to an $(N-1)$-dimensional discrete dynamical system, which can be obtained as a return map by considering its state each time when some of the phases reaches a fixed value, for example, $\varphi_1 = 2\pi$. Let us explain how this map appears. Without loss of generality, we may assume that the phases are ordered as

$$2\pi = \varphi_1 \geq \varphi_2 \geq \cdots \geq \varphi_N \tag{12.12}$$

at $t = 0$. We use the important property of Eqs. (12.9) and (12.10) that the oscillators do not overrun each other for all times if the system size N is sufficiently large. Indeed, since the inequality

$$\varphi_j + \frac{\varkappa}{N}Z(\varphi_j) \geq \varphi_{j+1} + \frac{\varkappa}{N}Z(\varphi_{j+1}) \tag{12.13}$$

holds for sufficiently large N, the order of oscillators is preserved during the spike. It is also evident that the order is preserved between the spikes as well. More exactly, the inequality $2\pi \geq \varphi_{1+l} \geq \varphi_{2+l} \geq \cdots \geq \varphi_{N+l} \geq 0$ holds for all t, where l is some shift and the indices are considered modulo N.

Let us introduce the map K_1, which maps the initial phases (Eq. (12.12)) into the phases at the moment when the oscillator φ_2 reaches the threshold, that is, $\varphi_2 = 2\pi$. It is easy to obtain that

$$K_1(\varphi_1, \varphi_2, \varphi_3, \ldots, \varphi_N) =$$
$$(2\pi - \mu(\varphi_2), 2\pi, \mu(\varphi_3) + 2\pi - \mu(\varphi_2), \ldots, \mu(\varphi_N) + 2\pi - \mu(\varphi_2)),$$

where

$$\mu(\varphi) := \varphi + \frac{\varkappa}{N}Z(\varphi). \tag{12.14}$$

In a similar way, the mapping K_2 exists, which maps the phases to the state, where the third oscillator is at the threshold, and so on. The composition of maps

$$K = K_N \circ K_{N-1} \circ \cdots \circ K_1 \tag{12.15}$$

gives the dynamical system on N-dimensional torus $\mathbb{T}^N$

$$(\varphi_1,\ldots,\varphi_N) \to K(\varphi_1,\ldots,\varphi_N), \tag{12.16}$$

which maps the initial state (Eq. (12.12)) into a new state after all N oscillators have crossed the threshold once and the first oscillator reaches again the threshold. We call the map K *return map*.

In this chapter, we will not use the explicit form of the mapping (Eq. (12.15)). For our purposes, it is important to conclude that the dynamics of system (Eqs. (12.9) and (12.10)) are indeed equivalent to some $(N-1)$-dimensional, discrete dynamical system on the N-dimensional torus. The smoothness of this system depends on the smoothness of its PRC function.

12.2
Numerical Results

In order to detect the appearance of one- or two-cluster states, we have numerically computed the order parameters

$$R_1(t) = \left| \frac{1}{N} \sum_{k=1}^{N} e^{i\varphi_k(t)} \right| \tag{12.17}$$

and

$$R_2(t) = \left| \frac{1}{N} \sum_{k=1}^{N} e^{i 2\varphi_k(t)} \right|. \tag{12.18}$$

A perfect one-cluster state is characterized by $R_1 = R_2 = 1$ and a perfect antiphase two cluster is characterized by $R_1 = 0$ and $R_2 = 1$. We present the results of simulations for $\varkappa = 0.5$, but qualitatively, we observe similar behavior for a broad range of $\varkappa > 0$.

As shown in Figure 12.6, we observe two qualitatively different types of behavior depending on parameter β. For $\beta < 0.5$, that is, when the maximum of the PRC is shifted toward the right (Figure 12.5), the one-cluster state seems to be the attractor; for $\beta > 0.5$ and the maximum of the PRC is shifted toward the left, a two-cluster state is attracting. We have chosen initial conditions in a vicinity of a two-cluster state in Figure 12.6a and b, therefore the initial values of the order parameters are $R_1 \approx 0$ and $R_2 \approx 1$. Figure 12.6b shows how the instability of the two-cluster state implies desynchronization transient, after which the system is attracted to a synchronous one-cluster state. Similar behavior occurs for other initial conditions also. Figure 12.6c and d illustrates the order parameters behavior for initial conditions close to the splay state (a state, where the phases are distributed). The initial values for the order parameters in the splay state are close to zero, but after a transient, they approach again the same asymptotic values as in Figure 12.6a and b.

A more complicated behavior occurs for the intermediate value of the parameter $\beta = 0.5$, that is, when the PRC is symmetric. In this case, the order parameters

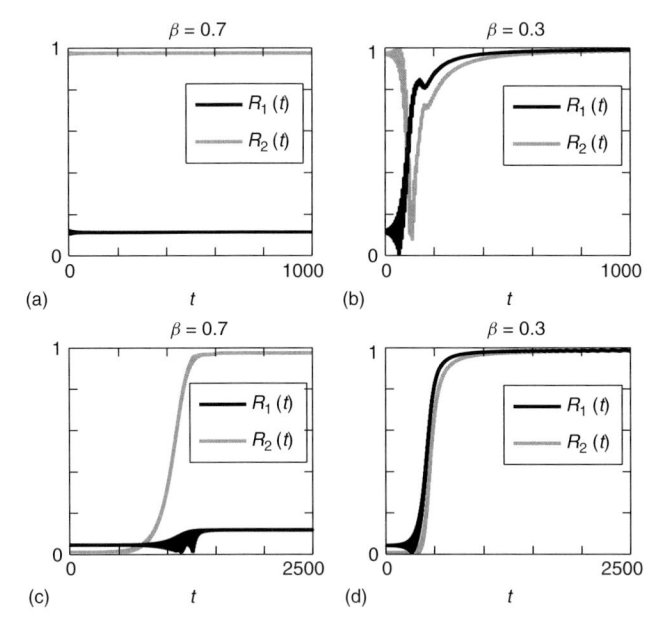

Figure 12.6 Behavior of the order parameters $R_1(t)$ and $R_2(t)$ for a trajectory starting in a vicinity of the two-cluster state for (a) and (b). The lower panel (c) and (d) corresponds to a trajectory starting in a vicinity of the splay state. Panels (a) and (c) correspond to the parameter value $\beta = 0.7$, where the two-cluster state is attracting and (b) and (d) to $\beta = 0.3$, where one-cluster state is attracting.

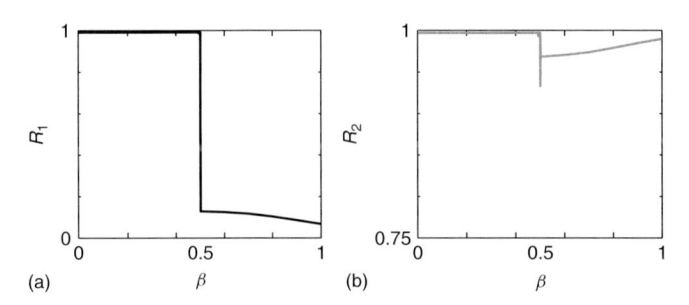

Figure 12.7 Dependence of the asymptotic values for the order parameter R_1 (a) and R_2 (b) on β. For the most values of β, except $\beta = 0.5$, the order parameters tend to some constant value, when initialized near the splay state or the symmetric two-cluster state.

$R_1(t)$ and $R_2(t)$ do not approach some asymptotic constant values, but remain periodic in time. As a result, the maximum asymptotic values of both R_1 and R_2 do not coincide with the corresponding minimum values. This type of behavior is observed for a very small parameter interval of order 10^{-3} around $\beta = 0.5$. We discuss it in Section 12.5 in more details. Figure 12.7 summarizes the behavior of the order parameters for different β.

12.3
Appearance and Stability Properties of One-Cluster State

In an ideal one-cluster synchronized state, all oscillators have the same phases $\varphi_j = \varphi_s$ for all j. This state is a fixed point of the map (Eq. (12.16)), because the PRC turns to zero at $\varphi = 2\pi$ and $\varphi = 0$. This means that the coupling vanishes for one-cluster state. More exactly, when an oscillator φ_j fires, that is, $\varphi_j = 2\pi$, all other oscillators have the phase 2π and do not obtain the spike. As a result, the period of this state is determined simply by the uncoupled dynamics and equals 2π.

12.3.1
Inadequacy of the Linear Stability Analysis

In order to obtain conditions for the stability of one-cluster state, one can examine the return map (Eq. (12.16)). The linearization of this return map around the one-cluster state gives then the corresponding multipliers, which determine its local linear stability. As it is expected, the local stability is governed by the properties of the PRC at $\varphi = 0$ and $\varphi = 2\pi$. This procedure has been done in [20]. Applying these results to our case, the resulting conditions for the local linear stability of one-cluster state is

$$\left(1 + \frac{\varkappa}{N} Z'(2\pi^-)\right)^l \left(1 + \frac{\varkappa}{N} Z'(0^+)\right)^{N-l} < 1, \quad l = 1, N-1. \tag{12.19}$$

We observe that the necessary condition for the linear stability is that the derivatives of $Z(\varphi)$ at the ends of the interval $[0, 2\pi]$ do not vanish. This is not the case for our PRC (Eq. (12.11)). Hence, all associated multipliers have modulus 1 and the linear stability analysis do not provide useful information about the stability of one-cluster state.

12.3.2
One-Cluster State is a Saddle Point

In this section, we show that one-cluster state is a saddle point, that is, there are some arbitrary small perturbations of this state, which grow with time. At the same time, some other small perturbations decay.

12.3.2.1 Existence of a Local Unstable Direction
First of all, let us show that one-cluster state is unstable with respect to the following special perturbation:

$$\varphi_1 = \varphi_s + \varepsilon, \quad \varphi_2 = \cdots = \varphi_N = \varphi_s \tag{12.20}$$

with arbitrary small $\varepsilon > 0$. During the period between spikes, the dynamics is monotonous $\varphi_j(t) = \varphi_s + t$ for $j = 2, \ldots, N$ and $\varphi_1(t) = \varphi_s + \varepsilon + t$, thus, the distance between the phases remain constant. Without loss of generality, we may assume that

$$\varphi_1(0^-) = 2\pi, \quad \varphi_2(0^-) = \cdots = \varphi_N(0^-) = 2\pi - \varepsilon.$$

12.3 Appearance and Stability Properties of One-Cluster State

After the first oscillator moves over the threshold and resetting occurs, the phases are as follows:

$$\varphi_1(0^+) = 0, \quad \varphi_2(0^+) = \cdots = \varphi_N(0^+) = 2\pi - \varepsilon + \frac{\varkappa}{N}Z(2\pi - \varepsilon) = \mu(2\pi - \varepsilon).$$

The next resetting occurs at time $t_1 = 2\pi - \varphi_2(0^+) = \varepsilon - \frac{\varkappa}{N}Z(2\pi - \varepsilon)$ when the group of $N - 1$ synchronous oscillators reaches the threshold. At this moment

$$\varphi_1(t_1^-) = \varepsilon - \frac{\varkappa}{N}Z(2\pi - \varepsilon) > 0, \quad \varphi_2(t_1^-) = \cdots = \varphi_N(t_1^-) = 2\pi.$$

Now the group of $N - 1$ synchronous oscillators is at the threshold. The correct definition of the firing rule for this case can be naturally obtained by extending it to the situation when all the oscillators $\varphi_2, \ldots, \varphi_N$ in the cluster have slightly different phases and then allowing the phases to converge them to the same value. This leads to the following resetting rule when passing the threshold by the $N - 1$ cluster:

$$\varphi_1(t_1^+) = \mu^{N-1}\left(\varphi_1(t_1^-)\right) = \mu^{N-1}\left(\varepsilon - \frac{\varkappa}{N}Z(2\pi - \varepsilon)\right), \quad (12.21)$$

$$\varphi_2(t_1^+) = \cdots = \varphi_N(t_1^+) = 0, \quad (12.22)$$

where μ^{N-1} denotes the superposition of $N - 1$ functions $\mu \circ \mu \circ \mu \circ \cdots \circ \mu$, where μ is defined by Eq. (12.14). The resetting Eq. (12.21) simply means that the function μ is applied $N - 1$ times (whenever an oscillator from the cluster $\varphi_2, \ldots, \varphi_N$ fires) in order to obtain the final position of φ_1.

In this way, we obtain a mapping, which maps the initial size of the perturbation ε at time $t = 0$ into its new size $Y_1(\varepsilon)$ at time t_1. The mapping is

$$\varepsilon \to Y_1(\varepsilon) = \mu^{N-1}\left(\varepsilon - \frac{\varkappa}{N}Z(2\pi - \varepsilon)\right). \quad (12.23)$$

It is clear that $Y_1(0) = 0$, what corresponds to the invariance of the one cluster, and the stability properties of the origin of Eq. (12.23) determine the stability of the one-cluster state with respect to the specific perturbation (Eq. (12.20)) chosen. Up to the linear level, the origin of Eq. (12.23) is neutrally stable, that is, $Y_1'(0) = 1$, which is clear, since the one-cluster state is linearly neutrally stable. The second derivative of Eq. (12.23) at $\varepsilon = 0$ is nontrivial

$$Y_1''(0) = \varkappa Z''(0) - \frac{\varkappa}{N}\left(Z''(2\pi) + Z''(0)\right)$$

and is positive for sufficiently large N since $Z''(0) > 0$ for $\beta \in (0, 1]$. Hence, for sufficiently large N, the origin of Eq. (12.23) is unstable, see Figure 12.8a. This leads to the *local instability of one-cluster state for all* $\beta \in (0, 1]$. Accordingly to this, the distance of the advanced oscillator φ_1 from the remaining cluster will grow, but this growth is not exponential.

12.3.2.2 Existence of a Local Stable Direction

Now let us show that the one-cluster state is locally stable with respect to perturbations of the form

$$\varphi_1 = \varphi_s - \varepsilon, \quad \varphi_2 = \cdots = \varphi_N = \varphi_s \quad (12.24)$$

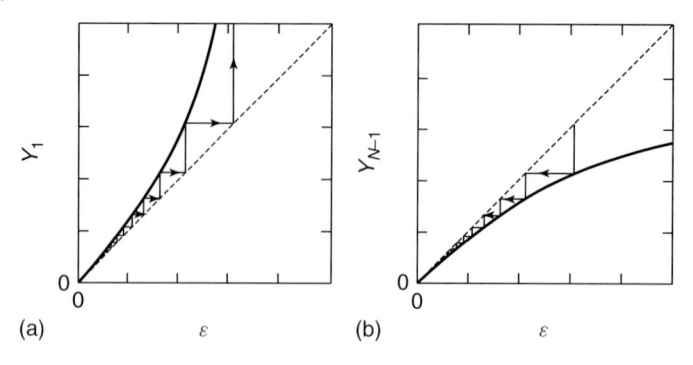

Figure 12.8 Local Cobweb diagram of the functions $Y_1(\varepsilon)$ and $Y_{N-1}(\varepsilon)$ around $\varepsilon = 0$. Iterations of these maps determine the behavior of special perturbations to the one-cluster state. (a) Small perturbations grow with time; (b) small perturbations decay.

with $\varepsilon > 0$. This can be shown similarly to the previous case by obtaining the discrete mapping, which describes the dynamics of the perturbation. In the case of perturbations (Eq. (12.24)), this mapping reads

$$\varepsilon \to Y_{N-1}(\varepsilon) = \mu\left(2\pi - \mu^{N-1}(2\pi - \varepsilon)\right) \tag{12.25}$$

and has the following properties

$$Y_{N-1}(0) = 0,$$
$$Y'_{N-1}(0) = 1,$$

and

$$Y''_{N-1}(0) = -\varkappa Z''(2\pi) + \frac{\varkappa}{N}\left(Z''(2\pi) + Z''(0)\right). \tag{12.26}$$

It implies that for sufficiently large N the second derivative is negative and the origin of the discrete mapping $\varepsilon \to Y_{N-1}(\varepsilon)$ is asymptotically stable (Figure 12.8b). Hence, the one-cluster state is stable with respect to perturbations of the form (Eq. (12.24)). This, together with the instability with respect to perturbations (Eq. (12.20)), implies that the one-cluster state is the saddle point in the phase space (see schematically Figure 12.9).

12.3.2.3 Other Stable and Unstable Local Directions

In general, the two-cluster perturbations of the one-cluster state are given by

$$\varphi_1(0) = \cdots = \varphi_{N_1}(0) = 2\pi, \tag{12.27}$$
$$\varphi_{N_1+1}(0) = \cdots = \varphi_N(0) = 2\pi - \varepsilon, \tag{12.28}$$

where $N_1 + N_2 = N$. This means, there are N_1 oscillators in the front group and the remaining N_2 oscillators in the backgroup. The corresponding discrete 1-D

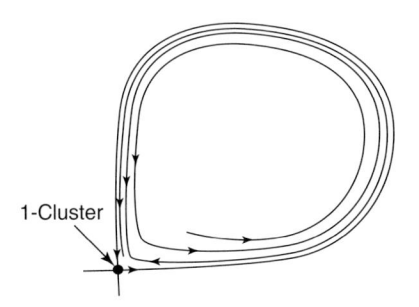

Figure 12.9 One-cluster state as a saddle point in the phase space with a homoclinic loop.

systems, which describe the dynamics of such perturbations are given by

$$\varepsilon \to Y_{N_1}(\varepsilon) \text{ and } \varepsilon \to Y_{N_2}(\varepsilon),$$

where $Y_j(0) = 0$, $\frac{d}{d\varepsilon} Y_j(0) = 1$ for $j = 1, \ldots, N-1$ and

$$\frac{d^2}{d\varepsilon^2} Y_{N_1}(0) = \frac{\varkappa}{N} \left(N_2 Z''(0) - N_1 Z''(2\pi) \right), \tag{12.29}$$

$$\frac{d^2}{d\varepsilon^2} Y_{N_2}(0) = \frac{\varkappa}{N} \left(N_1 Z''(0) - N_2 Z''(2\pi) \right). \tag{12.30}$$

The expressions (Eqs. (12.29) and (12.30)) may have different signs depending on the values of N_1, N_2, as well as the second derivatives $Z''(0)$ and $Z''(2\pi)$. This implies the existence of multiple unstable as well as stable directions to the one-cluster solution, for more details, see Section 12.4.

12.3.3
Stable Homoclinic Orbit to One-Cluster State

Let us first note that the two clusters of the form (Eqs. (12.27) and (12.28)) do not split with time. In geometric terms, this means, that the subspace corresponding to such solutions is invariant. In particular, the subspace that corresponds to $N_1 = 1$ and $N_2 = N - 1$ is invariant as well. Being restricted to this invariant subspace, the one-cluster state is a saddle point, as we have shown in the previous section. In section 12.7, we prove that there exists a homoclinic orbit in this subspace, which connects the both unstable and stable manifolds, see Figure 12.9. In fact, as will be shown in Section 12.4, the dynamics within the invariant subspace is given by the 1-D mapping shown in Figure 12.11b.

Numerical calculations further support this result and show that the invariant set, which is composed of a homoclinic loop and the fixed point is an attractor. Figure 12.10 shows how the width of the cluster changes as time evolves for some typical initial conditions. More specifically, we compute

$$\Delta(t) = \max_{1 \leq i, j \leq N} \left\{ |\varphi_i(t) - \varphi_j(t)| \right\}.$$

One can clearly observe that the width tends eventually to zero interrupted by some blowouts. The blowouts correspond to the events, during which the first oscillator

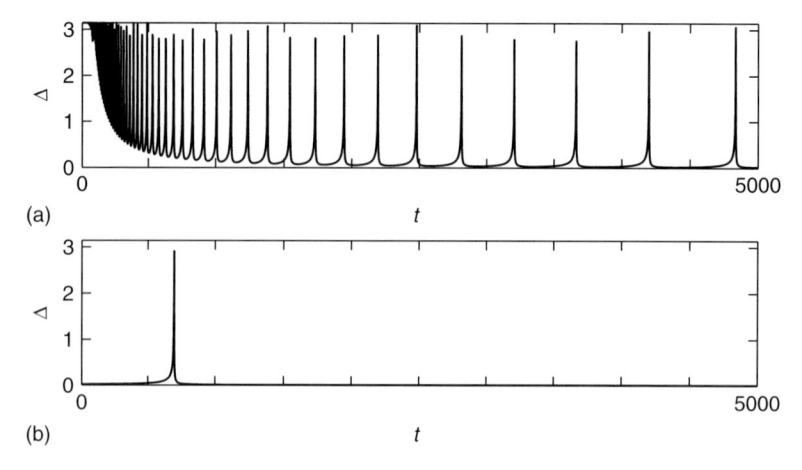

Figure 12.10 Width of the cluster $\Delta(t) = \max_{1 \leq i,j \leq N} \{|\varphi_i(t) - \varphi_j(t)|\}$ as a function of time. Panel (a) shows the behavior along the orbit started at an initial condition close to the splay state (far from the one cluster). Panel (b) shows the behavior along the orbit started close to the state (Eq. (12.20)). The behavior indicates the existence of a stable homoclinic orbit.

leaves behind the remaining cluster and makes a rotation in the phase. After the rotation, it joins again the cluster and becomes the "last" one. The time interval between such events grows unboundedly with time supporting the homoclinic nature of the attractor. Note that the width of the cluster should be nonzero in order to observe this phenomenon, that is, one should perturb the system slightly from the fixed point, see Figure 12.9.

Finally, we would like to remark that the same methods allow proving the existence of other homoclinic orbits, which correspond to two-cluster perturbations (Eqs. (12.27) and (12.28)) with $N_1 \ll N_2$. Hence, one should rather speak about an attracting family of homoclinic orbits.

12.4
Two-Cluster States

Two-cluster state appears when the oscillators split into two groups (Figure 12.1)

$$\varphi_1 = \cdots = \varphi_{N_1} := \psi_1, \quad \varphi_{N_1+1} = \cdots = \varphi_{N_1+N_2} := \psi_2. \tag{12.31}$$

Contrary to one-cluster state, the two-cluster state must not be a fixed point of the return map (Eq. (12.16)). Indeed, when two clusters appear, their relative behavior is then given by the following discrete return map (by assuming that the return map is computed for $\psi_1 = 2\pi$ and $\psi_2 < \psi_1$)

$$\psi_2 \to Y_{N_1}(\psi_2) := 2\pi - \mu^{N_2}\left(2\pi - \mu^{N_1}(\psi_2)\right). \tag{12.32}$$

This map has different properties depending on N_1, $N_2 = N - N_1$, as well as on β. All such maps have zero fixed point corresponding to the case when two clusters merge into one. One can obtain

$$Y_{N_1}(0) = 0, \quad Y_{N_1}(2\pi) = 2\pi,$$
$$Y'_{N_1}(0) = 1, \quad Y'_{N_1}(2\pi) = 1,$$

and

$$Y''_{N_1}(0) = \frac{\varkappa}{N}\left(N_2 Z''(0) - N_1 Z''(2\pi)\right),$$
$$Y''_{N_1}(2\pi) = \frac{\varkappa}{N}\left(N_2 Z''(2\pi) - N_1 Z''(0)\right).$$

Figure 12.11 shows typical maps for three different situations. (a) The map has an unstable fixed point inside the interval $[0, 2\pi]$ and the endpoints $x = 0$ and $x = 2\pi$ are asymptotically stable. Hence, within the corresponding subspace, the one-cluster state is asymptotically stable (similarly to Figure 12.8b). (b) The map has unstable fixed point at $x = 0$ and stable at $x = 2\pi$. This case corresponds exactly to the case, when the one-cluster state has a homoclinic orbit starting in $x = 0$ and ending at $x = 2\pi$ ($0 \sim 2\pi$ on the torus). (c) The map has a stable fixed point inside the interval $[0, 2\pi]$ and the endpoints $x = 0$ and $x = 2\pi$ are unstable. Hence, within the corresponding subspace, the one-cluster state is asymptotically unstable and the two-cluster stationary state is stable.

The fixed points of the map (Eq. (12.32)) give two-cluster stationary states:

$$\psi_2 = Y_{N_1}(\psi_2). \tag{12.33}$$

The condition for the merging of two cluster into one cluster is given by the condition for the existence of the double root of the function $Y_{N_1}(\psi)$ at $\psi = 0$ or $\psi = 2\pi$, that is, $Y''_{N_1}(0) = 0$ or $Y''_{N_1}(2\pi) = 0$. This results into

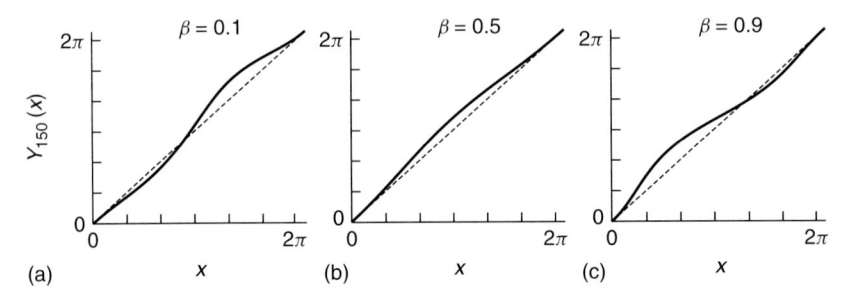

$$N_1 Z''(0) = N_2 Z''(2\pi). \tag{12.34}$$

Figure 12.11 Typical behavior of functions $Y_{N_1}(x)$ (Eq. (12.32)), which determine the behavior of two clusters, $N_1 = 150$ and $N = 500$.

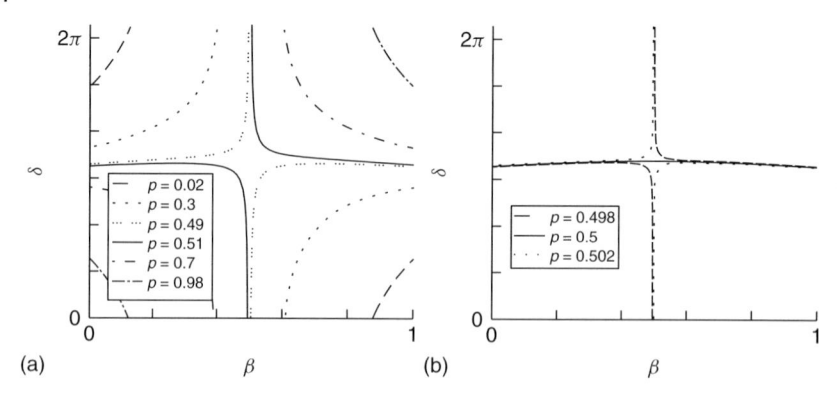

Figure 12.12 Positions of the two-cluster states $\delta = 2\pi - \psi_2$, where ψ_2 are fixed points of Eq. (12.33). Different lines correspond to different cluster splittings, that is, $N_1 = pN$, $N_2 = (1-p)N$. At $\delta = 0$ or $\delta = 2\pi$, the corresponding two cluster is merging into the one cluster.

Expression (Eq. (12.34)) determines also the moments when one-cluster state undergoes bifurcations. At such bifurcation, two different nonsymmetric two clusters bifurcate from the one-cluster state: one with $N_1 = pN$, $N_2 = (1-p)N$, and another with $N_1 = (1-p)N$, $N_2 = pN$. The bifurcation diagram in Figure 12.12 shows some of the branches of two clusters, which originate from $\psi_2 = 0$ or $\psi_2 = 2\pi$.

The bifurcations for $\beta < 0.5$ are subcritical. Namely, the two-cluster states are unstable and they merge into the one-cluster state. With increasing β the one-cluster state becomes more and more locally unstable by transforming stable directions into homoclinics (Figure 12.11). In spite of this fact, we observe numerically, that the invariant set, which is composed of the one-cluster state and homoclinic connections is still attracting in the phase space. All two-cluster states, which exist at this moment, are unstable. As a result, one computes high values of the order parameters R_1 and R_2 on the numerically obtained Figure 12.7 for $\beta < 0.5$.

12.4.1
Stability of Two-Cluster States

For $\beta > 0.5$, the invariant set composed of one-cluster state and homoclinic orbits losses its stability and two-cluster states emerge, which are asymptotically stable. Numerical results in Figure 12.13 show which two clusters are stable depending on the parameter β. In general, for β closer to 0.5, the symmetric clusters with $p \approx 0.5$ are stable. As β increases, the more asymmetric clusters stabilize as well. This implies that the PRCs with the maximum, which is shifted toward the left favor the coexistence of a large number of stable branches of two clusters.

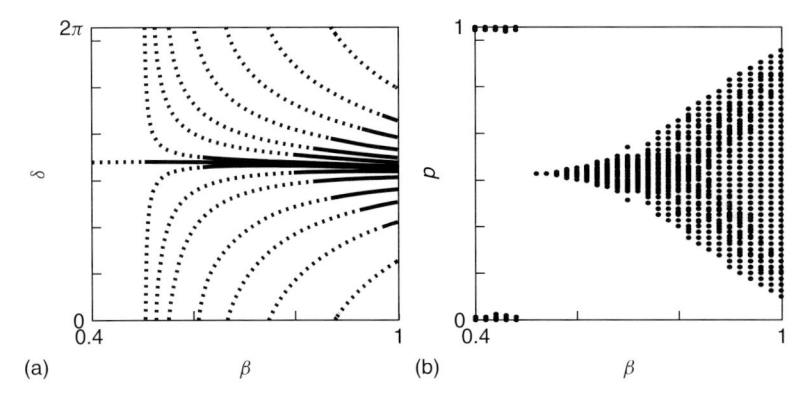

Figure 12.13 Stability and existence of two-cluster states. (a) Solid lines denote stable two-cluster stationary states and dashed unstable. The lines are shown only for selected values of $p = N_1/N$, while the dense set of branches for all possible p exist. Panel (b) shows which two clusters are stable in dependence on β (obtained numerically). $p = 0.5$ corresponds to the symmetric cluster and $p \neq 0.5$ to nonsymmetric clusters.

12.5 Intermediate State for Symmetric PRC with $\beta = 0.5$

The case of symmetric PRC for $\beta = 0.5$ is degenerate. When increasing β through 0.5, the homoclinic sets including the one-cluster state become unstable and a two-cluster state becomes stable as it is described in the previous section. The numerical calculations for $\beta = 0.5$ show nonstationary dependence of the order parameters on time, see Figure 12.14. One observes periods of time, when two clusters persist. These periods are characterized by almost constant order parameters. The periodic blowouts of the order parameters correspond to the behavior, during which the oscillators from the advancing cluster spread over a big

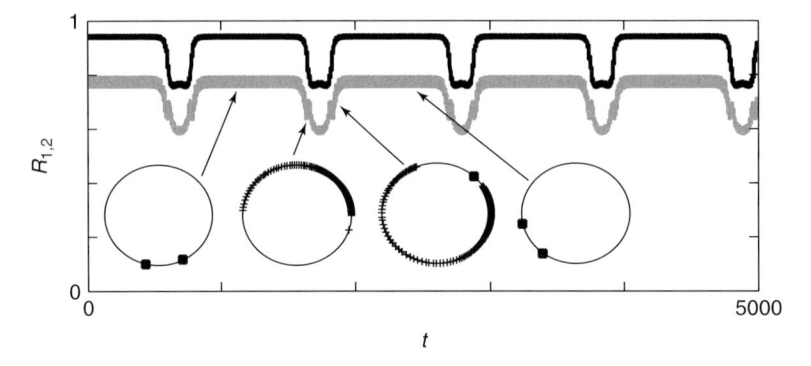

Figure 12.14 Nonstationary behavior of the order parameters R_1 and R_2 with time for $\beta = 0.5001$. One observes periodic restructuring of two clusters.

part of the phase circle and finally form another cluster behind (see the inset in Figure 12.14).

12.6
Conclusions

In this chapter, we have studied the asymptotic behavior of a system of globally pulse-coupled phase oscillators (Eqs. (12.9) and (12.10)) with the phase-response function, which is positive, unimodal, and turns zero at the threshold together with its first derivative. In particular, we considered the question how the position of the maximum of the PRC influences the dynamics of the coupled system.

We have numerically observed that for the PRCs with the maximum shifted to the right (for our model, it corresponds to $\beta < 0.5$), a one-cluster state becomes apparently stable. More detailed analysis reveals that the one-cluster state is, in fact, asymptotically locally unstable, that is, a generic small perturbation will grow with time. Moreover, we show that trajectories of the system has a behavior, which is characterized by long-time intervals when the system stays close to the one-cluster state and long excursions away from the one-cluster state (Figure 12.10). The excursions become less and less frequent with time. This behavior is explained by the existence of the family of homoclinic orbits to the one-cluster state, which altogether form an attracting set in the phase space of the system.

In the case, when the maximum of the PRC is shifted to the left, that is, the oscillators are mostly sensitive to perturbations in the phase just after the threshold, the one-cluster state no more dominates the dynamics and various stationary two-cluster states become stable. These two-cluster states bifurcate from the one-cluster state as parameter β increases. First, at $\beta = 0.5$, there appears a symmetric two cluster with equal number of oscillators in each cluster. With further increasing β more and more asymmetric clusters appear and become stable leading to the increasing coexistence of stable two clusters.

12.7
Appendix: Existence of a Homoclinic Orbit

Theorem *For $\beta \in (0, 1)$, there exists N_0, such that for populations of size $N > N_0$, system (Eq. (12.16)) possesses a homoclinic trajectory, which connects the one-cluster stationary state. The homoclinic trajectory has the form*

$$2\pi = \varphi_2(n) = \cdots \varphi_N(n) \neq \varphi_1(n) \tag{12.35}$$

where $\lim_{n \to -\infty} \varphi_1(n) = 0^+$ and $\lim_{n \to +\infty} \varphi_1(n) = 2\pi^-$.

Proof. Fix $\beta \in (0, 1)$. We will consider

$$Y_1(N, x) = 2\pi - \mu\left(N, 2\pi - \mu^{N-1}(N, x)\right)$$

12.7 Appendix: Existence of a Homoclinic Orbit

where $\mu^j(N, x)$ denotes the jth iteration of

$$x \mapsto \mu(N, x) = x + \frac{x}{N} Z_\beta(x).$$

∎

The map $Y_1(N, x)$ describes the evolution of the distance $x \in (0, 2\pi)$ during a time interval in which all oscillators of a population $\varphi_1 = x;\ \varphi_2 = \cdots = \varphi_N = 2\pi$, emit exactly one spike. Homoclinicity then is equivalent to

$$Y_1^k(N, x) \to 2\pi, \text{ for all } x \in (0, 2\pi), \text{ as } k \to \infty$$

where $Y_1^k(N, x)$ denotes the kth iteration of $x \mapsto Y_1(N, x)$. Analogously to the analysis of Section 12.4, we find that

$$Y_1(N, 0) = 0, \quad Y_1(N, 2\pi) = 2\pi,$$
$$Y_1'(N, 0) = Y_1'(N, 2\pi) = 1,$$
$$Y_1''(N, 0) > 0, \quad Y_1''(N, 2\pi) > 0.$$

Here and in the following, primes denote the derivatives with respect to the second argument (phase). For fixed N, there exists a rejecting region $(0, \varepsilon_N)$, where $Y_1''(N, x) > 0$, for $x \in (0, \varepsilon_N)$ and an attracting region $(2\pi - \varepsilon_N, 2\pi)$ with $Y_1''(N, x) > 0$, for $x \in (2\pi - \varepsilon_N, 2\pi)$. This gives

$$Y_1^{k_0}(N, x) > \varepsilon_N,$$

for $x \in (0, \varepsilon_N)$ and some finite $k_0 = k_0(N, x) \in \mathbb{N}$, and

$$Y_1^k(N, x) \to 2\pi,$$

for $k \to \infty$ and $x \in (2\pi - \varepsilon_N, 2\pi)$. Our goal is to show, that there exists a uniform $\varepsilon_0 > 0$, such that for all $N > N_0$:

$$Y_1''(N, x) > 0, \quad \text{for } x \in (0, \varepsilon_0) \quad \text{and}$$
$$Y_1''(N, x) > 0, \quad \text{for } x \in (\varepsilon_0, 2\pi - \varepsilon_0),$$

and such that for all $N > N_0$ and all $x \in [\varepsilon_0, 2\pi - \varepsilon_0]$:

$$Y_1(N, x) > x + \Delta_N,$$

with

$$\Delta_N := \min_{x \in [\varepsilon_N, 2\pi - \varepsilon_N]} Y_1(N, x) - x > 0.$$

Thus, any $x \in (0, 2\pi)$ will reach the attracting region $(2\pi - \varepsilon_0, 2\pi)$ within a finite number of iterations of $x \mapsto Y_1(N, x)$. Let us write

$$Y_1(N, x) = \tilde{Y}_1(x) + \frac{1}{N} w(N, x),$$

where $\tilde{Y}_1(x) = x + x Z_\beta(x)$ is independent of N. For $\tilde{Y}_1$ we have

$$\tilde{Y}_1(x) > x, \quad \text{for } x \in (0, 2\pi),$$
$$\tilde{Y}_1(0) = 0, \ \tilde{Y}_1(2\pi) = 2\pi,$$
$$\tilde{Y}_1'(0) = \tilde{Y}_1'(2\pi) = 1,$$

This implies for
$$w(N, x) = N\left(Y_1(N, x) - \tilde{Y}_1(x)\right)$$
that
$$w(N, 0) = w(N, 2\pi) = 0,$$
$$w'(N, 0) = w'(N, 2\pi) = 0.$$

We will show that the region $[0, \varepsilon_N]$ may be chosen as $[0, \varepsilon_0]$, independently on large N. The analysis for the other region $[2\pi - \varepsilon_0, 2\pi]$ can be done similarly. Around $x = 0$, we have the following representation of $Y_1(N, x)$:

$$Y_1(N, x) = Y_1(N, 0) + Y_1'(N, 0) x + \frac{x^2}{2} Y_1''(N, \xi_N)$$
$$= \tilde{Y}_1(0) + \tilde{Y}_1'(0) x + \frac{x^2}{2} \tilde{Y}_1''(\xi_N)$$
$$+ \frac{1}{N}\left(w(N, 0) + w'(N, 0) x + \frac{x^2}{2} w''(N, \xi_N)\right)$$
$$= x + \frac{x^2}{2}\left(\tilde{Y}_1''(\xi_N) + \frac{1}{N} w''(N, \xi_N)\right),$$

for some $\xi_N \in [0, \varepsilon]$. Further it holds $\tilde{Y}_1''(0) > 0$. This means, there exists an $\varepsilon_0 > 0$, such that for $x \in [0, \varepsilon_0]$, $\tilde{Y}_1''(x) > 0$. Now, we construct an N-independent lower bound for $w''(N, x)$ in $x \in [0, \varepsilon_0]$, where ε_0 will be further altered in the analysis without always choosing a new notation. In other words, we claim that there exists $c_0 \in \mathbb{R}$ with

$$\liminf_{N \to \infty} \left(\min_{x \in [0, \varepsilon_0]} w''(N, x)\right) > c_0. \tag{12.36}$$

We have
$$w(N, \varepsilon) = N\left(Y_1(N, x) - \tilde{Y}_1(x)\right)$$
$$= N\left(2\pi - \mu\left(2\pi - \mu^{N-1}(N, x)\right) - x - x Z_\beta(x)\right)$$
$$= N\left(\mu^{N-1}(N, x) - \frac{x}{N} Z_\beta\left(2\pi - \mu^{N-1}(N, x)\right) - x - x Z_\beta(x)\right)$$
$$= N\left(\frac{x}{N}\sum_{j=0}^{N-2} Z_\beta\left(\mu^j(N, x)\right) - \frac{x}{N} Z_\beta\left(2\pi - \mu^{N-1}(N, x)\right) - x Z_\beta(x)\right)$$
$$= x \underbrace{\sum_{j=0}^{N-2}\left(Z_\beta\left(\mu^j(N, x)\right) - Z_\beta(x)\right) - x Z_\beta\left(2\pi - \mu^{N-1}(N, x)\right) - x Z_\beta(x)}_{\equiv I}.$$

Since part I, as well as its derivatives, is obviously uniformly bounded in N and x, we restrict us to establish Eq. (12.36) for

$$\tilde{w}(N, x) = \sum_{j=0}^{N-2}\left[Z_\beta\left(\mu^j(N, x)\right) - Z_\beta(x)\right].$$

We have

$$\tilde{w}'(N,x) = \sum_{j=0}^{N-2}\left[Z'_\beta\left(\mu^j(N,x)\right)\left(\mu^j(N,x)\right)' - Z'_\beta(x)\right],$$

$$\tilde{w}''(N,x) = \sum_{j=0}^{N-2}\left[Z''_\beta\left(\mu^j(N,x)\right)\left(\left(\mu^j(N,x)\right)'\right)^2 \right. \quad (12.37)$$

$$\left. + Z'_\beta\left(\mu^j(N,x)\right)\left(\mu^j(N,x)\right)'' - Z''_\beta(x)\right].$$

To handle this, we need some uniformity properties of $\mu^j(N,x)$. Elementary calculations give

$$\left(\mu^j(N,x)\right)' = \prod_{k=0}^{j-1}\mu'\left(N,\mu^k(N,x)\right) = \prod_{k=0}^{j-1}\left(1 + \frac{\varkappa}{N}Z'_\beta\left(\mu^k(N,x)\right)\right),$$

$$\left(\mu^j(N,x)\right)'' = \sum_{l=0}^{j}\prod_{k=0, k\neq l}^{j-1}\left[\left(1 + \frac{\varkappa}{N}Z'_\beta\left(\mu^k(N,x)\right)\right)\right]$$

$$\times \frac{\varkappa}{N}Z''_\beta\left(\mu^l(N,x)\right)\left(\mu^l(N,x)\right)'.$$

This implies that the following inequality

$$0 < \left(\mu^j(N,x)\right)' < \exp(\varkappa\zeta'), \quad \text{where} \quad \zeta' \equiv \max_{x\in[0,2\pi]}\left|Z'_\beta(x)\right| \quad (12.38)$$

holds for all large enough N. This again yields

$$x \leq \mu^j(N,x) = \mu^j(N,0) + \int_0^x \left(\mu^j(N,y)\right)' dy$$

$$\leq x + x\exp(\varkappa\zeta'). \quad (12.39)$$

Using this upper bound, we get some N-independent ε_0, such that for $x \in [0,\varepsilon_0]$

$$Z''_\beta\left(\mu^k(N,x)\right) > 0.$$

This gives N-independent monotonicity of

$$x \mapsto Z'_\beta\left(\mu^k(N,x)\right) > 0 \quad \text{for} \quad x \in (0,\varepsilon_0).$$

Further, we can use Eq. (12.39) to improve the bounds (Eq. (12.38)) for $\left(\mu^j(N,x)\right)'$ in $x \in [0,\varepsilon_0]$ to

$$1 \leq \left(\mu^j(N,x)\right)' < \exp(\varkappa\zeta'). \quad (12.40)$$

This implies

$$\mu^j(N,x) < x \cdot \exp(\varkappa\zeta').$$

We find

$$0 < \left(\mu^j(N,x)\right)'' \leq \frac{j\varkappa\zeta''}{N}\exp(2\varkappa\zeta') \leq \varkappa\zeta''\exp(2\varkappa\zeta')$$

where

$$\zeta'' \equiv \max_{x \in [0, 2\pi]} \left| Z_\beta''(x) \right|.$$

Now observe that

$$Z_\beta''(0) + Z_\beta'''(0) = \left(4 - \frac{8}{\pi}\right) \beta^2 + \frac{2}{\pi} \beta + \frac{1}{\pi} > 0,$$

that is, eventually further decreasing of $\varepsilon_0 > 0$ gives, with $\tilde{\varepsilon}_0 = \varepsilon_0 \cdot \exp(\varkappa \zeta')$:

$$\min_{y \in [0, \tilde{\varepsilon}_0]} Z_\beta'''(y) > - \min_{y \in [0, \tilde{\varepsilon}_0]} Z_\beta''(y). \tag{12.41}$$

Hence, for $x \in [0, \varepsilon_0]$:

$$\tilde{w}''(N, x) = \sum_{j=0}^{N-2} \left(Z_\beta''(\mu^j) \left((\mu^j)'\right)^2 + \underbrace{Z_\beta'(\mu^j) (\mu^j)''}_{\geq 0} - Z_\beta''(x) \right)$$

$$\geq \sum_{j=0}^{N-2} \left(Z_\beta''(\mu^j) \left((\mu^j)'\right)^2 - Z_\beta''(\mu^j) + \int_x^{\mu^j} Z_\beta'''(y) \, dy \right)$$

$$= \sum_{j=0}^{N-2} \left(Z_\beta''(\mu^j) \left(\left((\mu^j)'\right)^2 - 1\right) + \int_x^{\mu^j} Z_\beta'''(y) \, dy \right)$$

$$\geq \sum_{j=0}^{N-2} \left(\min_{y \in [0, \tilde{\varepsilon}_0]} Z_\beta''(y) \left(\left((\mu^j)'\right)^2 - 1\right) + \min_{y \in [0, \tilde{\varepsilon}_0]} Z_\beta'''(y) (\mu^j - x) \right)$$

where we have omitted the arguments (N, x) of μ for brevity. Using Eq. (12.41), we continue the estimations

$$\cdots \geq \sum_{j=0}^{N-2} \left(\min_{y \in [0, \tilde{\varepsilon}_0]} Z_\beta''(y) \left(\left((\mu^j)'\right)^2 - 1 - (\mu^j - x)\right) \right)$$

$$= \min_{y \in [0, \tilde{\varepsilon}_0]} Z_\beta''(y) \sum_{j=0}^{N-2} \left(\underbrace{\left((\mu^j)'\right)^2}_{\geq 1} - 1 - \int_0^x \underbrace{\left((\mu^j)' - 1\right)}_{\leq \mu^j} dy \right)$$

$$\geq \min_{y \in [0, \tilde{\varepsilon}_0]} Z_\beta''(y) \sum_{j=0}^{N-2} \left(\left((\mu^j)'\right) - 1 - x \left((\mu^j)' - 1\right) \right)$$

$$\geq \min_{y \in [0, \tilde{\varepsilon}_0]} Z_\beta''(y) \sum_{j=0}^{N-2} (1 - x) \left((\mu^j)' - 1\right) \geq 0.$$

This establishes Eq. (12.36) and hence $Y_1(N, x) > x \in [0, \varepsilon_0]$ for large enough N.

References

1. Pikovsky, A., Rosenblum, M., and Kurths, J. (2001) *Synchronization: A Universal Concept in Nonlinear Sciences*, Cambridge University Press.
2. Strogatz, S.H. (2001) Exploring complex networks. *Nature*, **410**, 268–276.
3. Strogatz, S.H., Abrams, D.M., McRobie, A., Eckhardt, B., and Ott, E. (2005) Theoretical mechanics: Crowd synchrony on the millennium bridge. *Nature*, **438** (7064), 43–44.
4. Tass, P. (1999) *Phase Resetting in Medicine and Biology: Stochastic Modelling and Data Analysis*, Springer Series in Synergetics, Springer.
5. Timme, M., Geisel, T., and Wolf, F. (2006) Speed of synchronization in complex networks of neural oscillators: analytic results based on random matrix theory. *Chaos*, **16**, 015 108.
6. Popovych, O.V., Hauptmann, C., and Tass, P.A. (2006) Control of neuronal synchrony by nonlinear delayed feedback. *Biol. Cybern.*, **95** (1), 69–85.
7. Wünsche, H.J., Bauer, S., Kreissl, J., Ushakov, O., Korneyev, N., Henneberger, F., Wille, E., Erzgräber, H., Peil, M., Elsäßer, W., and Fischer, I. (2005) Synchronization of delay-coupled oscillators: a study of semiconductor lasers. *Phys. Rev. Lett.*, **94**, 163 901-1–163 901-4.
8. Fischer, I., Vicente, R., Buldú, J.M., Peil, M., Mirasso, C.R., Torrent, M.C., and Garcia-Ojalvo, J. (2006) Zero-lag long-range synchronization via dynamical relaying. *Phys. Rev. Lett.*, **97**, 123902.
9. Yanchuk, S., Schneider, K.R., and Recke, L. (2004) Dynamics of two mutually coupled semiconductor lasers: instantaneous coupling limit. *Phys. Rev. E*, **69**, 056 221-1–056 221-12.
10. Yanchuk, S., Stefanski, A., Kapitaniak, T., and Wojewoda, J. (2006) Dynamics of an array of coupled semiconductor lasers. *Phys. Rev. E*, **73**, 016209.
11. Cuomo, K.M. and Oppenheim, A.V. (1993) Circuit implementation of synchronized chaos with applications to communications. *Phys. Rev. Lett.*, **71** (1), 65–68. DOI: 10.1103/PhysRevLett.71.65.
12. Kanter, I., Kopelowitz, E., and Kinzel, W. (2008) Public channel cryptography: Chaos synchronization and hilbert's tenth problem. *Phys. Rev. Lett.*, **101** (8), 084 102.
13. Perlikowski, P., Stefanski, A., and Kapitaniak, T. (2008) 1:1 mode locking and generalized synchronization in mechanical oscillators. *J. Sound Vibrat.*, **318** (1–2), 329–340.
14. Perlikowski, P., Yanchuk, S., Wolfrum, M., Stefanski, A., Mosiolek, P., and Kapitaniak, T. (2010) Routes to complex dynamics in a ring of unidirectionally coupled systems. *Chaos*, **20**, 013 111.
15. Elble, R. and Koller, W. (1990) *Tremor*, The John Hopkins University Press.
16. Tass, P.A. (2003) A model of desynchronizing deep brain stimulation with a demand-controlled coordinated reset of neural subpopulations. *Biol. Cybern.*, **89** (2), 81–88.
17. Mirollo, R. and Strogatz, S. (1990) Synchronization of pulse-coupled biological oscillators. *SIAM J. Appl. Math.*, **50** (6), 1645–1662.
18. Bottani, S. (1996) Synchronization of integrate and fire oscillators with global coupling. *Phys. Rev. E*, **54** (3), 2334–2350.
19. Tsodyks, M., Mitkov, I., and Sompolinsky, H. (1993) Pattern of synchrony in inhomogeneous networks of oscillators with pulse interactions. *Phys. Rev. Lett.*, **71** (8), 1280–1283.
20. Goel, P. and Ermentrout, B. (2002) Synchrony, stability, and firing patterns in pulse-coupled oscillators. *Phys. D: Nonlin. Phenomena*, **163** (3-4), 191–216.
21. LaMar, M.D. and Smith, G.D. (2010) Effect of node-degree correlation on synchronization of identical pulse-coupled oscillators. *Phys. Rev. E*, **81** (4), 046 206.
22. Guardiola, X., Díaz-Guilera, A., Llas, M., and Pérez, C.J. (2000) Synchronization, diversity, and topology of networks of integrate and fire oscillators. *Phys. Rev. E*, **62** (4), 5565–5570.
23. Bressloff, P.C., Coombes, S., and de Souza, B. (1997) Dynamics of

a ring of pulse-coupled oscillators: Group-theoretic approach. *Phys. Rev. Lett.*, **79** (15), 2791–2794.

24. Ernst, U., Pawelzik, K., and Geisel, T. (1995) Synchronization induced by temporal delays in pulse-coupled oscillators. *Phys. Rev. Lett.*, **74** (9), 1570–1573.

25. Zumdieck, A., Timme, M., Geisel, T., and Wolf, F. (2004) Long chaotic transients in complex networks. *Phys. Rev. Lett.*, **93** (24), 244103.

26. Zillmer, R., Livi, R., Politi, A., and Torcini, A. (2007) Stability of the splay state in pulse-coupled networks. *Phys. Rev. E*, **76** (4), 046102.

27. Abbott, L.F. and van Vreeswijk, C. (1993) Asynchronous states in networks of pulse-coupled oscillators. *Phys. Rev. E*, **48** (2), 1483–1490.

28. Olmi, S., Livi, R., Politi, A., and Torcini, A. (2010) Collective oscillations in disordered neural networks. *Phys. Rev. E*, **81** (4), 046119.

29. Brown, E., Moehlis, J., and Holmes, P. (2004) On the phase reduction and response dynamics of neural oscillator populations. *Neural Comput.*, **16**, 673–715.

30. Izhikevich, E.M. (2005) *Dynamical Systems in Neuroscience: The Geometry of Excitability and Bursting*, The MIT Press.

31. Hoppensteadt, F. and Izhikevich, E. (1997) *Weakly Connected Neural Networks*, Springer-Verlag, New York.

32. Ermentrout, B. (1996) Type i membranes, phase resetting curves, and synchrony. *Neural Comput.*, **8** (5), 979–1001.

13
Broadband Chaos
Kristine E. Callan, Lucas Illing, and Daniel J. Gauthier

13.1
Introduction

The study of chaotic dynamics has been an active area of interdisciplinary research since the 1970s. Today, researchers are interested in practical applications of chaos, such as communications [1, 2], ranging [3], and ultra-wide-band (UWB) sensor networks [4], which require simple devices that produce complex and high-speed dynamics. To produce the high-dimensional chaos required for applications, a nonlinear system needs to have a high-dimensional phase space. One way to achieve this effect in relatively simple devices is to incorporate time-delayed feedback, as depicted in Figure 13.1. Furthermore, since all physical signals travel at finite speeds, it is important to understand how inherent time delays in both natural and man-made systems interact with nonlinearities to influence their behavior.

Time-delayed feedback systems obey delay differential equations (DDEs), rather than ordinary differential equations (ODEs). A DDE is an equation in which the state of a dynamic variable at a given time depends on the values of the dynamic variables at both current and previous times, unlike ODEs where only values at current times matter [5]. An example of a generic DDE with a single time delay τ is given by

$$\dot{x}(t) = F[x(t), x(t-\tau)] \tag{13.1}$$

where F is an arbitrary function of the current and delayed variables.

The phase space corresponding to a DDE is infinite-dimensional, allowing for the possibility of the previously mentioned high-dimensional chaotic solutions. Therefore, systems with sufficiently long time-delayed feedback can often be comprised of a small number of simple components and yet can still give rise to rich dynamics (including chaos) because of their infinite-dimensional phase space.

Studying experimental time-delayed feedback systems has provided much-needed insight about the solutions and properties of particular classes of DDEs. Since DDEs are commonly used to model the behavior of many types of systems (i.e., physiological diseases [6], population dynamics [7], neuronal networks [8], as

Nonlinear Laser Dynamics: From Quantum Dots to Cryptography, First Edition. Edited by Kathy Lüdge.
© 2012 Wiley-VCH Verlag GmbH & Co. KGaA. Published 2012 by Wiley-VCH Verlag GmbH & Co. KGaA.

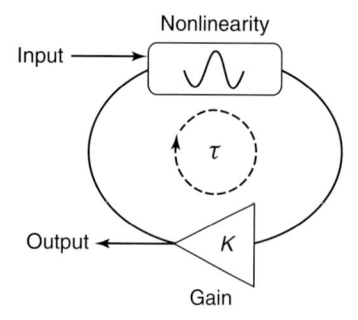

Figure 13.1 Schematic of a simple nonlinear time-delay system with feedback gain K and time delay τ.

well as nonlinear optical devices [9]), these results have important implications for many different fields of study.

One example of such a result is the broadband chaos we observe in a particular nonlinear time-delayed feedback system: an optoelectronic oscillator. The spectra of typical chaotic devices are broadband, yet they often contain several sharp peaks that stand out above the broad background. These features correspond to weakly unstable periodic orbits that comprise the backbone of the strange attractor. The fact that the power spectra for typical chaotic devices are not featureless limits their utility in the applications mentioned above.

In contrast, we will show in this chapter that our optoelectronic oscillator displays high-speed chaos with an essentially featureless power spectrum for certain choices of parameter values, as reported in [10]. The flat nature of the spectrum makes it difficult to distinguish from white noise, which could be attractive for use in applications where one wants there to be a low probability of detecting the deterministic signal.

Additionally, we find that the chaotic behavior coexists with a linearly stable quiescent state. If the system starts in this state, a finite perturbation of sufficient amplitude can force the system to the chaotic state. Furthermore, the transition between the two states takes the form of a train of ultrafast pulses that overlap and merge to eventually give rise to the chaotic solution. We will explain how these observations motivate a nonlinear stability analysis of the steady state, which yields excellent agreement with our experimental results.

13.2
Optoelectronic Oscillators

Optoelectronic oscillators have frequently been used as a benchtop tool for studying nonlinear time-delayed feedback, with their origins dating back to the seminal work of Ikeda [9]. The essential ingredients for such a system include a constant intensity optical power source, a nonlinear device to modulate the optical signal, an element to provide gain to compensate for any losses, and a feedback delay line with a timescale longer than the characteristic timescales of the resulting dynamics. The finite propagation time necessary for light to traverse the loop and its nonlinear interaction with the modulator results in new types of instabilities. In particular, Ikeda showed that multiple stable steady states and periodic states can coexist for the same parameter values. Ikeda also showed numerically that, as the feedback gain is slowly increased, the steady state becomes unstable and subsequently undergoes a period-doubling bifurcation to chaos. Shortly after Ikeda's prediction in 1979, this behavior was first observed experimentally by Gibbs et al. [11] in 1981.

After the pioneering work of Ikeda and others, several more experiments were designed in order to investigate the behavior of nonlinear time-delayed feedback systems. One reason these devices became so popular is that the generated chaos could be of arbitrarily high dimension: Farmer showed that the dimension of chaotic attractor increases as a function of the delay [12]. Additionally, the speed of these systems began to increase with advances in technology, making them even more attractive for certain applications. Along with the increase in speed, however, came components that were ac-coupled, meaning that signals below a certain frequency (f_h) were blocked. This led to a new class of DDEs, which is used to model modern high-speed optoelectronic oscillators.

The dynamics of this new class of high-speed optoelectronic oscillator were recently studied by Peil et al. [13]. They showed both experimentally and numerically that their device was capable of producing a variety of rich behaviors, including fast square-wave solutions, low-frequency periodic solutions, breathers [14], multipulse dynamics, and chaos. In addition to its wide range of dynamics, the utility of this device has also been successfully demonstrated in the realm of secure chaos communication [15]. The chaos generated by the optoelectronic oscillator was used to encode a message, and the resulting signal was transmitted over 120 km of optical fiber using the metropolitan area network of Athens, Greece. The message was then retrieved using chaos synchronization with an identical device at the end of the line. The transmission rates were on the order of gigabits per second. See Chapter 14 by Kanter and Kinzel for a more detailed discussion of secure chaos communication.

Our optoelectronic oscillator is similar to the one studied in [13]. In greater detail, and as shown in Figure 13.2, the beam generated by a continuous-wave semiconductor laser (wavelength 1.55 μm) is injected into a single-mode optical fiber, passes through a polarization controller and optical attenuator, and is injected into a Mach–Zehnder modulator (MZM). Light exiting the modulator passes through an additional piece of single-mode fiber (length ~5 m) serving as a delay line and is incident on a photodetector. Half of the resulting signal, denoted by V, is amplified by an inverting modulator driver (gain $g_{MD} = -22.6$) and fed back to the MZM via the ac-coupled input port. The other half of the signal is directed to a high-speed oscilloscope (8 GHz analog bandwidth, 40 GS s^{-1} sampling rate). The gain in the feedback loop, the bias voltage applied to the MZM, and the length of the time delay are all easily accessible parameters that determine the dynamics of the measured voltage.

To model the dynamics of the optoelectronic system, one needs to consider the nonlinear transmission functions of the MZM and modulator driver, the finite bandwidth of the system components, and the amount of time it takes the signal

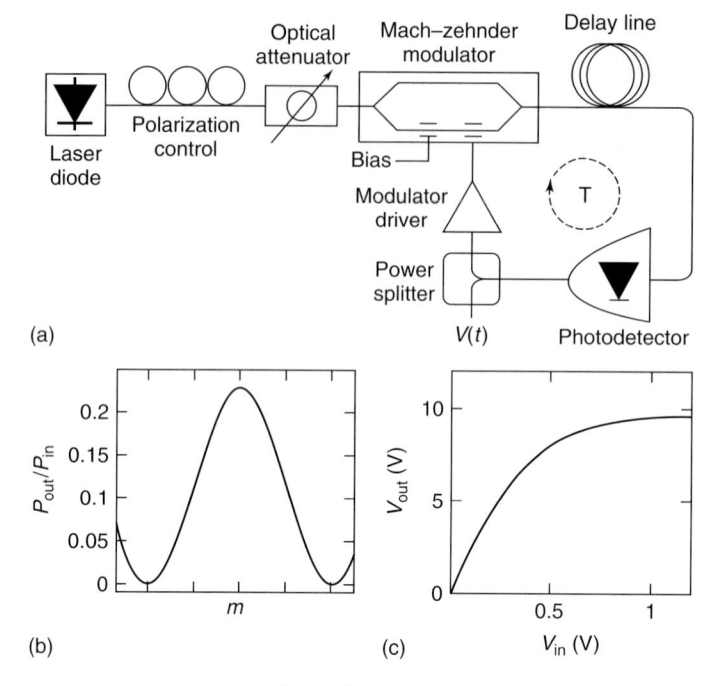

Figure 13.2 (a) Schematic of optoelectronic oscillator. Nonlinear transmission functions of MZM (b) and modulator driver (c).

to propagate from the output of the MZM back to the radio frequency (rf) input of the MZM.

The MZM modulates the intensity of an incident optical signal by exploiting Pockels electro-optic effect in a lithium niobate crystal in one arm of a Mach–Zehnder interferometer. When the signals from each arm of the interferometer are recombined at the output, their resulting interference depends on a constant bias voltage (V_B) and a fluctuating rf voltage ($V_{in}(t)$) applied to two electrodes across the crystal. The optical power (P_{out}) transmitted through the devices is given by

$$P_{out} = P_{in} \cos^2\left[\frac{\pi}{2}\left(\frac{V_B}{V_{\pi,dc}} + \frac{V_{in}}{V_{\pi,rf}}\right)\right] \tag{13.2}$$

where P_{in} is the power incident on the MZM, and $V_{\pi,dc}$ and $V_{\pi,rf}$ characterize the widths of the interference fringe ($V_{\pi,rf} = 7.4$ V, $V_{\pi,dc} = 7.7$ V). The interference fringe obtained by slowly varying V_B is shown in Figure 13.2b.

The modulator driver also has a nonlinear response: it saturates at high voltage with a saturation voltage of $V_{sat} = 9.7$ V. We model this saturation with a hyperbolic tangent function, as shown in Figure 13.2c. We find it essential to take into account this additional nonlinearity, as it limits the extent to which we can access multiple fringes of the MZM interference curve.

The high-speed components in our device are bandpass-coupled so that feedback of both low and high frequencies are suppressed. This differs from Ikeda's original model, which only incorporated low-pass filtering, and it has been shown that the inclusion of a high-pass filter results in fundamentally different dynamics [16]. We use a two-pole bandpass filter to approximate the effects of bandpass coupling, with low- (high-) frequency cutoff $\omega_- = 1.5 \times 10^5$ s^{-1} ($\omega_+ = 7.5 \times 10^{10}$ s^{-1}), center frequency $\omega_0 = \sqrt{\omega_- \omega_+} = 1.1 \times 10^8$ s^{-1}, and bandwidth $\Delta = \omega_+ - \omega_- = 7.5 \times 10^{10}$ s^{-1}.

Finally, we measure the time delay of the feedback loop to be approximately 24 ns. Thus, our oscillator has three widely separated timescales: the time delay of the feedback (on the order of 10 ns), the high-pass filter response time (on the order of microseconds), and the low-pass filter response time (on the order of 10 ps). Other researchers have found that these timescales play a prominent role in the dynamics they observe [13].

By combining the effects of the nonlinearities, bandpass filtering and time delay, we derive an integro-delay differential equation describing the fluctuating voltage $V(t)$ [17]

$$V(t) + \frac{1}{\Delta}\frac{dV(t)}{dt} + \frac{\omega_0^2}{\Delta}\int_0^t V(l)dl$$
$$= G\cos^2\left\{\frac{\pi V_B}{2V_{\pi,dc}} + \frac{\pi V_{sat}}{2V_{\pi,rf}}\tanh\left[\frac{g_{MD} V(t-T)}{V_{sat}}\right]\right\} \quad (13.3)$$

where G characterizes the gain in the feedback loop (proportional to the injected optical power) in units of volts and all other variables have previously been defined. We can then rewrite this integro-delay differential equation as two coupled DDEs given by

$$\frac{1}{\Delta}\frac{dV(t)}{dt} = -V(t) - U(t) + G\cos^2\left\{m + d\tanh\left[\frac{g_{MD}V(t-T)}{V_{sat}}\right]\right\}, \quad (13.4)$$

$$\frac{1}{\Delta}\frac{dU(t)}{dt} = \frac{\omega_0^2}{\Delta^2}V(t) \quad (13.5)$$

where $m = \pi V_B/2V_{\pi,rf}$ is the dimensionless operating point of the nonlinearity and $d = \pi V_{sat}/2V_{\pi,rf}$ characterizes the saturation of the modulator driver. By defining dimensionless variables

$$x = \frac{g_{MD}}{V_{sat}}V, \quad (13.6)$$

$$y = \frac{g_{MD}}{V_{sat}}U - K\cos^2 m, \quad (13.7)$$

and rescaling time ($s = t\Delta$), we obtain two coupled dimensionless DDEs

$$\dot{x}(s) = -x(s) - y(s) + F[x(s - \tau)], \quad (13.8)$$
$$\dot{y}(s) = \epsilon x(s). \quad (13.9)$$

Here, the overdot denotes the derivative with respect to the dimensionless time s, K is the dimensionless feedback loop gain, $\tau = T\Delta$ is the dimensionless time delay,

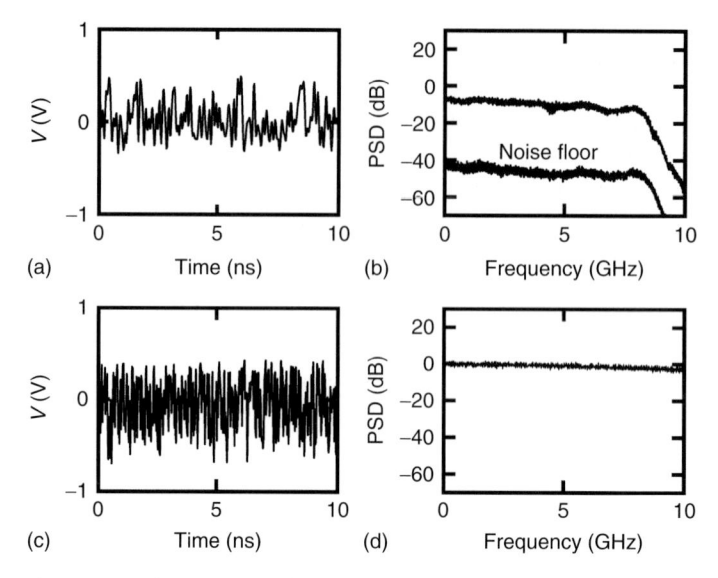

Figure 13.3 The experimental time series (a) and power spectral density (b) of the broadband chaotic behavior in the optoelectronic oscillator for $m = 0.063$ and $K = 3.47$ (upper trace). We find that the spectrum is flattest (i.e., the small peaks corresponding to T nearly vanish) for $m \gtrsim 0$. The power spectral density of the noise floor obtained below threshold (lower trace) is also shown. Theoretically predicted time series (c) and power spectral density (d) for $m = 0.063$ and $K = 3.47$. The numerical time series contains higher frequency components than the experimental time series, since the bandwidth of the oscilloscope affects the experimental time series.

$\epsilon = \omega_0^2/\Delta^2$ characterizes the bandpass filter, and the nonlinear delayed feedback term is

$$F[x] = K\cos^2(m + d\tanh x) - K\cos^2 m. \tag{13.10}$$

In our experiments, three parameters are held fixed ($d = 2.1$, $\tau = 1820$, and $\epsilon = 2.0 \times 10^{-6}$), while K can range from 0–5 by adjusting the injected optical power with an attenuator and m ranges from $-\pi/2$ to $\pi/2$. For future reference, note that V and its dimensionless analog x have opposite signs because $g_{MD} < 0$.

One important distinction of our work is that we bias the MZM near the top of an interference fringe ($m \approx 0$), which, as we will show in the next section, is where the quiescent state of the system is the most linearly stable. It is in this regime where we obtain the broadband chaotic behavior shown in Figure 13.3a. The one-sided power spectral density (PSD) of the experimental chaotic time series with a resolution bandwidth of 8 MHz is shown in Figure 13.3b. One can see that the power spectrum is essentially "featureless," as it is roughly flat up to the cutoff frequency of the oscilloscope used to measure the dynamics (8 GHz). More precisely, the spectrum is contained with a range of 12 dB with a standard deviation of 2 dB for frequencies below 8 GHz.

To further emphasize the flatness of the chaotic spectrum, we compare our results to the case where the oscillator is in the quiescent state, just below the instability threshold to be discussed in the next section. As seen in Figure 13.3b, the PSD is at least 40 dB below the PSD of the chaotic state and is qualitatively consistent with the noise floor of the overall system. Quantitatively, the noise floor is contained within a range of 18 dB with a standard deviation of 2 dB. Comparing the statistics for both spectra shows that the spectrum of the broadband chaos is nearly as featureless as the spectrum of the system noise.

One can obtain a qualitatively similar time series and spectrum by integrating the (noise-free) Eqs.(13.8) and (13.9), as shown in Figure 13.3c,d, indicating that the flat, broad spectrum is due to the deterministic dynamics rather than experimental noise. We also determine a positive largest Lyapunov exponent of $\sim 0.03 \, \text{ns}^{-1}$, showing that the trajectory is indeed chaotic.

In the following sections, we explain how this broadband chaotic solution can be accessed with either experimental noise or a controlled perturbation, despite the fact that we are operating in the regime where the quiescent state is linearly stable.

13.3
Instability Threshold

As a starting point for understanding the dynamics of the oscillator, we first study the linear stability of the single fixed point of Eqs. (13.8) and (13.9). This type of analysis provides insight as to how the system will respond to small perturbations. Chapter 7 by Krauskopf and Walker also discusses the stability and bifurcation properties of DDEs.

The fixed point $(x^*, y^*) = (0, 0)$ is found by setting both derivatives equal to zero and corresponds to the quiescent state of the oscillator. If we then Taylor-series expand the nonlinear term $F[x(t - \tau)]$ about $x^* = 0$ and assume a perturbation of the form $\delta y = e^{\lambda s}$, we obtain the resulting characteristic equation

$$\lambda^2 + \lambda + \epsilon + b\lambda e^{-\lambda \tau} = 0 \tag{13.11}$$

where $b = -Kd\sin(2m)$ is the effective slope of the nonlinearity in the vicinity of the fixed point. Here, λ represents the infinite number of eigenvalues whose real parts determine the stability of the solution. The quiescent state becomes linearly unstable when $\Re[\lambda]$ becomes positive, corresponding to exponential growth away from the steady-state solution. Thus, by setting $\Re[\lambda] = 0$ and $\Im[\lambda] = \Omega$, we determine the instability threshold of the quiescent state from

$$(i\Omega)^2 + i\Omega + \epsilon + b(i\Omega)e^{-i\Omega\tau} = 0. \tag{13.12}$$

Separating the real terms from the imaginary terms gives the following set of equations for the instability threshold

$$-\Omega^2 + \epsilon - b\Omega \sin(\Omega\tau) = 0, \tag{13.13}$$

$$1 - b\cos(\Omega\tau) = 0. \tag{13.14}$$

Note that these equations remain unchanged for $\Omega \to -\Omega$ and, for $\epsilon > 0$, there is no solution for $\Omega = 0$. This implies that the eigenvalues cross the imaginary axis in

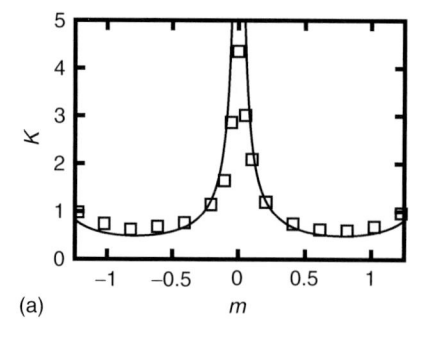

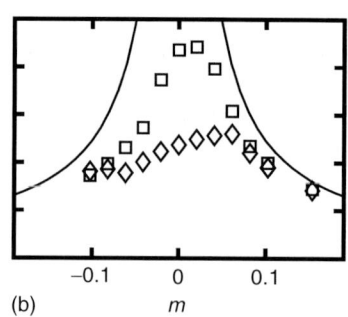

Figure 13.4 Observed values of K, for which the system transitions from quiescent state to oscillatory or pulsing behavior as a function of m, with K_H superimposed (solid line). The squares in (a) and (b) indicate low experimental noise, while the diamonds in (b) indicate a higher noise level, as shown in [10]. (Copyright (2010) by the American Physical Society.)

complex conjugate pairs, which is the signature of a Hopf bifurcation. With τ and ϵ set to the values appropriate for our experimental setup, Eqs. (13.13) and (13.14) can be used to determine the values of Ω and b that give rise to a Hopf bifurcation. While the frequency of the oscillatory motion at the onset of a Hopf bifurcation (Ω) is often of interest, here we are mainly concerned with finding b since, for a given m, it determines the gain for which a Hopf bifurcation occurs according to

$$K_H = -\frac{b}{d\sin(2m)}. \tag{13.15}$$

For the parameter values corresponding to our experimental setup, we find that $b \approx 1$ for $m < 0$ and $b \approx -1$ for $m > 0$. One can see that, for $m = 0$, which corresponds to the operating point at the top of the interference fringe, K_H diverges and the quiescent state of the model is linearly stable for all values of K, as shown by the solid lines in Figure 13.4.

Experimentally, however, we find that the situation is much more complicated than linear stability analysis predicts. In particular, as we increase K near $m = \pm\pi/4$, we find excellent agreement between the value of K for which the quiescent state is destabilized and K_H, as shown in Figure 13.4a. Near $m = 0$, however, we find that the quiescent state is destabilized well before K_H is reached, as shown with the squares in Figure 13.4b. Interestingly, it is also near $m = 0$ where we observe broadband chaos. One can also notice a slight asymmetry in the experimentally determined instability threshold about $m = 0$, which is not predicted by linear stability analysis.

Next, we investigate the influence of noise on the instability threshold by using an erbium-doped fiber amplifier, in succession with an attenuator, to add more noise to the system while keeping the total optical power the same. We find that, by increasing the root-mean-square noise in V by a factor of 2.3 (over a bandwidth from dc to 8 GHz), we observe a substantial decrease in the instability threshold, as shown by the diamonds in Figure 13.4b. In addition, the asymmetry in the threshold is also more pronounced than in the low-noise case. Our findings indicate that the presence of experimental noise in our system, due to laser relaxation oscillations and detector dark and shot noise, is responsible for the deviation from the linear theory. As shown in Section 13.5, these features can be understood with a global (nonlinear) stability analysis of the model.

13.4
Transition to Broadband Chaos

To further explore the region in parameter space where broadband chaos is observed and the dynamics appear to deviate from the linear theory, we look at how the system leaves the quiescent state at the instability threshold for $m \approx 0$. A representative time series of this transient behavior is shown in Figure 13.5. One can see that at around 50 ns, a small pulselike perturbation (due to noise) in V appears. At a time T later, this pulse is regenerated, but with a greater amplitude. Subsequent pulses (with a full width at half maximum of ~200 ps) continue to be generated each T and grow in amplitude until they begin to fold over the nonlinearity around 0.4 V and finally saturate at a maximum amplitude of about

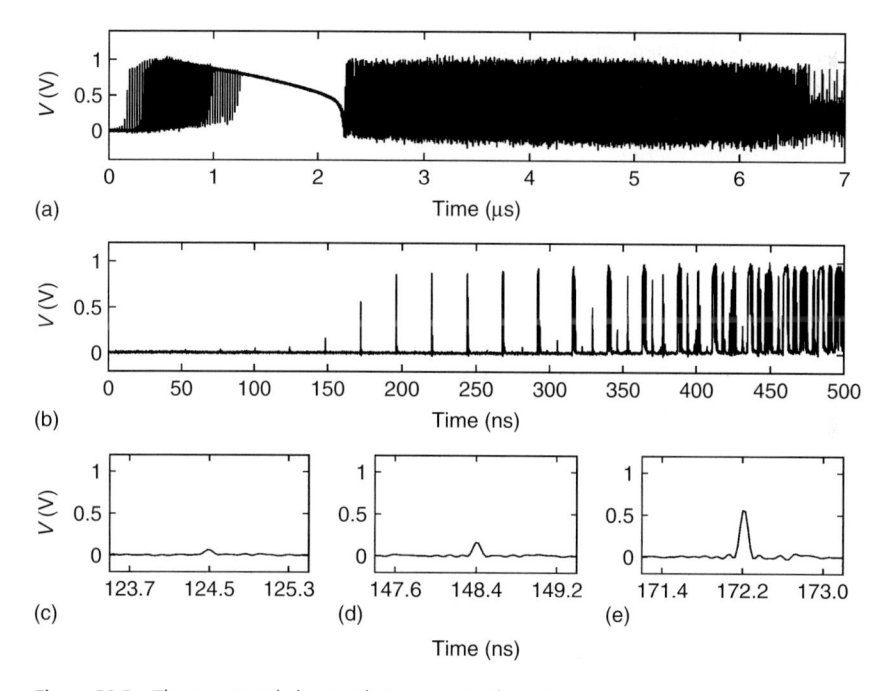

Figure 13.5 The transient behavior that occurs in the optoelectronic oscillator when the quiescent state first loses stability for parameter values $m = 0$ and $K = 4.36$. The initial 7 μs of data (a) shows the complex breather-like behavior. A zoom in of the first 500 ns (b) shows the growing pulse trains, with three of the pulses shown in (c–e).

1 V (corresponding to the input saturation voltage of the amplifier). Since noise spikes occur at random times, there can be more than one of these growing pulse trains contributing to the transient behavior (e.g., notice the spike around 280 ns). In our experiments, this pulsing transient eventually gives rise to the broadband chaos we are interested in. This behavior is also verified in noise-free numerical simulations of Eqs. (13.8) and (13.9).

However, the perturbations necessary to drive the system away from the linearly stable quiescent state do not have to originate from experimental noise. One can also apply a controlled perturbation and study its effect on the system's dynamics, both experimentally and numerically. By injecting 200 ps long electrical pulses of varying amplitudes into the feedback loop for values of K below the instability threshold, we find that, in general, a single input pulse will generate a train of pulses spaced in time approximately by T. For a small initial pulse amplitude, the subsequent pulse train will decay back to the quiescent state. For sufficiently large initial pulse amplitude, however, the subsequent pulse train grows and the steady-state solution is lost. For sufficiently large K, the system transitions to the chaotic state in a manner quite similar to the transient observed when noise was providing the initial perturbation. An example of both a decaying and growing pulse train are shown in Figure 13.6. We observe similar results numerically when we integrate the DDEs using a Gaussian pulse with the same width as in the experiment. The amplitudes V_{th} at which the transition between growth and decay occurs in the experiment (triangles) and simulation (stars) as a function of K are shown in Figure 13.9.

The features of these pulse trains will be exploited in the next section to understand how a small perturbation can be used to switch from the linearly stable quiescent state to broadband chaos.

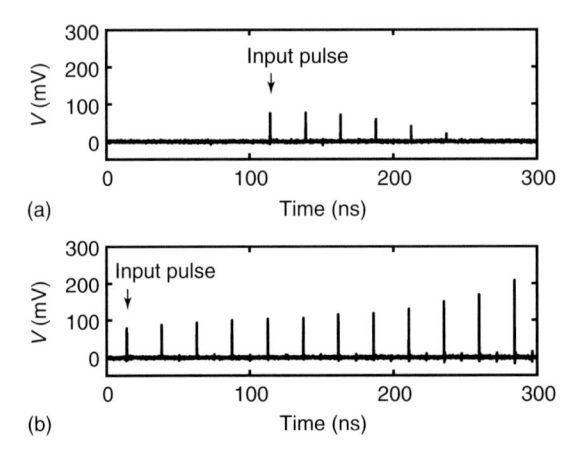

Figure 13.6 Two pulse trains generated by injecting pulses with amplitudes of (a) 75.1 and (b) 78.7 mV into the feedback loop of the optoelectronic oscillator.

13.5
Asymptotic Analysis

Our experimental observations show that, near $m = 0$, the system transitions from steady-state to non-steady-state behavior (sometimes in the form of broadband chaos) if seeded with a pulselike perturbation of sufficient amplitude. To better understand this observation, we consider the phase portrait for Eqs. (13.8) and (13.9) with $m = 0$. The time-delay term $F[x(s - \tau)]$ vanishes when the system is in the quiescent state, leaving us with a two-dimensional ODE. Solving for the nullclines of the system under these conditions gives

$$\dot{x} = 0 \implies y = -x, \tag{13.16}$$
$$\dot{y} = 0 \implies x = 0. \tag{13.17}$$

The intersection of the nullclines at the origin corresponds to the quiescent state of the oscillator. As ϵ is small, motion is slow in the y-direction and trajectories that start away from the stable fixed point are approximately horizontal until they reach the $y = -x$ nullcline.

Now consider what happens to the nullclines if x is perturbed with a short pulse at time $s = 0$. The feedback term $F[x(s - \tau)]$ will become nonzero for a short time in the vicinity of $s = \tau$, because of the pulse from τ earlier. This will effectively shift the $\dot{x} = 0$ nullcline to $y = -x + K\cos^2(d \tanh x_0) - K$, where x_0 is the amplitude of the initial perturbation. Trajectories that start near the origin will be drawn horizontally toward the shifting nullcline in an attempt to reach the new fixed point at (x^*, y^{**}), as shown in Figure 13.7. For a sufficiently short initial pulse, the trajectories will not have enough time to move appreciably in the y-direction

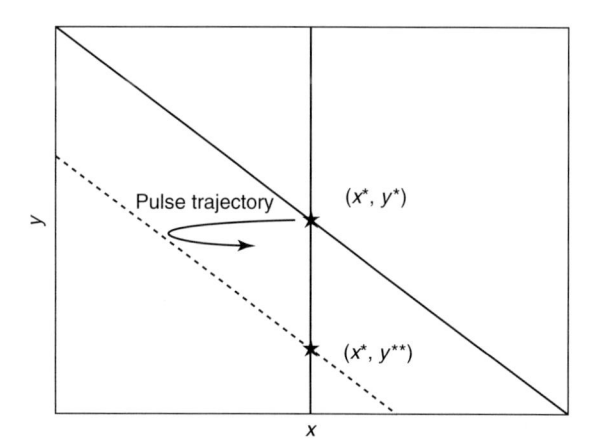

Figure 13.7 The nullclines with and without the presence of a pulse. The $\dot{y} = 0$ nullcline remains unchanged under the influence of a pulse, but the $\dot{x} = 0$ nullcline at $y = -x$ (solid line) is shifted to $y = -x - y^{**}$ (dashed line) when the pulse reaches its maximum amplitude. Trajectories originating near (x^*, y^*) approximately follow the shifting nullcline, but are unable to reach (x^*, y^{**}) since motion parallel to the y-axis is slow.

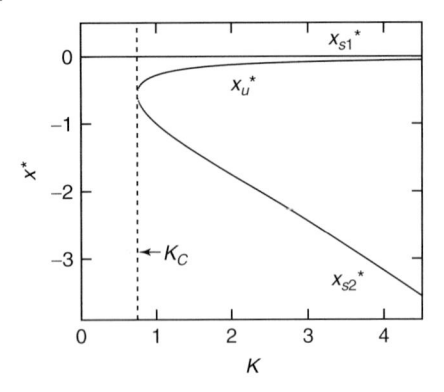

Figure 13.8 Fixed points of the one-dimensional map derived to approximate the system's pulsing behavior. For $K > K_C$ there are three fixed points (x_{s1}^*, x_u^*, and x_{s2}^*).

before the nullcline shifts back. While the motion in the y-direction is negligible, the out-and-back motion in the x-direction approximately reproduces the original pulse, but with a possibly different amplitude. As shown in Figure 13.7, this pulse will be in the negative x-direction, which corresponds to the positive V-direction, regardless of the direction of the initial pulse. The pulse regeneration will continue as described with each τ as long as y does not grow appreciably.

The phase-portrait analysis presented above explains how the system operating near $m = 0$ can produce equally spaced pulses with negative amplitudes in x (positive in V) if first seeded with a pulse. The discrete nature of the trajectories in time serves as motivation to investigate a one-dimensional map of the form

$$x_{n+1} = F(x_n) \tag{13.18}$$

where x_n corresponds to the amplitude of a pulse at time $n\tau$, F is the nonlinear feedback term defined in Eq. (13.10), and the slowly changing variable y has been neglected. One should keep in mind, however, that the map given by Eq. (13.18) only gives approximate predictions of the dynamics of the physical system, as reducing the DDE to a map neglects all effects of the bandpass filter.

For $m = 0$, Eq. (13.18) can have either one or three fixed points, depending on the value of K. The numerical solutions for the fixed points as a function of K are shown in Figure 13.8. Using standard stability analysis we find that fixed point at the origin x_{s1}^* is always stable and corresponds to the quiescent state of the oscillator. When the other two fixed points exist, they are both negative. However, the fixed point with the smaller magnitude x_u^* is unstable, while the fixed point with the greater magnitude x_{s2}^* is stable. Thus, x_u^* forms a separatrix between the quiescent state x_{s1}^* and the pulsating state x_{s2}^* and can be used to give an approximate value for the critical amplitude of a pulse needed to generate a train of pulses with increasing amplitudes. The agreement between x_u^* (when converted to physical units) and the critical amplitude found in the experiment and simulation (as discussed in the previous section) is shown in Figure 13.9.

13.5 Asymptotic Analysis

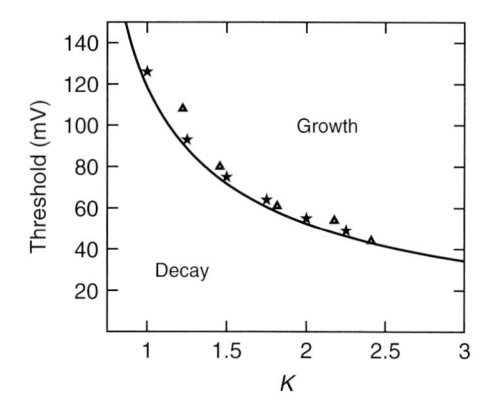

Figure 13.9 The pulse amplitudes V_{th} at the border between growth and decay as a function of K in the experiment (triangles) and simulation (stars) with the unstable fixed point of the map x_u^* superimposed as a solid curve.

Note that, for $m = 0$, the minimum perturbation size predicted by the map decreases asymptotically to zero as a function of K. Thus, if any noise is present, there exists a sufficiently large K such that the system will leave the quiescent state near $m = 0$.

For all values of m, one can also determine the threshold gain K_{th} required to destabilize the quiescent state for a given noise intensity $D = \sqrt{2 < x^2 >}$, where $< x^2 >$ is the variance of x in the stochastic analog of Eqs. (13.8) and (13.9) without feedback. First, consider determining the fixed points of the map by setting $x_{n+1} = x_n$. This gives

$$x^* = K \cos^2[m + d \tanh(x^*)] - K \cos^2(m), \tag{13.19}$$

which can be rewritten as

$$x^* = -K \sin[2m + d \tanh(x^*)] \sin[d \tanh(x^*)]. \tag{13.20}$$

For the unstable fixed point x_u^*, we can use the following approximation

$$x_u^* \approx -K \sin(2m + dx_u^*) dx_u^*, \tag{13.21}$$

because $x_u^* \ll 1$. Next, we set x_u^* equal to $< x^2 >$ to obtain

$$K_{th} \approx \frac{1}{d \sin \frac{d}{\sqrt{2}} D - 2m}. \tag{13.22}$$

As mentioned previously, the separatrix x_u^* and pulsating state x_{s2}^* only exist in a certain region of parameter space. For a given value of m, we can determine the value of K_C where the fixed points x_u^* and x_{s2}^* coalesce, as shown in Figure 13.8. Since x_{s2}^* represents the pulsating state, the transient pulse trains that we observe are only possible for $K > K_C$. We determine K_C numerically and find that there

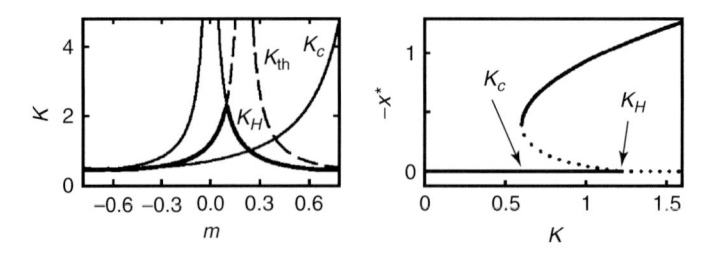

Figure 13.10 Instability thresholds using $D = 0.28$ in K_{th} (left). Stable (solid) and unstable (dashed fixed points versus K for $m = -0.2$ (right).

is a strong asymmetry about $m = 0$, as shown in Figure 13.10, indicating the pulsing behavior is least likely around $m = \pi/4$. Also shown in Figure 13.10 are K_H (Hopf) and K_{th} (noise threshold) for one value of the noise intensity. For $K_{th} < K_H$ ($\sim -\pi/4 < m \lesssim 0.1$), the quiescent state will be destabilized by a pulsing instability. For $K_H < K_{th}$, $K \sim K_H$, and small noise, the quiescent state will be destabilized by the Hopf bifurcation. Thus, we predict that the quiescent state will be unstable for $K > \min[K_H, K_{th}]$. We see that there is qualitative agreement between $\min[K_H, K_{th}]$ highlighted in Figure 13.10 with a thick line, and the high-noise experimental measurements (Figure 13.4b).

13.6
Summary and Outlook

In summary, we have investigated the dynamics of an optoelectronic oscillator operated in a regime where the quiescent state is linearly stable. We find experimentally and numerically that, for sufficiently high feedback gain, the system exhibits high-speed chaos with a featureless power spectrum extending beyond 8 GHz. By analyzing the experimental behavior when the system switches between these two dynamical regimes, we construct a nonlinear discrete map and find that we can predict the amplitude of a perturbation necessary to destabilize the quiescent state.

The broadband, featureless chaos generated by this device may find use in applications such as private chaos communication [15] or chaotic lidar [3], as its noiselike characteristics could improve security. In addition, the switching effect we report might also be useful for these types of applications.

Furthermore, coexisting states are common in time-delay systems, and the existence of such states could influence these systems' stability and performance. For example, optoelectronic microwave oscillators [18], synchronized neuronal networks [19], synthetic gene networks [20], and controlled chaotic systems [21, 22] may have their noise sensitivity or stability adversely affected if a coexisting chaotic state exists and internal or external perturbations to the system are large enough that this state can be accessed.

Acknowledgments

Professor Eckehard Schöll has been an invaluable collaborator on this work, and we have benefited greatly from his extensive knowledge of nonlinear dynamical systems. In particular, the asymptotic analysis described in Section 13.5 of this chapter was largely the result of his work with us during his sabbatical visit to Duke University in Spring 2008. His insights allowed us to develop a simple and intuitive picture of the transient dynamics of the optoelectronic oscillator. We are honored to have our work in this collection that celebrates his contribution to the field of nonlinear dynamics.

DJG and KEC also gratefully acknowledge the financial support of the US Office of Naval Research (N00014-07-1-0734). We thank Lauren Shareshian for helping to determine the Lyapunov exponent.

References

1. Pecora, L.M. and Carroll, T.M. (1990) Synchronization in chaotic systems. *Phys. Rev. Lett.*, **64**, 821–824.
2. Cuomo, K.M. and Oppenheim, A.V. (1993) Circuit implementation of synchronized chaos with applications to communications. *Phys. Rev. Lett.*, **71**, 65–68.
3. Lin, F.Y. and Liu, J.M. (2004) Chaotic lidar. *IEEE J. Sel. Top. Quantum Eletron.*, **10**, 991–997.
4. Li, J., Fu, S., Xu, K., Wu, J., Lin, J., Tang, M., and Shum, P. (2008) Photonic ultrawideband monocycle pulse generation using a single electro-optic modulator. *Opt. Lett.*, **33**, 288–290.
5. Erneux, T. (2009) *Applied Delay Differential Equations*, Springer, New York.
6. Mackey, M.C. and Glass, L. (1997) Oscillation and chaos in physiological control systems. *Science*, **197**, 287–289.
7. Kuang, Y. (1993) *Delay Differential Equations with Applications in Population Dynamics*, Academic Press, San Diego, CA.
8. Wilson, H.R. (1999) *Spikes, Decisions, and Actions: The Dynamical Foundations of Neuroscience*, Oxford University Press, Oxford.
9. Ikeda, K. (1979) Multiple-valued stationary state and its instability of the transmitted light by a ring cavity system. *Opt. Commun.*, **30**, 257–261.
10. Callan, K.E., Illing, L., Gao, Z., Gauthier, D.J., and Schöll, E. (2010) Broadband chaos generated by an optoelectronic oscillator. *Phys. Rev. Lett.*, **104**, 1113901.
11. Gibbs, H.M., Hopf, F.A., Kaplan, D.L., and Shoemaker, R.L. (1981) Observation of chaos in optical bistability. *Phys. Rev. Lett.*, **46**, 474–477.
12. Farmer, J.D. (1982) Chaotic attractors of an infinite-dimensional dynamical system. *Physica D*, **4**, 366–393.
13. Peil, M., Jacquot, M., Chembo, Y., Larger, L., and Erneux, T. (2009) Routes to chaos and multiple time scale dynamics in broadband bandpass nonlinear delay electro-optic oscillators. *Phys. Rev. E*, **79**, 026208.
14. Kouomou, Y.C., Colet, P., Larger, L., and Gastaud, N. (2005) Chaotic breathers in delayed electro-optical systems. *Phys. Rev. Lett.*, **95**, 203903.
15. Argyris, A., Syvridis, D., Larger, L., Annovazzi-Lodi, V., Colet, P., Fischer, I., García-Ojalvo, J., Mirasso, C.R., Pesquera, L., and Shore, K.A. (2005) Chaos-based communications at high bit rates using commercial fibre-optic links. *Nature*, **438**, 343–346.
16. Illing, L. and Gauthier, D.J. (2005) Hopf bifurcations in time-delay systems with band-limited feedback. *Physica D*, **210**, 180–202.

17. Udaltsov, V.S., Larger, L., Goedgebuer, J., Lee, M.W., Genin, E., and Rhodes, W.T. (2002) Bandpass chaotic dynamics of electronic oscillator operating with delayed nonlinear feedback. *IEEE Trans. Circuit. Syst. I*, **49**, 1006–1009.
18. Chembo, Y., Larger, L., and Colet, P. (2008) Nonlinear dynamics and spectral stability of optelectronic microwave oscillators. *IEEE J. Quantum Electron.*, **44**, 858–866.
19. Vicente, R., Gollo, L., Mirasso, C., Fischer, I., and Pipa, G. (2008) Dynamical relaying can yield zero time lag neuronal synchrony despite long conduction delays. *Proc. Natl. Acad. Sci. U.S.A.*, **105**, 17157–17162.
20. Weber, W., Stelling, J., Rimann, M., Keller, B., Baba, M.D.E., Weber, C., Aubel, D., and Fussenegger, M. (2007) A synthetic time-delay circuit in mammalian cells and mice. *Proc. Natl. Acad. Sci. U.S.A.*, **104**, 2643–2648.
21. Schöll, E. and Schuster, H. (eds) (2008) *Handbook of Chaos Control*, 2nd edn, Wiley-VCH Verlag GmbH, Weinheim.
22. Dahms, T., Hövel, P., and Schöll, E. (2008) Stabilizing continuous-wave output in semiconductor lasers by time delayed feedback. *Phys. Rev. E*, **78**, 056213.

14
Synchronization of Chaotic Networks and Secure Communication
Ido Kanter and Wolfgang Kinzel

14.1
Introduction

Chaos synchronization is a counterintuitive phenomenon. On one hand, a chaotic system is unpredictable, since its trajectory is extremely sensitive to its initial state [1]. On the other hand, two identical chaotic units that are coupled to each other may synchronize to a common chaotic trajectory [2–4]. The system is still chaotic, but after a transient time the two chaotic trajectories are locked to each other in finite precision. This coupling may be unidirectional so that one sender is driving a receiver and is then called master–slave configuration. It may also be bidirectional, with both units influencing each other.

The phenomenon of chaos synchronization is attracting a lot of research. It is a fundamental problem of nonlinear dynamics, which has many interdisciplinary aspects. For example, synchronization plays an important role in biological systems [5]. Hence, investigating the cooperative behavior of chaotic networks may help to understand the functioning of biological networks [6]. Furthermore, chaotic networks have the potential to be applied for novel secure communication systems [7–12]. In this regard, synchronization between coupled chaotic lasers is vitally important [13–15]. In fact, a private key secure communication over a distance of 120 km in a public fiber-optic communication network has recently been demonstrated with chaotic semiconductor lasers in a master–slave configuration [16].

Note that in this scenario, the chaotic signal of the lasers is the carrier of the information. Chaos may also be used to construct a secret message itself [17]. But in this article we consider only chaotic signals that may be modulated to create a secret message.

When a message is added in the master–slave configuration with a tiny amplitude to the chaotic carrier of the sender Alice [1]. it is not easy – if possible at all – to

---

1) In this report, we commonly refer to the communicating units (sender) A and (receiver) B as Alice and Bob and to the eavesdropping unit E as Eve.

Nonlinear Laser Dynamics: From Quantum Dots to Cryptography, First Edition. Edited by Kathy Lüdge.
© 2012 Wiley-VCH Verlag GmbH & Co. KGaA. Published 2012 by Wiley-VCH Verlag GmbH & Co. KGaA.

extract it from the transmitted laser beam unless its undisturbed chaotic signal is known [10–12]. The undisturbed chaotic signal, however, can only be known from a system that synchronizes with the sender, for example, the receiver Bob. Thus, chaos synchronization opens the possibility for a private key secure communication with high bit rates of the order of gigahertz.

However, this communication protocol is only secure if the two partners Alice and Bob have agreed on identical private laser parameters. If an attacker Eve can use identical equipment, she can synchronize as well, and extract the message by subtracting her laser output from the transmitted signal. Therefore, public cryptography is not possible with a unidirectional configuration. Is it possible to generalize this concept to the realm of public channel communication?

In fact, it was suggested that dynamical systems with bidirectional couplings [18, 19] may be able to realize a secret communication over a public channel [9]. When the two lasers of Alice and Bob are interacting, they may have an advantage over an attacker Eve, who is only recording the signal [20].

Thus, applications for public cryptography are related to the following general problems of nonlinear dynamics: For which conditions can networks of chaotic units synchronize to a common chaotic trajectory? For which systems is it not possible to synchronize an attacking unit, which is recording the exchanged signals and which cannot interact with the network?

For typical applications in neurobiology as well as for lasers, the transmitted signals have a time delay which is comparable or even much longer than the internal time scales of the single units. Thus, we have to consider networks with time-delayed couplings. Delay times may lead to new phenomena. Chaos can be either generated or controlled with time-delayed feedback, neural oscillators change their properties when they are coupled with delay [21], and the dynamics of the system becomes infinite dimensional [22]. Thus, many recent studies consider the cooperative properties of networks with time-delayed couplings [23].

In this article, we give an introduction and an overview of chaos synchronization with emphasis on our recent work. For pedagogical reasons, we demonstrate the main results with networks of chaotic Bernoulli units. But these results are compared with experiments on semiconductor lasers and with numerical simulations of Lang–Kobayashi equations, which are used to describe chaotic semiconductor lasers [24, 25].

14.2
Unidirectional Coupling

Most of the phenomena that have been observed for synchronized chaotic lasers have been described by a simple mathematical model: coupled Bernoulli maps [26–28]. Hence, in this overview, we restrict the mathematics to this simple case. But we keep in mind that many details of coupled lasers depicted by rate equations, in particular the Lang–Kobayashi equations for semiconductor lasers [24, 25], are the same as for the Bernoulli case. An isolated laser is not chaotic, chaos stems

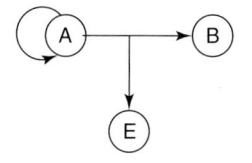

Figure 14.1 Scheme of two unidirectional coupled units A and B with an attacker E recording the transmitted signal.

from time-delayed feedback of the laser itself or from its partners. In the simplest case, a laser becomes chaotic by feeding its beam back to its resonator by an external mirror. When this chaotic laser beam is inserted into a second laser it may synchronize it. This unidirectional coupling, sketched in Figure 14.1, has the corresponding equations for the Bernoulli maps with $f(x) = (\alpha x) \mod 1$, which is chaotic for $\alpha > 1$.

$$a_t = (1 - \varepsilon) f(a_{t-1}) + \varepsilon f(a_{t-\tau})$$
$$b_t = (1 - \varepsilon) f(b_{t-1}) + \varepsilon f(a_{t-\tau}). \qquad (14.1)$$

The Bernoulli system allows an analytic calculation of the stability of the synchronization manifold [26, 28–32]. Here, we find that

1) The unit A is chaotic for all parameters $\varepsilon \in [0, 1]$ and
2) B synchronizes to A for $(1 - \varepsilon)\alpha < 1$.

When B synchronizes to A, the isolated units A and B without coupling and feedback are not chaotic. In this case, chaos is generated by the feedback, and only a nonchaotic unit can be synchronized by a signal from its partner. Note that although the signal is transmitted with a time-delayed signal with an arbitrary large delay time τ, B synchronizes to A without any time shift, $a_t = b_t$. This holds because we have used identical delay times for feedback and coupling, otherwise the trajectories would be synchronized with a time shift.

For shorter coupling delays, even anticipating chaos is possible: B can predict the chaotic trajectory of A [33]. Chaos synchronization with unidirectional coupling has been demonstrated in numerous experiments on semiconductor lasers and electronic circuits [13].

14.3
Transmission of Information

How can Alice hide and transmit her secret message with synchronized laser beams? In principle, there are two possibilities: (i) The message is modulating only the transmitted signal (chaos masking or chaos pass filter) and (ii) the message is modulating the sending unit as well as the transmitted signal (chaos modulation) [10–12, 34]. These two principles may be demonstrated with our Bernoulli systems. The first case corresponds to adding a message m_t to the transmitted signal:

$$b_t = (1 - \varepsilon) f(b_{t-1}) + \varepsilon f(a_{t-\tau} + m_{t-\tau}). \qquad (14.2)$$

Hence, the unit of Bob is driven by the signal plus the message and, therefore, Alice cannot completely synchronize with Bob. Surprisingly, Bob can recover the message by subtracting the incoming signal from its own time-shifted trajectory:

$$\tilde{m}_t = a_t + m_t - b_t. \tag{14.3}$$

It seems that the dynamic of B filters out the message; therefore, the name chaos pass filter was coined [35, 36]. The message is a perturbation that drives the unit B away from the synchronization manifold.

It turns out that this mechanism is more complex than that simple explanation [37]. The tiny perturbation m_t is amplified by the chaotic dynamics. In fact, the distribution of $\tilde{m}_t$ can have power-law tails, extremely large excursions away from the synchronization manifold are possible. But if the message is encoded with bits, $m_t = \pm 1$ the bit error rate (BER), the probability of $\tilde{m}_t m_t < 0$ can be very small. In fact, BERs of 10^{-7} are reported for the laser demonstration [16].

For the second scenario, chaos modulation, even error-free transmission is possible. One example is feeding the message back to the sender:

$$a_t = (1 - \varepsilon) f(a_t) + \varepsilon f(a_{t-\tau} + m_{t-\tau})$$
$$b_t = (1 - \varepsilon) f(b_t) + \varepsilon f(a_{t-\tau} + m_{t-\tau}). \tag{14.4}$$

It is immediately obvious that $a_t = b_t$ is still a solution of these equations; hence, from Eq. (14.3) one finds $\tilde{m}_t = m_t$. Another possibility of chaos modulations is to modulate one parameter of the sending unit, for example, the pump current of the laser of Alice [13]. This may be easier than modulating the beam, but it generates bit errors. In this case, one can even transmit signals with a chain of identical units with arbitrary transmission delays, and extract the message just by the difference of the trajectories of the last two units. However, the length of the chain is limited by convective instabilities [38].

14.4
Bidirectional Coupling

As mentioned before, unidirectional configurations allow any attacking unit to synchronize as well, if the attacker uses an identical dynamical unit. For bidirectional coupling, however, this is not obvious; hence, in this section, we discuss chaos synchronization with mutually interacting units.

For lasers, chaos can be generated by coupling two lasers via their time-delayed laser beams. But they will not synchronize until one includes a self-feedback or another coupling with certain constraints to the delay times [39, 40]. For the inclusion of a self-feedback, a schemata can be seen in Figure 14.2, and the corresponding Bernoulli system has the equations

$$a_t = (1 - \varepsilon) f(a_{t-1}) + \varepsilon \kappa f(a_{t-\tau}) + \varepsilon(1 - \kappa) f(b_{t-\tau})$$
$$b_t = (1 - \varepsilon) f(b_{t-1}) + \varepsilon \kappa f(b_{t-\tau}) + \varepsilon(1 - \kappa) f(a_{t-\tau}). \tag{14.5}$$

14.4 Bidirectional Coupling

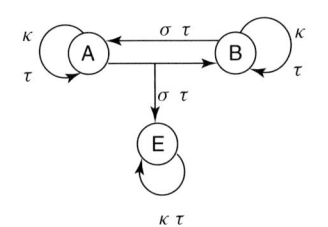

Figure 14.2 Scheme of two bidirectional coupled units A and B with an attacker E recording both transmitted signals.

For all parameters $\varepsilon, \kappa \in [0, 1]$ the system is chaotic. Complete synchronization, $a_t = b_t$, is a solution of these equations, but its stability has to be calculated. The analytic solution for large values of the delay time τ is shown in Figure 14.3. In fact, the stability of synchronization is determined by the roots of a polynomial of degree τ, and general symmetry considerations do not allow synchronization without feedback, $\kappa = 0$. This is different for a triangular configuration of three units or they can synchronize without feedback [28].

These results have been derived for a simple Bernoulli network. But, in fact, complete synchronization for semiconductor lasers with feedback and mutual coupling has been demonstrated experimentally [39]. The phase diagram is similar to Figure 14.3. Of course, there are differences when compared to iterated maps: if we are operating close to the threshold current, the laser will generate quasiperiodic spike patterns, as can be seen in Figure 14.4, interrupted by sudden intensity breaks, known as low frequency fluctuations. Almost complete synchronization of the intensity has been observed on a picosecond time scale [41], corresponding to 10 ns delay time, and even optical phase synchronization on a femtosecond scale has been measured [42].

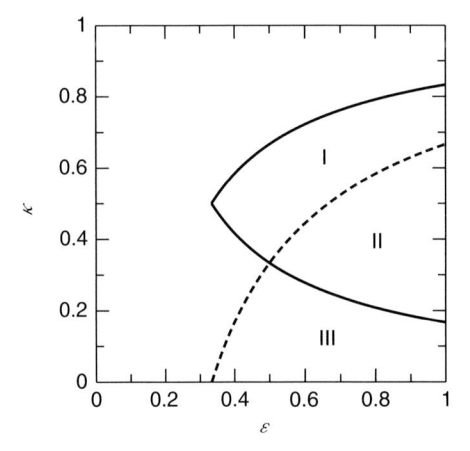

Figure 14.3 Phase diagram for Bernoulli units. Regions II + III: synchronization of A and B for unidirectional coupling, regions I + II: synchronization for bidirectional coupling. From [26].

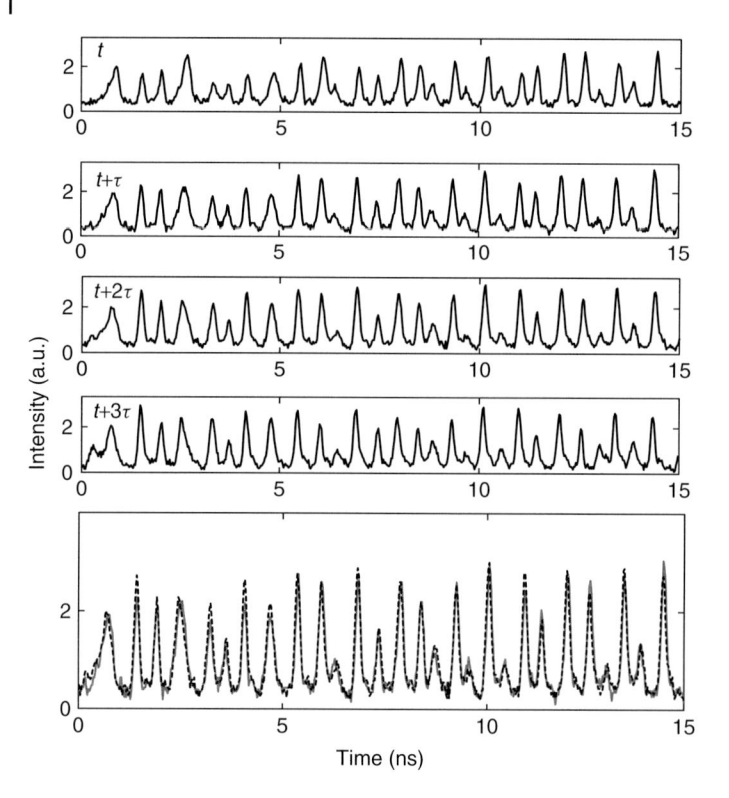

Figure 14.4 A trace of 15 ns duration of the intensity of one laser followed by plots of the same laser intensity after a time τ, 2τ and 3τ with $\tau = 23.55$ ns. The bottom panel shows the intensity traces at time $t+\tau$ and at time $t+2\tau$, demonstrating the slowly decaying periodicity of the spiking pattern. From [41].

For both systems, iterated Bernoulli maps and chaotic semiconductor lasers usually described by ordinary differential equations, it turns out that synchronization is extremely sensitive to a careful adjustment of delay times. The previous results have been derived with identical delay and coupling times. For the Bernoulli system, an analytic calculation shows that already for the minimal possible difference of one time step between delay and feedback time, synchronization is destroyed for large delay times. Analogous to that, a careful adjustment of delay times of less than the coherence time (few picoseconds), which means that less than a millimeter mismatch among distances in the laser experiments are necessary. However, recently we have shown how to avoid this sensitivity of the coupling delay. If the two partners use identical multiple feedback delays, the system synchronizes in broad intervals of the coupling delay [31, 32].

Without feedback, a signal with a single delay time cannot synchronize two nonlinear units to a common chaotic trajectory. However, when the signal is transmitted with two delay times, chaos synchronization can be achieved. In this case, the ratio of the two delays $\tau_1/\tau_2 = u/v$ has to be a ratio of odd relatively prime integers [31, 32]. Experiments with semiconductor lasers have confirmed this theoretical result [43].

Is there any difference between bi- and unidirectional coupling? Our simple model shows (Figure 14.3) that the phase diagrams are different. Two identical units with bidirectional coupling synchronize in regions I and II, whereas with a unidirectional coupling they synchronize in regions II and III. Thus, with identical units, Alice and Bob can select their parameters in region I, and Eve does not synchronize her unit in the Bernoulli system as well as in the lasers. It was shown, however, that Eve can always find parameters for her unit, which lead to synchronization [27].

Nevertheless, in the following we suggest that secret communication is possible for two interacting chaotic units. In the following section, we show that in the scenario of Figure 14.3 the bit error of Alice and Bob is much smaller than the one of Eve. Using methods from information theory, such a difference can be used for secure communication. In Section 14.6, we suggest that private filters be used for public cryptography with chaos synchronization.

14.5
Mutual Chaos Pass Filter

Random bit generators to be applied to cryptography have twofold requirements. The random bits have to be generated at as high a speed as possible [44] while maintaining high quality, as measured by the unpredictability of the bit string produced. The second requirement is to find an effective key-exchange protocol over a public channel, such that the communicating parties will hold the same random bit string that cannot be revealed by a powerful computational eavesdropper [45].

Until recently, the only realistic way of generating random bit strings at high data rates was to use deterministic algorithms to generate pseudorandom number sequences, for they are only limited by computational hardware speed. However, their unpredictability is limited by the very nature of their deterministic origin [45]. It is widely accepted that the core of any true random bit generator must be an intrinsically nondeterministic physical process, such as measuring thermal noise from a resistor, shot noise from a Zener diode or a vacuum tube [46, 47], and measuring radioactive decay from a radioactive source [48]. Owing to low signal levels, the generation rate from such sources is typically less than 100 Mb s^{-1} [44]. An intriguing possibility for a physical system for random bit generators is a semiconductor laser in the presence of external feedback, whose output consists of a large chaotic intensity fluctuations, characterized by pulses with a typical width of 100 ps [44, 49–51]. Indeed, great progress has been made recently in demonstrating a generation rate from such lasers, from few gigabits per second [49, 50] toward terabits per second [51, 52], where the randomness of the bit strings is verified by standard statistical tests [53, 54].

A secure key-exchange protocol between two parties over a public channel was discovered in 1976 by Diffie and Hellman based on number theory, and paved the road for modern cryptography [45]. Alternative physical mechanisms based on quantum mechanics have been suggested more recently for a secure key-exchange protocol, with the important and unique ability of the two communicating parties to detect the presence of any third party trying to gain knowledge of the key [55–58]. The first layer of the quantum protocol is based on quantum ingredients such as entangled pairs of photons and results in correlated keys for both partners. The second, classical layer, consists of an error-correcting code, information reconciliation, and privacy amplification. These result in identical keys for the communicating pair, while leakage of information to an eavesdropper is eliminated. Significantly, the quantum protocol relies on classical ingredients, such as random bit generators, error-correcting code, and source coding, which govern the security of the entire protocol.

The main focus of this section is to securely synchronize two random bit generators with high bandwidth and fidelity over a public channel using a classical mechanism – zero lag synchronization (ZLS) of two mutually coupled chaotic lasers [52]. The ZLS mechanism is not sufficiently secure in its simple form to act as a key-exchange protocol [20], and it serves only as an information carrier to generate correlated random bit sequences. Identical random bit sequences can be constructed from the correlated sequences using our proposed protocol, together with information reconciliation and privacy amplification [57, 58]. Furthermore, the presented mechanism allows the secure generation of a synchronized random bit string amongst a small network of communicating parties. We have numerically investigated the scenario of Figure 14.8, where two mutually coupled lasers, A and B, are subject to both optical feedback and mutual coupling in a symmetric configuration [52]. The optical self-feedback time delay, τ_f and the mutual coupling time delay, τ_m, were both selected to be equal, $\tau = 10$ ns in the examples below. The self-feedback strength and the mutual coupling strength are denoted by κ and σ, respectively. The injection current to the threshold current ratio is selected to be 1.5, so that the lasers are operating in the coherence collapse regime [59]. The Lang–Kobayashi equations are known to be a good model for the intensity dynamics of coupled semiconductor lasers [25], and the equations for the scenario of Figure 14.8 are very similar to reference [39]. For each point in the phase space, (κ, σ), the cross correlation at zero time lag was measured over a window of 20 ns and averaged over 1 μs. The formula that was used to calculate the cross correlation for unit A and B is

$$\text{Corr} = \frac{\sum_i \left[(a_i - \bar{a}) \cdot (b_i - \bar{b})\right]}{\sqrt{\sum_i (a_i - \bar{a})^2} \cdot \sqrt{\sum_i (b_i - \bar{b})^2}} \tag{14.6}$$

where $\bar{a}$ is the mean of the trajectory of unit A and $\bar{b}$ is the mean of the trajectory of unit B. There are mainly two phases as depicted in Figure 14.5. For small $\kappa + \sigma$, A and B are not synchronized, whereas for larger values, ZLS emerges as the cross correlation gradually increases toward one.

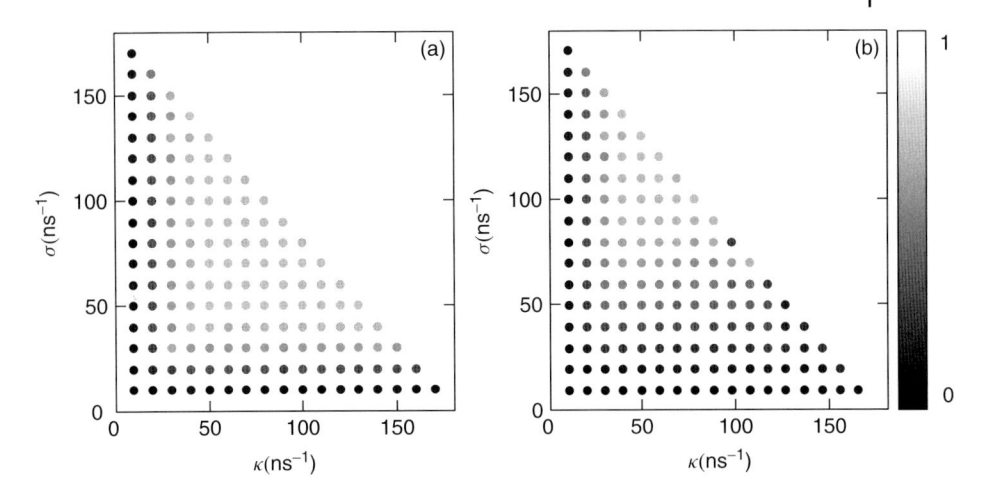

Figure 14.5 Cross correlation for mutual and unidirectional coupling taken from [52]. (a) Cross correlation at zero time lag between two mutually coupled semiconductor lasers for a range of parameter values: κ, feedback strength, and, σ, coupling strength. (b) Cross correlation at zero time lag between a third semiconductor laser coupled unidirectionally to one of the parties using identical κ and σ.

In the case that a third laser, C, is coupled unidirectionally to the transmitted signal of laser A with the same delays and coupling strengths, it is clear that ZLS of C with A and B is a possible solution of the chaotic dynamics, but its stability is questionable. Figure 14.5b depicts the cross correlation at zero time lag between C and either A or B when all lasers use the same parameters (κ, σ). A comparison between Figure 14.5a and 14.5b indicates that ZLS of mutually coupled chaotic lasers is superior to the unidirectional coupling of laser C in a large fraction of the phase space (κ, σ) [9, 34]. Typically, laser C can achieve the same level of synchronization as the mutually coupled lasers by amplifying the coupling signal, while maintaining its total input $\kappa_c + \sigma_c \sim \kappa + \sigma$. A central aspect of the proposed cryptographic protocol is the advantage of ZLS of mutually coupled lasers over unidirectionally coupled lasers with identical coupling and feedback strengths. In what follows, we first describe the utilization of ZLS as a carrier synchronizing the two random bit generators of the communicating parties and then we analyze the security of the channel. In the first step, each partner encodes a random binary sequence by modulating the chaotic intensity of its laser. The modulated intensity is thus M^2, where $M = 1$ corresponds to the transmission of bit 1, while $M = M_0$ corresponds to the transmission of bit-1, where in simulations below M_0 is set to 0.9 with a bandwidth of 1 Gb s^{-1}. Our simulations indicate that the ZLS between the communicating pair remains robust even in the presence of such independent modulation by each of the parties. B, for instance, decodes the message transmitted from A by dividing the intensity received from A with its own synchronized laser output, prior to his modulation, $<I_{AR}>/<I_B>$, where the average, $<>$, is over a predetermined duration of one bit transmission time. If this fraction is

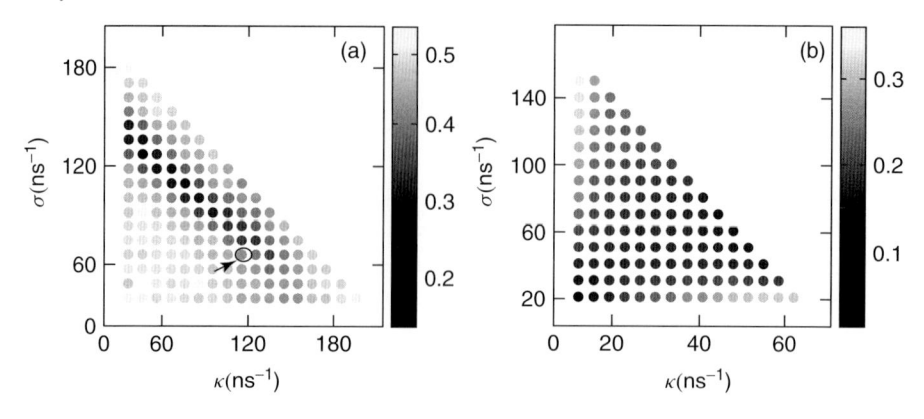

Figure 14.6 The bit error rate (BER) of the attacker and the parties taken from [52]. (a) The BER of laser E for the setup in Figure 14.8 as a function of (κ, σ), when the parties are operating with $\kappa = 90\,\text{ns}^{-1}$ and $\sigma = 40\,\text{ns}^{-1}$ (indicated by the arrow). (b) The BER among the parties in the setup of Figure 14.8 as a function of (κ, σ). The BER for each (κ, σ) is averaged over 1 μs and the modulation bandwidth is 1 Gb s^{-1}.

larger than $(1 + M_0^2)/2$, then the estimated received bit is 1, otherwise −1. The encoding/decoding procedures are implemented simultaneously at both lasers, and are known as a mutual chaos pass filter mechanism [34]. The average BER as a function of (κ, σ) in Figure 14.6b.

14.5.1
Protocol

Parties A and B encode different random bit sequences; hence, the decoded bits are uncorrelated and independent of BER level. An identical random binary sequence is obtained using the following protocol: (i) The two partners start with an identical public random binary sequence of length L, $S_A = S_B = S$. (ii) A compares his estimated received bit at time interval m, $R_A(m)$, to his random transmitted bit at the same time interval, $T_A(m)$. If $R_A(m) = T_A(m)$, $S_A(m)$ is set equal to $R_A(m)$, otherwise $S_A(m)$ remains unchanged. Similarly, in the event $R_B(m) = T_B(m)$, $S_B(m)$ is set equal to $R_B(m)$. (iii) At the end of the mutual chaos pass filter procedure, the average fraction of identical bits between S_A and S_B is given by $1 - p + 0.5p^2$, where p stands for the BER of the mutual chaos pass filter procedure and is calculated using symbolic mathematics [52]. The meaning of $p = 0$ is that A and B acted identically on the initial vectors S and $P_{AB} = 1$, whereas for $p = 1$ only when the partners send different bits, S is altered differently by the two partners, hence $P_{AB} = 0.5$. For simplicity of discussion, we assume statistically independent errors in the decoding procedure of A and B. However, it is expected that both decoders are correlated, since in the event the two lasers are temporally desynchronized the probability for an error bit for both of them increases in comparison to time slots of enhanced synchronization, as was indeed observed in simulations and is analyzed in [52]. The two partners now possess correlated bit sequences. Identical random bit sequences can be constructed from the correlated sequences using our proposed protocol together with information reconciliation procedure, a form of error-correcting code, as for protocols of quantum cryptography [60]. At the end of this procedure, the two partners hold identical random bit sequences. Inevitably, leakage of information occurs during the information reconciliation procedure and is eliminated by a privacy amplification procedure, which is also utilized in quantum cryptography for similar reasons [55–57]. The identical random bit sequences can serve as a common key generated over a public channel. The main question is whether a passive, unidirectionally coupled attacker, C, is capable of deducing the key, when all details of the protocol are publicly known.

Figure 14.6 indicates that it is possible to select sets of parameters (κ, σ) such that the ZLS of the parties, A and B, is superior to the ZLS of the attacker, C. For instance, for $\kappa = 90\,\text{ns}^{-1}$ and $\sigma = 40\,\text{ns}^{-1}$, the cross correlation at zero time lag between the parties is much higher, ~ 0.94, than the correlation between the attacker and the parties ~ 0.5. An attacker using the same set of parameters as A and B would obtain a very high BER in his chaos pass filter mechanism, $q \sim 0.4$ in our simulations, in comparison to $p \sim 0.07$ for A and B. In order to minimize his BER, the attacker can amplify σ_C while decreasing κ_C so that $\kappa_C + \sigma_C \sim \kappa + \sigma$. Figure 14.6 indicates that the minimum BER for the attacker, $q \sim 0.15$, is obtained for $\kappa = 40\,\text{ns}^{-1}, \sigma_C = 90\,\text{ns}^{-1}$, while the parties are operating with $\kappa = 90\,\text{ns}^{-1}$, $\sigma = 40\,\text{ns}^{-1}$. Although this is a much lower BER then E would obtain without the use of amplification, it remains more than twice as high as the BER of A and B.

The mutual chaos pass filter procedure is based on the synchronization of lasers A and B on the unmodulated portion of the signal shared between them. The modulated part of the shared signal can be considered as "noise" for the synchronization process. The noise-to-signal ratio for A and B is given by

$$\frac{\sigma M^2}{I(\sigma + \kappa)} \tag{14.7}$$

and this is larger for the attacker

$$\frac{\sigma_c M^2}{I(\sigma_C + \kappa_C)} \tag{14.8}$$

since $\sigma_C > \sigma$, and as a result $q > p$. However, higher BER for C does not necessarily indicate that the fraction of identical bits between S_C and S_A is reduced in comparison to the fraction of identical bits between S_A and S_B. One can show, using symbolic mathematics [52], that the average fraction of identical bits between S_C and S_A is given by

$$P_{AB} = 1 - 0.5p - q(1 - 1.5p) + q^2(0.5 - p). \tag{14.9}$$

The regions where $P_{AC} < P_{AB}$ are indicated by the black and gray colored region in Figure 14.7a. Note that for $p = q$ and also for a limited region where $p < q$, one finds $P_{AC} < P_{AB}$.

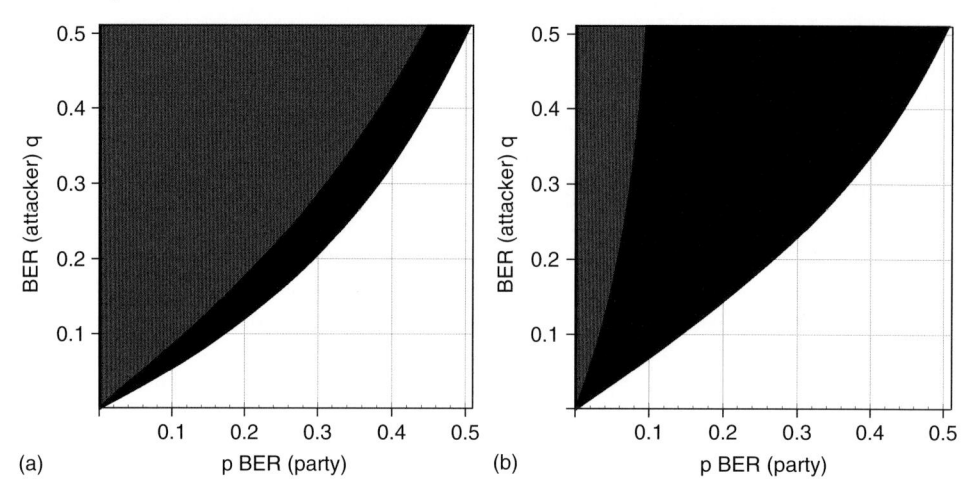

Figure 14.7 Secure regions for two and three synchronized random bit generators [52]. (a) Two mutually coupled lasers as in schematic Figure 14.8. (b) Three mutually coupled lasers as in reference [52]. The BER of the mutual chaos pass filter procedure between a pair of parties, p, and between a party and the attacker, q, where we assume statistically uncorrelated decoded bits by the parties and by the attacker. The colored regions (colored in either black or gray) indicate a necessary condition for the failure of an attacker before the reconciliation procedure, $P_{AC} < P_{AB}$. The region where an attacker cannot succeed in recovering the random bits sequence of the parties, even when using the leakage of information of the reconciliation procedure, is indicated in gray.

A reconciliation procedure sets $P_{AB} = 1$, resulting in identical random bit sequences for A and B. The leakage of information in the reconciliation procedure for the case $P_{AC} < P_{AB}$, can also be expected to be usable for enhancing P_{AC}, but it cannot be boosted to one. The exact bound for when A and B can be considered secure from attack by C is given by

$$I(S_A, S_B) > I(S_C, S_A) + I(S_C, S_B | S_A) \tag{14.10}$$

where $I(S_C, S_B)$ and $I(S_C, S_B | S_A)$ stand for the mutual information and the conditional mutual information, respectively, and S_A, S_B and S_C stand for the binary sequences before the reconciliation procedure. The above equation states that in case the minimum required exchange of information for the reconciliation procedure, $1 - I(S_A, S_B)$, is less than the total missing information C possesses about S_A and S_B, $1 - I(S_C, S_A) - I(S_C, S_B | S_A)$, the attacker fails to recover the random bits sequence. This condition as a function of p and q is depicted by the gray region of Figure 14.7a, where the details of symbolic mathematics calculation are given in [52]. Figure 14.7b depicts a similar scenario as Figure 14.7a but for the case of three mutually coupled lasers as discussed in details in [52]. The above-mentioned example ($p = 0.07$, $q = 0.15$) lies in the gray region of Figure 14.7a and thus indicates that a secure synchronization of two random bit generators over a public channel is achieved.

14.6 Private Filters

Let us repeat the initial problem related to public cryptography: Alice and Bob want to synchronize their dynamical systems to a common chaotic trajectory. Eve has as much knowledge about Bob's systems as Alice has and vice versa. Eve can record and manipulate the transmitted signals between Alice and Bob, but she cannot influence their dynamics. How can Alice and Bob ensure that Eve does not synchronize as well?

For this problem, we focus on the chaos pass filter and not on chaos modulation. It turns out that we can add a new mechanism to the configuration of the previous section to not only raise the BER of Eve but also to prevent her from synchronization: secret commutative filters [20, 27] as can be seen in Figure 14.8. Both Alice and Bob transmit their signals via a private filter, which is not known to their partner. Since these filters commute, both receive an identical signal if they are synchronized. Eve only knows the two transmitted signals after passing through one filter and thus cannot synchronize as a hardware attacker.

Let us specify this principle for the Bernoulli systems. The dynamical equations of Alice and Bob are

$$a_t = (1 - \varepsilon)f(a_{t-1}) + \varepsilon \kappa f(a_{t-\tau}) + \varepsilon(1 - \kappa)f[F_A(F_B(b_t))]$$
$$b_t = (1 - \varepsilon)f(b_{t-1}) + \varepsilon \kappa f(b_{t-\tau}) + \varepsilon(1 - \kappa)f[F_B(F_A(a_t))]. \quad (14.11)$$

$F_B(b_t)$ and $F_A(a_t)$ are the signals transmitted from Alice and Bob, respectively. They consist of the variables that are filtered with private kernels K_r^A and K_r^B:

$$F_A(a_t) = \sum_r K_r^A a_{t-\tau}$$
$$F_B(b_t) = \sum_r K_r^B b_{t-\tau}. \quad (14.12)$$

If $a_t = b_t$, then the two driving signals $F_A(F_B(b_t))$ and $F_B(F_A(a_t))$ are identical, since convolutions commute. In fact, for randomly chosen filter parameters, Alice and Bob can still synchronize the units. But Eve, as a hardware attacker, receives only $F_A(a_t)$ and $F_B(b_t)$; thus, she cannot synchronize her unit. Any hardware attack fails.

We now analyze the situation with an idealized software attacker, which can record the signal with infinite precision. In principle, when enough information is transmitted, it may be possible to calculate the private filters K_r^A and K_r^B for $r = 1, \ldots, N$ with N being the number of filter parameters. This is due to the counting argument as, with each time step, the attacker gets two additional

Figure 14.8 Scheme of two bidirectional coupled units A and B, each with a private commutative filter. From [27].

equations for $F_A(a_t) F_B(a_t)$ and only one additional unknown new variable a_t. To avoid this, an additional protocol has been suggested:

1) When the number of time steps is about the number of unknowns $a_t = b_t$, K_r^A, K_r^B, the transmission is interrupted, and we have a period of silence and Alice and Bob select new filter parameters.
2) The transmitted signals as well as the filter parameters are discretized to integers.
3) A nonlinear term of the past signal is added to the transmitted signal; its effect vanishes as the partners are synchronized.

The first point avoids an exact calculation of the filter parameters. The second point maps the problem to the solution of equations with integers (Diophantine problems). These problems belong to the class of NP problems; hence, in principle it should be impossible to solve these equations with a number of calculations, which increase with a power of N, the number of filter parameters, only.

14.7
Networks

In the previous sections we have considered chaos synchronization for two interacting chaotic units, for example, two semiconductor lasers interacting by their mutual laser beams with a long transmission time. Now we want to extend this configuration to a chaotic network of nonlinear units. In fact, secure communication in networks of users is a challenging problem in information theory and applications. Of course, it is possible to couple these users pairwise. But collective chaos synchronization of networks may lead to new communication protocols, and, in addition, understanding the cooperative behavior of interacting nonlinear units may give some insight into the functioning of neuronal networks.

As shown before, ZLS is a necessary condition for secure communication over public channels. Therefore, we want to discuss ZLS in general networks of identical nonlinear units interacting by a function of their time-delayed variables. For simplicity, we consider the simple model of a network of N Bernoulli units with variables x_t^i, which follow the equations

$$x_t^i = (1-\varepsilon) f(x_{t-1}^i) + \varepsilon \sum_{j \neq i} G_{ij} f(x_{t-\tau}^j). \tag{14.13}$$

The row sum of the coupling matrix is unity, $\sum_j G_{i,j} = 1$; thus, the synchronized trajectory $x_t^i = s_t$ is a solution of these equations:

$$s_t = (1-\varepsilon) f(s_{t-1}) + \varepsilon f(s_{t-\tau}). \tag{14.14}$$

To analyze the stability of the synchronization manifold, we use the powerful method of the master stability function [61], which has been extended to time-delayed systems [29, 31, 62–64]. Linearizing Eq. (14.13) in the vicinity of the synchronization manifold yields a system of N perturbation modes, whose amplitudes $\xi_{k,t}$ follow the equations

$$\xi_{k,t} = (1-\varepsilon)f'(s_{t-1})\xi_{k,t-1} + \varepsilon\gamma_k f'(s_{t-\tau})\xi_{k,t-\tau}. \tag{14.15}$$

γ_k are the eigenvalues of the coupling matrix G. Thus, a system of N coupled equations reduces to a one-dimension equation with an additional variable γ. For $\gamma = 1$, this equation determines the stability inside the synchronization manifold. All other modes of perturbation determine the stability perpendicular to the manifold.

In the case of Bernoulli units, one has constant coefficients of the linear Eq. (14.15), $f'(s_t) = \alpha > 1$. Therefore, with $\xi_t = z^t$, this equation reduces to determining the roots of the polynomial

$$z^\tau = (1-\varepsilon)\alpha z^{\tau-1} + \varepsilon\alpha\gamma_k. \tag{14.16}$$

If one root of this polynomial lies outside the unit circle, then the corresponding perturbation mode $\xi_{k,t}$ is unstable. Thus, if the matrix G of a network has an eigenvalue γ_k (except $\gamma_0 = 1$) for which Eq. (14.16) is unstable, then the network cannot synchronize to a common chaotic trajectory.

In the limit of large delay times, that is, when τ is much larger than all internal time scales of the system, the root of the polynomial and the corresponding conditional Lyapunov exponents can be calculated analytically. We find a simple relation that determines the stability of the synchronization manifold:

$$|\gamma_1| < \exp(-\tau\lambda_{max}). \tag{14.17}$$

γ_1 is the eigenvalue of G with the second largest modulus and λ_{max} is the largest Lyapunov exponent of the network, which, in our case, is identical to the one of the synchronization manifold (Eq. (14.14)).

This equation is exact for Bernoulli networks, but numerical simulations of several networks of semiconductor lasers have shown that this relation holds for this case, as well [30]. Thus, the eigenvalue gap $1 - |\gamma_1|$ is responsible for chaos synchronization of general networks.

Equation (14.17) has interesting consequences. For example, for a pair of units without self-feedback and for a general bipartite network, one has $\gamma_1 = -1$. Therefore, these networks cannot be synchronized to a chaotic trajectory. For a triangle, however, one has $\gamma_1 = -1/2$; hence, a triangle can be synchronized if chaos is sufficiently weak. For an all-to-all network, the eigenvalue $\gamma_1 = -1/(N-1)$ decreases to zero with the size of the network. Hence, any chaotic network can be synchronized for sufficiently large-size N.

Eq. (14.17) is also true in the limit of $\tau \to \infty$. If the single isolated unit is chaotic, the Lyapunov exponent λ_{max} is of the order of 1 and the network cannot synchronize [62]. However, if the unit is not chaotic, then λ_{max} is of the order of $1/\tau$ and the network can synchronize to a common chaotic trajectory. In this case, chaos is generated by the time-delayed couplings, as in the case of lasers that become chaotic either by self-feedback or by interacting with other lasers.

A challenging test of Eq. (14.17) are networks with directed couplings without self-feedback. In this case, the eigenvalue γ_1 usually is a complex number and according to Eq. (14.17) only the modulus determines synchronization. In fact, numerical simulations of the laser rate equations have shown that the master stability function depends on the modulo of γ, only [64].

As a consequence of Eq. (14.17), any ring of units cannot be synchronized since one has $\gamma_1 = 1$. But a general oriented graph can have an eigenvalue gap [29, 30]. For example, a square with a diagonal shown in Figure 14.9 has a complex eigenvalue, which is largest for $\rho = 5/8$ with $\gamma_1 = \sqrt{3}/2$. ρ and $1 - \rho$ are the input weights in the network shown in Figure 14.9. According to Eq. (14.17), this network can synchronize without time shift when $\lambda_{max}\tau < 0.15$.

Figure 14.10 shows $\lambda_{max}\tau$ for semiconductor lasers obtained from numerical simulations of the corresponding rate equations [30]. Combing this result of a single laser with the eigenvalues gap of the network of Figure 14.9, we find ZLS [29] and the phase diagram of Figure 14.11, which is in agreement with the one obtained from simulating the complete network (see the cross correlations shown in Figure 14.11.

The analysis with the polynomial Eq. (14.16) can be extended to networks with multiple time delays [29]. The simplest system is a pair of units coupled with two

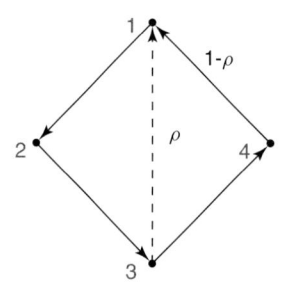

Figure 14.9 Directed ring with a diagonal of coupling strength ρ.

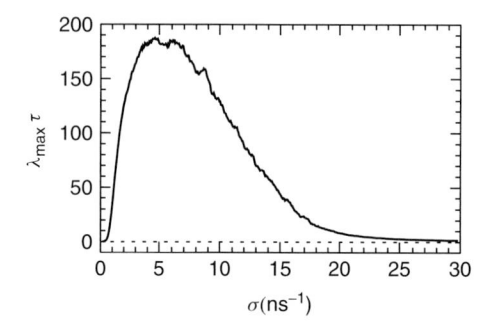

Figure 14.10 Maximal Lyapunov exponent for a single semiconductor laser with one self-feedback as a function of its self-feedback strength from [30], representing the synchronization manifold of an arbitrary network. The parameters of the numerical simulations of the laser rate equations can be found in [30].

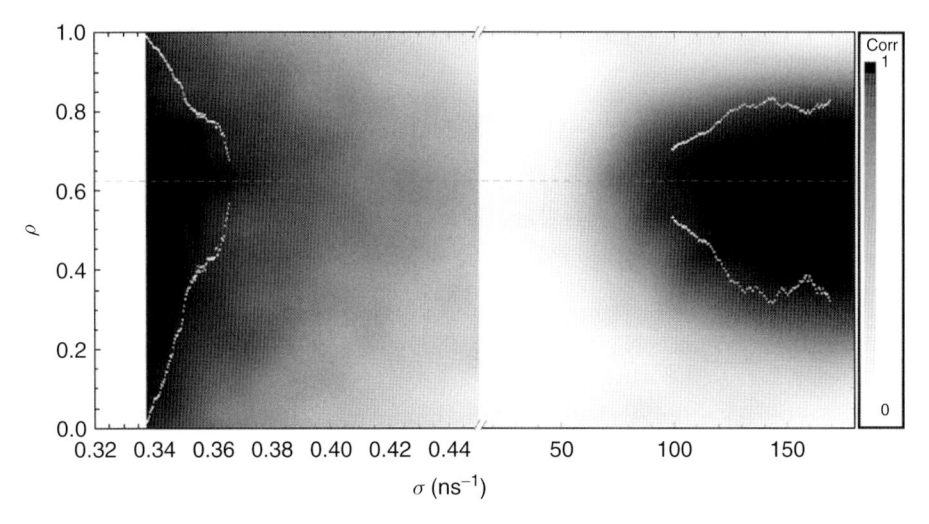

Figure 14.11 Cross correlations of semiconductor lasers of the network of Figure 14.9. The white line shows the transition to chaos synchronization predicted by Eq. (14.17) and Figure 14.10.

delay times τ_1 and τ_2. Note that for a single delay, time synchronization is not possible. However, multiple delay times can enforce synchronization [29, 31, 32]. The symmetries of the corresponding polynomial exclude synchronization when the ratio $\tau_2/\tau_1 = u/v$ is a ratio of two odd integers u and v, which are relatively prime [31, 32, 43]. For other ratios synchronization is possible. For $\tau_2 = 2\tau_1$, the parameter region for synchronization is largest. In fact, for this case, chaos synchronization has been demonstrated with experiments on semiconductor lasers [43]. For odd ratios, the cross correlation is high, whereas for other ratios of small integers a high cross correlation has been measured.

For general networks with multiple delay times, an analytic solution has not been found, yet. However, in this case, a self-consistent mixing argument has been given, which determines the kind of synchronization that is possible in the limit of weak chaos [29]. The argument is based on mixing information: each unit has a chaotic trajectory that is transmitted to its connected partners. This information is mixed with the signals from the other units after some time. When all the information from all units are finally spread over the complete network, this network can synchronize. When only some subnets have mixed information form the nodes of the subnet, the system shows cluster synchronization. It turns out that the greatest common divisor (GCD) of all loop lengths of the graph determines the kind of synchronization. If GCD $= 1$, complete synchronization occurs in the limit of weak chaos. If GCD $= m$, then the network synchronizes to m clusters or sublattices.

14.8
Outlook

Key-exchange protocols based on number theory are fundamentally limited to only two users. The key-exchange protocol presented here, however, can be generalized to a secure synchronization of random bit generators among a small network of three mutually coupled lasers [52], although we assume that an advanced attacker may be able to amplify the transmitted signal without introducing additional noise. For synchronizing more than three random bit generators, it is expected that the secure region in the BER of the mutual chaos pass filter procedure (p,q) will become smaller, since the superior information of a party (knowledge of its transmitted bit) relative to the attacker is reduced. Sufficiently low BER values, p, are within the secure region for any number of synchronized random bit generators, and the remaining question is whether sufficiently low p values can be achieved. It is experimentally expected that by implementing a high pass filter on the mutually transmitted signals, p can be significantly reduced to the order of 10^{-7}, as was observed in a unidirectional configuration and modulation bandwidth of 1Gb s^{-1}. Consequently, it is expected that the presented secure synchronization of two random bit generators can be extended to a larger network of communicating chaotic lasers. An important property of the proposed key-exchange protocol is to detect the presence of an active attacker trying to interfere with the mutual transmitted signals in order to enhance his ZLS or to prevent the parties from achieving an identical key. A possible solution for the parties is to communicate part of their encoded and estimated decoded bits over a public channel, where each encoded/decoded configuration has a given probability which is a function of p. An active attacker cannot imitate these probabilities, since its knowledge of the transmitted information is inferior in comparison to the parties. This additional feature of the key-exchange protocol, based on synchronization of chaotic lasers, is similar to known features of the quantum key-exchange protocol and certainly deserves further research.

References

1. Schuster, H.G. and Just, W. (2005) *Deterministic Chaos*, Wiley-VCH Verlag GmbH, Weinheim.
2. Pikovsky, A., Rosenbluh, M., and Kurths, J. (2001) *Synchronization, A Universal Concept in Nonlinear Sciences*, Cambridge University Press, Cambridge.
3. Pecora, L.M. and Carroll, T.L. (1990) Synchronization in chaotic systems. *Phys. Rev. Lett.*, **64** (8), 821–824.
4. Pikovsky, A.S. (1984) On the interaction of strange attractors. *Z. Phys. B: Condens. Matter*, **55**, 149–154. DOI: 10.1007/BF01420567.
5. Buszaki, G. (2006) *Rhythms of the Brain*, Oxford University Press, Oxford.
6. Hövel, P., Dahlem, M., and Schöll, E. (2010) Control of synchronization in coupled neural systems by time-delayed feedback. *Int. J. Bifurcat. Chaos*, **20**, 813.
7. Boccaletti, S., Kurths, J., Osipov, G., Valladares, D.L., and Zhou, C.S. (2002) The synchronization of chaotic systems. *Phys. Rep.*, **366**, 1–2.

8. Klein, E., Mislovaty, R., Kanter, I., and Kinzel, W. (2003) Public channel cryptography by synchronization of networks and chaotic maps. *Phys. Rev. Lett.*, **91**, 118701.
9. Klein, E., Mislovaty, R., Kanter, I., and Kinzel, W. (2003) Public-channel cryptography using chaos synchronization. *Phys. Rev. E*, **72**, 016214.
10. Kinzel, W., Kanter, I., and Schuster, H. (2008) Secure communication with chaos synchronization, in *Handbook of Chaos Control*, 2nd edn (ed. E. Scho), Wiley-VCH Verlag GmbH, Weinheim.
11. Cuomo, K.M. and Oppenheim, A.V. (1993) Circuit implementation of synchronized chaos with applications to communications. *Phys. Rev. Lett.*, **71** (1), 65–68.
12. Kocarev, L. and Parlitz, U. (1995) General approach for chaotic synchronization with applications to communication. *Phys. Rev. Lett.*, **74** (25), 5028–5031.
13. Uchida, A., Rogister, F., Garcja-Ojalvo, J., and Roy, R. (2005) Synchronization and communication with chaotic laser systems. *Prog. Opt.*, **48**, 203–341.
14. VanWiggeren, G.D. and Roy, R. (1998) Communication with chaotic lasers. *Science*, **279**, 1198–1200.
15. Colet, P. and Roy, R. (1994) Digital communication with synchronized chaotic lasers. *Opt. Lett.*, **19**, 2056–2058.
16. Argyris, A., Syvridis, D., Larger, L., Annovazzi-Lodi, V., Colet, P., Fischer, I., García-Ojalvo, J., Mirasso, C.R., Pesquera, L., and Shore, K.A. (2005) Chaos-based communications at high bit rates using commercial fibre-optic links. *Nature*, **437**, 343–346.
17. Baptista, M.S. (1998) Cryptography with chaos. *Phys. Lett.*, **A240**, 50.
18. Klein, E., Gross, N., Rosenbluh, M., Khaykovich, L., and Kanter, I. (2006) Stable isochronization of mutually coupled chaotic lasers. *Phys. Rev. E*, **73**, 066214.
19. Kanter, I., Gross, N., Klein, E., Kopelowitz, E., Yoskovits, P., Khaykovich, L., Kinzel, W., and Rosenbluh, M. (2007) Synchronization of mutually coupled lasers in the presence of a shutter. *Phys. Rev. Lett.*, **98**, 154101.
20. Kanter, I., Kopelowitz, E., and Kinzel, W. (2008) Public Channel Cryptography: Chaos Synchronization and Hilbert's Tenth Problem. *Phys. Rev. Lett.*, **101** (8), 084 102. DOI: 10.1103/PhysRevLett.101.084102.
21. Schöll, E., Hiller, G., Hövel, P. and Dahlem, M. (2009) Time-delayed feedback in neurosystems. *Phil. Trans. Roy. Soc.*, **267**, 1079.
22. Schöll, E. and Schuster, H. (eds) (2008) *Handbook of Chaos Control*, 2nd edn, Wiley-VCH Verlag GmbH, Weinheim.
23. Arenas, A., Díaz-Guilera, A., Kurths, J., Moreno, Y., and Zhou, C. (2008), Synchronization in complex networks, DOI: 10.1016/j.physrep.2008.09.002.
24. Lang, R. and Kobayashi, K. (1980) External optical feedback effects on semiconductor injection laser properties. *IEEE J. Quantum Electron.*, **16**, 347–355. DOI: 10.1109/JQE.1980.1070479.
25. Ahlers, V., Parlitz, U. and Lauterborn, W. (1998) Hyperchaotic dynamics and synchronization of external-cavity semiconductor lasers. *Phys. Rev. E*, **58**, 7208–7213. DOI: 10.1103/PhysRevE.58.7208.
26. Kestler, J., Kopelowitz, E., Kanter, I., and Kinzel, W. (2008) Patterns of chaos synchronization. *Phys. Rev. E*, **77** (4), 046209.
27. Kanter, I., Kopelowitz, E., Kestler, J., and Kinzel, W. (2008) Chaos synchronization with dynamic filters: two-way is better than one-way. *Europhys. Lett.*, **83**, 50005. DOI: 10.1209/0295-5075/83/50005.
28. Kestler, J., Kinzel, W., and Kanter, I. (2007) Sublattice synchronization of chaotic networks with delayed couplings. *Phys. Rev. E*, **76**, 035202.
29. Kanter, I., Zigzag, M., Englert, A., Geissler, F., and Kinzel, W. (2011) Synchronization of unidirectional time delay chaotic networks and the greatest common divisor. *Europhys. Lett.*, **93**, 60003.
30. Englert, A., Heiligenthal, S., Kinzel, W., and Kanter, I. (2011) Synchronization of chaotic networks with time-delayed couplings: an analytic study. *Phys. Rev. E*, **83**, 046222

31. Zigzag, M., Butkovski, M., Englert, A., Kinzel, W., and Kanter, I. (2009) Emergence of zero-lag synchronization in generic mutually coupled chaotic systems. *Europhys. Lett.*, **85**, 60005.
32. Zigzag, M., Butkowski, M., Kanter, A., Kinzel, W., and Kanter, I. (2010) Zero-lag synchronization and multiple time delays in two coupled chaotic systems. *Phys. Rev. E*, **81**, 036215.
33. Voss, H. (2000) Anticipating chaotic synchronization. *Phys. Rev. E*, **61**, 5115–5119. DOI: 10.1103/PhysRevE.61.5115.
34. Klein, E., Gross, N., Kopelowitz, E., Rosenbluh, M., Khaykovich, L., Kinzel, W., and Kanter, I. (2006) Public-channel cryptography based on mutual chaos pass filters. *Phys. Rev. E*, **74**, 046201.
35. Murakami, A. and Shore, K.A. (2005) Chaos-pass filtering in injection-locked semiconductor lasers. *Phys. Rev. A*, **72**, 053810.
36. Fischer, I., Liu, Y., and Davis, P. (2000) Synchronization of chaotic semiconductor laser dynamics on subnanosecond time scales and its potential for chaos communication. *Phys. Rev. A*, **62**, 011801.
37. Kinzel, W., Kestler, J., and Kanter, I. (2008) Chaos pass filter: linear response of synchronized chaotic systems, http://arxiv.org/abs/0806.4291.
38. Schmitzer, B., Kinzel, W., and Kanter, I. (2009) Pulses of chaos synchronization in coupled map chains with delayed transmission. *Phys. Rev. E*, **80**, 047203.
39. Klein, E., Gross, N., Rosenbluh, M., Kinzel, W., Khaykovich, L., and Kanter, I. (2006) Stable isochronal synchronization of mutually coupled chaotic lasers. *Phys. Rev. E*, **73** (6), 066214.
40. Fischer, I., Vicente, R., Buldu, J., Peil, M., Mirasso, C., Torrent, M., and García-Ojalvo, J. (2006) Zero-lag long-range synchronization via dynamical relaying. *Phys. Rev. Lett.*, **97** (12), 123902.
41. Rosenbluh, M., Aviad, Y., Cohen, E., Khaykovich, L., Kinzel, W., Kopelowitz, E., Yoskovits, P., and Kanter, I. (2007) Spiking optical patterns and synchronization. *Phys. Rev. E*, **76** (4), 046207.
42. Aviad, Y., Reidler, I., Kinzel, W., Kanter, I., and Rosenbluh, M. (2008) Phase synchronization in mutually coupled chaotic diode lasers. *Phys. Rev. E*, **78** (2), 025204. DOI: 10.1103/PhysRevE.78.025204.
43. Englert, A., Kinzel, W., Aviad, Y., Butkovski, M., Reidler, I., Zigzag, M., Kanter, I., and Rosenbluh, M. (2010) Zero lag synchronization of chaotic systems with time delayed couplings. *Phys. Rev. Lett.*, **104**, 114102.
44. Murphy, T. and Roy, R. (2008) Chaotic lasers: The world's fastest dice. *Nat. Photon.*, **2**, 714.
45. Stinson, D.R. (1995) *Cryptography: Theory and Practice*, CRC Press, Boca Raton, FL.
46. Vincent, C. (1971) Precautions for accuracy in generation of truly random binary numbers. *J. Phys. E Sci. Instrum.*, **4**, 825.
47. Maddocks, R., Vincent, C., Walker, E., and Matthews, S. (1972) Compact and accurate generator for truly random binary digits. *J. Phys. E Sci. Instrum.*, **5**, 542.
48. Stefanov, A., Gisin, N., Guinnard, O., Guinnard, L., and Zbinden, H. (2000) Optical quantum random number generator. *J. Mod. Opt.*, **47**, 595.
49. Uchida, A., Amano, K., Inoue, M., Hirano, K., Naito, S., Someya, H., Oowada, I., Kurashige, T., Shiki, M., Yoshimori, S., Yoshimura, K., and Davis, P. (2008) Fast physical random bit generation with chaotic semiconductor lasers. *Nat. Photon.*, **2**, 728.
50. Reidler, I., Aviad, Y., Rosenbluh, M., and Kanter, I. (2009) Ultrahigh-speed random number generation based on a chaotic semiconductor laser. *Phys. Rev. Lett.*, **103**, 024102.
51. Kanter, I., Aviad, Y., Reidler, I., Cohen, E., and Rosenbluh, M. (2010) An optical ultrafast random bit generator. *Nat. Photon.*, **4**, 58.
52. Kanter, I., Butkovski, M., Peleg, Y., Zigzag, M., Aviad, Y., Reidler, I., Rosenbluh, M., and Kinzel, W. (2010) Synchronization of random bit generators based on coupled chaotic lasers and application to cryptography. *Opt. Express*, **18**, 18292.

53. Nist statistical tests suite, http://csrc.nist.gov/groups/ST/toolkit/rng/documentation_soft.
54. Marsaglia, G. Diehard: a battery of tests of randomness, http://www.stat.fsu.edu/pub/diehard/.
55. Bennett, C.H., Hand, C., and Brassard, G. (1984) Quantum cryptography: public key distribution and coin tossing. *Proceedings of the IEEE International Conference on Computers, Systems, and Signal Processing, Bangalore*, p. 174.
56. Ekert, A. (1991) Quantum cryptography based on bell's theorem. *Phys. Rev. Lett.*, **67**, 661.
57. Gisinand, N., Ribordy, G., Tittel, W., and Zbinden, H. (2002) Quantum cryptography. *Rev. Mod. Phys.*, **74**, 145.
58. Scarani, V., Bechmann-Pasquinucci, H., Cerf, N., Dusek, M., Lutkenhaus, N., and Peev, M. (2009) The security of practical quantum key distribution. *Rev. Mod. Phys.*, **81**, 1301.
59. Jones, R., Spencer, P., Lawrence, J., and Kane, D. (2001) Influence of external cavity length on the coherence collapse regime in laser diodes subject to optical feedback. *Optoelectron. IEEE Proc.*, **148**, 7.
60. Asmussen, S. and Glynn, P. (2007) *Stochastic Simulation: Algorithms and Analysis*, Springer-Verlag, New York.
61. Pecora, L.M. and Carroll, T.L. (1998) Master stability functions for synchronized coupled systems. *Phys. Rev. Lett.*, **80** (10), 2109–2112. DOI: 10.1103/PhysRevLett.80.2109.
62. Kinzel, W., Englert, A., Reents, G., Zigzag, M., and Kanter, I. (2009) Synchronization of networks of chaotic units with time-delayed couplings. *Phys. Rev. E*, **79** (5), 056207.
63. Choe, C., Dahms, T., Hövel, P., and Schöll, E. (2010) Controlling synchrony by delay coupling in networks. *Phys. Rev. E*, **81**, 025205(R).
64. Flunkert, V., Yanchuk, S., Dahms, T., and Schoell, E. (2010) Synchronizing distant nodes: a universal classification of networks. *Phys. Rev. Lett.*, **105**, 254101.

15
Desultory Dynamics in Diode-Lasers: Drift, Diffusion, and Delay[†]

K. Alan Shore

Eckehard Schoell and the present author shared a PhD supervisor: Professor Peter Landsberg. However, our studies were coincident in neither space nor time. Eckehard Schoell worked with Peter Landsberg at Southampton University, while the present author did so at the then University College, Cardiff – now Cardiff University. Our trajectories intersected when Eckehard Schoell organized a meeting to celebrate the sixtieth birthday of Peter Landsberg at Gregynog Hall in Wales in 1982.

Spatial linkages between Peter Landsberg and Eckehard Schoell were enhanced when Professor Schoell moved to Berlin – Peter Landsberg's birth place. Peter Landsberg left us in February 2010 at the age of 87 – having continued publishing into his eighty-fourth year. It remains to be seen whether any of his students can emulate that longevity.

15.1
Introduction

The title of this chapter is chosen to convey a technical concept as well as a personal sentiment. Dealing with the latter first, this author notes that his research career has not been meticulously planned but has rather been subject to a series of unexpected stimuli, which has resulted in the execution of a Brownian-like motion through a number of research areas related to semiconductor optoelectronic devices. In this sense, the author's research career is more desultory than deterministic and perhaps a little chaotic.

Turning to the technical concept, which is intended to be encapsulated in the title, it is seen that a number of physical mechanisms, and notably electron diffusion and drift and optical feedback time delays, can have significant impact on the dynamics of semiconductor lasers and thereby often engendering nonlinear behaviors and definitely including chaotic dynamics. These technical themes are developed in this chapter in a manner which interleaves somewhat with the personal dimension treated above.

[†] A previous 60 years.

Nonlinear Laser Dynamics: From Quantum Dots to Cryptography, First Edition. Edited by Kathy Lüdge.
© 2012 Wiley-VCH Verlag GmbH & Co. KGaA. Published 2012 by Wiley-VCH Verlag GmbH & Co. KGaA.

In Section 15.1, the author recalls very early work on electron diffusion in devices that were then called the *injection lasers* and are now generally termed *semiconductor lasers* or *diode lasers*. This work gives a linkage to previous work by Landsberg as well as to an earlier resident of Berlin. Effects of electron diffusion resurfaced later when attention began to be paid to transverse mode instabilities in semiconductor lasers including laser arrays.

In Section 15.2, attention is given to electron drift dynamics in intersubband lasers. This theoretical work predates the spectacular experimental work performed by Capasso, Faist, and coworkers in the invention of the quantum cascade laser. The work owes its inclusion in this chapter to the fact that it exemplifies the serendipitous nature of the present author's research and also provides an organic link to the subject of this birthday tributary volume. The relevant work was stimulated by attending a presentation by a diploma student of Eckehard Schoell at the Technical University of Berlin. That work focused carrier transport in quantum well structures, but discussions with Schoell and hard work by Wai MunYee resulted in the joint publication on prospects for intersubband optical gain. That paper laid the foundation for subsequent work, which appears to be the first work on the modulation properties of intersubband lasers. In turn, that work resulted in collaborative work with Pesquera *et al.* in Santander, Spain.

The Santander connection provides a natural bridge to the work in Section 15.3 in which the role of carrier diffusion and stimulated emission in determining the modal properties of vertical cavity surface emitting lasers (VCSELs) is described. This work on the spatiotemporal dynamics of VCSELs had been initiated during an extended stay at Bath University by Angel Valle from Santander. A particular aspect of VCSEL modal behavior, which has retained its currency, is the selection of polarization modes.

Section 15.4, giving particular emphasis to experimental work, offers a discussion on how delayed optical feedback can be used to impact the dynamics of VCSELs and also to influence modal selection. Attention is also given to the impact of optical injection on VCSEL behavior. Both delayed optical feedback and optical injection may cause semiconductor lasers, including VCSELs, to exhibit optical chaos.

Section 15.5 treats optical chaos generation, optical chaos synchronization, and optical chaos communications using semiconductor lasers, with particular exemplars being taken from experimental work on VCSELs. Optical chaos communications provide a rich context in which to explore the nonlinear dynamics of semiconductor lasers. One particular feature of chaos synchronization for communications is the impact of the time of flight arising to the physical separation between the synchronized chaotic lasers, which function as transmitter and receiver in chaos communications. The expectation is that a time delay associated with the time of flight will arise in chaos synchronization.

Section 15.6 treats an early experiment that identified conditions under which the time of flight delay could be eliminated. Inherent in this work is the use of lasers that are rendered chaotic by delayed optical feedback. Knowledge of that feedback delay time is important for reconstructing the laser dynamics and hence may affect the security of chaos communications.

The final technical section of the chapter treats some practical aspects of chaos communications schemes. First, it is shown that conditions for high-quality chaos communications do not necessarily coincide with those that provide the optimum condition for message transmission and extraction. The second topic concerns the security of such systems, pointing out when the presence of an eavesdropper can be detected. A brief conclusion terminates the chapter.

15.2
Carrier Diffusion in Diode Lasers

This author began his association with the semiconductor laser in 1972 when the semiconductor laser was 10 years of age, just 2 years younger than its sibling the ruby laser, and had become a device that could be operated fairly reliably at room temperature. Tracing the early development of the semiconductor laser is enabled by the special issues of the *IEEE Journal of Quantum Electronics* associated with the *IEEE Semiconductor Laser Conference*. That conference was first held in Las Vegas, Nevada, between 29 November 1967 and 1 December 1967, and J.I. Pankove served as the first Guest Editor of the relevant special issue published in April 1968. In his editorial, Pankove referred to the challenge of achieving continuous-wave room temperature operation of semiconductor lasers. Even when the second meeting of the series was held in Mexico City in December 1969, that challenge had not been met, although the journal guest editor J.E. Ripper in June 1970 was cautiously optimistic that the introduction of double heterostructure lasers "revived hope of soon achieving continuous operation at room temperature." However, by the time delegates convened in Boston in May 1972 for the third *IEEE Semiconductor Laser Conference*, the *IEEE Journal of Quantum Electronics* Guest Editor, L.A. D'Asaro was able to indicate in February 1973 that "continuous operation of junction lasers at room temperature has become a widely achieved fact." That performance capability owed a great deal to the ingenuity and persistence of Professor Zhores Alferov and Professor Herbert Kroemer who championed the

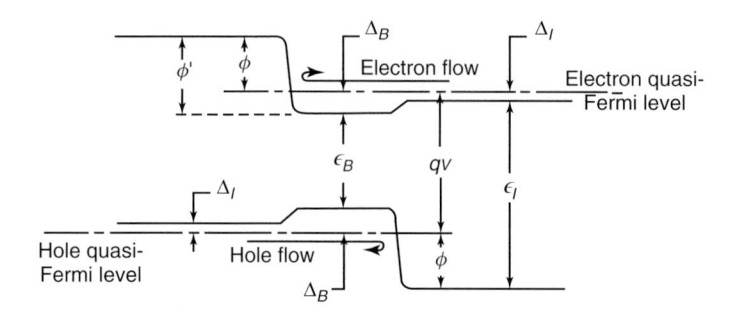

Figure 15.1 Double heterostructure proposed by Kroemer. (Reprinted Figure 15.8 from [1]. Copyright 2001 by the American Physical Society; see also *Proc. IEEE*, **51**, 1782, 1963.)

use of semiconductor heterostructure technology (Figure 15.1) in semiconductor laser design. Appropriate recognition of their contributions came with the 2000 Nobel Prize for Physics being awarded to them.

T.L. Paoli acted as Guest Editor for the July 1975 *IEEE Journal of Quantum Electronics* associated with the *fourth IEEE Semiconductor Laser Conference*, which was held in Atlanta, Georgia in November 1974 (the first such meeting attended by the present author). In his editorial, Paoli was able to highlight another critical issue in a rather wry manner. He remarked, "Although the reliability of junction lasers remained a primary concern of most workers, the nervous fear that such devices might be limited to relatively short CW lives was somewhat alleviated by the informal report by Bell Laboratories of a continuous lifetime in excess of 14 000 h for *at least one* junction laser." (This author's emphasis.)

It was clear then, and several subsequent decades of research have confirmed, that the development of semiconductor lasers was far from complete when the present author joined the Department of Applied Mathematics and Mathematical Physics at University College, Cardiff to begin work with Professor Peter Landsberg and Dr. Mike Adams toward a PhD concerning the "Double-Hetero-structure Injection Laser." The focus of attention was the transport of electrons and holes across the laser heterostructure and thence to determine in a self-consistent manner the optical wave-guiding properties of the laser. In studying these effects, both carrier drift and carrier diffusion were taken into account. Some care was needed in defining the diffusion coefficient, D, and electron mobility, μ, for use in the simulations. It is widely known that these parameters are interlinked by the Einstein relation $D/\mu = kT/e$. Early work on rectifier theory [2] had shown that such a simple relation is not always applicable but should be replaced by a generalized Einstein relation of the form:

$$D/\mu = \alpha kT/e \tag{15.1}$$

where $\alpha = F_{1/2}(\eta)/F_{-1/2}(\eta)$; the F's are Fermi integrals and $\eta = E_f/kT$.

Then, in the case of the carrier degeneracy, which obtains under lasing conditions, the ratio between the diffusion coefficient and mobility depends on the carrier density or equivalently, the Fermi level, E_f.

Measurements of the current voltage (I–V) characteristics of double heterostructure lasers had shown that they took the form $I = I_0 \exp(eV/\alpha kT)$ with $\alpha > 2$. It was shown in [3] that an association could be made between the parameter α included in Eq. (15.1) and the nonideality factor of the I–V characteristics. This use of an α parameter was later superseded by a much wider usage in relation to the line-width enhancement and antiguiding properties of semiconductor lasers. On the other hand, it appears that the carrier degeneracy effects at the heart of [3] continue to have some currency in work on quantum well semiconductor lasers [4].

As the design of semiconductor laser continued to evolve, issues of carrier transport and specifically of carrier diffusion gained increasing attention in relation to the stability of transverse optical modes [5]. Also at this time, wider interest was also beginning to be shown in understanding general optical instabilities including optical bistability and chaos. It is, of course, now widely appreciated that such phenomena are ubiquitous in physical and nonphysical systems, but optical systems provided many early practical demonstrations of such effects. From the perspective of semiconductor laser development, such phenomena are generally unwelcome. On the other hand, from a scientific viewpoint, optical systems, and specifically semiconductor lasers, provide a convenient laboratory in which to explore such phenomena. Those viewpoints can be reconciled when studies of such nonlinear optical phenomena open up new functionality by means of device designs for controlling instabilities [6]. Both perspectives were adopted in work, which first examined optical bistability and optical instability in semiconductor lasers [7–13]. Specifically, twin-stripe lasers were conceived as a means for controlling the instabilities and thereby accessing, for example, beam-steering capabilities [14]. In the process, it was found useful to utilize methodologies and terminologies from nonlinear dynamics – perhaps the first occasion when such explicit use was made of such approaches in the context of semiconductor lasers [15–17]. The significance of such phenomena was fully recognized when attention was directed at the development of high-power semiconductor laser arrays [18]. The prediction of the onset of chaos in such devices [19] underlined the need to understand in detail the underlying physics of such devices and specifically the nonlinear dynamical properties of these lasers.

Having first been of interest for the determination of heterostructure carrier transport and then becoming key to understanding optical mode instabilities in semiconductor lasers and laser arrays, carrier diffusion played yet another prominent role when the modal properties of VCSELs came under scrutiny. Such aspects are considered in Section 15.3.

As progress sometimes requires retracing previous paths, it is noted that carrier transport across heterojunctions has returned to prominence, particularly in treating the behavior of quantum well laser structures. A particular example of such transport phenomena is treated in the next section.

15.3
Intersubband Laser Dynamics

For the first 20 years or so of semiconductor laser usage, the semiconductor material optical bandgap was seen as the sole determinant of the lasing wavelength. Thus, in order to develop semiconductor lasers that could offer a span of operating wavelengths – mainly from the visible into the near-infrared – attention was given to a variety of ternary and quaternary heterostructure materials whose bandgap corresponded to the target emission wavelength. That "bandgap slavery" was relaxed, if not totally abolished, with the development of the so-called low-dimensional semiconductors and specifically, quantum well and quantum dot materials. Using such materials, lasing transitions were no longer principally determined by quasi-continuous energy states in the conduction and valence bands of traditional or "bulk" semiconductors. Rather, such transitions occurred between quantized electron and hole energy levels. Those levels were in turn determined largely by the physical dimensions of the material layers used in the device construction. Realization of low-dimensional structures placed considerable demands on semiconductor growth technology, which rapidly and effectively met that challenge.

As is discussed in more detail elsewhere in this volume [20], electron transport in quantum well and quantum dot materials is a rich area for fundamental studies and also has significant applications. Focusing on conduction band electrons for definiteness, one may consider a number of mechanisms by which carriers are transported across, for example, a multilayered structure formed in semiconductor quantum well materials. Depending on the exact structure utilized, several configurations of energy levels may arise, but here, attention is drawn to the case in which a number of discrete energy levels appear in the conduction and valence bands. Interest then is narrowed to the opportunities which emerge to build lasers that utilize transitions between, for example, the discrete electron energy states in the conduction band. It is noted for consistency that when one focuses attention on transitions involving one species of charge carrier, for example, electrons, then one cannot refer to the resultant emitter as being a diode laser – no use is made of a p–n junction. It is hoped that the reader is prepared to allow the present author some latitude to treat this topic despite the alliterative title of the chapter.

Typically, in as much as several quantized energy levels may exist in such a structure, it is natural to consider whether an opportunity exists to achieve optical gain via electron transitions between such intersubband energy levels. A positive answer to that question was obtained in the joint work with Schoell [21]. It is emphasized that such transitions are unipolar in nature, and hence the lasing device is a semiconductor laser but not, as mentioned above, a diode laser.

A salient feature of such intersubband transitions is that the energy transition is small relative to typical semiconductor laser bandgaps. Put another way, the laser emission from such transitions is at a much longer wavelength than the typical 1 μm wavelength obtained from interband semiconductor lasers. Intersubband lasers offer opportunities for producing compact laser sources with wavelengths one or two orders of magnitude greater than those provided by conventional lasers (Figure 15.2).

It took the Herculean efforts of Faist and coworkers, and notably Sivco and Cho, to realize this concept in the form of the quantum cascade laser (QCL) [23]. A crucial aspect of the device design was the use of a semiconductor superlattice. A proposal for the use of such superlattices to achieve electromagnetic amplification had been made within the very first decade of the history of semiconductor lasers [24]. Further considerations of this device are offered elsewhere in this volume [25].

The practical realization of the quantum cascade laser, based on intersubband transitions, has opened yet another new chapter in the development of semiconductor lasers. The QCL and its variants open the opportunity to provide a convenient source of terahertz radiation with actual and prospective applications in medical imaging, trace gas detection, and security screening. In the context of increased attention being given to the need for scientific research to be "applicable'" it is observed that such capabilities were probably not at the forefront of the minds of

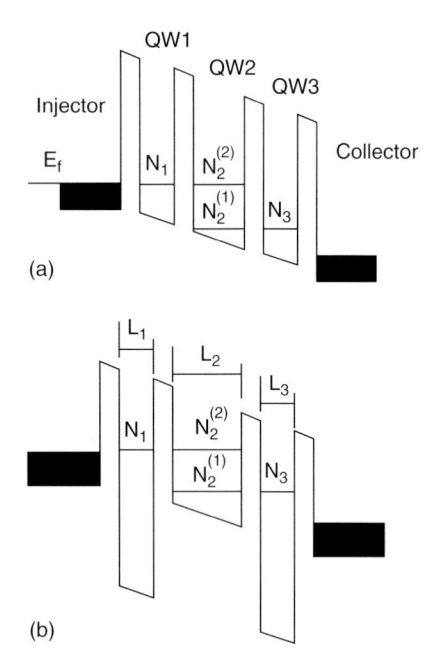

Figure 15.2 Intersubband laser structure. (Reprinted with permission of the IET, Figure 15.1 from [22].)

those who first explored the transport properties of electrons in semiconductors. (Figure 15.2)

The model assembled in Berlin [21] was refined and utilized in subsequent explorations of the modulation performance of intersubband lasers. Here, the key parameter is the electron lifetime that in typical intersubband structures, is of the order of 1 ps, about three orders of magnitude smaller than that for typical interband transitions. The direct consequence is the expectation arising of terahertz modulation bandwidths in intersubband lasers [22, 26, 27] (Figure 15.3).

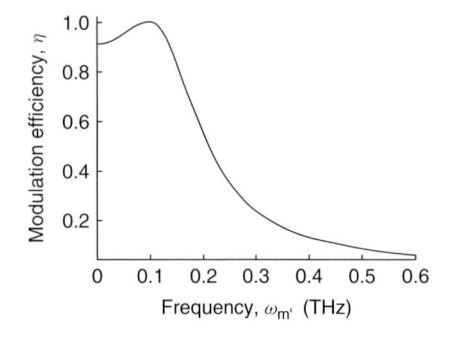

Figure 15.3 Predicted modulation response of intersubband laser. (Reprinted with permission of the IET, Figure 15.2 from [22].)

Here, carrier transport across the structures is of key importance [28, 29]. Such dynamical properties are beginning to be exploited [30].

Intersubband transitions also offer opportunities for tailoring nonlinear optical properties to enable, for example, third harmonic generation [31–35]. Coupled quantum well structures offer further opportunities for studying transport phenomena [36, 37].

15.4
Carrier Diffusion Effects in VCSELs

As signaled above, carrier diffusion effects regain importance when consideration is given to the properties and behavior of VCSELs. Here, the principal consideration is the interplay between the two-dimensional carrier density profile and the optical modes of the laser cavity. Typically, VCSELs have a circular cross-section and usually emit linearly polarized light. However, because of their circular symmetry there is no a priori defined direction for that linear polarization. Consideration of the diffusion of charged carriers in VCSELs reveals much about the basic physics of mode selection. In this section, we consider three aspects of VCSEL performance that are strongly impacted by diffusion effects. Sections 15.4.1 and 15.4.2 are exclusively concerned with theoretical analysis of such effects. In Section 15.4.3, attention is drawn to a nascent area of activity where again, carrier diffusion effects have a significant role.

15.4.1
Transverse Mode Competition and Secondary Pulsations

Theoretical exploration of VCSEL dynamics requires a treatment of both carrier diffusion and the interaction of the lasing mode with the gain profile. Early models treating such aspects were reported in [38] and [39]. Here emphasis was given to the competition between transverse modes a stand-alone VCSEL. The fundamental transverse mode, having a maximum intensity at the center of the laser, is normally excited close to the lasing threshold. This is because the mode profile is well matched to the two-dimensional carrier profile derived from the current injection into the VCSEL. As the laser drive current is increased, the optical power of the lasing mode generally increases. In turn, this increases the stimulated emission, which in turn preferentially enhances the carrier depletion in the regions where the mode intensity is maximum, that is, at the center of the laser. This causes spatial hole burning in the gain [39]. Depending on the rate of carrier diffusion, such spatial hole burning can be ameliorated somewhat, but typically, the gain profile exhibits a strong depletion. In this case, a higher order transverse mode with an optical profile that matches the gain profile begins to lase. Further increase in the drive current with concomitant changes in the gain profile can cause further changes in the transverse mode structure.

Carrier diffusion also can play an interesting role when the laser is switched off. In this case, with a cessation of lasing, carrier diffusion is the main determinant of the carrier profile. In appropriate circumstances, the restoration of the gain profile can result in laser emission even after the laser is nominally switched off. Such predicted secondary pulsations [40] have subsequently been observed.

One of the consequences of transverse mode switching is a change of the far-field emission of the laser and specifically, a manifestation as a high-frequency beam-steering behavior [41]. Transverse mode selection in VCSELs can be aided by the use of optical feedback when the laser is operated in an external cavity configuration [42]. Attention is also drawn to statistical effects that impact transverse mode selection in laser turn-on [43].

15.4.2
VCSEL Polarization Selection

As is made clear in Section 15.4.1, modal selection in VCSELs is highly dependent on the injected carrier profile. The carrier profile also contributes to the selection of the linear polarization mode of VCSELs. Spatial hole burning has been shown to impact polarization dynamics in VCSELs [44]. Nominally, circularly symmetric VCSELs often exhibit a degree of asymmetry in material composition or structure, which results in some birefringence that assists in polarization mode selection [45] including in multi transverse-mode VCSELs [46]. Additional dimensions to the polarization mode behavior arise in laser modulation [47] and when operated in external cavities [48].

15.4.3
Nanospin VCSELs

In the foregoing discussion, the emphasis has been given to determination of the direction of linear polarization in VCSELs. Consideration has also been given to the opportunity to generate circularly polarized light from VCSELs. The basic requirement in this case is that appropriate quantum mechanical transitions are accessed. Such general requirements were highlighted in a seminal paper by San Miguel et al. [49], which has been widely applied to the study of the polarization dynamics of VCSELs. The model is often referred to as the *SFM model*, an acronym which both reflects the authorship of the paper and the underlying physical mechanism considered in the paper: the spin-flip mechanism.

Increasing attention has been given, in recent years, to the exploitation of electron spin in a range of electronic devices, giving rise to a new discipline spin electronics or spintronics.

In the case of semiconductor lasers, a particular attraction of using spin effects is to access a potential 50% reduction of the laser threshold current [50]. It has been shown that optically pumped VCSELs can generate circularly polarized light emission [51], and electrically pumped LEDS and semiconductor lasers have been developed to exploit spin effects [52–54]. The challenge is to deliver an electrical spin-injected semiconductor laser having operability at and above room temperature.

Another thrust of activity, motivated by developments in nanotechnology, is to miniaturize semiconductor lasers to provide the so-called nanolasers [55]. One approach to semiconductor laser miniaturization is to utilize metal-clad structures. In such structures, lasing can occur in optical modes, which are located at the interface between the metal and the semiconductor dielectric, surface plasmon modes [56]. Such structures are thus linked to the burgeoning area of research activity known as *plasmonics* [57]. Metal-clad VCSELs are strong candidates to provide electrically injected nanospin semiconductor lasers [58, 59]. In such devices, the issue of diffusion of the electrically injected spin electrons can be expected to play a significant role in device performance [60].

15.5
Delayed Feedback and Control of VCSEL Polarization

A motif of the present chapter is the persistence of certain underlying physical effects such as carrier drift and carrier diffusion in determining the properties of successive generations of semiconductor lasers. The exploration of optical injection into lasers in general and semiconductor lasers in particular [61] is consistent with that motif. The study of optical feedback effects in diode lasers has perhaps a longer pedigree [62] than optical injection and is particularly associated with a truly seminal publication [63] whose continued use over 30 years is noteworthy. A remarkable feature of the Lang–Kobayashi model [63] is that appears to provide quite accurate predictions even outside the domain of its precise validity. A relatively complete summary of the state of the art of optical feedback effects in semiconductor lasers is available [64]. Other aspects of this configuration are presented elsewhere in this volume [65, 66].

In this section, attention is directed at some of the impacts of optical injection and optical feedback on the behavior of VCSELs. Exploration of these aspects continues apace, and hence further insights in this behavior are anticipated. Experimentally observed phenomena include bistability in various forms [67, 68], and theoretical studies have highlighted pertinent dynamical properties [69].

Early theoretical studies of optical feedback effects on VCSELs considered the noise properties of VCSELs subject to optical feedback [70, 71] including the effects of multiple external cavity reflections [72] and strong coherent optical feedback [73]. Modeling activity was extended to include modal dynamics in modulated VCSELs subject to optical feedback [74]. Complementary experimental work investigated polarization-resolved noise properties of VCSELs subject to optical feedback [75].

The attention paid in [75] to the polarization aspect is of considerable significance for practical applications of VCSELs. Often, VCSELs have circular apertures for which the direction of the polarization of the lasing mode in VCSELs is generally neither defined a priori nor fixed during laser operation. Near the lasing threshold, VCSELs normally support one dominant linearly polarized fundamental transverse mode with a suppressed, nearly degenerate, orthogonally polarized fundamental mode. When the laser bias current is increased, the polarization often switches from its near-threshold direction to the orthogonal polarization. Such polarization switching will impact the use of VCSELs in polarization-sensitive systems, and hence effort has been expended to control polarization switching in VCSELs by a careful design of VCSEL structure [76–78].

Despite their very high facet reflectivity and because of their very short cavity lengths and large emitting area, VCSELs are very sensitive to the effects of optical feedback. The response depends in particular on the polarization properties of the reflected light and specifically on whether the optical feedback is polarization preserving [79–83] or polarization selective [84–87]. An experimental comparison of the effects of polarization-preserving and polarization-selective optical feedbacks on polarization switching in VCSELs has been reported [88].

When the bias current is increased, the polarization often switches to orthogonal polarization. Such polarization switching adds to the richness of VCSEL dynamics [89–94]. Anticorrelation dynamics between the polarization states of VCSELs has been reported in several articles [95–98]. There has also been some work on the quantitative analysis of the cross correlation between the two orthogonal polarizations near threshold in free-running VCSELs [91] and in the regime of low-frequency fluctuations (LFFs) in the VCSEL with isotropic feedback [94, 98]. Tabaka et al. [99] also reported the cross correlation between the two orthogonal polarizations in VCSELs with a short external cavity. An experimental study of the effect of optical feedback on the magnitude of polarization dynamics anticorrelation in VCSELs operating at a number of bias currents relative to the polarization switching current has also been reported [100].

In order to further emphasize the scope for intervention in VCSEL dynamics, reference is made to work on optical injection into multimode VCSELs [101], the control VCSEL polarization by optical injection [102], and the appearance of instabilities and chaos in optically injected VCSELs [103]. Polarization dynamics in such configurations has also been investigated in detail both experimentally [104, 105] and theoretically [106] where interest was focused on effects in modulated lasers. The dynamics of polarization switching in modulated VCSELs has also been examined experimentally [107, 108]. Optical injection has been experimentally shown to be capable of effecting polarization switching in VCSELs [108] and as a means for providing an optical flip-flop [109].

As shown above, both optical injection and optical feedback can cause VCSELs to exhibit nonlinear dynamics including chaos. In the next section, the use of VCSELs in optical chaos communications is discussed.

15.6
VCSEL Chaos and Synchronization and Message Transmission

The key concepts of chaos communications will be rehearsed in detail elsewhere in this volume [110], and other summaries of these concepts have appeared elsewhere

[111, 112]. In this chapter, only a brief summary is presented in order to motivate relevant experiments performed with VCSELs.

The general concept of synchronization has a long historical tradition, and its modern manifestations have been summarized in a major work [113]. The general context for chaos synchronization in external cavity semiconductor lasers has been delineated by Sivaprakasam and Masoller Ottieri [114] to which attention is directed for further references. In the following sections, we outline experimental work on chaos synchronization in external cavity lasers from the perspective of implementing optical chaos communications. The generic experimental configuration includes a transmitter or master laser and a receiver or slave laser. Unidirectional optical coupling between the transmitter and slave laser enables synchronization of the dynamics of the two lasers. (Such unidirectional coupling is effected by the deployment of optical isolators between the lasers.) The transmitter laser is driven into chaos by the application of optical feedback from an external mirror – an external cavity laser. Two options arise in respect of the receiver laser: (i) it may be configured as an external cavity laser and enter the chaotic regime via optical feedback from the external cavity mirror or (ii) it may be a stand-alone laser whose dynamics is affected via the optical coupling from the chaotic transmitter laser. The former case is generally termed a *closed-loop configuration*, while the latter is referred to as an *open-loop configuration*. In such experiments, several regimes of nonlinear dynamics can be accessed. A specific regime that is convenient for careful examination of laser dynamics is the so-called LFFs in which the laser output power exhibits a sudden drop-out followed by a relatively long recovery time. A sudden power drop-out recurs and the cycle is repeated, although it should be emphasized that the repetition occurs without simple periodicity.

In essence, the approach, illustrated in Figure 15.4, is straightforward. It is intended to transmit information between two lasers using a chaotic optical carrier.

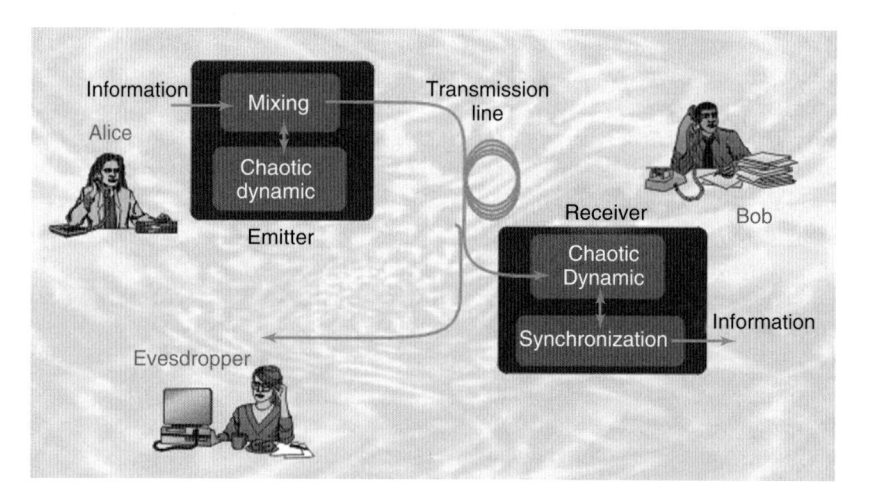

Figure 15.4 Schematic chaos communications system. (Prepared by L.Larger; reprinted by permission.)

The basic requirement for this is that optical chaos synchronization is effected between the transmitter laser where the information originates and a receiver laser which is the destination of the information. The driver for this work is the aim of providing a secure communication where unauthorized recipients – eavesdroppers – are unable to extract the transmitted message specifically because they cannot reproduce the operating conditions to allow them to synchronize an eavesdropper laser with the transmitter laser. In the case of an authorized recipient who is able to synchronize to the transmitter chaos, the information added at the transmitter can be subtracted at the receiver. Such an approach was first demonstrated by Wiggeren and Roy using fiber laser [115]. A considerable research effort has been made over the following decade to refine the basic approach. Here, emphasis is given to the use of VCSELs for chaos communications.

VCSELs have many impressive characteristics, such as a low threshold current, single-longitudinal operation, circular output-beam profile, and wafer-scale integrability. It can be anticipated that because of the relatively low output powers of VCSELs, chaos synchronization may require operation of the lasers well above the threshold. A salient feature of VCSELs is their tendency to exhibit changes in the emission polarization due to changes, for example, of bias current and operating temperature. Polarization effects may therefore be expected to be of some importance in VCSEL synchronization. Interest in implementing chaotic communications using VCSELs is indicated by previous theoretical and experimental work performed on chaotic synchronization in VCSELs. Spencer et al. [116] investigated theoretically the synchronization of chaotic VCSELs. Later, Fujiwara et al. [117] reported an experimental observation of chaotic synchronization in mutually coupled stand-alone VCSELs.

Chaos synchronization was achieved experimentally in unidirectionally coupled external cavity VCSELs operating in an open-loop regime [92]. Synchronization was observed when the polarization of the transmitter is perpendicular to the polarization (x-polarization) of the free-running receiver. The transmitter output versus the x-polarized receiver output power shows normal (positive-slope) synchronization. However, inverse (negative-slope) synchronization was found to arise between the transmitter output and the y-polarized receiver output power.

In this section, a relatively detailed discussion is presented of an experimental demonstration of chaos synchronization in unidirectionally coupled VCSELs. The transmitter was rendered chaotic by optical feedback, while the receiver was a stand-alone VCSEL. One polarization component of the transmitter was coupled into the receiver. Synchronization is observed when the polarization of the transmitter is perpendicular to the polarization (x-polarization) of the free-running receiver. In the experiment, the transmitter laser was biased so that almost all the output power was in the y-polarized lasing mode. The laser was subjected to sufficient optical feedback to enter a regime of chaotic dynamics. The x-polarized component was excited and showed antiphase dynamics with the y-polarized component. For the receiver laser, the laser bias was such that most of the output power was contained in the x-polarized mode. The polarization of the injected laser beam was rotated to perpendicular to that of the RL. Figure 15.5a

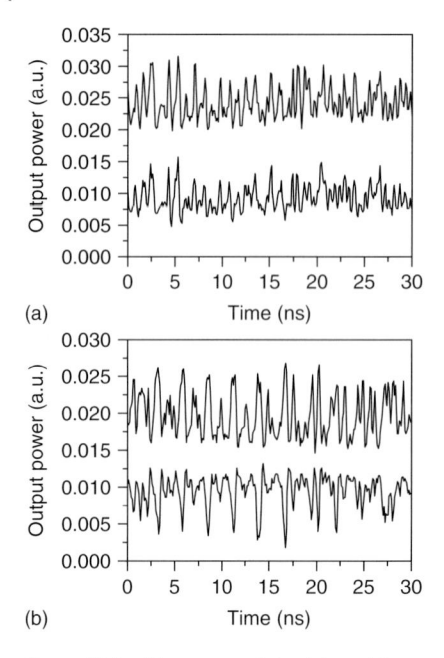

Figure 15.5 Time trace of the injected beam and the receiver laser (a) in the y-polarization and (b) in the x-polarization. (Reprinted, with permission of the Optical Society of America, Figure 15.2 from [92].)

shows the time traces of the injected beam and the RL in the y-polarization. The time trace of the injected beam has been displaced vertically for clarity. These time traces show evidence of synchronization between the injected beam and y-polarization component of the RL. Figure 15.5b shows the time traces of the injected beam and the RL in the x-polarization taken at a different time from that in Figure 15.5a. The fluctuations in the time trace of the injected beam and the RL time trace in the x-polarization are seen to be in antiphase, which is consistent with previous observations of antiphase dynamics of orthogonal polarizations in unstable VCSELs.

Synchronization is readily demonstrated by plotting the instantaneous injected power against the receiver output power at the same time.

Figure 15.6a shows the injected power versus the y-polarization component of the receiver. It demonstrates that good synchronization is obtained between the injected beam and the y-polarization component of the receiver. Figure 15.6b shows the injected power as a function of the x-polarization component of the receiver. The injected power and the x-polarization component of the receiver also show good synchronization; however, the gradient of the synchronization is negative. Such synchronization has been termed *"inverse synchronization."* The absolute correlation coefficient is 0.768, which is lower than that of chaotic synchronization between injected beam and the y-polarization component of the receiver. The reason for this poorer correlation is that the y-polarization mode of the receiver

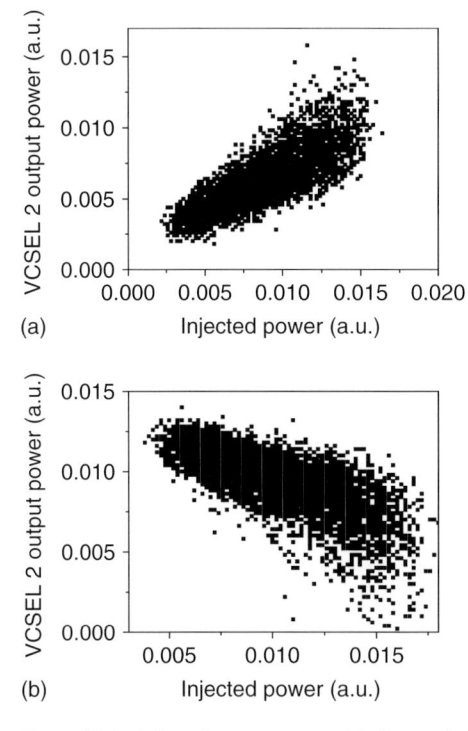

Figure 15.6 Injected power versus (a) the y-polarization component of the receiver and (b) the x-polarization component of the receiver. (Reprinted with permission of the Optical Society of America, Figure 15.3 from [92].)

was locked to the transmitter's frequency when the perpendicular polarization injected beam with proper injection power and frequency detuning excited the y-polarization mode of the receiver. The y-polarization output power of the receiver and the injected beam show normal (positive-slope) chaos synchronization.

Having successfully demonstrated that VCSELs could be robustly chaos synchronized, work was undertaken to show that chaotic message transmission could be achieved [118]. Taking advantage of the polarization properties of VCSELs, a means for enhancing the quality of chaos synchronization using polarization-preserved injection was demonstrated [119]. Subsequent work with a higher-frequency VCSEL showed that gigahertz message transmission could be accomplished using a chaotic carrier generated in an external cavity VCSEL [120].

15.7
Delay Deletion: Nullified Time of Flight

The foregoing work has illustrated the use of optical chaos synchronization in order to enable message transmission using a chaotic optical carrier. It was pointed out that a sine qua non for such an accomplishment is the achievement of optical chaos synchronization between the identified transmitter and receiver lasers. In any practical configuration, such devices will be physically separated and usually over a significant distance. In this case, there is a nonnegligible time of flight for the light emitted by the transmitter to reach the receiver. That time flight will normally result in a lag between the dynamics of the receiver laser relative to that of the transmitter laser. Account must be taken of that lag in order to ensure high-quality chaos synchronization.

Having explored synchronization between pairs of lasers, some effort was directed at obtaining conditions wherein several lasers could be synchronized. Such efforts were principally motivated by a wish to provide additional functionality in chaos communications including chaotic message broadcasting [121] and chaotic message relay [122] (Figure 15.7).

Chaos relay utilized cascaded synchronization of three chaotic semiconductor lasers [123]. In exploring this configuration, it was appreciated that means were available for affecting the lag time and effect a transition from lag to lead [124]. In that transition, a regime was accessed where there was no delay between the dynamics of the first and third laser in the cascade. Such dynamics were termed *isochronous*. Recalling that the lag in dynamics arises principally due to the physical separation between the devices, the achievement of a zero delay can be expressed as an elimination of time of flight effects [125, 126]. The underlying effects

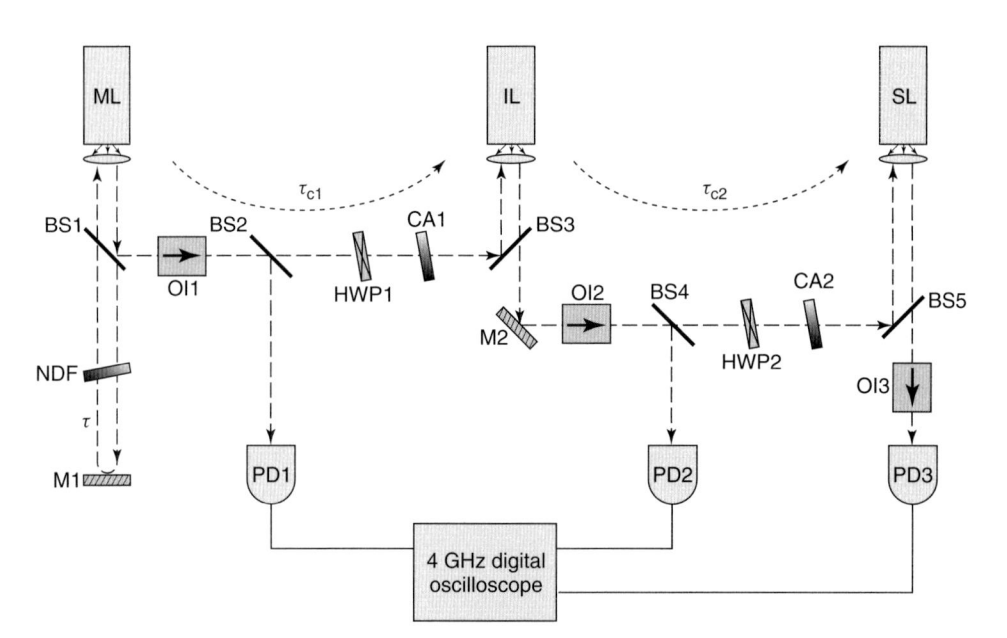

Figure 15.7 Schematic diagram of the experimental setup. ML, master laser; IL, intermediate laser; SL, slave laser. (Reprinted with permission of the Optical Society of America, Figure 15.1 from [126].)

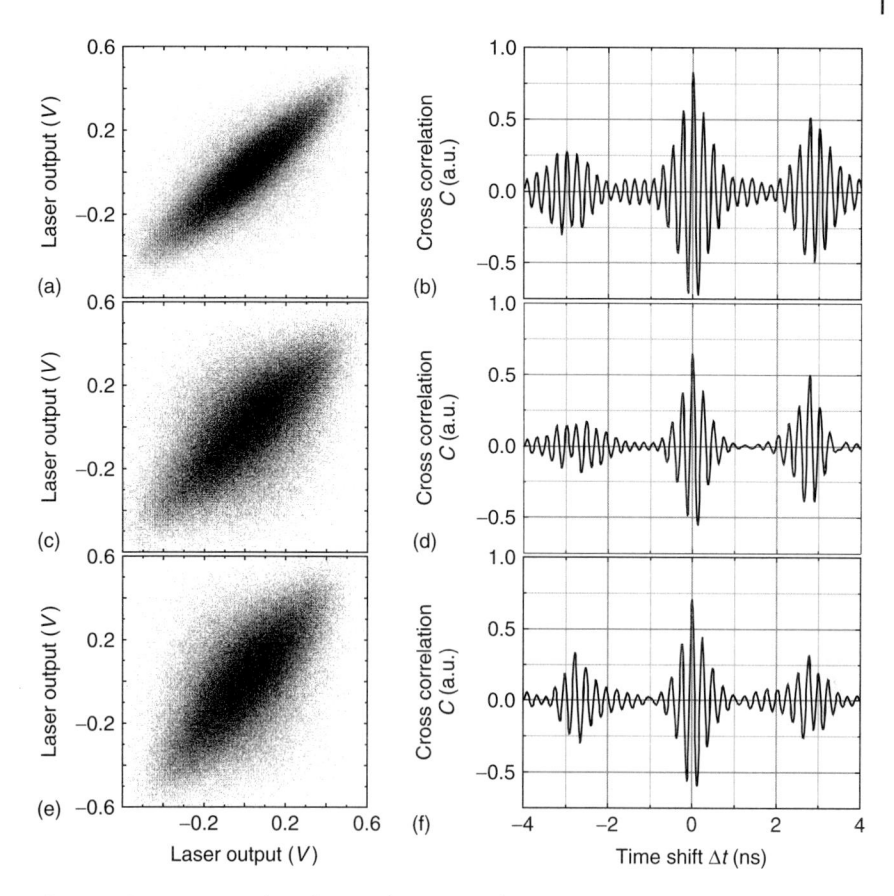

Figure 15.8 Experimental results. Synchronization diagrams: (a) ML versus IL, (c) ML versus SL, and (e) IL versus SL; cross-correlation diagrams: (d) ML and SL, and (f) IL and SL. (Reprinted with permission of the Optical Society of America, Figure 15.3 from [126].)

are related to other dynamical phenomena including the so-called anticipatory synchronization, which is described in detail elsewhere in this volume [65].

Other approaches to the elimination of time of flight time-lags have subsequently been reported in the literature [127] (Figure 15.8).

15.8
Chaos Communications: Optimization and Robustness

In the previous two sections, it was noted that high-quality chaos synchronization is required for successful optical chaos communications. Here, a brief discussion is offered of recent work concerned with the determining the optimal operating conditions for chaos communications. Also, attention is given to the impact on the quality of synchronization of power losses in chaotic optical communications systems. The latter topic touches on the central issue of the security of message transmission using chaos.

Careful theoretical work has been undertaken to determine the optimum time delay for chaos synchronization [128]. In experimental work, for a given external cavity length, optimization of synchronization is most easily achieved by varying either the laser bias current or the laser operating temperature or both. Using the laser drive current and operating temperature as control parameters, a detailed experimental investigation has been carried out on the optimum conditions for chaos synchronization and also to locate the optimum conditions for message transmission [129]. The principal conclusion from this work is that that optimizing chaos synchronization may not always ensure optimized message transmission. This has some implications for the engineering application of this approach. Specifically, it will be necessary to allow for adjustment of the operating conditions of the transmitter and receiver lasers in situ. It should be emphasized, however, that good-quality message extraction can be obtained at the conditions for optimized chaos synchronization. In the case that the available message quality is sufficient for a given communications system, the need for in situ adjustments of the lasers can be obviated.

A persistent issue in the utilization of chaos in communications is the quantification of the security of transmission. In early experiments, it was shown that chaos synchronization could not be achieved between rather dissimilar lasers [130]. The outstanding issue is a precise determination of the dissimilarity required between lasers in order to ensure robustness against eavesdropper attack. One signature of an eavesdropper attack is the attenuation of light coupled between the transmitter and receiver lasers. A drop in the optical power coupled to the receiver laser could also arise because of a physical failure in the communications without intervention by an eavesdropper. Experimental work has been undertaken to determine the robustness of a chaos communications to such power losses whether for benign or malign reasons. In this way, identification has been made of the power loss at which the system fails to operate due to a loss of chaos synchronization [131]. It remains to be explored whether an eavesdropper can successfully extract a transmitted message without causing power losses to the extent that causes noticeable reductions in the quality of chaos synchronization. These results offer further stimulus for exploring the nonlinear dynamics of semiconductor lasers.

15.9
Conclusion

Semiconductor lasers have been in existence for almost half a century. From the earliest days of the laser, there has been an interest in their dynamical behavior. Despite the apparent maturity of the field, the continued vitality of research into semiconductor laser dynamics can be attributed to the remarkable innovations that have been made, and continue to be made, in the design and functionality of laser diodes. From a research viewpoint, appreciation of the nonlinear optical and nonlinear dynamical behavior of semiconductor lasers has opened one of the richest veins for exploration. It is confidently predicted that challenging research problems will continue to emerge in respect of such aspects of semiconductor lasers. It is to be hoped that the celebration of a landmark birthday by one of the younger contributors to this book may provide an occasion on which to review progress made in that direction.

Acknowledgments

The first acknowledgement is to Professor M.J. Adams, Essex University. Mike was a research fellow working with Peter Landsberg at University College, Cardiff – they had published an influential article on the theory of injection lasers in 1971 – when this author began his PhD. Mike obtained support for this author to undertake his research from the then Post Office Research Department, Dollis Hill, London which, during the period of the PhD transmuted into the British Telecom Research Laboratories (BTRL), Martlseham Heath, Ipswich. It was Mike who, without being a formal supervisor, provided day-to-day support for the PhD. However, by coincidence, both Mike and Peter Landsberg departed Cardiff for new locations on the Solent before the completion of the relevant thesis. Nevertheless, Mike continued his support and ensured a successful completion of the thesis. It may be worth noting for younger readers that at this time e-mail was nonexistent. It is expected that somewhere in the archives of BTRL are the cyclostyled quarterly reports prepared and produced by this author during the three years of the PhD.

This article was written during the tenure of a Japanese Society for the Promotion of Science (JSPS) invitation fellowship at the Graduate School of Materials in the Nara Institute of Science and technology (NAIST). Grateful acknowledgment is made of the support from JSPS as well as the hospitality received from Professors H. Kawaguchi, K. Ikeda, and S. Koh of the Ultrafast Photonics Group in the Graduate School of Materials and many other staff and students at NAIST.

Finally, acknowledgment is made of the work of a host of colleagues and collaborators who have made contributions to the work cited here. Such a list would be long and would present a high probability that this author commits the sin of omission in respect of one or more coworkers. Holding that discretion is the better part of valor, it is hoped that they will accept the explicit recognition of their contributions through cited publications without having all their names listed here.

References

1. Kroemer, H. (2001) *Rev. Mod. Phys.*, **73**, 783–793.
2. Landsberg, P.T. (1952) On the diffusion theory of rectification. *Proc. R. Soc. Lond., Series A: Math. Phys. Sci.*, **213**, 226.
3. Shore, K.A. and Adams, M.J. (1976) The effects of carrier degeneracy on

transport properties of the double-heterostructure injection laser. *Appl. Phys.*, **9**, 161–164.
4. Eliseev, P.G. et al. (2005) Anomalous differential resistance change at the oscillation threshold in quantum -well laser diodes. *IEEE J. Quant. Electron.*, **41**, 9–14.
5. Hakki, B.W. (1975) GaAs double hetero-structure lasing behaviour along the junction plane. *J. Appl. Phys.*, **46**, 292–302.
6. Shore, K.A. and Rozzi, T.E. (1981) Near field control in multi-stripe-geometry injection lasers. *IEEE J. Quant. Electron.*, **QE-17**, 718–722.
7. Shore, K.A. and Hartnett, P.J. (1982) Diffusion and waveguiding effects in twin-stripe injection lasers. *Opt. Quant. Electron.*, **14**, 169–176.
8. Shore, K.A. (1982) Optically induced spatial instability in twin-stripe geometry lasers. *Opt. Quant. Electron.*, **14**, 177–181.
9. Shore, K.A. (1982) Semiconductor laser bistable operation with an adjustable trigger. *Opt. Quant. Electron.*, **14**, 321–326.
10. White, I.H., Carroll, J.E., and Plumb, R.G. (1982) Closely coupled twin stripe lasers. *IEE Proc. Part I: Solid-State Electron Devices*, **129**, 291–296.
11. White, I.H. and Carroll, J.E. (1983) New mechanism of bistable operation of closely coupled twin stripe lasers. *Electon. Lett.*, **19**, 337–339.
12. Carroll, J.E. and White, I.H. (1984) Optical bistability in semiconductor lasers. *Phil. Trans. R. Soc. Lond., Ser. A: Math. Phys. Eng. Sci.*, **313**, 333–340.
13. Shore, K.A. and Rozzi, T.E. (1984) Transverse switching due to Hopf bifurcation in semiconductor lasers. *IEEE J. Quant. Electron.*, **QE-20**, 246–255.
14. Shore, K.A. (1984) Radiation patterns for optically steered twin-stripe laser beam scanner. *Appl. Opt.*, **23**, 1386–1390.
15. Shore, K.A. (1985) Stability of self pulsing due to Hopf bifurcation in semiconductor lasers. *IEEE J. Quant. Electron.*, **QE-21**, 1249–1256.
16. Shore, K.A. (1988) Amplification properties of dynamic instabilities: Hopf bifurcation in semiconductor lasers. *J. Opt. Soc. Am. B: Opt. Phys.*, **5**, 1211–1215.
17. Rozzi, T.E. and Shore, K.A. (1985) Spatial and temporal instabilities in multi-stripe semiconductor lasers. *J. Opt. Soc. Am. B: Opt. Phys.*, **2**, 237–249.
18. Wiecorek (this volume); See also e.g. Cross, P.S. et al. (1987) Ultra-high power semiconductor laser arrays. *Science*, **237**, 1305–1309.
19. Winful, H.G. and Wang, S.S. (1988) Stability of phase-locking in semiconductor laser arrays. *Appl. Phys. Lett.*, **53**, 1894–1896.
20. Eaves, L., Balanov, A., Janson, N., and Fromhold, M. (2011) Charge domains in superlattices, this volume, Chapter 6, Wiley-VCH Verlag GmbH, Weinheim.
21. Yee, W.M., Shore, K.A., and Schoell, E. (1993) Carrier transport and inter-sub-band population inversion in coupled quantum wells. *Appl. Phys. Lett.*, **63**, 1089–1091.
22. Cheung, C.Y.L., Spencer, P.S., and Shore, K.A. (1997) Modulation bandwidth optimisation for unipolar inter-subband semiconductor lasers. *IEE Proc. Optoelectron.*, **144**, 44–47.
23. Faist, J., Capasso, F., Sivco, D.L., Sirtori, C., Hutchinson, A.L., and Cho, A.Y. (1994) Quantum cascade laser. *Science*, **264**, 553–556.
24. Kasarinov, R.F. and Suris, R.A. (1971) Possibility of amplification of electromagnetic waves in a semiconductor with a superlattice. *Sov. Phys. Semicond.*, **5**, 707–709.
25. Wacker, A. Quantum Cascade Laser: an emerging technology covering two decades of the optical spectrum, this volume, Chapter 5, Wiley-VCH Verlag GmbH, Weinheim.
26. Cheung, C.Y.L. and Shore, K.A. (1998) Self-consistent analysis of the direct-current modulation response of unipolar semiconductor lasers. *J. Mod. Opt.*, **45**, 1219–1229.
27. Mustafa, N., Pesquera, L., Cheung, C.Y.L., and Shore, K.A. (1999) THz

bandwidth prediction for amplitude modulation response of unipolar intersubband semiconductor lasers. *IEEE Photonics Technol. Lett.*, **11**, 527–529.

28. Kalna, K., Cheung, C.Y.L., Pierce, I., and Shore, K.A. (2000) Self-consistent analysis of carrier transport and carrier capture dynamics in quantum cascade intersubband semiconductor lasers. *IEEE Trans. Microw. Theory Tech.*, **48**, 639–644.

29. Kalna, K., Cheung, C.Y.L., and Shore, K.A. (2001) Electron transport process in quantum cascade intersubband semiconductor lasers. *J. Appl. Phys.*, **89**, 2001–2005.

30. Gkortsas, V.M. (2010) Dynamics of mode-locked quantum cascade lasers. *Opt. Express*, **18**, 13616–13630.

31. Shore, K.A., Chen, X., and Blood, P. (1996) Frequency doubling and sum frequency generation in semiconductor optical waveguide devices. *Prog. Quant. Optoelectron.*, **20**, 181–218.

32. Banerjee, S., Spencer, P.S., and Shore, K.A. (2006) Tunable quantum cascade lasers with phase-matched third harmonic generation. *Appl. Phys. Lett.*, **89**, 05113.

33. Banerjee, S. and Shore, K.A. (2003) MIR and NIR nonlinear optical processing using intersubband $\chi^{(3)}$ in triple quantum well structures. *Semicond. Sci. Technol.*, **18**, 655–660.

34. Banerjee, S., Shore, K.A., Mitchell, C.J., Sly, J.L., and Missous, M. (2005) Current-voltage and light-current characteristics in highly strained InGaAs/InAlAs quantum cascade laser structures. *IEE Proc. Circuits, Devices Syst.*, **152**, 497–501.

35. Banerjee, S., Spencer, P.S., and Shore, K.A. (2006) Design of a tunable quantum cascade laser with enhanced optical non-linearities. *IEE Proc. Optoelectron.*, **153**, 40–42.

36. Sahu, T. and Shore, K.A. (2009) Multi-interface roughness effects on electron mobility in $Ga_{0.5}In_{0.5}GaAs$ multisubband coupled quantum well structures. *Semicond. Sci. Technol.*, *Semicond. Sci. Technol.*, **24**, 095021.

37. Sahu, T. and Shore, K.A. (2010) Effect of electric field on low temperature multisubband electron mobility in a coupled $Ga_{0.5}In_{0.5}P/GaAs$ quantum well structure. *J. Appl. Phys.*, **107**, 113708.

38. Valle, A., Sarma, J., and Shore, K.A. (1995) Dynamics of transverse mode competition in vertical cavity surface emitting laser diodes. *Opt. Commun.*, **115**, 297–302.

39. Valle, A., Sarma, J., and Shore, K.A. (1995) Spatial hole burning effects on the dynamics of vertical cavity surface emitting laser diodes. *IEEE J. Quant. Electron.*, **QE-31**, 1423–1431.

40. Valle, A., Sarma, J., and Shore, K.A. (1995) Secondary pulsations driven by spatial hole burning in vertical cavity surface emitting laser diodes. *J. Opt. Soc Am. B*, **12**, 1741–1746.

41. Valle, A., Rees, P., Pesquera, L., and Shore, K.A. (1999) High-frequency beam steering induced by transverse mode switching in VCSELs: optical gain and waveguiding effects. *J. Opt. Soc. Am. B*, **16**, 2045–2054.

42. Dellunde, J., Valle, A., and Shore, K.A. (1996) Transverse mode selection in external cavity VCSELs. *J. Opt. Soc. Am. B*, **13**, 2477–2483.

43. Dellunde, J., Torrent, M.C., Sancho, J.M., and Shore, K.A. (1997) Statistics of transverse mode turn-on dynamics in VCSELs. *IEEE J. Quant. Electron.*, **33**, 1197–1204.

44. Mueller, R., Klehr, A., Valle, A., Sarma, J., and Shore, K.A. (1996) Spatial hole-burning effects on polarisation dynamics in edge-emitting and vertical cavity surface emitting laser diodes. *IOP Semicond. Sci. Technol.*, **11**, 587–596.

45. Valle, A., Shore, K.A., and Pesquera, L. (1996) Polarisation selection in birefringent vertical cavity surface emitting lasers. *IEEE Lightwave Technol.*, **LT-14**, 2062–2068.

46. Valle, A., Pesquera, L., and Shore, K.A. (1997) Polarisation behaviour of birefringent multi-transverse mode VCSELs. *IEEE Photonics Technol. Lett.*, **9**, 557–559.

47. Valle, A., Pesquera, L., and Shore, K.A. (1998) Polarisation modulation dynamics of birefringent vertical – cavity laser

48. Valle, A., Pesquera, L., and Shore, K.A. (1998) Polarisation selection and sensitivity of external cavity VCSELs. *IEEE Photonics Technol. Lett.*, **10**, 639–641.
49. San Miguel, M., Feng, Q., and Moloney, J.V. (1995) Light polarization dynamics in vertical cavity surface-emitting lasers. *Phys. Rev. A*, **52**, 1728–1739.
50. Rudolph, J. et al. (2003) Laser threshold reduction in a spintronic device. *Appl. Phys. Lett.*, **82**, 4516–4518.
51. Fujino, H., Koh, S., Iba, S., Fujimoto, T., and Kawaguchi, H. (2009) Circularly polarized lasing in a (110) oriented quantum well VCSEL under optical spin injection. *Appl. Phys. Lett.*, **94**, 131108.
52. Holub, M., Shin, J., Chakrabati, S., and Bhatacharya, P.K. (2005) Electrically injected spin-polarized vertical cavity surface emitting lasers. *Appl. Phys. Lett.*, **87**, 091108.
53. Hoevel, S. et al. (2005) Spin controlled optically pumped vertical cavity surface emitting laser. *Electron. Lett.*, **41**, 251–253.
54. Hoevel, S. et al. (2007) Spin-controlled LEDs and VCSELs. *Phys. Status Solidi A*, **204**, 500–507.
55. Ning, C.Z. (2010) Semiconductor nanolasers. *Phys. Status Solidi B*, **247**, 774–788.
56. Welford, K. (1991) Surface plasmon-polaritons and their uses. *Opt. Quant. Electron.*, **23**, 1–27.
57. Dragoman, M. and Dragoman, D. (2008) Plasmonics: applications to nanoscale terahertz and optical devices. *Prog. Quant. Electron.*, **32**, 1–41.
58. Chang, S.-W. and Chuang, S.L. (2009) Fundamental formulation for plasmonic nanolasers. *IEEE J. Quant. Electron.*, **45**, 1004–1013.
59. Li, D.B. and Ning, C.Z. (2009) Giant modal gain, amplified surface plasmon-polariton propagation and slowing down of energy velocity in a metal-semiconductor-metal structure. *Phys. Rev. B*, **80**, 153304.
60. Basu, D., Saha, D., and Bhatacharya, P.K. (2009) Optical polarization modulation and gain anisotropy in an electrically injected spin laser. *Phys. Rev. Lett.*, **102**, 093904.
61. Kobayashi, S. (1981) Injection locking in AlGaAs semiconductor lasers. *IEEE J. Quant. Electron.*, **17**, 681–688.
62. (a) Broom, R.F., Mohn, E., Risch, C., and Salathe, R. (1970) Microwave self-modulation of a diode laser coupled to an external cavity. *IEEE J. Qunat. Electron.*, **6**, 328–334; (b) Spencer, M.B. and Lamb, W.E. Jr. (1972) Laser with a transmitting mirror. *Phys. Rev. A*, **5**, 884–892.
63. Lang, R. and Kobayashi, K. (1980) External optical feedback effects on semiconductor injection-laser properties. *IEEE J. Quant. Electron.*, **16**, 347–355.
64. Kane, D.M. and Shore, K.A. (2005) (eds) *Unlocking Dynamical Diversity*, John Wiley & Sons, Ltd., pp. 1–339.
65. Masoller, C. and Zamora-Munt, J. Exploiting noise, nonlinearities and polarization bistability, this volume, Chapter 4, Wiley-VCH Verlag GmbH, Weinheim.
66. Krauskopf, B. and Walker, J. Delayed optical feedback on a laser with saturable absorber, this volume, Chapter 7, Wiley-VCH Verlag GmbH, Weinheim.
67. Hong, Y., Shore, K.A., Larsson, A., Ghisoni, M., and Halonen, J. (2000) Pure frequency-polarization bistability in a VCSEL subject to optical injection. *Electron. Lett.*, **36**, 2019–2020.
68. Hong, Y., Spencer, P.S., and Shore, K.A. (2001) Power and frequency dependence of hysteresis in optically bistable injection-locked vertical cavity surface emitting semiconductor lasers. *Electron. Lett.*, **37**, 569–570.
69. Torre, M.S., Masoller, C., and Shore, K.A. (2004) Numerical study of optical injection dynamics of vertical-cavity surface-emitting lasers. *IEEE J. Quant. Electron.*, **40**, 25–30.
70. Langley, L.N., Shore, K.A., and Mork, J. (1994) Dynamics and noise properties of semiconductor lasers subject to

strong optical feedback. *Opt. Lett.*, **19**, 2137–2139.

71. Langley, L.N. and Shore, K.A. (1997) The effect of optical feedback on noise properties of vertical cavity surface emitting lasers. *IEE Proc. Optoelectron.*, **144**, 34–38.

72. Dellunde, J., Valle, A., Pesquera, L., and Shore, K.A. (1999) Noise properties of VCSELs subject to multiple external cavity reflections. *J. Opt. Soc Am. B*, **16**, 2131–2139.

73. Ju, R., Spencer, P.S., and Shore, K.A. (2004) Relative intensity noise of semiconductor lasers subject to strong coherent optical feedback. *J. Opt. B: Quant. Semiclass. Opt.*, **6**, S775–S779.

74. Torre, M.S., Masoller, C., Mandel, P., and Shore, K.A. (2004) Transverse-mode dynamics in directly modulated vertical-cavity surface-emitting lasers with optical feedback. *IEEE J. Quant. Electron.*, **40**, 620–628.

75. Sivaprakasam, S., Bandyopadhyay, S., Hong, Y., Spencer, P.S., and Shore, K.A. (2004) Polarisation resolved relative intensity noise measurements of a vertical cavity surface emitting laser subjected to strong optical feedback. *IEEE Photonics Technol. Lett.*, **16**, 9–11.

76. Mukaihara, T., Ohnoki, N., Hayashi, Y., Hatori, N., Koyoma, F., and Iga, K. (1995) Polarization control of vertical cavity surface-emitting lasers using a birefringent metal/dielectric polarizer loaded on top distributed Bragg reflector. *IEEE J. Select. Top. Quant. Electron.*, **1**, 667–672.

77. Badilita, V., Carlin, J.-F., Ilegems, M., Brunner, M., Verschaggelt, G., and Panajotov, K. (2004) Control of polarization switching in vertical-cavities surface emitting lasers. *IEEE Photon. Technol. Lett.*, **16**, 365–367.

78. Augustin, L.M., Smalbrugge, E., Choquette, K.D., Karouta, F., Strijbos, R.C., Verschaffelt, G., Geluk, E.-J., van de Roer, T.G., and Thienpont, H. (2004) Controlled polarization switching in VCSELs by means of asymmetric current injection. *IEEE Photon. Technol. Lett.*, **16**, 708–710.

79. Jiang, S., Pan, Z., Dagenais, M., Morgan, R.A., and Kojima, K. (1994) Influence of external optical feedback on threshold and spectral characteristics of vertical-cavity surface-emitting lasers. *IEEE Photon. Technol. Lett.*, **6**, 34–36.

80. Russell, T.H. and Milster, T.D. (1997) Polarization switching control in vertical cavity surface emitting lasers. *Appl. Phys. Lett.*, **70**, 2520–2522.

81. Besnard, P., Chares, M.L., Stephan, G., and Robert, F. (1999) Switching between polarized modes of a vertical-cavity surface-emitting laser by isotropic optical feedback. *J. Opt. Soc. Am. B*, **16**, 1059–1063.

82. Ackemann, T., Sondermann, M., Naumenko, A., and Loiko, N.A. (2003) Polarization dynamics and low-frequency fluctuations in vertical-cavity surface-emitting lasers subjected to optical feedback. *Appl. Phys. B: Lasers Opt.*, **77**, 739–746.

83. Sciamanna, M., Panajotov, K., Thienpont, H., Veretennicoff, I., Megret, P., and Blondel, M. (2003) Optical feedback induces polarization mode hopping in vertical-cavity surface-emitting lasers. *Opt. Lett.*, **28**, 1543–1545.

84. Robert, F., Besnard, P., Chares, M.L., and Stephan, G.M. (1995) Switching of the polarization state of a vertical-cavity surface-emitting laser using polarized feedback. *Opt. Quant. Electron.*, **27**, 805–811.

85. Kuksenkov, D.V. and Temkin, H. (1997) Polarization related properties of vertical-cavity surface-emitting lasers. *IEEE Select. Top. Quant. Electron.*, **3**, 390–395.

86. Wilkinson, C.I., Woodhead, J., Frost, J.E.F., Roberts, J.S., Wilson, R., and Lewis, M.F. (1999) Electrical polarization control of vertical-cavity surface-emitting lasers using polarized feedback and a liquid crystal. *IEEE Photon. Technol. Lett.*, **11**, 155–157.

87. Loiko, N.A., Naumenko, A.V., and Abraham, N.B. (2001) Complex polarization dynamics in a VCSEL with external polarization-selective feedback.

88. Hong, Y., Spencer, P.S., and Shore, K.A. (2004) Suppression of polarization-switching in vertical-cavity surface-emitting lasers using optical feedback. *Opt. Lett.*, **29**, 2151–2153.
89. Robert, F., Besnard, P., Chares, M.L., and Stephan, G.M. (1997) Polarization modulation dynamics of vertical-cavity surface-emitting lasers with an extended cavity. *IEEE J. Quant. Electron.*, **33**, 2231–2238.
90. Giacomelli, G., Marin, F., and Romanelli, M. (2003) Multi-time-scale dynamics of a laser with polarized optical feedback. *Phys. Rev. A*, **67**, 053809.
91. Sondermann, M., Weinkath, M., Ackemann, T., Mulet, J., and Balle, S. (2003) Two-frequency emission and polarization dynamics at lasing threshold in vertical-cavity surface-emitting lasers. *Phys. Rev. A*, **68**, 033822.
92. Hong, Y., Lee, M.W., Spencer, P.S., and Shore, K.A. (2004) Synchronization of chaos in unidirectionally coupled vertical-cavity surface-emitting semiconductor lasers. *Opt. Lett.*, **29**, 1215–1217.
93. Hong, Y., Ju, R., Spencer, P.S., and Shore, K.A. (2005) Investigation of polarization bistability in vertical-cavity surface-emitting lasers subjected to optical feedback. *IEEE J. Quant. Electron.*, **41**, 619–624.
94. Sondermann, M. and Ackemann, T. (2005) Correlation properties and drift phenomena in the dynamics of vertical-cavity surface-emitting lasers with optical feedback. *Opt. Express*, **13**, 2707–2715.
95. Giudici, M., Balle, S., Ackemann, T., Barland, S., and Tredicce, J.R. (1999) Polarization dynamics in vertical-cavity surface-emitting lasers with optical feedback: experiment and model. *J. Opt. Soc. Am. B*, **11**, 2114–2123.
96. Bandyopadhyay, S., Hong, Y., Spencer, P.S., and Shore, K.A. (2002) Experimental observation of anti-phase polarisation dynamics in VCSELs. *Opt. Commun.*, **202**, 145–154.
97. Hong, Y. and Shore, K.A. (2005) Influence of optical feedback time-delay on power-drops in vertical-cavity surface-emitting lasers. *IEEE J. Quant. Electron.*, **41**, 1054–1057.
98. Naumenko, A.V., Loiko, N.A., Sondermann, M., and Ackemann, T. (2003) Description and analysis of low-frequency fluctuations in vertical-cavity surface-emitting lasers with isotropic optical feedback by a distant reflector. *Phys. Rev. A*, **68**, 033805.
99. Tabaka, A., Peil, M., Sciamanna, M., Fischer, I., Elsäßer, W., Thienpont, H., Veretennicoff, I., and Panajotov, K. (2006) Dynamics of vertical-cavity surface-emitting lasers in the short external cavity regime: pulse packages and polarization mode competition. *Phys. Rev. A*, **73**, 013810.
100. Hong, Y., Paul, J., Spencer, P.S., and Shore, K.A. (2006) Optical feedback dependence of anti-correlation polarization dynamics in vertical-cavity surface-emitting lasers. *J. Opt. Soc. Am. B*, **23**, 2285–2290.
101. Hong, Y., Spencer, P.S., Rees, P., and Shore, K.A. (2002) Optical injection dynamics of two-mode vertical cavity surface emitting semiconductor lasers. *IEEE J. Quant. Electron.*, **38**, 274–279.
102. Bandyopadhyay, S., Hong, Y., Spencer, P.S., and Shore, K.A. (2003) VCSEL polarisation control by optical injection. *IEEE J. Lightwave Technol.*, **21**, 2395–2404.
103. Hong, Y., Spencer, P.S., Bandyopadhyay, S., Rees, P., and Shore, K.A. (2003) Polarization resolved chaos and instabilities in a VCSEL subject to optical injection. *Opt. Commun.*, **216**, 185–189.
104. Hong, Y., Paul, J., Spencer, P.S., and Shore, K.A. (2006) The effects of polarization-resolved optical feedback on the relative intensity noise and polarization stability of vertical-cavity surface-emitting lasers. *IEEE J. Lightwave Technol.*, **24**, 3210–3216.
105. Paul, J., Masoller, C., Hong, Y., Spencer, P.S., and Shore, K.A. (2007) Impact of orthogonal optical feedback on the polarisation switching of

vertical-cavity surface-emitting lasers. *J. Opt. Soc. Am. B*, **24**, 1987–1994.
106. Masoller, C., Torre, M.S., and Shore, K.A. (2007) Polarization dynamics of current-modulated vertical-cavity surface-emitting lasers. *IEEE J. Quant. Electron.*, **43**, 1074–1082.
107. Hong, Y., Paul, J., Spencer, P.S., and Shore, K.A. (2008) Influence of low-frequency modulation on polarization switching of VCSELs subject to optical feedback. *IEEE J. Quant. Electron.*, **44**, 30–35.
108. Paul, J., Masoller, C., Mandel, P., Hong, Y., Spencer, P.S., and Shore, K.A. (2008) Experimental and theoretical study of dynamical hysteresis and scaling laws in the polarization switching of vertical-cavity surface-emitting lasers. *Phys. Rev. A*, **77**, 043803.
109. Jeong, K.H., Kim, K.H., Lee, S.H., Lee, M.H., Yoo, B.-S., and Shore, K.A. (2008) Optical injection-induced polarization switching dynamics in 1.5 μm wavelength single-mode vertical cavity surface emitting lasers. *IEEE Photonics Technol. Lett.*, **20**, 779–781.
110. Kinzel, W. and Kanter, I. Chaos synchronization and secure communication, this volume, Chapter 16, Wiley-VCH Verlag GmbH, Weinheim.
111. Shore, K.A., Spencer, P.S., and Pierce, I. (2008) in *Chaos Control Handbook* (eds E. Schoell and H.G. Schuster), Wiley-VCH Verlag GmbH, pp. 369–376.
112. Shore, K.A., Spencer, P.S., and Pierce, I. (2008) in *Recent Advances in Laser Dynamics: Control and Synchronization* (ed. A.N. Pisarchik), Research Signpost, Kerala, India, pp. 79–104.
113. Pikovsky, A., Rosenblum, M., and Kurths, J. (2001) *Synchronisation: A Universal Concept in Nonlinear Sciences*, Cambridge University Press.
114. Sivaprakasam, S. and Masoller Ottieri, C. (2005) in *Unlocking Dynamical Diversity*, Chapter 6 (eds D.M. Kane and K.A. Shore), John Wiley & Sons Inc., New York.
115. van Wiggeren, G.D. and Roy, R. (1998) Communication with chaotic lasers. *Science*, **279**, 1198–1200.
116. Spencer, P.S., Mirasso, C.R., Colet, P., and Shore, K.A. (1998) Modelling of optical synchronisation of chaotic VCSELs. *IEEE J. Quant. Electron.*, **34**, 1673–1679.
117. Fujiwara, N., Takiguchi, Y., and Ohtsubo, J. (2003) Observation of the synchronization of chaos in mutually injected vertical cavity surface emitting lasers. *Opt. Lett.*, **28**, 1677–1679.
118. Lee, M.W., Hong, Y., Spencer, P.S., and Shore, K.A. (2004) Experimental demonstration of VCSEL-based chaotic optical communications. *IEEE Photonics Tech. Lett.*, **16**, 2392–2394.
119. Hong, Y., Paul, J., Lee, M.W., Spencer, P.S., and Shore, K.A. (2008) Enhanced Chaos Synchronization in Unidirectionally Coupled VCSELs with Polarization-Preserved Injection. *Opt. Lett.*, **33**, 587–589.
120. Hong, Y., Lee, M.W., Paul, J., Spencer, P.S., and Shore, K.A. (2009) GHz bandwidth message transmission using chaotic vertical-cavity surface-emitting lasers. *IEEE J. Lightwave Technol.*, **22**, 5099–5105.
121. Lee, M.W. and Shore, K.A. (2004) Chaotic message broadcasting using DFB laser diodes. *Electron. Lett.*, **40**, 614–615.
122. Lee, M.W. and Shore, K.A. (2006) Demonstration of a chaotic optical message relay using DFB laser diodes. *IEEE Photon. Technol. Lett.*, **18**, 169–171.
123. Sivaprakasam, S. and Shore, K.A. (2001) Cascaded synchronization of external cavity laser diodes. *Opt. Lett.*, **26**, 253–255.
124. Sivaprakasam, S., Spencer, P.S., Rees, P., and Shore, K.A. (2002) Transition between anticipating and lag synchronization in chaotic external cavity diode lasers. *Opt. Lett.*, **27**, 1250–1252.
125. Sivaprakasam, S., Paul, J., Spencer, P.S., Rees, P., and Shore, K.A. (2003) Nullified time-of-flight lead-lag in synchronisation of chaotic external cavity laser diodes. *Opt. Lett.*, **28**, 1397–1399.
126. Lee, M.W., Paul, J., Masoller, C., and Shore, K.A. (2006) Observation of cascade complete chaos synchronisation

with zero time lag in laser diodes. *J. Opt. Soc. Am. B*, **23**, 846–851.
127. Fisher, I. et al. (2006) Zero-lag long-range synchronisation via dynamical relaying. *Phys. Rev. Lett.*, **97**, 123902.
128. Peters-Flynn, S., Spencer, P.S., Pierce, I., and Shore, K.A. (2006) Identification of the optimum time-delay for chaos synchronisation regimes of semiconductor lasers. *IEEE J. Quant. Electron.*, **42**, 427–434.
129. Hong, Y., Yee, M.W., and Shore, K.A. (2010) Optimised message extraction in laser diode based optical chaos communications. *IEEE J. Quant. Electron.*, **46**, 253–257.
130. Jones, R.J., Sivaprakasam, S., and Shore, K.A. (2000) Integrity of semiconductor laser chaotic communications to naive eavesdroppers. *Opt. Lett.*, **25**, 1663–1665.
131. Hong, Y. and Shore, K.A. (2010) Power loss resilience in laser diode based optical chaos communications systems. *IEEE J. Lightwave Technol.*, **28**, 270–276.

Further Reading

Lee, S.H., Jung, H.W., Kim, K.H., Lee, M.H., Yoo, B.-S., Roh, J., and Shore, K.A. (2010) 1-GHz All-Optical flip-flop operation of conventional cylindrical-shaped single-mode VCSELs under low power optical injection. *IEEE Photonics. Technol. Lett.*, **22**, 1759–1761.

Index

a
accumulation layer   116
active region   4, 92
adiabatic elimination   17
– of atomic polarization   189
all-optical memory element   257
all-optical signal processing   246
all-to-all coupling   246
amplitude noise   203
amplitude–phase coupling   274
arbitrary phase   140
asymmetry parameter   47
asymptotic analysis   21, 144, 327
asymptotically stable   304
attractor   167, 271, 285, 287, 319
auger scattering   5, 6, 20
autocorrelation function   220

b
back injection   224
backward wave oscillators   131
bandpass filter   321
Bernoulli network   337, 347
Bernoulli units   335–337, 345, 346
bidirectional coupling   334, 336, 339
bidirectional transmission   239
bifurcation
– Andronov–Hopf   82, 83, 119, 145, 166, 170, 175, 189, 206, 255, 283, 324, 330
– Bogdanov–takens   165
– codimension-three   176
– codimension-two   167
– of limit cycles   263
– homoclinic loop   166, 176, 305, 306
– Hopf bifurcation bridge   64
– Hopf–Hopf bifurcation point   171
– inverse Andronov–Hopf   191
– inverse period-doubling   261
– Neimark–Sacker   203, 205, 210
– period doubling cascade   79
– period-doubling bifurcation   258, 319
– pitchfork   272
– saddle-node bifurcation   166, 169, 175, 191, 255
– saddle-node bifurcation of limit cycles   166, 172, 175, 260
– saddle transition of Hopf curves   177
– stochastic d-bifurcation   271, 279, 280, 283
– torus bifurcation   83, 259
– transcritical   166, 169, 205, 255, 262
bifurcation analysis   79, 82, 161, 163, 164, 272
bifurcation diagram   172, 173, 175, 281, 308
bipartite network   247
bistability   42, 76, 80, 173, 256
bit error rate   336
Bloch gain   102
Bloch oscillations   118, 129
blowouts   305
Bogdanov–takens point   175
Boltzmann equation   6, 100
Born approximation
– first order   252
– second order   5
boundary condition   115, 185
– Ohmic boundary conditions   115
– periodic boundary conditions   187
broadband chaos   317, 318, 324
bubbling   231
bunching   129

c
capture rate   6, 148
carrier diffusion   358, 362
carrier heating   14
carrier lifetimes   13, 16, 27, 140
carrier transport   358

Nonlinear Laser Dynamics: From Quantum Dots to Cryptography, First Edition. Edited by Kathy Lüdge.
© 2012 Wiley-VCH Verlag GmbH & Co. KGaA. Published 2012 by Wiley-VCH Verlag GmbH & Co. KGaA.

carrier–carrier scattering 5, 143
center manifold theory 273
chaos 111, 317, 359, 365
– period-doubling route to chaos 260
– quantum chaos 111
chaos communications 365, 366, 371
chaos modulations 336
chaos relay 370
chaos synchronization 239, 333, 334, 366, 372
chaotic attractor 287, 319
chaotic currents 115
chaotic dynamics 224, 277
chaotic electron transport 111
chaotic intensity dynamics 218
chaotic invariant sets 286
chaotic message transmission 369
chaotic network 333, 346
chaotic optical carrier 369
chaotic regime 193, 261, 285
chaotic spectrum 322
chaotic trajectory 230, 335
characteristic equation 146, 150, 155, 169, 170, 191, 255, 272, 302, 323
charge conservation 26
charge domain 111, 123–125, 127, 130
charge domain dynamics 115
chirp 193
circularly polarized light 363
class B lasers 23, 272
cluster state 293, 300, 302, 304, 308, 349
codimension-two points 175
coherence properties 74
coherent effects 3, 103
collapse of optical coherence 219
common noise 236
common optical external forcing 275
communication process 239
conductance 95, 101
confined QD levels 5
continuation method 82, 145, 164
continuous-time perturbation 287
convective instabilities 336
correlation scaling 219
Coulomb matrix elements 7
Coulomb scattering 4
coupled lasers
– bidirectionally coupled lasers 218, 220, 226
– coupled two-mode lasers 245
– delay-coupled lasers 217

– mutually coupled lasers 217, 341
coupled phase oscillators 298
coupling path 224
critical feedback strength 139, 140
critical slowing down 42, 43
cross-correlation function 222, 224, 225, 230, 236, 340, 343
cryptographic protocol 341
current continuity equation 113
current modulation 42, 231
– asymmetric triangular 44
– external pump modulation with different frequencies 232
– high-frequency modulation 20
– periodic modulation 231
current oscillations 120, 121, 128, 130
current–voltage characteristics 119
cutoff frequency 21
cyclotron frequency 118

d
damping rate of ROs 23, 24, 139–158
delay differential equations 141, 148ff, 164, 184, 317, 334, 346
– integro-delay differential equation 321
delay-coupled lasers 217
delay-induced instabilities 219
delayed optical feedback 60, 68, 69, 72, 85, 139, 141, 161, 162, 218, 226, 231, 364
delocalization 111, 118
density matrix theory 101, 103
detailed balance 8
dimensionless variables 148, 153, 321
dipole matrix element 94, 105
discrete mapping 304
distributed feedback lasers 59
doped carrier reservoir 26
doping concentrations 26, 117
drift velocity 114, 118, 124, 130
diffusion 114, 355
drift–diffusion model 114
dynamical hysteresis 42
dynamical instability 218

e
eavesdropper 239, 339, 367, 372
eigenvalues see characteristic equation
Einstein coefficient 10
Einstein relation 358
elastic scattering time 114
electric field envelope 186
electron-scattering time 114, 361
electronic subbands 92
encoding scheme 50

encrypted communication 238
encrypted key 239
energy confinement 15
energy flux density 105
energy losses 204
entrainment 231
equilibrium
– globally attracting 167
Esaki–Tsu approach 114
Esaki–Tsu domain 124, 126
Esaki–Tsu peak 129
excitability threshold 167
excitable SLSA 163, 171
excitatory response 298
excited states 9, 84
external cavity modes 149, 155
external forcing 279
external modulation 183, 231
external noise 235, 271
eye pattern diagrams 21, 22

f
Fabry–Perot laser 58, 248
Fabry–Perot modes 250
far-field emission 363
feedback 206, 226
feedback loop 163, 321
Fermi integrals 358
Fermi's golden rule 93, 104
filter 345
firing rule 303
fixed point 328
Floquet exponents 272, 273, 284
Floquet multipliers 205
Fourier transformed field 185
frequency splitting 69

g
gain
– modal gain 93, 95, 105
– nonlinear gain 257, 258
– threshold gain 68
– unsaturated gain 187
gain profile 253, 363
gain spectrum 101, 102
gain window 196
Gaussian processes 275
Gaussian pulse 326
Gaussian white noises 40
ghost stochastic resonance 233
Ginzburg–Landau equation 208
global constraint 115, 116
Green's function 103

h
Haus master equation 208
Haus model 184
– generalized Haus model 208
heating 98
heterostructure 4, 92, 357, 358
Hodgkin–Huxley neuron model 296
homoclinic loop 166, 176, 305, 306
Hopf normal form 273
horseshoes 272, 286
hysteresis 200
Hopf see bifurcation

i
Ikeda map 187
implicit function theorem 252
in-phase oscillation 128
index perturbation 251
induced emission 10
inhomogeneous broadening 20
initial conditions 173, 300
injection locking 73, 80, 83
injection-locking instabilities 217
injector region 92
instability boundaries 205
instability threshold 211, 324
integrate-and-fire (IF) oscillators 294
intensity fluctuations 220, 278
intensity–current characteristic 40
interband transitions 361
interdropout probability distribution 233
interface conditions 185
interminiband laser 96
intermittency 193
internal noise 235
interspike time 173
intersubband phonon scattering 97
intersubband transitions 360
invariant sets 285
isochronal solution 223, 235, 370
isochrone 274, 275

j
Jacobi matrix 255

k
key exchange protocol 238, 340
Kramers law 69
Kramers' hopping 36

l
Landau–Stuart model 271, 273
– with white noise external forcing 283

Lang–Kobayashi (LK) model   60, 222, 236, 334, 340, 364
laser linewidth   19, 75
laser Relay   228
lasing condition   249
lasing threshold   11, 42, 46, 205
laterally coupled laser arrays   217
lattice temperature   14, 114
lattice vibrations   98
leader–laggard dynamics   224
leading edge stability boundaries   202
leading order equations   23, 153
limit cycle   258, 272, 273
– overdamped limit cycle   272
– stable hyperbolic   286
linear gain   10
linear gain parameter   190
linear stability analysis *see* charcteristic equation
linear threshold line   190, 201
linewidth enhancement factor   29, 64, 139, 148, 185, 193, 203, 223, 254, 272, 274
link   247
instability (local)   303
locked cavity modes   197
locking   75, 85, 197, 260
logic operations   50
logic stochastic resonance   36
logical operators   38
long delay regime   218
longitudinal modes   19, 58, 183
Lorentzian line shape   187
loss of noise synchrony   271
loss of synchrony   277
low-frequency dropouts   231
low-frequency fluctuation (LFF)   61, 218, 231, 337, 365
– type I LFF   63
– type II LFF   63
lumped element approach   184, 185
Lyapunov exponent   231, 279, 284, 285, 323, 347, 348

m

Mach–Zehnder modulator   319
magnetic field   111
map   299, 328
– Bernoulli map   334
– 1-D map   328
– 3-D map   205
– 4-D map   210
– return map   299, 306
master stability function   346
master–slave configuration   333

message decryption   239
message transmission   372
microscopically based rate equation model   3, 142, 153
miniband   93, 97, 112
miniband transport   112
minimum linewidth mode   140
mirror relay   230
missing fundamental illusion   233
mode beating   64
mode competition   57, 58, 61, 67
mode hopping   69
mode locked laser   183
– fundamental ML solution   191, 193
– harmonic ML   191
– passively mode-locked laser   183, 184, 212
– Q-switching   161, 189, 193
– synchronously pumped actively ML lasers   196
mode-locking stability boundaries   200
modulation asymmetry   46
modulation response   16, 18, 28
monochromatic external forcing   276
monostability   40
Monte-Carlo simulations   100
multimode laser   57, 61, 188, 246
multimode optical injection   75
multipulse dynamics   319
mutual chaos pass filter   342
mutual injection   235
mutually coupled lasers   217, 341

n

nanolasers   364
network theory   247
network topology graph   247
networks   293
– chaotic network   333, 346
– complex networks   245, 246, 248
– excitable networks   233
New's model   187
– generalized New's model   203
New's stability criterion   193, 197, 201, 205, 210
nodes   247
noise   35, 44, 48, 51–53, 67, 72, 162, 179, 235, 271, 275, 326
noise correlation time   236, 237
noise floor   323
noise modulation   235
noise realizations   282
noise-induced pulses   168
noise-to-signal ratio   343
non-KAM chaos   112

nonequilibrium Green's functions   101
nonexcitable SLSA   173
nonlinear delayed feedback   322
nonlinear stability analysis   318, 325
nonlinear transmission functions   319
nonlinear units   346
nullcline   327, 328
numerical continuation   179

o

optical bistability   62, 359
optical buffer memory   54
optical coupling   231
optical data communication   5
optical feedback see delayed optical feedback
optical injection   74, 77, 83, 85, 235, 253, 364, 365
optical-fiber coupling   246
optimal noise level   235
optoelectronic oscillator   318, 325
order parameter   142, 275, 300, 301
oscillating dynamics   258

p

parabolic dispersion   93
parabolic gain dispersion   207
Pauli blocking factor   141
Pauli matrices   249, 251
periodic blowouts   309
periodic ML solution   210
periodic orbits   318
periodic spiking   296
perturbation   197, 304
– index perturbation   251
– pulselike perturbation   325
perturbation mode   347
phase oscillators   298
phase-response function   293, 295, 297
phase-space stretching   275
phonon emission   97
Poisson's equation   115
polarization bistability   35, 36
polarization dynamics   77
polarization mode hopping   69
polarization rotating optical feedback   64
polarization switching   36, 40, 60, 68, 365
polarized light   362
power dropouts   61, 85, 220, 231, 366
power spectrum   121, 122, 126, 318, 322
– featureless   318
private filter   345
private key   334
probability distribution function   232, 233, 278

pseudorandom digital messages   239
pseudorandom number   339
public channel communication   239, 334, 339, 340
public cryptography   345
pullback convergence   280, 282
pulsating state   329
pulse energy   203, 204
pulse group velocity   196
pulse package   67
pulse triggering   49
pulse-coupled oscillators   293, 294, 298

q

Q-switching instability   189, 202
Q-switching instability boundary   206, 210
quantized energy levels   360
quantum cascade laser   64, 91, 356, 360
quantum dot (QD) laser   3–30, 84, 139–142
quantum kinetic calculations   103
quantum protocol   340
quasi periodic dynamics   224, 259
quasi-Fermi distribution   7
quasiperiodicity   224
quiescent state   318, 323
quantum well (QW) laser model   29, 146

r

radiative recombination   143
random bit generators   339
random bit sequences   340, 342
random chaotic attractor   287
random dynamical system   279
random sink   279, 280, 282
random strange attractor   279, 280, 283
rate equations
– for QD laser subject to optical feedback   141
– for solitary QW laser   146
– laser with saturable absorber (Yamada)   163, 188
– QD laser (3 variables)   29, 148
– QD laser (microscopic)   3, 153
– two mode VCSEL (spin flip model)   39, 71
refractory period   167
relaxation cascade   17
relaxation oscillations (ROs)   4, 23, 24, 42, 44, 139, 147, 273
relaxation processes   7, 14
reliability   52
resetting rule   303
residence time   69, 71

resonant behavior 233
return map 299, 306
ring cavity 184
ring configuration 221
ring of unidirectionally coupled lasers 219
ring of units 348
rotational symmetry 272, 273
round-trip time 141
Routh–Hurwitz stability criteria 255
RPP dynamics 65
Runge–Kutta scheme 113

s

saddle-node *see* bifurcation
saddle point 302, 304
saturable absorber 161, 163, 183, 188
saturable gain 198, 203
saturation parameter 188
scattering
– elastic impurity scattering 100
– electron–electron scattering 5–8, 100
– optical phonon scattering 98
– phonon scattering 100, 114
scattering rates 24, 143
scattering time 23, 94, 97, 117
secret message 335
secure chaos communication 319
secure communication 333, 339
self-pulsations 167
self-sustained current oscillations 111
self-sustained oscillations 162, 173
semiclassical approach 103
semiconductor Bloch equations 4
separatrix 328
shear 275
short delay regime 218
short external cavity 64, 65, 140
side mode suppression ratios 253
single-mode laser 59
size distribution 20
slow saturable absorber 188
slow saturable absorber approximation 197
slow–fast system 163
spatial hole burning 362
spatiotemporal electron charge dynamics 122
spectral filtering 185, 187, 204, 209, 211
spectral filtering bandwidth 200
spectral properties 19
spiking events 295
spin-flip model 39, 82, 363
splay state 301
spontaneous emission 10, 48, 52
spontaneous emission noise 275, 276

square-wave solutions 319
stability analysis 41, 60, 128, 150, 151, 168, 231, 302, 308, 328, 347
stability boundary 139, 205
stability condition 157
stable manifold 167, 305
stimulated processes 10, 105
stochastic bifurcation 271
stochastic forcing 272, 279
stochastic hopping 41
stochastic logical operator 49
stochastic resonance 49, 72
stochastic-web 111, 118
strange attractor 285, 318
stretch-and-fold action 286, 287
success probability 52, 53
superlattice 92, 111, 112
switching dynamics 18
symmetry breaking 224
synchronization 72, 219, 366, 367
– chaos synchronization 239, 333, 334, 366, 372
– cluster synchronization 349
– complete synchronization 295, 337
– generalized synchronization 222
– identical synchronization 230
– incomplete synchronization 280
– inverse synchronization 368
– lack of synchronization 276
– lag synchronization 226
– leader–laggard synchronization 222
– localized synchronization 217
– noise synchronization 271, 275, 278, 280, 282
– out-of-phase synchronized 225
– partial synchronization 276
– secure synchronization 344
– trivial synchronization 276
– white noise synchronization 277
– zero-lag synchronization 228, 230, 236, 238, 340
synchronization manifold 231, 336, 346, 348
synchronization transitions 280
synchronized power dropouts 226

t

terahertz radiation 360
thermal energy 97
thermal noise 339
thermodynamic equilibrium 8
thermodynamic phase transition 276
third harmonic generation 362
time of flight 370

time-delayed feedback   317, 335
time-delayed systems *see* delay differential equations
timescale separation   24, 27, 61, 163
timing jitter   162
trailing edge instability boundaries   211
transcendental characteristic equation   169
transfer matrix   249–251
transition probability   6
transverse mode   59, 78, 362
transversely stable   231
traveling wave model   185
tunneling injection   98, 99
tunneling resonance   95
turn-on dynamics   11, 12
two-mode lasing   15, 84, 85, 245

u
unidirectional coupling   224, 339, 366

v
variational approach   209
vertical-cavity surface-emitting laser (VCSEL)   35, 59, 362, 364
voltage drop   116

w
wave-mixing   75
waveguide losses   95
weak chaos   349
white noise external forcing   277, 280

y
Yamada rate equations   163

z
zero-lag synchronization   228, 230, 236, 238, 340

References in Nonlinear Dynamics and Complexity

Schuster, H. G. (ed.)

Reviews of Nonlinear Dynamics and Complexity

Volume 1

2008
ISBN: 978-3-527-40729-3

Schuster, H. G. (ed.)

Reviews of Nonlinear Dynamics and Complexity

Volume 2

2009
ISBN: 978-3-527-40850-4

Schuster, H. G. (ed.)

Reviews of Nonlinear Dynamics and Complexity

Volume 3

2010
ISBN: 978-3-527-40945-7

Grigoriev, R. and Schuster, H.G. (eds.)

Transport and Mixing in Laminar Flows

From Microfluidics to Oceanic Currents

2011
ISBN: 978-3-527-41011-8

Klages, R., Just, W., Jarzynski, Ch., Schuster, H.G. (eds.)

Nonequilibrium Statistical Physics of Small Systems

Fluctuation Relations and Beyond

2013
ISBN: 978-3-527-41094-1

Niebur, E., Plenz, D., Schuster, H.G. (eds.)

Criticality in Neural Systems

2013
ISBN: 978-3-527-41104-7